Handbook of Secondary Fungal Metabolites

VOLUME II

Handbook of Secondary Fungal Metabolites

VOLUME II

RICHARD J. COLE
Albany, Georgia

MILBRA A. SCHWEIKERT
National Peanut Research Laboratory
Dawson, Georgia

ACADEMIC PRESS
An imprint of Elsevier Science
Amsterdam Boston London New York Oxford Paris
San Diego San Francisco Singapore Sydney Tokyo

Cover images: Photograph by Dr. Bruce Horn and design by Brian E. Cole. The cover is a collage of various fungi, some presented in pure culture and others in their natural forms, i.e. mshrooms, which are the easily recognizable fruiting structures of some fungi, superimposed on these fungi are the chemical structures of some representative secondary fungal metabolites.

Academic Press Rapid Manuscript Reproduction

This book is printed on acid-free paper. ∞

Academic Press
An imprint of Elsevier Science
525 B Street, Suite 1900, San Diego, California 92101-4495, USA
http://www.academicpress.com

Academic Press
84 Theobald's Road, London WC1X 8RR, UK
http://www.academicpress.com

Library of Congress Catalog Card Number: 2003103019

International Standard Book Number: 0-12-179460-1 (Set)
International Standard Book Number: 0-12-179461-X (Volume 1)
International Standard Book Number: 0-12-179462-8 (Volume 2)
International Standard Book Number: 0-12-179463-6 (Volume 3)

PRINTED IN THE UNITED STATES OF AMERICA
03 04 05 06 07 8 7 6 5 4 3 2 1

Contents

Preface

The "Handbook of Secondary Fungal Metabolites" is presented in three volumes and is comprehensive to the extent that all major groups of secondary fungal metabolites are included. The format is similar to that presented in the "Handbook of Toxic Fungal Metabolites" with the major exception that actual spectra are not included; however, spectral data are included where available. Also included in these volumes are the methods used by the authors to isolate and purify metabolites. Another major difference is that the appropriate references are presented with each metabolite, negating the need to turn to the end of each group to find the appropriate references. Each volume contains four indexes: secondary metabolite index, molecular formula index, molecular weight index, and fungal/plant source index. In a few instances, plant sources are included when the metabolites are closely related to fungal metabolites or the source of precursors may be fungal; i.e., the baccharins, which are found in extracts from *Baccharis megapotamica*. These metabolites are closely related to the macrocyclic trichothecenes found in extracts of fungi such as *Myrothecium* spp. and *Stachybotrys* spp. Also, metabolites from the fungal symbiont of lichens are sometimes presented. To aid in the interpretation of NMR data, the numbering system presented in the literature is included for the major representative fungal metabolite and, at times, for several related metabolites. Fungal sources are given as reported in the original references. It is recognized that the taxonomy in several cases has been revised, perhaps more than once. It is beyond the scope of these volumes to deal with what is "currently accepted taxonomy" because this is a dynamic science that, in many cases, is as yet undefined.

The "Handbook" has been divided into sections, and the placement of metabolites is based on chemical relationships. One section of each volume contains a miscellaneous section to accommodate metabolites difficult to place into one of the sections. The miscellaneous section of Volume III contains some metabolites related to those that appear in Volumes I and II. This occurred when related metabolites were discovered after Volumes I and II were completed.

It is hoped that this compilation of data on secondary fungal metabolites will aid investigators in the identification of known or related fungal metabolites. Because fungal metabolites represent a wide diversity of chemical species, these volumes will be useful to scientists interested in correlations of structural features with various spectral and biological characteristics. The known biological activity of metabolites is presented, which may aid in future studies related to the identification of new uses for fungal metabolites.

Richard J. Cole
Milbra A. Schweikert

Acknowledgments

The authors thank the following investigators for their assistance in producing the "Handbook of Secondary Fungal Metabolites." Their contributions made this compilation of data on fungal metabolites possible.

Wayne L. Bryden
University of Queensland
Gotton, Queensland 4343
Australia

Brian E. Cole
Images, Inc.
Snellville, Georgia

Horace G. Cutler
Department of Pharmacology
Mercer University
Atlanta, Georgia

Jens C. Frisvad
Department of Biotechnology
Technical University of Denmark
DK-2800 Lyngby
Denmark

Bruce Horn
USDA-ARS
National Panut Research Laboratory
Dawson, Georgia

William Norred
Toxicology and Mycotoxin Research Unit
USDA-ARS
Athens, Georgia

James K. Porter
Toxicology and Mycotoxin Research Unit
USDA-ARS
Athens, Georgia

Ronald T. Riley
Toxicology and Mycotoxin Research Unit
USDA-ARS
Athens, Georgia

Victor S. Sobolev
USDA-ARS
National Peanut Research Laboratory
Dawson, Georgia

Bruce B. Jarvis
Department of Chemistry and Biochemistry
University of Maryland
College Park, Maryland

1

C_{27} Sterols

Cholest-5-en-3β-ol (Cholesterol)
Cholesta-5,24-dien-3β-ol (Desmosterol)
Cholest-7-en-3β-ol (Lathosterol)
Cholesta-5,22-dien-3β-ol (2-Dehydrocholesterol)
Cholest-8-en-3β-ol
Cholesta-5,7-dien-3β-ol (7-Dehydrocholesterol)
Cholesta-7,24-dien-3β-ol
Cholesta-8,24-dien-3β-ol (Zymosterol)
Cholesta-5,7,24-trien-3β-ol
Cholesta-5,8,24-trien-3β-ol
Cholesta-7,22,24-trien-3β-ol
Cholesta-8,22,24-trien-3β-ol
Cholesta-5,7,22,24-tetraen-3β-ol
Cholesta-5,8,22,24-tetraen-3β-ol

Common/Systematic Name

Cholesterol; Cholesterin
Cholest-5-en-3β-ol

Molecular Formula/Molecular Weight

$C_{27}H_{46}O$; MW = 386.35487

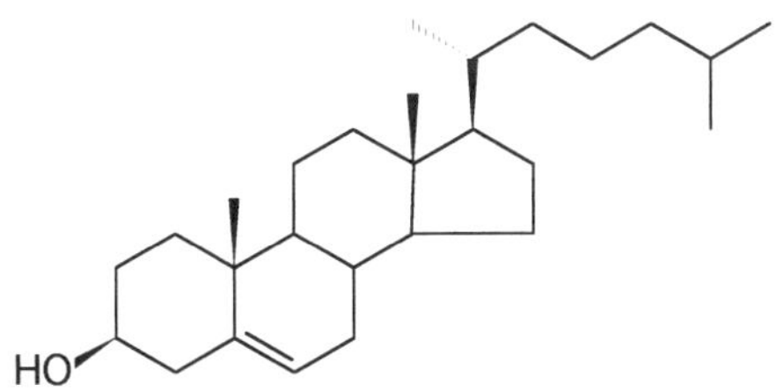

General Characteristics

Monohydrate, pearly leaflets or plates from dilute alcohol; becomes anhydrous at 70-80°C. When anhydrous, mp. 148.5°C. Sublimes as orthorhombic needles. $[\alpha]_D^{20}$ -31.5° (*c*=2, in ether); $[\alpha]_D^{20}$ -39.5° (*c*=2, in $CHCl_3$). Practically insoluble in water (about 0.2mg/100ml H_2O). Slightly soluble in alcohol (1.29%, w/w at 20°C), more soluble in hot alcohol (100 g of saturated 96% alcoholic solution contains 28g at 80°C). One gram dissolves in 2.8ml ether, in 4.5ml chloroform, in 1.5ml pyridine. Also soluble in benzene, petroleum ether, oils, and fats. Soluble in aqueous solutions of bile salts: Gives intense red color with rosaniline in chloroform solution. Methyl ether, crystals from acetone; mp., 84°C ; $[\alpha]_D^{20}$ -45.8°C (*c*=1.2, in $CHCl_3$). Acetate, needles from acetone; mp., 115-116°C; $[\alpha]_D^{20}$ -47.4° (*c*=2, in $CHCl_3$). Benzoate; mp., 145.5°C (the melt becomes clear at 180°C); and $[\alpha]_D^{25}$ -13.7° (*c*=0.9, in $CHCl_3$).

Fungal Source

Acaulospora laevis, Achlya bisexualis, Allomyces macrogynus, Allomyces spp., *Amoebidium parasiticum, Aspergillus oryzae, Blastocladia ramosa, Colletotrichum dematium, Dipsacomyces acuminosporus, Glomus caledonius, G. mosseae, Gnomonia leptostyla, Leptosphaeria typhae, Linderina pennispora, Monoblepharella* sp., *Nectria galligena, Phycomyces blakesleeanus, Rhizophlyctis rosea, Smittium* sp., and *Ustilago maydis.*

Isolation/Purification

Extracted with chloroform-methanol (3:1, v/v); the extract was saponified with ethanolic KOH and the non-saponifiable fraction extracted with *n*-hexane. The non-saponifiable fraction was fractionated on alumina (activity III). The nonpolar components were eluted with hexane, while diethyl ether eluted the sterol fraction. The free sterols were acetylated and separated on silica gel impregnated with silver nitrate. Cholesteryl acetate was identified using GLC.

Also, see R. C. Heupel; Isolation and Primary Characterization of Sterols, In Analysis of Sterols and Other Biologically Significant Steroids, W. D. Nes, and E. J. Parish (eds.), Academic Press Inc., New York NY, 341 pp. (1989).

Spectral Data

UV:
End absorption.

^{13}C NMR:
($CDCl_3$) C-1, 37.2; C-2, 31.5; C-3, 71.4; C-4, 42.2; C-5, 140.5; C-6, 121.3; C-7, 31.8; C-8, 31.8; C-9, 50.0; C-10, 36.4; C-11, 21.0; C-12, 39.7; C-13, 42.2; C-14, 56.6; C-15, 24.2; C-16, 28.2; C-17, 56.1; C-18, 11.8; C-19, 19.2; C-20, 35.7; C-21, 18.7; C-22, 36.1; C-23, 23.8; C-24, 39.4; C-25, 27.9; C-26, 22.5; and C-27, 22.7ppm.

Mass Data:
M^+ 386.64*m/e*, also 343, 301, 275, 247, 260, and 273*m/e*; Acetate derivative, 428, 413, 368(100), 353(28), 301, 299, 315, 273, 255, and 213*m/e*. C, 83.37%; H, 11.99%; O, 4.14%.

References

W. D. Nes, and E. J. Parish (eds.); Analysis of Sterols and Other Biologically Significant Steroids; Academic Press Inc., New York, NY, 341 pp. (1989).

W. B. Turner, and D. C. Aldridge; Fungal Metabolites II; Academic Press, New York, NY, 631 pp. (1983).

J. D. Weete; Structure and Function of Sterols in Fungi, Advances in Lipid Research, Vol. 23, pp. 115-167(1989).

Common/Systematic Name

Desmosterol; 24-Dehydrocholesterol

Cholesta-5,24-dien-3β-ol

Molecular Formula/Molecular Weight

$C_{27}H_{44}O$; MW = 384.33922

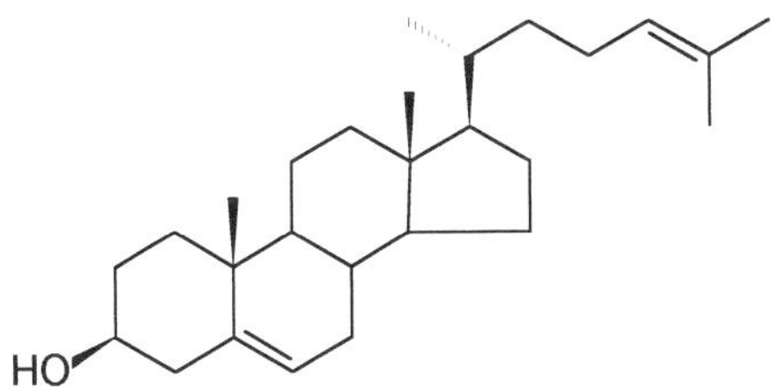

General Characteristics

Platelets from methanol; mp., 121.5°C; also reported as 119-119.5° C; $[\alpha]_D^{27}$ - 41.0° (*c*=1, in $CHCl_3$).

Fungal Source

Smittium culisetae, S. culicis, S. simulii, S. mucronatum, S. spp., and *Aplanopsis* spp.

Isolation/Purification

See R. C. Heupel; Isolation and Primary Characterization of Sterols, In Analysis of Sterols and Other Biologically Significant Steroids; W. D. Nes, and E. J. Parish (eds.), Academic Press Inc., New York, NY, 341 pp.(1989).

Spectral Data

^{13}C NMR:

($CDCl_3$) C-15, 24.4; C-16, 28.3; C-17, 56.2; C-18, 12.0; C-20, 36.2; C-21, 18.7; C-22, 35.7; C-23, 24.8; C-24, 125.5; C-25, 130.9; C-26, 17.7; and C-27, 25.7ppm (most other chemical shifts were omitted because these chemical shifts were consistent with those of cholesterol).

Mass Spectrum:

Acetate: 426(M^+), 411, 366(100), 356, 351(14), 315, 313, 296(4), 281, 273, 259, 255(8), 253(36), 229, 213(16), and *m/e*.

References

W. D. Nes and E. J. Parish (eds.); Analysis of Sterols and Other Biologically Significant Steroids; Academic Press Inc., New York, NY, 341 pp. (1989).

W. B. Turner and D. C. Aldridge; Fungal Metabolites II; Academic Press, Inc., New York, NY, 631 pp. (1983).

J. D. Weete; Structure and Function of Sterols in Fungi, Advances in Lipid Research, Vol. 23, pp. 115-167(1989).

Common/Systematic Name

Lathosterol

Cholest-7-en-3β-ol

Molecular Formula/Molecular Weight

$C_{27}H_{46}O$; MW = 386.35487

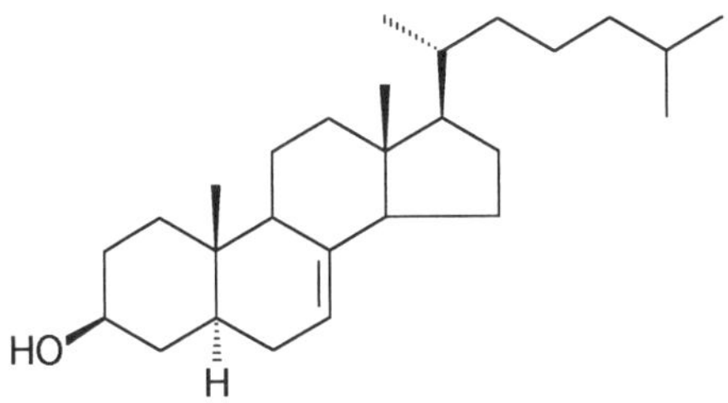

Fungal Source

Blastocladia ramosa, *Nectria galligena*, and *Dictyuchus monosporus*.

Isolation/Purification

See R. C. Heupel; Isolation and Primary Characterization of Sterols, In Analysis of Sterols and Other Biologically Significant Steroids; W. D. Nes, and E. J. Parish (eds.), Academic Press Inc., New York, NY, 341 pp.(1989).

Spectral Data

UV:

End absorption.

^{13}C NMR:

($CDCl_3$) C-1, 37.1; C-2, 31.3; C-3, 70.7; C-4, 37.8; C-5, 40.2; C-6, 29.6; C-7, 117.2; C-8, 139.3; C-9, 49.4; C-10, 34.1; C-11, 21.5; C-12, 39.5; C-13, 43.2; C-14, 54.9; C-15, 22.9; C-16, 27.9; C-17, 56.1; C-18, 11.8; C-19, 12.9; C-20, 36.1; C-21, 18.8; C-22, 36.1; C-23, 23.9; C-24, 39.4; C-25, 27.9; C-26, 22.5; and C-27, 22.7ppm.

References

W. D. Nes and E. J. Parish (eds.); Analysis of Sterols and Other Biologically Significant Steroids; Academic Press Inc., New York, NY, 341 pp. (1989).

W. B. Turner and D. C. Aldridge; Fungal Metabolites II; Academic Press, Inc., New York, NY, 631 pp. (1983).

J. D. Weete; Structure and Function of Sterols in Fungi, Advances in Lipid Research, Vol. 23, pp. 115-167(1989).

Common/Systematic Name

22-Dehydrocholesterol
Cholesta-5,22-dien-3β-ol

Molecular Formula/Molecular Weight

$C_{27}H_{44}O$; MW = 384.33922

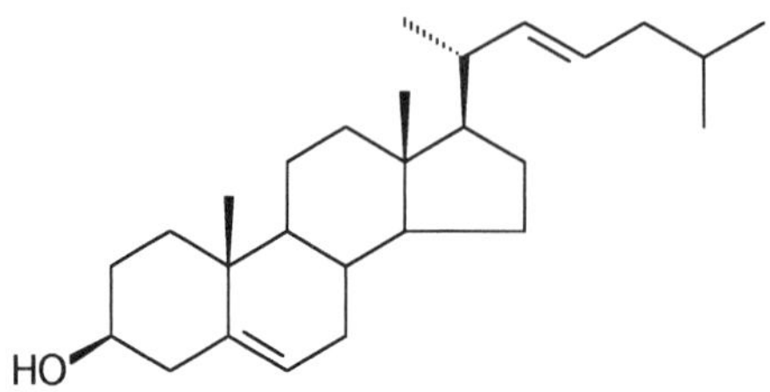

Fungal Source

Physarum polycephalum, and *Rhizophlyctis rosea*.

Isolation/Purification

See R. C. Heupel; Isolation and Primary Characterization of Sterols, In Analysis of Sterols and Other Biologically Significant Steroids; W. D. Nes, and E. J. Parish (eds.), Academic Press Inc., New York, NY, 341 pp.(1989).

Spectral Data

UV:
End absorption.

^{13}C NMR:
($CDCl_3$) (22*E*-dehydrocholesterol) C-15, 24.3; C-16, 28.6; C-17, 56.0; C-18, 12.1; C-20, 40.1; C-21, 20.9; C-22, 138.1; C-23, 126.2; C-24, 42.0; C-25, 28.6; C-26, 22.3; and C-27, 22.3ppm; (22*Z*-dehydrocholesterol) C-15, 24.3; C-16, 28.1; C-17, 56.3; C-18, 12.2; C-20, 34.3; C-21, 20.7; C-22, 137.2; C-23, 125.5; C-24, 36.8; C-25, 28.8; C-26, 22.4; and C-27, 22.6ppm.

References

W. D. Nes and E. J. Parish (eds.); Analysis of Sterols and Other Biologically Significant Steroids; Academic Press Inc., New York, NY, 341 pp. (1989).

W. B. Turner and D. C. Aldridge; Fungal Metabolites II; Academic Press, Inc., New York, NY, 631 pp. (1983).

J. D. Weete; Structure and Function of Sterols in Fungi, Advances in Lipid Research, Vol. 23, pp. 115-167(1989).

Common/Systematic Name

Cholest-8(9)-en-3β-ol

Molecular Formula/Molecular Weight

$C_{27}H_{46}O$; MW = 386.35487

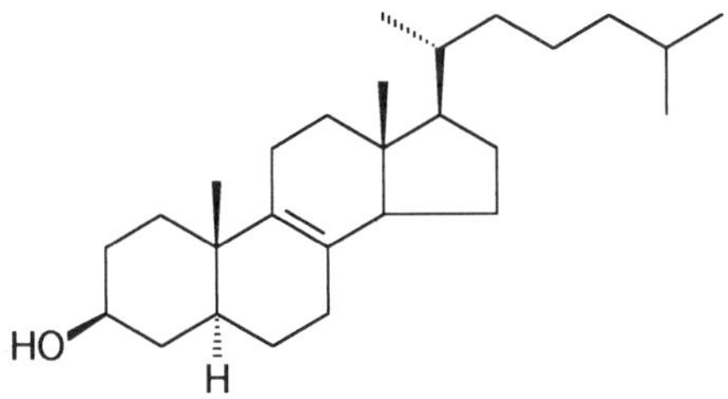

Fungal Source

Blastocladia ramosa.

Isolation/Purification

See R. C. Heupel; Isolation and Primary Characterization of Sterols, In Analysis of Sterols and Other Biologically Significant Steroids; W. D. Nes, and E. J. Parish (eds.), Academic Press Inc., New York, NY, 341 pp. (1989).

Spectral Data

UV:

End absorption.

^{13}C NMR:

($CDCl_3$) C-1, 35.1; C-2, 31.5; C-3, 70.9; C-4, 38.2; C-5, 40.7; C-6, 25.4; C-7, 27.1; C-8, 128.0; C-9, 134.8; C-10, 35.6; C-11, 22.7; C-12, 36.9; C-13, 42.0; C-14, 51.8; C15, 23.9; C-16, 28.7; C-17, 54.8; C-18, 11.2; C-19, 17.8; C-20, 36.2; C-21, 18.7; C22, 36.1; C-23, 23.7; C-24, 39.4; C-25, 27.9; C-26, 22.4; and C-27, 22.7ppm.

References

W. D. Nes and E. J. Parish (eds.); Analysis of Sterols and Other Biologically Significant Steroids; Academic Press Inc., New York, NY, 341 pp. (1989).

W. B. Turner and D. C. Aldridge; Fungal Metabolites II; Academic Press, Inc., New York, NY, 631 pp. (1983).

J. D. Weete; Structure and Function of Sterols in Fungi, Advances in Lipid Research, Vol. 23, pp. 115-167(1989).

Common/Systematic Name

7-Dehydrocholesterol; Provitamin D_3
Cholesta-5,7-dien-3β-ol

Formula/Molecular Weight

$C_{27}H_{44}O$; MW = 384.33922

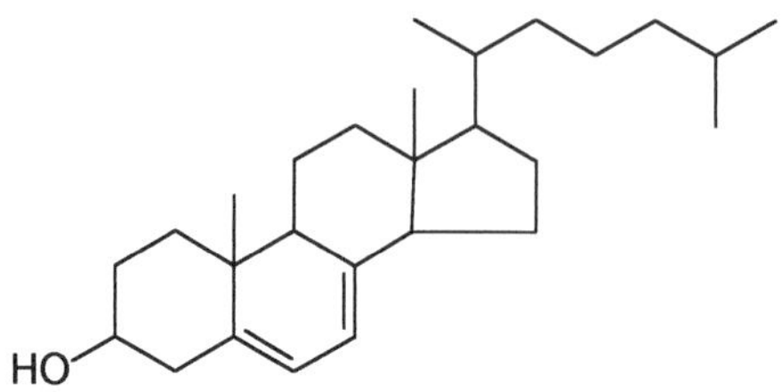

General Characteristics

Hydrated plates from ethyl ether-methanol; mp., 150-151°C (anhydrous); $[\alpha]_D^{20}$ = -113.6° (c=1.0, in $CHCl_3$); $[\alpha]_D^{20}$ - 127.1° (in benzene). Insoluble in water; soluble in usual organic solvents; gives pink to blue color with antimony trichloride, blue with chloral hydrate and red with 90% trichloroacetic acid. Acetate, crystals from methanol; mp., 129-130°C; $[\alpha]_D^{20}$ = -85.3° (c=1.2, in benzene).

Fungal Source

Lactarius sp., *Nectria galligena, Russula aeruginosa,* and *R. decolorans.*

Isolation/Purification

7-Dehydrocholesterol was isolated from the neutral lipid fraction by column chromatography on neutral alumina (grade IV) using 10%, 20%, 50% benzene/*n*-hexane and benzene. The sterol containing fraction was analyzed by silica gel-G TLC (rhodamine-impregnated plates, 0.1%) or acetylated with pyridine-acetic anhydride and purified by argentation TLC chromatography (20% $AgNO_3$ -impregnated plates).

Also see R. C. Heupel: Isolation and Primary Characterization of Sterols, In Analysis of Sterols and Other Biologically Significant Steroids; W. D. Nes, and E. J. Parish (eds.); Academic Press Inc., New York, NY, pp. 1-27(1989).

Biological Activity

Converted to vitamin D_3 by UV irradiation.

Spectral Data

UV:

λ_{max}^{MeOH} 260, 271, 281.5, and 293.5nm.

Mass Spectrum:
LREIMS: 384(M^+), 369(M^+ - methyl), 366(M^+ - H_2O), 299(M^+ - H_2O - C_5H_7), and 271*m/e* (M^+ - side chain).

References

B. Dehorter, M. F. Brocquet, L. Lacoste, J. Alais, A. Lablache-Combier, A. Maquestiau, Y. van Harverbeke, R. Flammang, and H. Mispreuve; Phytochemistry, Vol. 19, pp. 2311-2316(1980).

R. C. Heupel; Isolation and Primary Characterization of Sterols, In Analysis of Sterols and Other Biologically Significant Steroids; W. D. Nes, and E. J. Parish (eds.); Academic Press Inc., New York, NY, pp. 1-27(1989).

L. G. Jayko, T. I. Baker, R. D. Stubblefield, and R. F. Anderson; Nutrition and Metabolic Products of *Lactarius* species; Canad. J. Microbiol., Vol. 8, pp. 361-371(1962).

A. M. Pierce, A. M. Unrau, A. C. Oehlschlager, and R. A. Woods; Azasterol Inhibitors in Yeast. Inhibition of the Δ^{24} -Sterol Methyltransferase and the 24-Methylene Sterol $\Delta^{24(28)}$-Reductase in Sterol Mutants of *Saccharomyces cerevisiae*; Canad. J. Biochem., Vol. 57, pp. 201-208(1979).

W. B. Turner and D. C. Aldridge; Fungal Metabolites II; Academic Press, New York, N.Y., p. 307(1983).

Common/Systematic Name

Cholesta-7,24-dien-3β-ol

Molecular Formula/Molecular Weight

$C_{27}H_{44}O$; MW = 384.33922

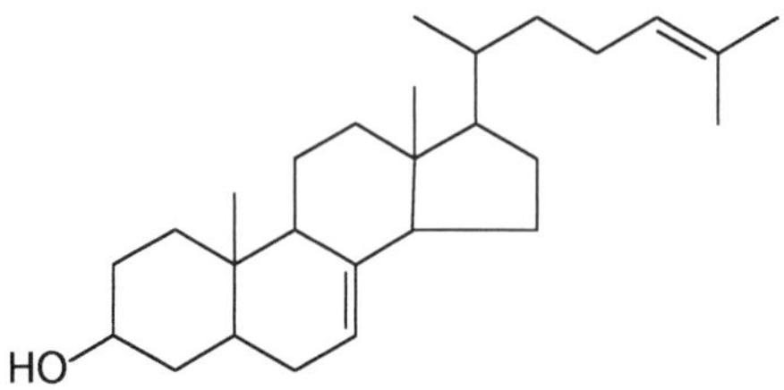

Fungal Source

Saccharomyces cerevisiae.

Isolation/Purification

Sterols from the fungal extract were purified by preparative TLC (plates were coated with silica gel GF_{254} impregnated with 45% silver nitrate and 0.25% rhodamine 6G and developed with cyclohexane-ethyl acetate, 1:1, v/v) of their acetate derivatives. The free sterol was purified via the benzoate.

Also see Heupel, R. C.; Isolation and Primary Characterization of Sterols, In Analysis of Sterols and Other Biologically Significant Steroids; W. D. Nes, and E. J. Parish (eds.); Academic Press Inc., New York, NY, 341 pp, (1989).

Spectral Data

^{1}H NMR:

($CDCl_3$) 0.58(s, 3H, 18-H_3); 0.84(s, 3H, 19-H_3); 0.97(d, 3H, $J_{20,21}$=6.0Hz, 21-H); 1.64 and 1.72(2s, 6H, 26-H_3 and 27-H_3); 3.68(m, 1H, H-3); and 5.16ppm (m, 2H, H-7,24).

Mass Spectrum:

LREIMS: 384*m/e* (M^+, 31%) (calcd for $C_{27}H_{44}O$, 384), 369(M^+ - CH_3, 29), 330(17), 300(17), and 271*m/e* (100) (M^+ - side chain); acetate, 422*m/e* (M^+) calcd for $C_{29}H_{42}O_2$.

References

L. Avruch, S. Fischer, H. Pierce, Jr., and A. C. Oehlschlager; The Induced Biosynthesis of 7-Dehydrocholesterol in Yeast: Potential Sources of New Provitamin D_3 Analogs; Can. J. Biochem., Vol. 54, pp. 657-665(1976).

W. D. Nes and E. J. Parish (eds.); Analysis of Sterols and Other Biologically Significant Steroids; Academic Press Inc., New York, NY, 341 pp.(1989).

W. B. Turner and D.C. Aldridge; Fungal Metabolites II; Academic Press, Inc., New York, NY, 631 pp. (1983).

J. D. Weete; Structure and Function of Sterols in Fungi, Advances in Lipid Research; Vol. 23, pp. 115-167(1989).

Common/Systematic Name

Zymosterol; Cholesta-8,24-dien-3β-ol

Formula/Molecular Weight

$C_{27}H_{44}O$; MW = 384.33922

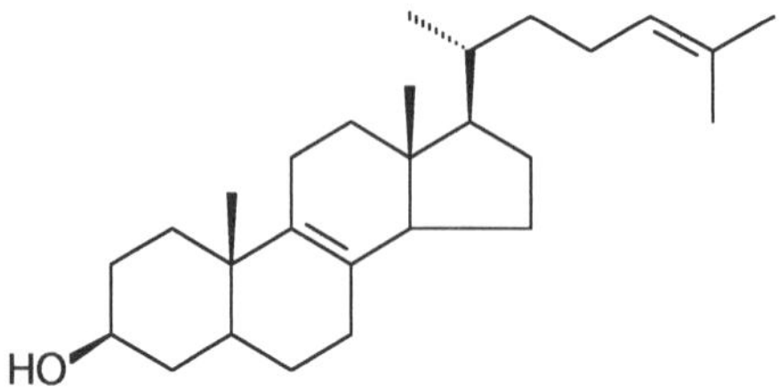

General Characteristics

Plates from chloroform-methanol and acetone-methanol; mp., 110-113°C; $[\alpha]_D^{25}$ + 53.7° (*c*=1.55, in $CHCl_3$); benzoate, prisms from chloroform-methanol; mp., 126-128°C; $[\alpha]_D^{27}$ = + 44.8° (*c*=1.3, in $CHCl_3$).

Fungal Source

Saccharomyces cerevisiae.

Isolation/Purification

Sterol residues with most of the ergosterol removed were chromatographed by column chromatography on neutral alumina using benzene and various combinations of benzene-*n*-hexane, benzene-acetone and benzene-methanol. The sterol was finally purified as the benzoate by silica gel GF_{254} TLC or by argentation TLC chromatography (20% $AgNO_3$ impregnated plates) and fractional crystallization.

Also see R. C. Heupel: Isolation and Primary Characterization of Sterols, In Analysis of Sterols and Other Biologically Significant Steroids; W. D. Nes, and E. J. Parish (eds.); Academic Press Inc., New York, NY, pp. 1-27(1989).

Spectral Data

UV:

λ_{max}^{MeOH} End absorption.

IR:

(KBr) 890 and 1640cm^{-1}.

^{1}H NMR:

($CDCl_3$) 9.00 and 9.3 8τ (singlet methyl chemical shifts); 4.92τ (24-H).

Mass Spectrum:
Acetate, LREIMS: 426(M^+, 100%), 411(M^+ - methyl, 41), 366(M^+ - ROH, 19), 351(M^+- Me - ROH), and 313*m/e* (M^+- side chain - 2H).

References

D. H. R. Barton, U. M. Kempe, and D. A. Widdowson; Investigations on the Biosynthesis of Steroids and Terpenoids. Part VI. The Sterols of Yeast; J. Chem. Soc., Perkin Trans. I, pp. 513-522(1972).

R. C. Heupel; Isolation and Primary Characterization of Sterols, In Analysis of Sterols and Other Biologically Significant Steroids; W. D. Nes, and E. J. Parish (eds.) Academic Press Inc., New York, NY, pp. 1-27(1989).

W. B. Turner and D. C. Aldridge; Fungal Metabolites II; Academic Press, New York, N.Y., p. 308 (1983).

Common/Systematic Name

Cholesta-5,7,24-trien-3β-ol

Molecular Formula/Molecular Weight

$C_{27}H_{42}O$; MW = 382.32357

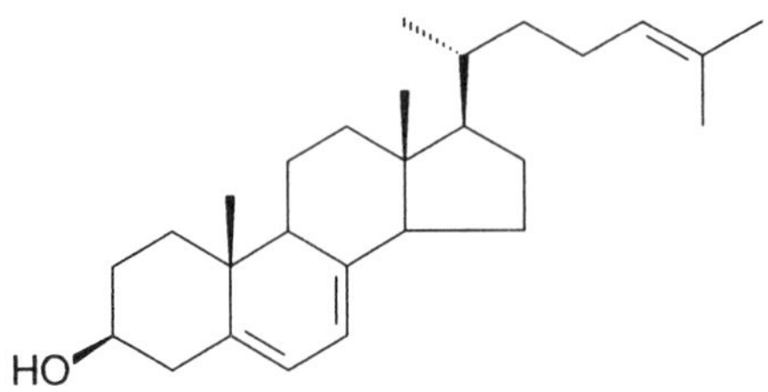

General Characteristics

Needles from methanol; mp., 103.5-105°C; $[\alpha]_D$ - 118.9° (*c*=0.35, in $CHCl_3$) mp., 102-102.5°C; $[\alpha]_D$ - 111.0°; benzoate, plates from chloroform-methanol; mp., 131-134°C; $[\alpha]_D$ - 55.0° (*c*=0.39, in $CHCl_3$).

Fungal Source

Saccharomyces cerevisiae.

Isolation/Purification

Yeast cells were refluxed in 10% methanolic KOH, methanol removed *in vacuo*, and ether soluble material converted into benzoate derivatives. The benzoates were separated with preparative silica gel GF_{254} TLC followed by silver nitrate-impregnated TLC.

Spectral Data (Benzoate)

UV:

λ_{max} 229(ε = 18,200), 252sh (7,400), 262sh (10,800), 272(15,100), 282(15,300), and 294nm (8,200).

Mass Data:

Found: C, 84.0; H, 9.4; calcd for $C_{34}H_{46}O_2$; C, 83.9; H, 9.5%.

References

D. H. R. Barton, J. E. T. Corrie, D. A. Widdowson, M. Bard, and R. A. Woods; Biosynthesis of Terpenes and Steroids. Part IX. The Sterols of Some Mutant Yeasts and Their Relationship to the Biosynthesis of Ergosterol; J. Chem. Soc. Perkin Trans I, pp. 1326-1333(1974).

D. H. R. Barton, A. A. L. Gunatilaka, T. R. Jarman, D. A. Widdowson, M. Bard, and R. A. Woods; Biosynthesis of terpenes and steroids. Part X. The Sterols of Some Yeast

Mutants Doubly Defective in Ergosterol Biosynthesis; J. Chem. Soc. Perkin Trans I, pp. 88-92 (1975).

Common/Systematic Name

Cholesta-5,8,24-trien-3β-ol

Molecular Formula/Molecular Weight

$C_{27}H_{42}O$; MW = 382.32357

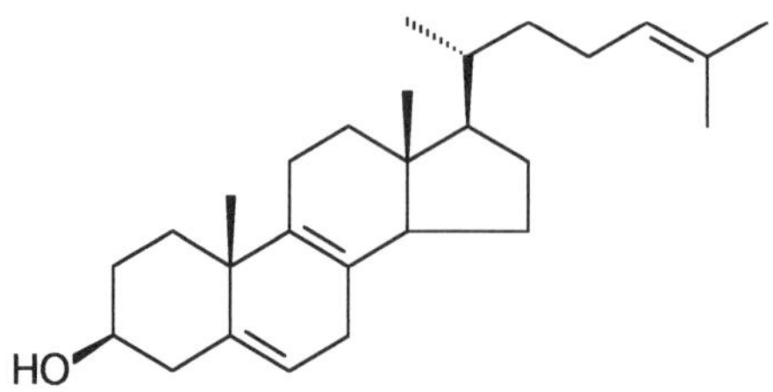

Fungal Source

Saccharomyces cerevisiae.

Isolation/Purification

The nonsaponifiable fraction was acetylated and the sterol was purified by preparative argentation TLC (silica gel GF_{254} coated with 25% $AgNO_3$) of the acetate derivative using benzene as mobile phase. 3β-Acetoxycholesta-5,8,24(28)-triene eluted together with a small amount of 3β-acetoxycholesta-5,8,22,24(28)-tetraene from a band at R_f 0.31-0.38. The fraction was further purified with three consecutive developments in benzene-hexane 3:2, 1:2, and 1:5 (v/v); two major bands were obtained at R_f 0.17-0.26 and 0.26-0.30. After further purification of both bands by TLC (silica gel GF_{254}, eluted with benzene), the upper band yielded 3β-acetoxycholesta-5,8,24-triene and the lower band yielded 3β-acetoxycholesta-5,8,22,24-tetraene.

Spectral Data

^{1}H NMR:

($CDCl_3$) (acetate) 0.66(3H, s, H3-18); 0.95(3H, d, $J_{20,21}$=6.OHz, H-21; 1.09(3h, d, H-21); 1.21(3H, s, H-19); 2.08(3H, s, CH_3CO-); 4.64(1H, m, H-3); 4.82(2H, m, H-28); 5.41(1H, m, H-6); 5.51(1H, dd, $J_{20,22}$=8.5Hz); and 5.92ppm (1H, d, $J_{22,23}$=15.3Hz).

Mass Spectrum:

Acetate, LREIMS: 436(M^+, 5.9%), 376(70.5), 361(100), 253(25.3), 251(23.2), 211(12.2), 157(31.2), 143(24.8), 123(17.2), and 81*m/e* (23.0).

Reference

A. M. Pierce, A. M. Unrau, A. C. Oehlschlager, and R. A. Woods; Azasterol Inhibitors in Yeast. Inhibition of the Δ^{24}-Sterol Methyltransferase and the 24-Methylene Sterol $\Delta^{24(28)}$-Reductase in Sterol Mutants of *Saccharomyces cerevisiae*; Can. J. Biochem., Vol. 57, pp. 201-208(1979).

Common/Systematic Name

Cholesta-7,22,24(28)-trien-3β-ol

Molecular Formula/Molecular Weight

$C_{27}H_{42}O$; MW = 382.32357

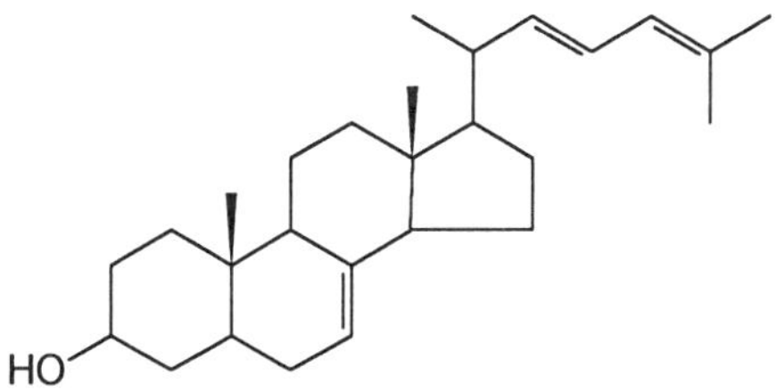

Fungal Source

Saccharomyces cerevisiae.

Isolation/Purification

See R. C. Heupel: Isolation and Primary Characterization of Sterols, In Analysis of Sterols and Other Biologically Significant Steroids, W. D. Nes, and E. J. Parish (eds.), Academic Press Inc., New York, NY, pp. 1-27(1989).

References

A. M. Pierce, A. M. Unrau, A. C. Oehlschlager, and R. A. Woods; Azasterol Inhibitors in Yeast. Inhibition of the Δ^{24}-Sterol Methyltransferase and the 24-Methylene Sterol $\Delta^{24(28)}$-Reductase in Sterol Mutants of *Saccharomyces cerevisiae*, Can. J. Biochem., Vol. 57, pp. 201-208(1979).

R. C. Heupel; Isolation and Primary Characterization of Sterols, In Analysis of Sterols and Other Biologically Significant Steroids, W. D. Nes and E. J. Parish (eds.) Academic Press Inc., New York, NY, pp. 1-27(1989).

Common/Systematic Name

Cholesta-8,22,24-trien-3β-ol

Molecular Formula/Molecular Weight

$C_{27}H_{42}O$; MW = 382.32357

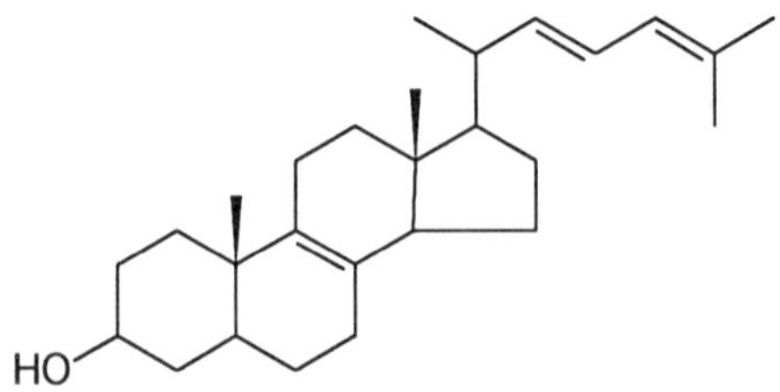

Fungal Source

Saccharomyces cerevisiae.

Isolation/Purification

See R. C. Heupel: Isolation and Primary Characterization of Sterols, In Analysis of Sterols and Other Biologically Significant Steroids; W. D. Nes, and E. J. Parish (eds.), Academic Press Inc., New York, NY, pp. 1-27(1989).

References

R. C. Heupel; Isolation and Primary Characterization of Sterols, In Analysis of Sterols and Other Biologically Significant Steroids; W. D. Nes, and E. J. Parish (eds.) Academic Press Inc., New York, NY, pp. 1-27(1989).

A. M. Pierce, A. M. Unrau, A. C. Oehlschlager, and R. A. Woods; Azasterol Inhibitors in Yeast. Inhibition of the Δ^{24}-Sterol Methyltransferase and the 24-Methylene Sterol $\Delta^{24(28)}$-Reductase in Sterol Mutants of *Saccharomyces cerevisiae*; Can. J. Biochem., Vol. 57, pp. 201-208(1979).

Common/Systematic Name

Cholesta-5,7,22,24-tetraen-3ß-ol

Molecular Formula/Molecular Weight

$C_{27}H_{40}O$; MW = 380.30792

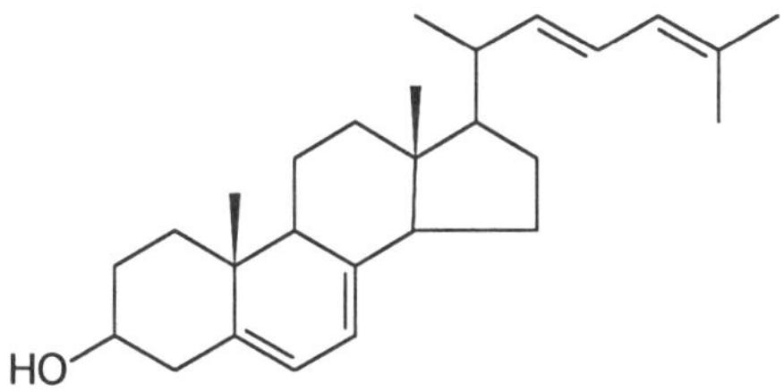

General Characteristics

Crystals from methanol; mp., 139-141°C (141-143°C); acetate, crystals from chloroform-methanol; mp., 137-138°C (sealed, evacuated capillary).

Fungal Source

Saccharomyces cerevisiae.

Isolation/Purification

Sterols from the fungal extract were purified by preparative TLC (plates were coated with silica gel GF_{254} impregnated with 45% silver nitrate and 0.25% rhodamine 6G and developed with cyclohexane-ethyl acetate, 1:1) of their acetate derivatives. The free sterol was purified via the benzoate.

Also see Heupel, R. C.; Isolation and Primary Characterization of Sterols, In Analysis of Sterols and Other Biologically Significant Steroids; W. D. Nes, and E. J. Parish (eds.); Academic Press Inc., New York, NY, 341 pp, (1989).

Spectral Data

UV:

λ_{max}^{EtOH} 232sh(ϵ = 27,170), 238(29,200), 247sh (21,900), 261(9,900), 271(12,800), 281(13,000), and 293nm (7,600); (acetate) 233sh (ϵ = 28,400), 239(30,200), 247sh (23,000), 261(10.100), 271(13,100), 282(13,400), and 293nm (7,680).

^{1}H NMR:

($CDCl_3$) 0.71(s, 3H, 18-H_3); 1.01(s, 3H, 19-H_3); 1.13(d, $J_{20,21}$=6.5Hz, 21-H); 1.81(s, 6H, 26-H_3 and 27-H_3); 3.65(m, 1H, H-3); 5.19-5.89(m, 4H, H-6, 7,22,24); and 6.19ppm (q, $J_{22,23}$=13.0Hz, $J_{23,24}$=10.0Hz, 1H); (Acetate) 0.72(s, 3H, 18-H_3); 1.02(s, 3H, 19-H_3); 1.12(d, $J_{20,21}$=6.5Hz, 21-H); 1.82(s, 6H, 26-H_3 and 27-H_3); 2.10(s, 3H, CH_3CO-); 4.72(m, 1H, H-3); 5.40(AB, $J_{6,7}$=5.5Hz, H-7); 5.44(dd, $J_{22,23}$=14.5Hz, $J_{21,22}$=8.2Hz, H-22); 5.60(1H, H-6); 5.78(d, m, 1H, $J_{23,24}$=10.5Hz, $J_{24,26,27}$=1Hz, H-24); and 6.18ppm (dd, 1H, H-24).

Mass Spectrum:

LREIMS: 380*m/e* (12%) (calcd for $C_{27}H_{40}O$, 380), 347(M^+ - H_2O - CH_3, 11), and 109*m/e* (100) (M^+ - side chain); acetate, 422*m/e* (M^+) calcd for $C_{29}H_{42}O_2$.

References

L. Avruch, S. Fischer, H. Pierce, Jr., and A. C. Oehlschlager; The Induced Biosynthesis of 7-Dehydrocholesterol in Yeast: Potential Sources of New Provitamin D_3 Analogs; Can. J. Biochem., Vol. 54, pp. 657-665(1976).

W. D. Nes and E. J. Parish (eds.); Analysis of Sterols and Other Biologically Significant Steroids; Academic Press Inc., New York, NY, 341 pp.(1989).

W. B. Turner and D.C. Aldridge; Fungal Metabolites II; Academic Press, Inc., New York, NY, 631 pp. (1983).

J. D. Weete; Structure and Function of Sterols in Fungi, Advances in Lipid Research; Vol. 23, pp. 115-167(1989).

Common/Systematic Name
Cholesta-5,8,22,24-tetraen-3β-ol

Molecular Formula/Molecular Weight
$C_{27}H_{40}O$; MW = 380.30792

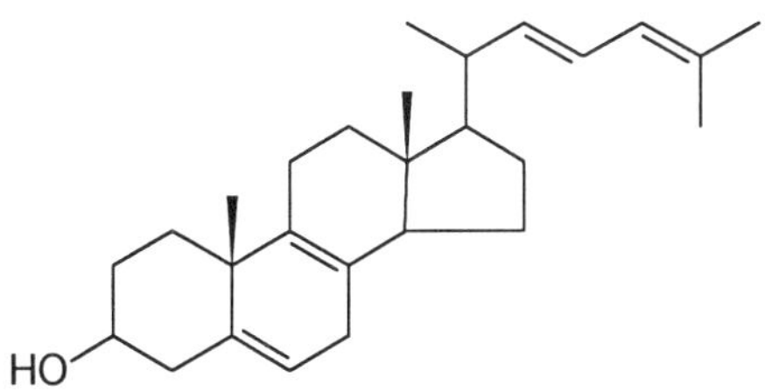

Fungal Source
Saccharomyces cerevisiae.

Isolation/Purification
The nonsaponifiable fraction was acetylated and the sterol was purified by preparative argentation TLC (silica gel GF_{254} coated with 25% $AgNO_3$) of the acetate derivative using benzene as mobile phase. 3β-Acetoxycholesta-5,8,24(28)-triene eluted together with a small amount of 3β-acetoxyergosta-5,8,22,24(28)-tetraene and 3β-acetoxy-cholesta-5,8,22,24(28)-tetraene from a band at R_f 0.31-0.38. The fraction was further purified with three consecutive developments in benzene:hexane 3:2, 1:2, and 1:5 (v/v); two major bands were obtained at R_f 0.17-0.26 and 0.26-0.30. After further purification of both bands by TLC (silica gel GF_{254}, benzene), the upper band yielded 3β-acetoxycholesta-5,8,24(28)-triene and the lower band yielded 3β-acetoxycholesta-5,8,22,24(28)-tetraene.

Spectral Data (Acetate)

UV:
$\lambda_{max}^{Hexanes}$ 232(sh), 238, and 247nm.

^{1}H NMR:
($CDCl_3$) 0.69(3H, s, H_3-18); 1.06(3H, d, $J_{20,21}$=6.3Hz, H-21); 1.21(3H, s, H-19); 1.75(6H, s, 26-CH_3 and 27-CH_3); 2.05(3H, s, CH_3-CO-); 4.61 (1H, m, H-3); 5.32(1H, dd, $J_{20,22}$=9.2Hz, H-22); 5.41(1H, m, H-6); 5.69(1H, d, $J_{23,24}$=12.0Hz, H-24); and 6.07ppm (1H, d, $J_{22,23}$=4.5Hz, H-23).

Mass Spectrum:
LREIMS: 422(M^+, 5.7%), 362(34), 347(55.3), 253(26.7), 251(16.1), and 109*m/e* (100).

Reference

A. M. Pierce, A. M. Unrau, A. C. Oehlschlager, and R. A. Woods; Azasterol Inhibitors in Yeast. Inhibition of the Δ^{24}-Sterol Methyltransferase and the 24-Methylene Sterol $\Delta^{24(28)}$-Reductase in Sterol Mutants of *Saccharomyces cerevisiae*; Can. J. Biochem., Vol. 57, pp. 201-208(1979).

C_{28} Sterols

2

(24*R*) 24-Methylcholest-5-en-3β-ol (Campesterol)
(24*R*) 24-Methylcholesta-5,22-dien-3β-ol (Brassicasterol, Ergosta-5,22-dien-ol)
(24*R*) 24-Methylcholestan-3β-ol (Campestanol)
(24*S*) 24-Methylcholesta-3β-ol (Ergostanol)
(24*S*) 24-Methylcholest-8-en-3β-ol (Ergost-8-en-3β-ol)
24*S*) 24-Methylcholesta-5,8-dien-3β-ol (Ergosta-5,8-dien-3β-ol)
24-Methylenecholesterol (Ergosta-5,24(28)-dien-3β-ol)
Ergosta-8,24(28)-dien-3β-ol (Fecosterol)
(24*S*) 24-Methylcholest-7-en-3β-ol (Fungisterol, Ergost-7-en-3β-ol)
(24S) 24-Methylcholesta-5,7,14-trien-3β-ol (Ergosta-5,7,14-trien-3β-ol)
Ergosta-5,7,22-trien-3β-ol (Ergosterol)
Ergosta-5,7-dien-3β-ol
Ergosta-7,22-dien-3β-ol (5,6-Dihydroergosterol)
Ergosta-7,24(28)-dien-3β-ol (Episterol)
Ergosta-5,7,24(28)-trien-3β-ol
Ergosta-8,14,22-trien-3β-ol
(24*S*)24-Methylcholest-5-en-3β-ol (Ergost-5-en-3β-ol)
(24*S*)24-Methylcholesta-5,8,22-trien-3β-ol (Lichesterol)
Ergosta-5,7-dien-3β-ol (22,23-Dihydroergosterol)
(24*S*) 24-Methylcholesta-7,16-dien-3β-ol
Ergosta-4,6,8(14),22-tetraen-3-one
(24*S*) 24-Methylcholesta-8,14-dien-3β-ol (Ignosterol)
(24*S*) 24-Methylcholesta-8,22-dien-3β-ol (Ergosta-8,22-dien-3β-ol)
Ergosta-7,22,24(28)-trien-3β-ol
Ergosta-8,14,24(28)-trien-3β-ol
Ergosta-8,22,24-trien-3β-ol
(24*S*)24-Methylcholesta-5,7,9,(11),22-tetraen-3β-ol (Ergosta-5,7,9(11),22-tetraen-3β-ol)
Ergosta-5,8,22,24-tetraen-3β-ol
Ergosta-4,7,22-trien-3-one
Ergosta-7,22-diene-3β,5α-diol
Ergosta-8,24-diene-3β,6α-diol
Portensterol (Hydroxy-*iso*-ergosterol)
Pyrocalciferol
Ergosta-5,7,22,24(28)-tetraen-3β-ol
Ergosta-7,22-dien-3-one
Ergosta-7,22-dien-5α,6β,7β-triol (Cerevisterol)
14α-Methylcholest-7-en-3β-ol
14α-Methylcholest-8-en-3β-ol
14α-Methylzymosterol
4α-Methylzymosterol

Common/Systematic Name

Campesterol; 24α-Methylcholesterol
(24*R*)24-Methylcholest-5-en-3β-ol

Molecular Formula/Molecular Weight

$C_{28}H_{48}O$; MW = 400.37052

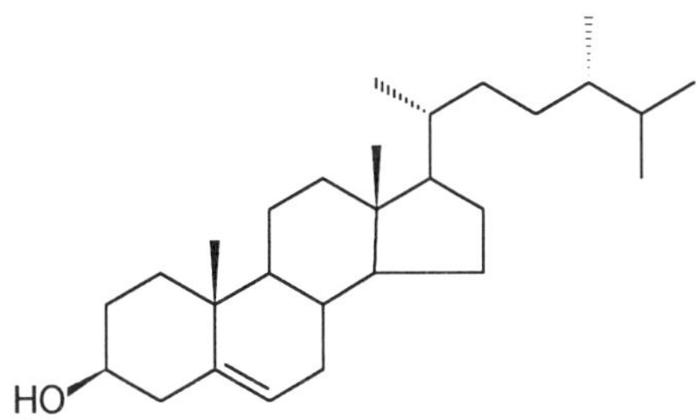

General Characteristics

Crystals from acetone; mp., 157-158°C; $[\alpha]_D^{23}$ -33° (22.5mg, in 5ml $CHCl_3$). Acetate: $C_{30}H_{50}O_2$, crystals from alcohol; mp., 137-138°C; $[\alpha]_D^{23}$ -35° (28.8mg, in 1ml $CHCl_3$).

Fungal Source

Allomyces macrogynus, Fusarium sporotrichioides, Glomus mosseae, Hyphochytrium catenoides, Rhizidiomyces apophysatus, Stachybotrys alternans, Spizellomyces punctatum, Monoblepharella sp., *Acaulospora laevis,* and *Physarum polycephalum.*

Isolation/Purification

See R. C. Heupel; Isolation and Primary Characterization of Sterols, In Analysis of Sterols and Other Biologically Significant Steroids; W. D. Nes and E. J. Parish (eds.), Academic Press Inc., New York, NY, 341 pp.(1989).

Spectral Data

UV:

End absorption.

^{13}C NMR:

($CDCl_3$) C-15, 24.4; C-16, 28.3; C-17, 56.2; C-18, 11.9; C-20, 36.0; C-21, 18.8; C-22, 33.8; C-23, 30.4; C-24, 38.9; C-25, 32.5; C-26, 20.3; C-27, 18.3; and C-28, 15.4ppm. (Skeletal carbon chemical shifts essentially similar to cholesterol).

Mass Data:

400(M^+), 385, 382, 367, 315, 289, 274, 273, 261, 255, 246, 229, 231, and 213*m/e*; C 83.93%. H 12.08%, O 3.99%.

References

M. Lenfant, M. F. Lecompte, and G. Farrugi; Identification des Stérols de *Physarum polycephalum*; Phytochemistry, Vol. 9, pp. 2529-2535(1970).

W. D. Nes and E. J. Parish (eds.); Analysis of Sterols and Other Biologically Significant Steroids; Academic Press Inc., New York, NY, 341 pp. (1989).

W. B. Turner and D. C. Aldridge; Fungal Metabolites II; Academic Press, Inc., New York, NY, 631 pp. (1983).

J. D. Weete; Structure and Function of Sterols in Fungi, Advances in Lipid Research, Vol. 23, pp. 115-167(1989).

Common/Systematic Name

Brassicasterol

(24*R*)24-Methylcholesta-5,22-dien-3β-ol

Molecular Formula/Molecular Weight

$C_{28}H_{46}O$; MW = 398.35487

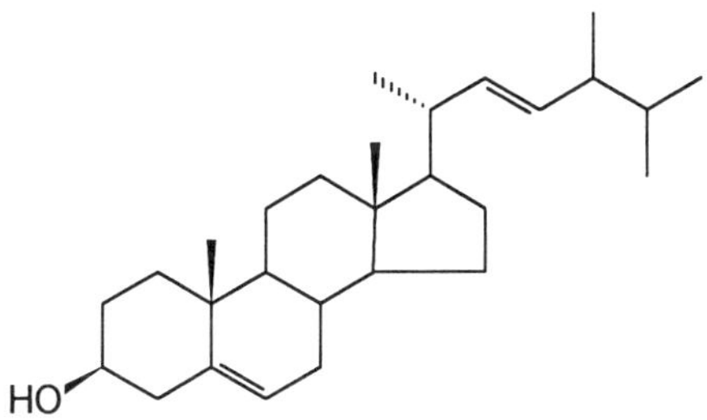

Fungal Source

Aspergillus oryzae, Cladonia gonecha, Stereocaulon tomentosum, Trichophyton rubrum, Tuber melanosporum, T. brumale, Terfezia sp., *Protomyces* sp., *Gnomonia leptostyla, Claviceps purpurea,* and *C. fusiformis.* Widespread in the genera *Protomyces* and *Taphrina.*

Isolation/Purification

See R. C. Heupel; Isolation and Primary Characterization of Sterols, In Analysis of Sterols and Other Biologically Significant Steroids; W. D. Nes, and E. J. Parish (eds.), Academic Press Inc., New York, NY, 341 pp.(1989).

Spectral Data

^{13}C NMR:

($CDCl_3$) C-15, 24.4; C-16, 28.6; C-17, 56.1; C-18, 12.1; C-20, 40.2; C-21, 21.1; C-22, 135.8; C-23, 131.8; C-24, 42.9; C-25, 33.2; C-26, 20.0; C-27, 19.7; and C-28, 17.7ppm.

Mass Spectrum:

Acetate derivative; 440(M^+), 425, 397, 380(100), 365, 337, 315, 313, 273, 255(47), 253, and 213*m/e* (13).

References

W. D. Nes and E. J. Parish (eds.); Analysis of Sterols and Other Biologically Significant Steroids; Academic Press Inc., New York, NY, 341 pp. (1989).

W. B. Turner and D. C. Aldridge; Fungal Metabolites II; Academic Press, Inc., New York, NY, 631 pp. (1983).

J. D. Weete; Structure and Function of Sterols in Fungi, Advances in Lipid Research, Vol. 23, pp. 115-167(1989).

Common/Systematic Name

Campestanol

(24*R*)-24-Methylcholestan-3β-ol

Molecular Formula/Molecular Weight

$C_{28}H_{50}O$; MW = 402.38617

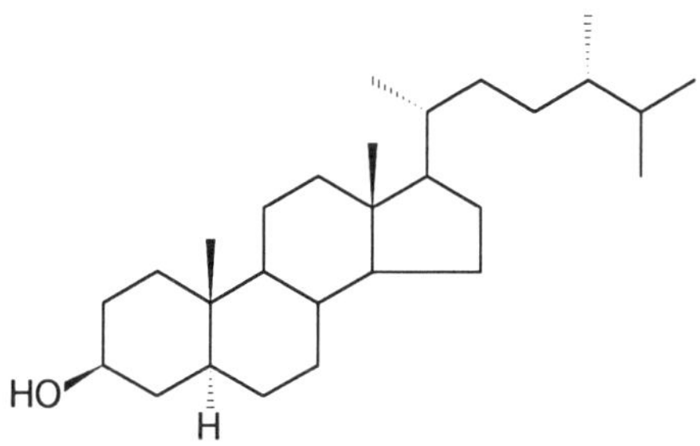

Fungal Source

Physarum polycephalum.

Isolation/Purification

See R. C. Heupel: Isolation and Primary Characterization of Sterols, In Analysis of Sterols and Other Biologically Significant Steroids; W. D. Nes, and E. J. Parish (eds.), Academic Press Inc., New York, NY, pp. 1-27(1989).

Spectral Data

UV:

End absorption.

Mass Spectrum:

LREIMS: (acetate) 444(M^+, 35%), 429(M^+ - methyl, 7), 384(M^+ - ROH, 34), 369(M^+ - Me - ROH, 23), 317(M^+- side chain), 275(M^+- side chain -42, 28), 257(M^+ - side chain- ROH, 16), and 215*m/e* (M^+- side chain - ROH -42).

References

M. Lefant, M. F. Lecompte, and G. Farugia; Identification Des Stérols de *Physarum polycephalum*; Phytochemistry, Vol. 9, pp. 2529(1970).

W. B. Turner and D. C. Aldridge; Fungal Metabolites II; Academic Press Inc., New York, New York, p. 308(1983).

Common/Systematic Name

Ergostanol

(24*S*)24-Methylcholestan-3β-ol

Molecular Formula/Molecular Weight

$C_{28}H_{50}O$; MW = 402.38617

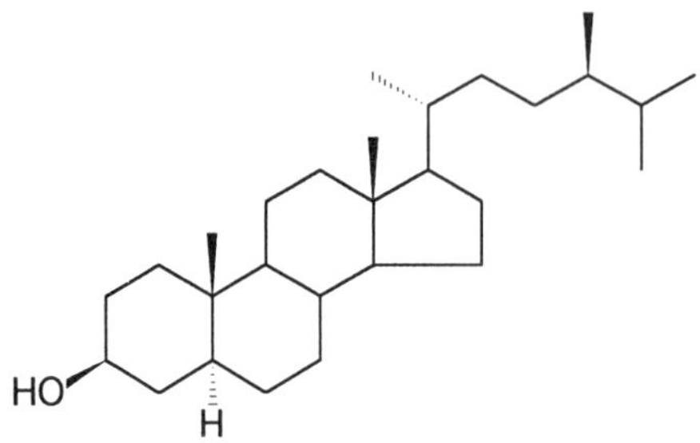

General Characteristics

Crystals; mp., 144-145°C; $[\alpha]_D^{20}$ +15.9° (*c*=1.8, in $CHCl_3$); acetate: $C_{30}H_{52}O_2$; mp., 145°C; $[\alpha]_D^{20}$ +6° (*c*=1.8, in $CHCl_3$); benzoate: $C_{35}H_{54}O_2$; mp., 163-165°C.

Fungal Source

Physarum polycephalum and *P. flavicomum.*

Isolation/Purification

See R. C. Heupel: Isolation and Primary Characterization of Sterols, In Analysis of Sterols and Other Biologically Significant Steroids; W. D. Nes, and E. J. Parish (eds.); Academic Press Inc., New York, NY, pp. 1-27(1989).

References

E. Bullock and C. J. Dawson; Sterol Content of the Myxomycetes *Physarum polycephalum* and *P. flavicomum*; J. Lipid Res., Vol. 17, p. 565(1976).

W. B. Turner and D. C. Aldridge; Fungal Metabolites II; Academic Press Inc., New York, New York, 631 pp.(1983).

Common/Systematic Name

(24*S*)24-Methylcholest-8-en-3β-ol; Ergost-8-en-3β-ol

Molecular Formula/Molecular Weight

$C_{28}H_{48}O$; MW = 400.37052

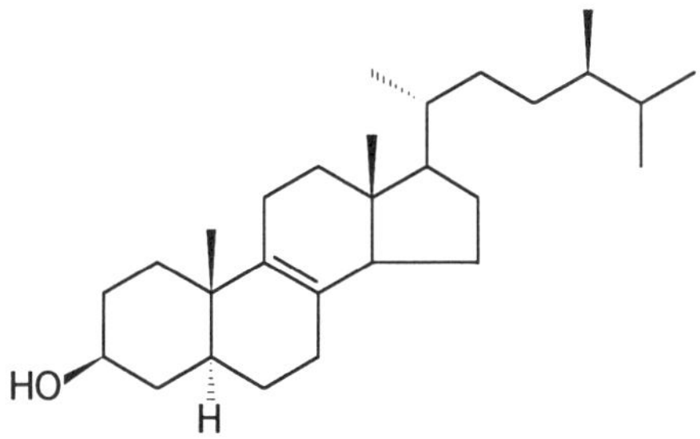

General Characteristics

Benzoate, plates from chloroform-methanol; mp., 152-155°C; $[\alpha]_D$ + 33.8° (*c*=0.60, in $CHCl_3$) (mp., 147-148°C; $[\alpha]_D$ + 34°).

Fungal Source

Candida albicans, C. utilis, Torulopsis glabrata, and *Saccharomyces cerevisiae.*

Isolation/Purification

Yeast cells were refluxed in 10% methanolic KOH, methanol was removed *in vacuo*, and ether-soluble material was converted into benzoate derivatives. The benzoates were separated with preparative silica gel GF_{254} TLC followed by silver nitrate-impregnated TLC.

Spectral Data

^{1}H NMR:

($CDCl_3$) 0.63(3H, s, 13β-Me); 1.01(3H, s, 10β-Me); and 4.9ppm (1H, m, 3α-H).

Mass Data:

Benzoate: LREIMS: 504(M^+, 100%), 489, 382, 377, 367, 335, 289, 255, 229, and 213*m/e*; found: C, 83.3%; H, 10.4; calcd for $C_{35}H_{52}O_2$; C, 83.3; H, 10.4.

References

D. H. R. Barton, J. E. T. Corrie, D. A. Widdowson, M. Bard, and R. A. Woods; Biosynthesis of Terpenes and Steroids. Part IX. The Sterols of Some Mutant Yeasts and their Relationship to the Biosynthesis of Ergosterol; J. Chem. Soc.,Trans Perkin I, pp. 1326-1333(1974).

D. H. R. Barton, A. A. L. Gunatilaka, T. R. Jarman, D. A. Widdowson, M. Bard, and R. A. Woods; Biosynthesis of Terpenes and Steroids. Part X. The Sterols of Some Yeast Mutants Doubly Defective in Ergosterol Biosynthesis; J. Chem. Soc.,Perkin Trans I, pp. 88-92(1975).

Common/Systematic Name

(24*S*)24-Methylcholesta-5,8-dien-3β-ol; Ergosta-5,8-dien-3β-ol

Molecular Formula/Molecular Weight

$C_{28}H_{46}O$; MW = 398.35487

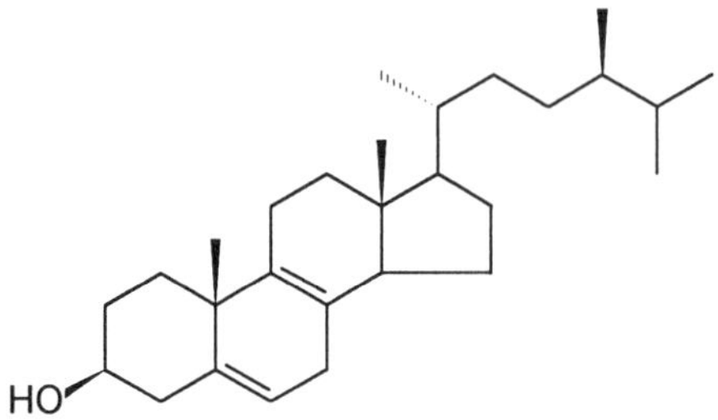

General Characteristics

Plates from acetone-methanol; mp., 119-122°C; $[\alpha]_D$ - 0.40° (*c*=0.2, in $CHCl_3$); benzoate, plates from acetone-methanol; mp., 136-139°C; $[\alpha]_D$ + 0.5°. (calcd + 2.0°)(*c*=0.25, in $CHCl_3$).

Fungal Source

Saccharomyces cerevisiae.

Isolation/Purification

Yeast cells were refluxed in 10% methanolic KOH, methanol was removed *in vacuo*, and ether-soluble material was converted into benzoate derivatives. The benzoates were separated with preparative silica gel GF_{254} TLC followed by silver nitrate-impregnated TLC.

Spectral Data (Benzoate)

UV:

λ_{max} 230nm (ϵ = 10,630)

^1H NMR:

($CDCl_3$) 0.67(3H, s, 13β methyl); 1.25(3H, s, 10β methyl); 4.90(1H, m, 3α-H); and 5.45ppm (1H, m, 6-H).

Mass Data:

LREIMS: 502(M^+), 380(100%), 365, 253, and 251*m/e*; found: C, 83.8%; H, 10.0; calcd for $C_{35}H_{50}O_2$; free alcohol, 398(M^+), 383, 380, 365(100%), and 271*m/e* (M+ - side chain); found: C, 84.0; H, 11.5; calcd for $C_{35}H_{50}O_2$; C, 84.3; H, 11.6.

<u>Reference</u>

D. H. R. Barton, A. A. L. Gunatilaka, T. R. Jarman, D. A. Widdowson, M. Bard, and R. A. Woods; Biosynthesis of Terpenes and Steroids. Part X. The Sterols of Some Yeast Mutants Doubly Defective in Ergosterol Biosynthesis; J. Chem. Soc.,Perkin Trans I, pp. 88-92 (1975).

Common/Systematic Name

24-Methylene cholesterol

24-Methylcholesta-5,24(28)-dien-3β-ol

Molecular Formula/Molecular Weight

$C_{28}H_{46}O$; MW = 398.35487

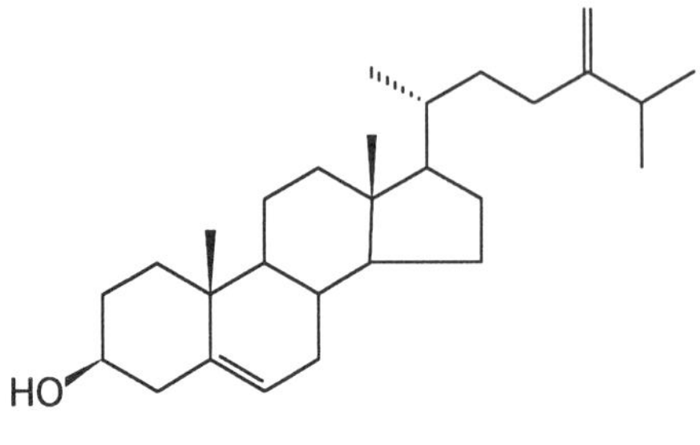

General Characteristics

Acetate crystallized from ethanol as plates; mp., 131-132°C.

Fungal Source

Achlya bisexualis, Glomus caledonius, Saprolegnia ferax, S. megasperma, Leptolegnia caudata, Pythiopsis cymosa, and *Rhizophydium sphaerotheca.*

Isolation/Purification

The mycelium from *Achlya bisexualis* was air dried, saponified with ethanolic KOH, and the lipids were extracted with hexane. The non-saponifiable material was chromatographed on alumina (activity III) and squalane and other hydrocarbons eluted with light petrol. Further elution with diethyl ether gave a sterol fraction which was subsequently acetylated and separated by preparative TLC on silica gel-impregnated with silver nitrate.

Also, see R. C. Heupel; Isolation and Primary Characterization of Sterols, In Analysis of Sterols and Other Biologically Significant Steroids; W. D. Nes and E. J. Parish (eds.), Academic Press Inc., New York, NY, 341 pp. (1989).

Spectral Data

UV:

End absorption.

^{13}C NMR:

($CDCl_3$) (Acetate derivative) C-15, 24.3; C-16, 28.2; C-17, 56.0; C-18, 11.9; C-20, 35.8; C-21, 18.7; C-22, 34.7; C-23, 31.0; C-24, 156.7; C-25, 33.8; C-26, 21.9; C-27, 22.0; and C-28, 105.9ppm.

(Skeletal carbon chemical shifts essentially similar to cholesterol).

TLC Data

TLC was carried out on silica gel G plates impregnated with $AgNO_3$ and rhodamine 6G. Plates were developed with benzene. R_f for 24-methylenecholesteryl acetate was 0.47-0.51.

GC Data

GC was performed using 1% QF-1. 24-Methylenecholesteryl acetate had relative retention time of 1.28 (relative to cholesteryl acetate).

References

M. Lenfant, M. F. Lecompte, and G. Farrugia; Identification des Stérols de *Physarum polycephalum*; Phytochemistry, Vol. 9, pp. 2529-2535(1970).

W. D. Nes and E. J. Parish (eds.); Analysis of Sterols and Other Biologically Significant Steroids; Academic Press Inc., New York, NY, pp. 341 (1989).

C. R. Popplestone and A. M. Uurau; Major Sterols of *Achlya bisexualis*; Phytochemistry, Vol. 12, pp. 1131-1133(1973).

W. B. Turner and D.C. Aldridge; Fungal Metabolites II; Academic Press Inc., New York, NY, 631 pp. (1983).

J. D. Weete; Structure and Function of Sterols in Fungi; Advances in Lipid Research, Vol. 23, pp. 115-167(1989).

Common/Systematic Name

Fecosterol; Ergosta-8,24(28)-dien-3β-ol
24-Methylcholesta-8,24(28)-dien-3β-ol

Molecular Formula/Molecular Weight

$C_{28}H_{46}O$; MW = 398.35482

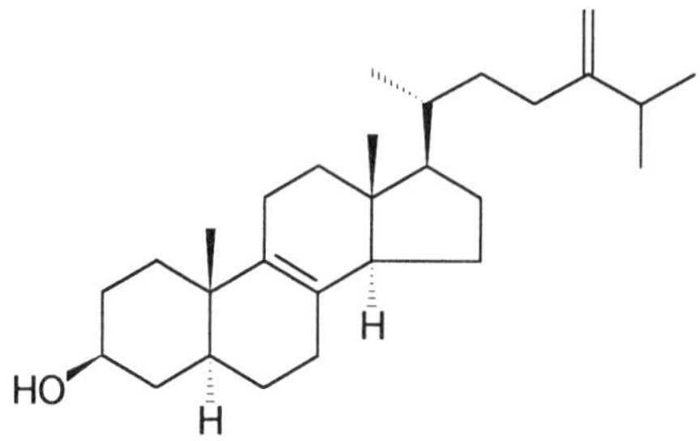

General Characteristics

Plates from chloroform-methanol; mp., 126-130°C; $[\alpha]_D^{24}$ + 44.9° (*c*=0.64, in $CHCl_3$); benzoate: plates from acetone-methanol; mp., 127-128°C; mp., 127-128°C + 43.2 (*c*=0.40) (mp., 127-129°C; $[\alpha]_D$ + 39.9°) ; acetate: plates from acetone-methanol; mp., 136-138°C; $[\alpha]_D$ + 34.7°.

Fungal Source

Botrytis cinerea, Candida albicans, C. utilis, Lobaria pulmonaria, L. scobiculata, Neurospora crassa, Pichia sp., *Saccharomyces carlsbergensis, S. cerevisiae, Torulopsis glabrata,* and *Usnea longissima.*

Isolation/Purification

See R. C. Heupel; Isolation and Primary Characterization of Sterols, In Analysis of Sterols and Other Biologically Significant Steroids; W. D. Nes and E. J. Parish (eds.), Academic Press Inc., New York, NY, 341 pp.(1989).

Spectral Data

UV:

End absorption.

IR:

(KBr) 3600, 1640, and 890cm^{-1}. Benzoate: 1710, 1640, and 890cm^{-1}. Acetate: 1710, 1640, and 890cm^{-1}.

^{1}H NMR:

($CDCl_3$) 5.35(2H, m, 28-H); 6.4(1H, m, 3α-H); 8.99(3H, d, *J*=7.0Hz, 26- and 27-Me); 9.05(3H, s, 10-Me); and 9.39τ (3H, s, 13-Me); benzoate: 5.1(1H, m, 3α-H); 5.3(2H, m, 28-H); 9.00(6H, d, *J*=8Hz, 26- and 27-Me); 9.02(3H, s, 10-Me); 9.38(3H, s, 13-Me); and 2.0 and 2.5τ (5H, m, aromatic-H). Acetate: 5.31(1H, m, 28-H); 5.35(1H, m, 3α-H); 8.0(3H, s, Ac); 8.98(6H, d, *J*=7.0Hz, 26- and 27-Me); 9.03(3H, s, 10-Me); and 9.38τ (3H, s, 13-Me).

Mass Spectrum:
(Acetate derivative) 440(M^+, 100%), 425, 380(64), 365, 356(7), 341, 315(10), 313(49), 296(3), 281, 273(32), 255(31), 253(30), and 213*m/e* (100).

References

D. H. R. Barton, J. E. T. Corrie, D. A. Widdowson, M. Bard, and R. A. Woods; Biosynthesis of Terpenes and Steroids. Part IX. The Sterols of Some Mutant Yeasts and their Relationship to the Biosynthesis of Ergosterol; J. Chem. Soc.,Perkin Trans I, pp. 1326-1333(1974).

D. H. R. Barton, A. A. L. Gunatilaka, T. R. Jarman, D. A. Widdowson, M. Bard, and R. A. Woods; Biosynthesis of Terpenes and Steroids. Part X. The Sterols of Some Yeast Mutants Doubly Defective in Ergosterol Biosynthesis; J. Chem. Soc.,Perkin Trans I, pp. 88-92(1975).

W. D. Nes and E. J. Parish (eds.); Analysis of Sterols and Other Biologically Significant Steroids; Academic Press Inc., New York, NY, 341 pp. (1989).

W. B. Turner and D. C. Aldridge; Fungal Metabolites II, Academic Press, New York, New York, 314 pp. (1983).

J. D. Weete; Structure and Function of Sterols in Fungi, Advances in Lipid Research, Vol. 23, pp. 115-167(1989).

Common/Systematic Name

Fungisterol; Ergost-7-en-3β-ol
(24*S*)24-Methylcholest-7-en-3β-ol

Molecular Formula/Molecular Weight

$C_{28}H_{48}O$; MW = 400.37052

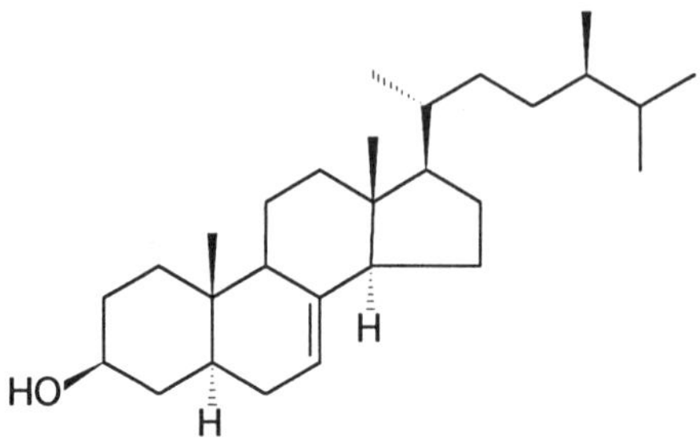

Fungal Source

Agaricus bisporus, Armillaria mellea, Boletus luridus, Candida albicans, C. utilis, C. tropicalis, Coriolus pargamenus, C. versicolor, Cronartium fusiforme, Cryptoderma citrinum, Fomitopsis pinicola, Gloeophyllum sepiarium, Glomus mosseae, Grifola frondosa, Hymenomycetes sp., Lenzites trabea, Linderina pennispora, Neurospora crassa, Penicillium claviforme, Pholiota aegerita, Phycomyces blakesleeanus, Puccinia graminis var. *triticil, P. striiformis, Rhizopus arrhizus, Russula senecis, Saccharomyces cerevisiae, Torulopsis glabrata, Ustilago maydis,* and *U. nuda.* The only C_{28} sterol detected in rusts and many Basidiomycetes.

Isolation/Purification

See R. C. Heupel; Isolation and Primary Characterization of Sterols, In Analysis of Sterols and Other Biologically Significant Steroids; W. D. Nes and E. J. Parish (eds.), Academic Press Inc., New York, NY, 341 pp. (1989).

Spectral Data

UV:

End absorption.

^{1}H NMR:

($CDCl_3$) 0.53; 0.80(each 3H, s, H_3-18, -19); 0.78(3H, d, *J*=5.2Hz); 0.83(3H, d, *J*=5.5Hz); 0.86(3H, d, *J*=5.5Hz); 0.92(3H, d, *J*=6.1Hz); 3.60(1H, m, H-3); and 5.18ppm (1H, m, H-7).

^{13}C NMR:

($CDCl_3$) 11.8(q, C-18); 13.0(q, C-19); 15.4(q, C-28); 17.6(q, C-27); 20.5(q, C-26); 19.0(q, C-21); 21.5(t, C-11); 22.9(t, C-15); 27.9(t, C-16); 29.6(t, C-6); 29.6(t, C-23);

30.7(t, C-2); 31.5(d, C-25); 33.6(t, C-22); 34.2(s, C-10); 36.6(d, C-20); 37.1(t, C-1); 38.0(t, C-4); 39.0(d, C-24); 39.5(t, C-12); 40.2(d, C-5); 43.3(s, C-13); 49.4(d, C-9); 55.1(d, C-14); 56.0(d, C-17); 71.0(d, C-37); 117.4(d, C-7); and 139.6ppm (s, C-8).

Mass Spectrum:

Acetate derivative: 442(M^+, 72%), 427(M^+ - Me, 15), 382(M^+ - AcOH, 36), 353(8), 315(11), 288(8), 273(16), 255(100), 229(26), and 213*m/e* (67); benzoate: 504(M^+), 489, 382, 377, and 255*m/e*.

References

D. H. R. Barton, J. E. T. Corrie, D. A. Widdowson, M. Bard, and R. A. Woods; Biosynthesis of Terpenes and Steroids, Part IX. The Sterols of Some Mutant Yeasts and Their Relationship to the Biosynthesis of Ergosterol; J. Chem. Soc.,Perkin Trans I, pp. 1326-1333(1974).

D. H. R. Barton, A. A. L. Gunatilaka, T. R. Jarman, D. A. Widdowson, M. Bard, and R. A. Woods; Biosynthesis of Terpenes and Steroids, Part X. The Sterols of Some Yeasts Mutants Doubly Defective in Ergosterol Biosynthesis; J. Chem. Soc.,Perkin Trans I, pp. 88-92(1975).

H. Fujimoto, K. Maeda, and M. Yamazaki; New Toxic Metabolites from a Mushroom, *Hebeloma vinosophyllum* III. Isolation and Structures of Three New Glycosides, Hebevinosides XII, XIII and XIV and Productivity of the Hebevinosides at Three Growth Stages of the Mushroom; Chem. Pharm. Bull., Vol. 39, pp. 1958-1961(1991).

W. D. Nes and E. J. Parish (eds.); Analysis of Sterols and Other Biologically Significant Steroids; Academic Press Inc., New York, NY, 341 pp. (1989).

W. B. Turner and D. C. Aldridge; Fungal Metabolites II; Academic Press, Inc., New York, NY, 312 pp. (1983).

J. D. Weete; Structure and Function of Sterols in Fungi; Advances in Lipid Research, Vol. 23, pp. 115-167(1989).

Common/Systematic Name

(24*S*)24-Methylcholesta-5,7,14-trien-3β-ol

Molecular Formula/Molecular Weight

$C_{28}H_{44}O$; MW = 396.33922

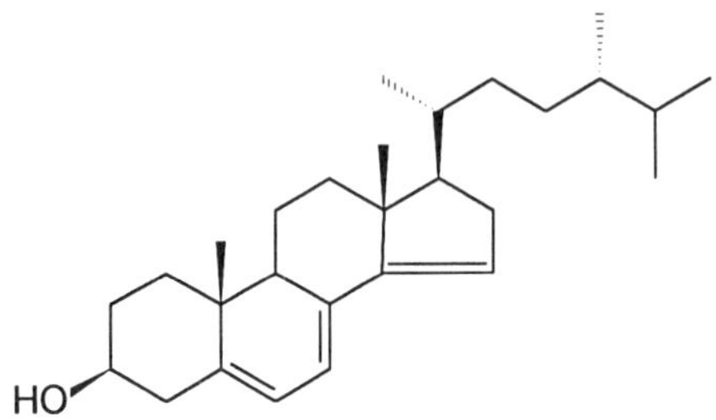

Fungal Source

Rhizopus arrhizus.

Isolation/Purification

See R. C. Heupel; Isolation and Primary Characterization of Sterols, In Analysis of Sterols and Other Biologically Significant Steroids; W. D. Nes and E. J. Parish (eds.), Academic Press Inc., New York, NY, 341 pp. (1989).

References

W. D. Nes and E. J. Parish (eds.); Analysis of Sterols and Other Biologically Significant Steroids; Academic Press Inc., New York, NY, 341 pp. (1989).

W. B. Turner and D. C. Aldridge; Fungal Metabolites II; Academic Press, Inc., New York, NY, 631 pp. (1983).

J. D. Weete; Structure and Function of Sterols in Fungi, Advances in Lipid Research, Vol. 23, pp. 115-167(1989).

Common/Systematic Name

Ergosterol

Ergosta-5,7,22-trien-3β-ol

Molecular Formula/Molecular Weight

$C_{28}H_{44}O$; MW = 396.33922

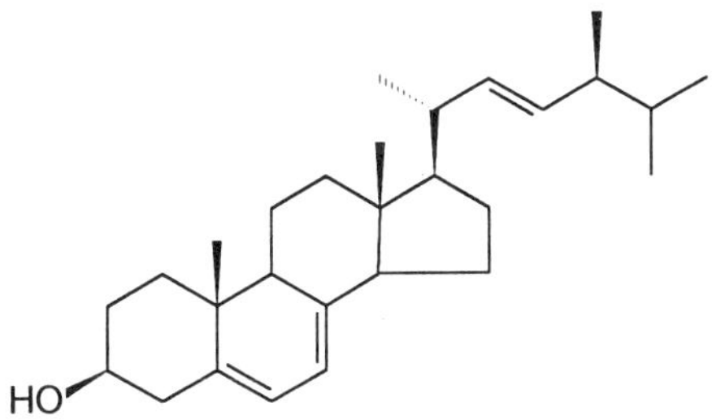

General Characteristics

Small hydrated plates from alcohol; hydrated needles from ether. The best crystallized form contains 1.5 moles of H_2O; mp., 168°C. Complete removal of water is almost impossible and results in an amorphous mass; melting range 166-183°C; $[\alpha]_D^{20}$ -135° (*c*=1.2, in $CHCl_3$ calcd as anhydrous). Practically insoluble in water. One gram dissolves in 660ml alcohol, in 45 ml boiling alcohol, in 70ml ether, in 39ml boiling ether, in 31ml chloroform. Precipitated by digitonin. Affected by light and air; turns yellow. Oxygen forms peroxides and hydrogen may form polyhydro compounds. Acetate, $C_{30}H_{46}O_2$; mp., 179°C, clear at 181°C. Benzoate, $C_{35}H_{48}O_2$; plates or needles; mp., 169-171°C; $[\alpha]_D^{23}$ -71° (*c*=1.1, in $CHCl_3$).

Isolation/Purification

See R. C. Heupel; Isolation and Primary Characterization of Sterols, In Analysis of Sterols and Other Biologically Significant Steroids; W. D. Nes, and E. J. Parish (eds.), Academic Press Inc., New York, NY, 341 pp. (1989).

Fungal Source

The major fungal sterol in many fungal species. The major sterol of most species of the family Mucoraceae, most Ascomycetes, and with the exception of rusts, most Basidiomycetes.

Agaricus campestris, Alternaria alternata, A. brassicicola, A. kikuchiana, A. tenuis, Amoebidium parasiticum, Amelaria mellea, Aspergillus niger, Boletus luridus, B. spp., *Candida albicans, C. guilliermondii, C. utilis, Candida* sp., *Cantharellus cibarius, Coriolus pargamenus, C. sangamenus, C. versicolor, Cryptoderma citrinum, Cyathus helenae, Fomes allardii, F. annosus, Fomitopsis cytisina, F. pinicola, F. pubertatis, Fusarium sporotrichioides, Ganoderma applanatum, Geotrichum flavo-brunneum, Gloeophyllum sepiarium, Glomus mosseae, Gnomonia leptostyla, Grifola frondosa, G. umbellata, Hemispora stellata, Leccinum aurantiacum, Leccinum* spp., *Lentinus lepideus,*

Lenzites trabea, Leptosphaeria typhae, Lobaria pulmonaria, L. scobiculata, Microporus flabelliformis, Monilinia fructigena, Mucor pusillus, Nectria galligena, Neurospora crassa, Penicillium claviforme, Phellinus gilvus, P. igniarius, P. robustus, Pholiota aegerita, Phycomyces blakesleeanus, Pichia sp., *Polyporus cretaceous, P. dryadeus, Poria cocos, Rhizopus arrhizus, Saccharomyces cerevisiae, Scleroderma aurantium, Sclerotinia fructicola, Smittium* sp., *Spicaria elegans, Suillus bovinus, S. variegatus, Torulopsis glabrata, Trametes feei, T. lilacino-gilva, Usnea longissima, Ustilago maydis, U. nuda,* and *Xanthoria parietina.*

Biological Activity

When irradiated with UV light, ergosterol develops powerful vitamin D_l (q.v.) activity.

Spectral Data

UV:

λ_{max}^{EtOH} 262, 271, 282, and 293nm.

^{13}C NMR:

($CDCl_3$) C-1, 38.5; C-2, 32.1; C-3, 70.5; C-4, 40.9; C-5, 139.8; C-6, 119.7; C-7, 116.4; C-8, 141.3; C-9, 46.3; C-10, 37.1; C-11, 21.1; C-12, 39.2; C-13, 42.9; C-14, 54.6; C-15, 23.0; C-16, 28.3; C-17, 55.9; C-18, 12.1; C-19, 16.3; C-20, 40.4; C-21, 21.2; C-22, 135.6; C-23, 132.1; C-24, 42.9; C-25, 33.1; C-26, 22.0; C-27, 19.7; and C-28, 17.6ppm.

Mass Spectrum:

396(M^+, 72%), 381(7), 378, 363(100), 271, 253, 229, 227, 211, 158, and 143*m/e*.
Acetate derivative; 438(M^+, 9%), 423, 378(100), 363, 313, 253, 211, and 143*m/e*.

References

W. D. Nes and E. J. Parish (eds.); Analysis of Sterols and Other Biologically Significant Steroids; Academic Press Inc., New York, NY, 341 pp. (1989).

W. B. Turner and D. C. Aldridge; Fungal Metabolites II; Academic Press, Inc., New York, NY, 631 pp. (1983).

J. D. Weete; Structure and Function of Sterols in Fungi, Advances in Lipid Research, Vol. 23, pp. 115-167(1989).

Common/Systematic Name

(24*S*)24-Methylcholesta-5,7-dien-3β-ol; Ergosta-5,7-dien-3β-ol

Molecular Formula/Molecular Weight

$C_{28}H_{46}O$; MW = 398.35487

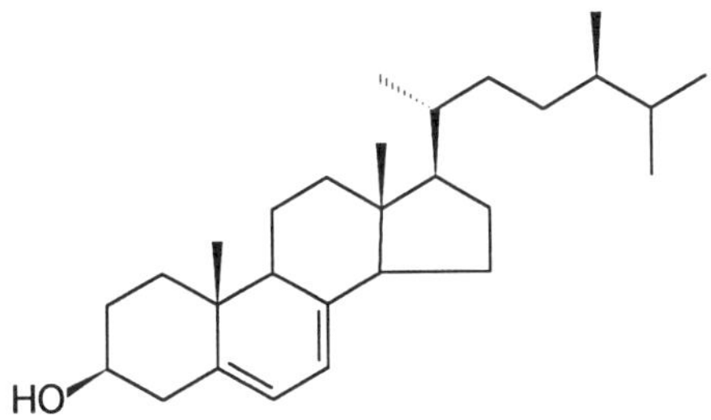

General Characteristics

Crystals from chloroform-methanol; mp., 148-150°C; $[\alpha]_D$ - 73.6° (*c*=0.41); acetate, crystals from acetone-methanol; mp., 153-155°C; $[\alpha]_D$ - 87.6° (*c*=0.45) (mp., 157-158°C; $[\alpha]_D$ - 74.8°).

Fungal Source

Found in several *Hymenomycetes* species. *Agaricus bisporus, Amelaria mellea, Aspergillus parasiticus, Boletus luridus, B.* spp., *Candida albicans, C. utilis, Coriolus pargamenus, C. sangamenus, C. versicolor, Daedalea quercina, Leccinum aurantiacum, L.* spp., *Microporus flabelliformis, Mucor pusillus, Pholiota aegerita, Phycomyces blakesleeanus, Pichia* sp., *Piptoporus betulinus, Polyporus paragamenus, Saccharomyces cerevisiae, S. carlsbergensis, Schizophyllum commune, Suillus bovinus, S. variegatus, Torulopsis glabrata, Tremella fuciformis,* and *Ustilago maydis.*

Isolation/Purification

Yeast cells were refluxed in 10% methanolic KOH, methanol was removed *in vacuo*, and ether soluble material was converted into benzoate derivatives. The benzoates were separated with preparative silica gel GF_{254} TLC followed by silver nitrate-impregnated TLC.

Spectral Data

UV:

(acetate) λ_{max} 262(ϵ = 8,400), 272(12,400), 282(13,600), and 294nm (7,400).

Mass Data:

(benzoate) Found: C, 84.1; H, 9.6; calcd for $C_{35}H_{48}O_2$; C, 83.9; H, 9.7%.

References

D. H. R. Barton, J. E. T. Corrie, D. A. Widdowson, M. Bard, and R. A. Woods; Biosynthesis of Terpenes and Steroids. Part IX. The Sterols of Some Mutant Yeasts and Their Relationship to the Biosynthesis of Ergosterol; J. Chem. Soc., Trans Perkin I, pp. 1326-1333(1974).

D. H. R. Barton, A. A. L. Gunatilaka, T. R. Jarman, D. A. Widdowson, M. Bard, and R. A. Woods; Biosynthesis of Terpenes and Steroids. Part X. The Sterols of Some Yeast Mutants Doubly Defective in Ergosterol Biosynthesis; J. Chem. Soc., Perkin Trans I, pp. 88-92(1975).

Common/Systematic Name

5,6-Dihydroergosterol; Ergosta-7,22-dien-3β-ol
24-Methylcholesta-7,22-dien-3β-ol

Molecular Formula/Molecular Weight

$C_{28}H_{46}O$; MW = 398.35487

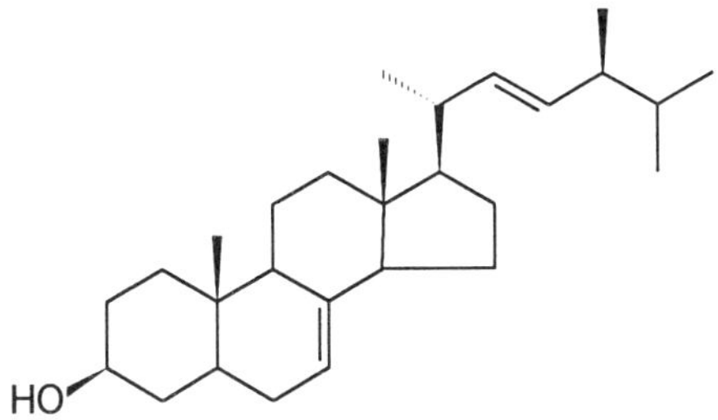

Fungal Source

Acremonium coenophialum, Amelaria mellea, Aspergillus fennelliae, A. oryzae, Boletus luridus, B. spp., *Candida albicans, C. utilis, C. tropicalis, Candida* sp., *Coriolus pargamenus, C. versicolor, Cryptoderma citrinum, Daedalea quercina, Flammulina velutipes, Fomes annosus, F. fomentarius, Fomitopsis cytisina, Gloeophyllum sepiarium, Leccinum aurantiacum, Leccinum* spp., *Hymenomycetes* sp., *Leptosphaeria typhae, Neurospora crassa, Phellinus gilvus, Pholiota aegerita, Phycomyces blakesleeanus, Piptoporus betulinus, Rhizopus arrhizus, Saccharomyces cerevisiae, Suillus bovinus, S. variegatus, Torulopsis glabrata,* and *Xanthoria parietina.* Many Basidiomycetes.

Isolation/Purification

See R. C. Heupel; Isolation and Primary Characterization of Sterols, In Analysis of Sterols and Other Biologically Significant Steroids; W. D. Nes and E. J. Parish (eds.), Academic Press Inc., New York, NY, 341 pp. (1989).

Spectral Data

UV:

End absorption.

^{1}H NMR:

($CDCl_3$) 0.54. 0.80(each 3H, s, H_3-18, -19); 0.78(3H, d, J=6.7Hz); 0.82(3H, d, J=6.7Hz); 0.91(3H, d, J=7.0Hz); 1.02(3H, d, J=6.7Hz); 3.60(1H, m, H-3); and 5.18ppm (3H, m, H-7, -22, -2).

^{13}C NMR:

($CDCl_3$) 12.1(q, C-18); 13.0(q, C-19); 17.6(q, C-28); 19.6(q, C-26); 19.9(q, C-27); 21.1(q, C-21); 21.5(t, C-11); 22.9(t, C-15); 28.1(t, C-16); 29.6(t, C-6); 31.5(t, C-2);

33.1(d, C-25); 34.2(s, C-10); 37.1(t, C-1); 38.0(t, C-4); 39.4(t, C-12); 40.2(d, C-5); 40.4(d, C-2); 42.8(d, C-24); 43.3(s, C-13); 49.4(d, C-9); 55.0(d, C-14); 56.0(d, C-17); 71.0(d, C-3); 117.4(d, C-7); 131.9(d, C-23); 135.6(d, C-22); and 139.6ppm (s, C-8).

Mass Spectrum:
EIMS: (acetate derivative) 440(M^+, 60%), 425(15), 397, 380(9), 365(9), 337, 315, 313(100), 273(12), 255(81), 253(15), and 213*m/e* (23).

TLC Data
R_f 0.33 with silica gel plates developed with benzene-diethyl ether (9:1, v/v).

HPLC Data
Relative retention time was 0.86 (relative to cholesterol) performed with C_{18} reversed phase column eluted with 4% aqueous methanol at ambient temperature with a flow rate of 1.6ml/min.

GLC Data
Relative retention time was 1.26 relative to cholesterol using 3% SE-30 at 245°C.

References

H. Fujimoto, K. Maeda, and M. Yamazaki; New Toxic Metabolites from a Mushroom, *Hebeloma vinosophyllum* III. Isolation and Structures of Three New Glycosides, Hebevinosides XII, XIII and XIV and Productivity of the Hebevinosides at Three Growth Stages of the Mushroom; Chem. Pharm. Bull., Vol. 39, pp. 1958-1961(1991).

W. D. Nes and E. J. Parish (eds.); Analysis of Sterols and Other Biologically Significant Steroids; Academic Press Inc., New York, NY, 341 pp. (1989).

W. B. Turner and D. C. Aldridge; Fungal Metabolites II; Academic Press, Inc., New York, NY, 631 pp. (1983).

J. D. Weete; Structure and Function of Sterols in Fungi, Advances in Lipid Research, Vol. 23, pp. 115-167(1989).

Common/Systematic Name

Episterol

Ergosta-7,24(28)-dien-3β-ol

Molecular Formula/Molecular Weight

$C_{28}H_{46}O$; MW = 398.35487

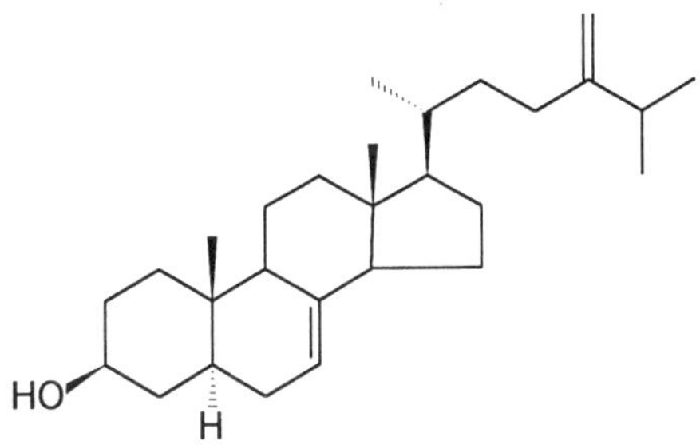

Fungal Source

Amanita caesaria, Coprinus atramentarius, Flammulina velutipes, Hygrocybe punicea, Lampteromyces japonicus, Lentinus edodes, Leucopaxillus giganteus, Russula foetens, R. nigricans, R. senecis, Stachybotrys atra, Dictyuchus monosporus, Agaricus campestris, Aspergillus oryzae IFO 4290 (occurs as acylsterol with linoleic acid), *Candida albicans, C. utilis, C. tropicalis, Inocybe macrosperma, Leptosphaeria typhae, Lobaria pulmonaria, L. scobiculata, Nectria galligena, Neurospora crassa, Phycomyces blakesleeanus, Saccharomyces cerevisiae, Torulopsis glabrata, Usnea longissima,* and *Xanthoria parietina.*

Isolation/Purification

The non-saponifiable lipid extract was purified by TLC on silica gel G developed with ethanol-chloroform (1:99, v/v) to give crude sterol fractions. These were acetylated, purified on TLC with silica gel G developed with cyclohexane-benzene (1:1, v/v) and analyzed by gas chromatography and gas chromatography-mass spectroscopy using 1% SE-30. Further purification was achieved with 10% $AgNO_3$-silica gel G TLC developed with cyclohexane-benzene (1:1, v/v) or alumina column chromatography (neutral grade III) using increasing amounts of ethyl ether in petroleum ether.

See R. C. Heupel; Isolation and Primary Characterization of Sterols, In Analysis of Sterols and Other Biologically Significant Steroids; W. D. Nes, and E. J. Parish (eds.), Academic Press Inc., New York, NY, 341 pp. (1989).

Spectral Data

UV:

End absorption.

Mass Spectrum:

LREIMS: (Acylsterol with linoleic acid) 660(M^+, 12.5%), 578, 563, 535(M^+ - side chain, 15.2), 533, 381(M^+ - RCOOH, 100), 380(27.3), 297, 255, 253, 228, and 213*m/e*.

References

J. R. Lenton, L. J. Goad, and T. W. Goodwin; Sterols of the Mycobiont and Phycobiont Isolated From The Lichen *Xanthoria parietina;* Phytochemistry, Vol. 13, pp. 2249-2253(1973).

W. D. Nes and E. J. Parish (eds.); Analysis of Sterols and Other Biologically Significant Steroids; Academic Press Inc., New York, NY, 341 pp. (1989).

M. Ohnishi and Y. Fujino; Principal Molecular Species of Acylsterols in *Aspergillus oryzae*; Agric. Biol. Chem., Vol. 44, pp. 681-683(1980).

W. B. Turner and D. C. Aldridge; Fungal Metabolites II; Academic Press, Inc., New York, NY, 631 pp. (1983).

J. D. Weete; Structure and Function of Sterols in Fungi, Advances in Lipid Research, Vol. 23, pp. 115-167(1989).

Common/Systematic Name

24-Methylcholesta-5,7,24(28)-trienol
Ergosta-5,7,24(28)-trien-3β-ol

Molecular Formula/Molecular Weight

$C_{28}H_{44}O$; MW = 396.33922

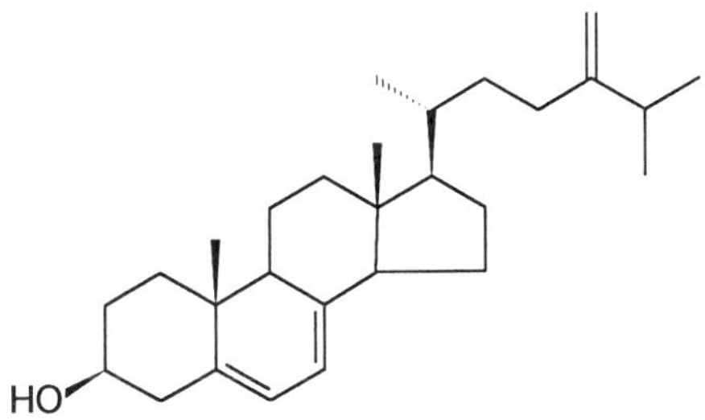

Fungal Source

Candida albicans, C. utilis, Phycomyces blakesleeanus, Leptosphaeria typhae, Saccharomyces cerevisiae, and *S. carlsbergensis.*

Isolation/Purification

See R. C. Heupel; Isolation and Primary Characterization of Sterols, In Analysis of Sterols and Other Biologically Significant Steroids; W. D. Nes and E. J. Parish (eds.), Academic Press Inc., New York, NY, 341 pp. (1989).

Spectral Data

UV:
Typical 5,7-diene absorption.

References

W. D. Nes and E. J. Parish (eds.); Analysis of Sterols and Other Biologically Significant Steroids; Academic Press Inc., New York, NY, 341 pp. (1989).

W. B. Turner and D. C. Aldridge; Fungal Metabolites II; Academic Press, Inc., New York, NY, 631 pp. (1983).

J. D. Weete; Structure and Function of Sterols in Fungi, Advances in Lipid Research, Vol. 23, pp. 115-167(1989).

Common/Systematic Name

(24*S*)24-Methylcholesta-8,14,22-trien-3β-ol; Ergosta-8,14,22-trien-3β-ol

Molecular Formula/Molecular Weight

$C_{28}H_{44}O$; MW = 396.33922

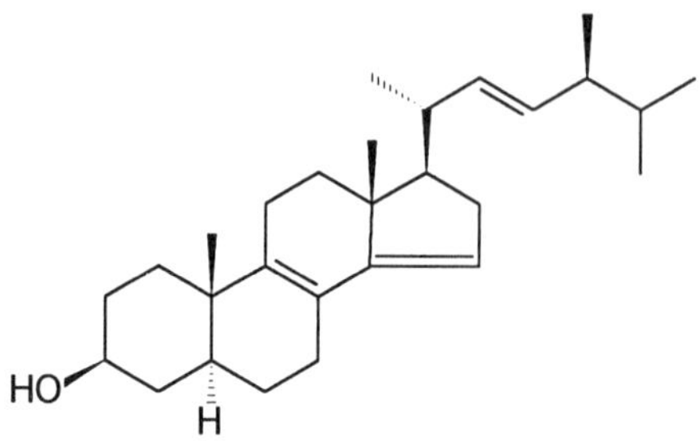

General Characteristics

Benzoate, plates from chloroform-methanol; mp., 140-142°C; $[\alpha]_D$ - 30.3° (*c*=0.89).

Fungal Source

Saccharomyces cerevisiae.

Isolation/Purification

Yeast cells were refluxed in 10% methanolic KOH, methanol was removed *in vacuo*, and ether-soluble material was converted into benzoate derivatives. The benzoates were separated with preparative silica gel GF_{254} TLC followed by silver nitrate-impregnated TLC.

Spectral Data (Benzoate)

UV:

λ_{max} 230(ϵ = 26,200) and 252nm sh (22,000).

Mass Data:

Found: C, 84.1; H, 9.6; calcd for $C_{35}H_{48}O_2$; C, 83.9; H, 9.7%.

References

D. H. R. Barton, J. E. T. Corrie, D. A. Widdowson, M. Bard, and R. A. Woods; Biosynthesis of Terpenes and Steroids. Part IX. The Sterols of Some Mutant Yeasts and their Relationship to the Biosynthesis of Ergosterol; J. Chem. Soc., Trans Perkin I, pp. 1326-1333(1974).

D. H. R. Barton, A. A. L. Gunatilaka, T. R. Jarman, D. A. Widdowson, M. Bard, and R. A. Woods; Biosynthesis of Terpenes and Steroids. Part X. The Sterols of Some Yeast Mutants Doubly Defective in Ergosterol Biosynthesis; J. Chem. Soc., Trans Perkin I, pp. 88-92(1975).

Common/Systematic Name

Ergost-5-en-3β-ol

(24*S*)24-Methylcholest-5-en-3β-ol

Molecular Formula/Molecular Weight

$C_{28}H_{48}O$; MW = 400.37052

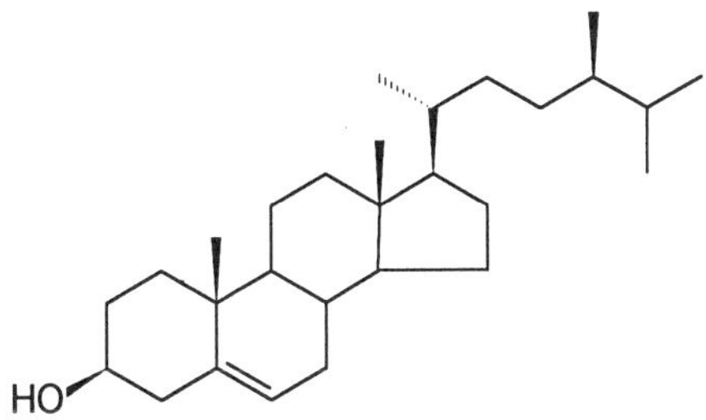

Fungal Source

Physarum polycephalum, P. flavicomum, Hyphochytrium catenoides, Trichophyton mentagrophytes, and *Fusarium culmorum.*

Isolation/Purification

See R. C. Heupel; Isolation and Primary Characterization of Sterols, In Analysis of Sterols and Other Biologically Significant Steroids; W. D. Nes and E. J. Parish (eds.), Academic Press Inc., New York, NY, 341 pp. (1989).

Spectral Data

UV:

End absorption.

References

W. D. Nes and E. J. Parish (eds.); Analysis of Sterols and Other Biologically Significant Steroids; Academic Press Inc., New York, NY, 341 pp. (1989).

W. B. Turner and D. C. Aldridge; Fungal Metabolites II; Academic Press, Inc., New York, NY, 631 pp. (1983).

J. D. Weete; Structure and Function of Sterols in Fungi, Advances in Lipid Research, Vol. 23, pp. 115-167(1989).

Common/Systematic Name

Lichesterol

(24*S*)24-Methylcholesta-5,8,22-trien-3β-ol; Ergosta-5,8,22-trien-3β-ol

Molecular Formula/Molecular Weight

$C_{28}H_{44}O$; MW = 396.33922

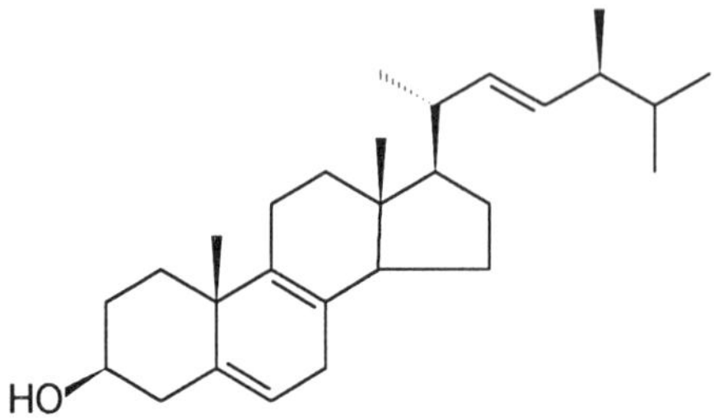

General Characteristics

Acetate, plates from acetone-methanol; mp., 138-140°C; $[\alpha]_D$ - 40.2° (*c*=0.39); benzoate, needles from chloroform-methanol; mp., 153-155°C; $[\alpha]_D$ - 6.1° (*c*=0.46).

Fungal Source

Candida albicans, Flammulina velutipes, Leptosphaeria typhae, Lobaria pulmonaria, L. scobiculata, Nectria galligena, Neurospora crassa, Saccharomyces cerevisiae, Usnea longissima, and *Xanthoria parietina.*

Isolation/Purification

Yeast cells were refluxed in 10% methanolic KOH, methanol was removed *in vacuo*, and ether-soluble material was converted into benzoate derivatives. The benzoates were separated with preparative silica gel GF_{254} TLC followed by silver nitrate-impregnated TLC.

Also, see R. C. Heupel; Isolation and Primary Characterization of Sterols, In Analysis of Sterols and Other Biologically Significant Steroids; W. D. Nes and E. J. Parish (eds.), Academic Press Inc., New York, NY, 341 pp. (1989).

Spectral Data

Mass Spectrum:

Acetate derivative, LREIMS: 438(M^+, 7%), 423, 378(70), 363(100), 313, 253(48), 251, 211(18), and 337*m/e* (10).

References

W. D. Nes and E. J. Parish (eds.); Analysis of Sterols and Other Biologically Significant Steroids; Academic Press Inc., New York, NY, 341 pp. (1989).

W. B. Turner and D. C. Aldridge; Fungal Metabolites II; Academic Press, Inc., New York, NY, 631 pp. (1983).

J. D. Weete; Structure and Function of Sterols in Fungi, Advances in Lipid Research, Vol. 23, pp. 115-167(1989).

Common/Systematic Name

22,23-Dihydroergosterol; Ergosta-5,7-dien-3β-ol

Molecular Formula/Molecular Weight

$C_{28}H_{46}O$; MW = 398.35487

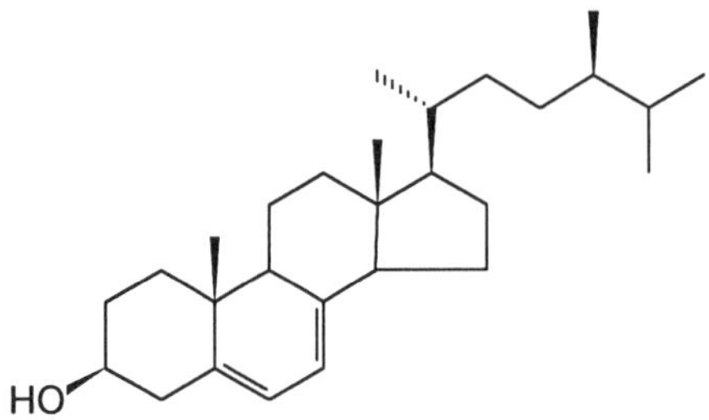

Fungal Source

Found in several *Hymenomycetes* species. *Agaricus bisporis, Amelaria mellea, Aspergillus parasiticus, Boletus luridus, B.* spp., *Candida albicans, C. utilis, Coriolus pargamenus, C. sangamenus, C. versicolor, Daedalea quercina, Leccinum aurantiacum, L.* spp., *Microporus flabelliformis, Mucor pusillus, Pholiota aegerita, Phycomyces blakesleeanus, Pichia* sp., *Piptoporus betulinus, Polyporus paragamenus, Saccharomyces cerevisiae, S. carlsbergensis, Schizophyllum commune, Suillus bovinus, S. variegatus, Torulopsis glabrata, Tremella fuciformis,* and *Ustilago maydis.*

Isolation/Purification

See R. C. Heupel; Isolation and Primary Characterization of Sterols, In Analysis of Sterols and Other Biologically Significant Steroids; W. D. Nes and E. J. Parish (eds.), Academic Press Inc., New York, NY, 341 pp. (1989).

Spectral Data

UV:

Typical 5,7-diene system (see ergosterol).

References

W. D. Nes and E. J. Parish (eds.); Analysis of Sterols and Other Biologically Significant Steroids; Academic Press Inc., New York, NY, 341 pp. (1989).

W. B. Turner and D. C. Aldridge; Fungal Metabolites II; Academic Press, Inc., New York, NY, 631 pp. (1983).

J. D. Weete; Structure and Function of Sterols in Fungi, Advances in Lipid Research, Vol. 23, pp. 115-167(1989).

Common/Systematic Name

(24*S*)24-Methyl-5α-cholesta-7,16-dien-3β-ol

Molecular Formula/Molecular Weight

$C_{28}H_{46}O$; MW = 398.35487

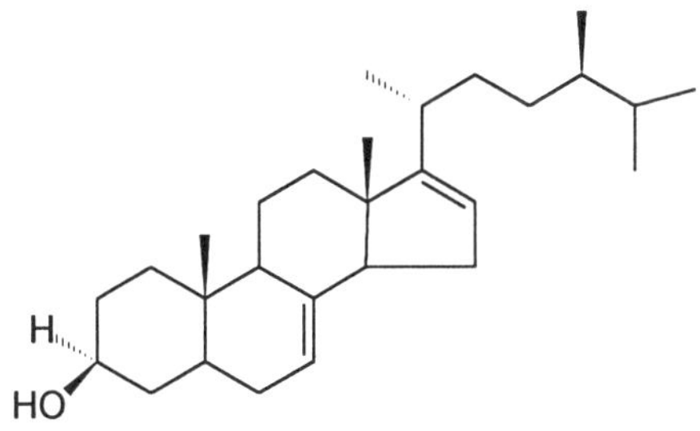

General Characteristics

Crystals from methanol; mp., 166-169.5°C; acetate; mp., 168-170°C.

Fungal Source

Ganoderma applanatum = *Fomes applanatus.*

Isolation/Purification

The ground fruit bodies of *G. applanatum* were extracted with methanol, evaporated to dryness and chromatographed on aluminum oxide (activity II) followed by chromatography on silica gel impregnated with silver nitrate using a gradient from hexane to ethyl acetate (2:1, v/v). The final product was crystallized from methanol.

Spectral Data

^{1}H NMR:

($CDCl_3$) 0.76(s, 19-CH_3); 0.56(s, 18-CH_3); and 5.08-5.2ppm (m, =CH).

Mass Spectrum:

LREIMS: 398(M^+) and 269*m/e* (base peak).

References

L. I. Strigina, Yu. N. Elkin, and G. B. Elyakov; Steroid Metabolites of *Ganoderma applanatum* Basidiomycete; Phytochemistry, Vol. 10, pp. 2361-2365(1971).

W. B. Turner and D.C. Aldridge; Fungal Metabolites II; Academic Press, Inc., New York, New York, 631 pp. (1983).

Common/Systematic Name

Ergosta-4,6,8(14),22-tetraen-3-one

Molecular Formula/Molecular Weight

$C_{28}H_{40}O$; MW = 392.30792

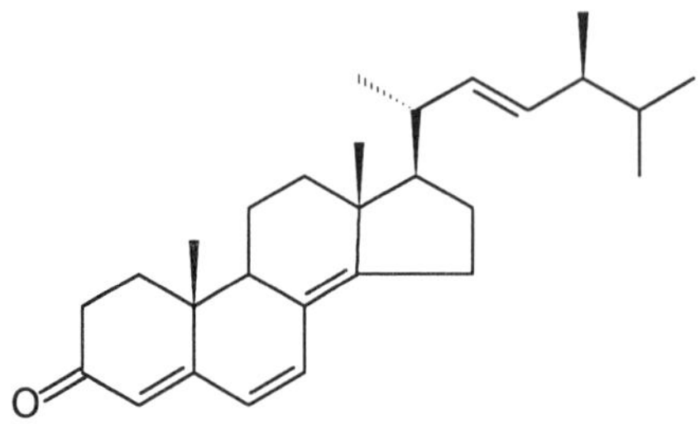

General Characteristics

Crystals; mp., 114-115°C; highly fluorescent.

Fungal Source

Alternaria alternata, Aspergillus flavus, A. fumigatus, A. ochraceus, Candida utilis, Fomes officinalis, F. applanatus, Fusarium moniliforme, Lampteromyces japonicus, Penicillium citrinum, P. patulum = P. griseofulvum, P. rubrum, and P. islandicum.

Isolation/Purification

Fungal cultures were extracted twice with methanol, saponified with ethanolic KOH, and extracted with *n*-hexane. The *n*-hexane extracts were reduced in volume, applied to a silica gel 60 column eluted in succession with hexane, ethyl acetate-hexane (5:95, v/v), and ethyl acetate-hexane (10:90, v/v). Ergosta-4,6,8(14),22-tetraen-3-one purification monitored with long-wave UV light was eluted in the 5% ethyl acetate-hexane fraction and recrystallized from petroleum ether.

Spectral Data

UV:

λ_{max}^{EtOH} 351(log ∊ 25,800), 282(5,800), and 239nm (4,100).

IR:

(KBr) 2940, 2860, 1650, 1562, 1445, 1347, 1320, 1265, 1217, 1191, 964, 946, 872, and 758cm^{-1}.

^{13}C NMR:

($CDCl_3$) 16.64; 17.62; 18.97; 19.44; 19.64; 19.96; 21.21; 25.34; 27.65; 33.05; 33.91; 34.09; 35.59; 36.70; 39.19; 42.82; 43.93; 44.31; 55.66; 122.83; 124.29; 132.35; 133.72; 155.73; 163.97; and 198.97ppm.

Mass Spectrum:
LREIMS: 392(M^+, 21%), 268(100), 267(55), 242(5), and 214*m/e* (22).

TLC Data

Brinkman SIL G-HR-25 precoated plates were activated for 1 hr at 110° C; solvent systems were chloroform-acetone (88:12, v/v) (R_f 0.87) and benzene-acetone (90:10, v/v) (R_f 0.68). The metabolite was detected as a blue fluorescent spot with UV light.

HPLC Data

SIL-X-I column, mobile phase was isooctane-ethyl acetate (92.5:7.5, v/v) at 2.5ml/min flow rate; detection at 350nm and elution at 2.5min.

References

R. G. Cooks, R. D. Daftary, and Y. Pomeranz; Changes in Mold-Damaged Wheat Flours Stored at Various Temperatures; J. Agric. Food Chem., Vol. 18, pp. 620-623(1970).

W. D. Nes and E. J. Parish (eds.); Analysis of Sterols and Other Biologically Significant Steroids; Academic Press Inc., New York, New York, p. 341(1989).

L. M. Seitz and J. V. Paukstelis; Metabolites of *Alternaria alternata*: Ergosterol and Ergosta-4,6,8(14),22-tetraen-3-one; J. Agric. Food Chem., Vol. 25, pp. 838-841(1977).

W. B. Turner and D.C. Aldridge; Fungal Metabolites II; Academic Press, Inc., New York, New York, 631 pp. (1983).

J. D. Weete; Structure and Function of Sterols in Fungi; Advances in Lipid Research, Vol. 23, pp. 115-167(1989).

Common/Systematic Name

Ergosta-8(9),14-dien-3β-ol; Ignosterol
(24*S*)24-Methylcholesta-8(9),14-dien-3β-ol

Molecular Formula/Molecular Weight

$C_{28}H_{46}O$; MW = 398.35487

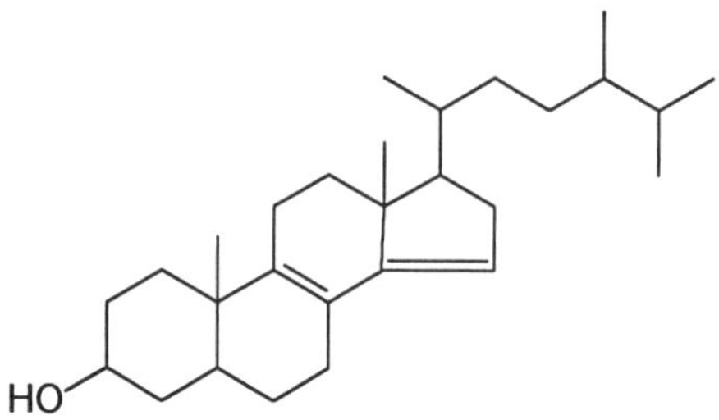

General Characteristics

Crystals from acetone-MeOH; mp., 133-134°C; $[\alpha]_D$ + 11° (*c*=1%, in $CHCl_3$).

Fungal Source

Saccharomyces cerevisiae strain 3701B and *Chlorella ellipsoidea.*

Isolation/Purification

Ignosterol was isolated from the neutral lipid fraction by column chromatography using Woelm grade II alumina and eluting with increasing amounts of ethyl ether in *n*-hexane. The sterol fraction was acetylated and further chromatographed by argentation chromatography eluting with increasing amounts of ethyl ether in *n*-hexane.

Spectral Data

UV:

λ_{max}^{MeOH} 251nm (ε = 18,000).

IR:

(KBr) 3050 and 795cm^{-1}.

^{1}H NMR:

The C-18 and C-19 methyl resonances appeared at 0.82 and 0.99ppm, respectively, which is in good agreement with the calculated shifts of 0.83 and 0.94ppm and compares favorably to those of ergosta-8,14,22-trien-3β-yl benzoate (0.85 and 0.94ppm). A broad one-proton absorption appeared at 5.33ppm for the olefinic hydrogen of the 8, 14 conjugation.

Mass Spectrum:
LREIMS: 398(M^+), 383, 365, and 371*m/e*.

GC Data
(Acetate derivative, relative retention time (RRT) relative to cholesteryl acetate) QF-1, RRT 1.24; Hi-Eff-8-BP, 1.44; PMPE, 1.38; and SE-30, 1.32.

References
L. G. Dickson, G. W. Patterson, C. F. Cohen, and S. R. Dutky; Two Novel Sterols from Inhibited *Chlorella ellipsoidea*; Phytochem., Vol. 11, pp. 3473-3477(1972).

P. R. Hays, L. W. Parks, H. D. Pierce, Jr., and A. C. Oehlschlager; Accumulation of Ergosta-8,14-dien-3β-ol by *Saccharomyces cerevisiae* Cultured with an Azasterol Antimycotic Agent; Lipids, Vol. 12, pp. 666-668(1977).

P. R. Hays, W. D. Neal, and L. W. Parks; Physiological Effects of an Antimycotic Azasterol on Cultures of *Saccharomyces cerevisiae*; Antimicrobial Agents and Chemotherapy, Vol. 12, pp. 185-191(1977).

Common/Systematic Name

(24*S*)24-Methylcholesta-8,22-dien-3β-ol; Ergosta-8(9),22-dien-3β-ol

Molecular Formula/Molecular Weight

$C_{28}H_{46}O$; MW = 398.35487

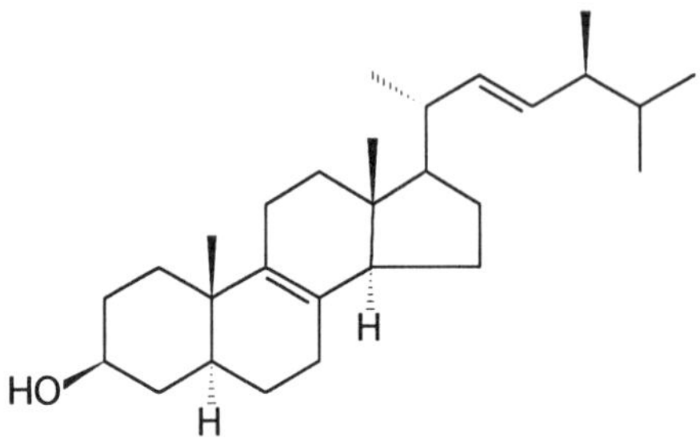

General Characteristics

Free alcohol, lathes from acetone; mp., 165-168°C; $[\alpha]_D$ + 27.6° (*c*=0.51, in $CHCl_3$) (mp., 166-169°C; $[\alpha]_D$ + 30°). Benzoate: plates from chloroform-methanol; mp., 152-153°C; $[\alpha]_D$ + 26.9° (*c*=0.42, in $CHCl_3$) (mp., 147-148°C; $[\alpha]_D$ + 34°).

Fungal Source

Candida albicans, C. utilis, Neurospora crassa, Saccharomyces cerevisiae, and *Torulopsis glabrata.*

Isolation/Purification

Yeast cells were refluxed in 10% methanolic KOH, methanol removed *in vacuo*, and ether soluble material converted into benzoate derivatives. The benzoates were separated with preparative silica gel GF_{254} TLC followed by silver nitrate-impregnated TLC.

The yeast cells were saponified by the alkaline pyrogallol procedure as modified for large volumes. The nonsaponifiable fraction was extracted into hexane and recovered by evaporating the solvent under nitrogen. The sterols were dissolved in ethanol, precipitated with digitonin, recovered by dimethyl sulfide cleavage and recrystallization from ethanol. The partially purified sterols were then separated by thin-layer chromatography on $AgNO_3$-silica gel G. To obtain large quantities of the sterol for detailed analyses by various spectroscopic procedures, the sterols were recrystallized three times from alcohol after digitonin precipitation. The recovered product was streaked onto plates of silica gel G (1.0mm), and the plates were developed in benzene-ethyl acetate (5:1, v/v). The sterol band was scraped from the plates and eluted with $CHCl_3$.

Spectral Data

^{1}H NMR:

($CDCl_3$) The spectrum showed the presence of two olefinic hydrogens (broadened triplet) at 5.14ppm and a single hydrogen on a carbon bearing oxygen as a broadened multiplet at 3.56ppm. In addition, there were clearly six distinct methyl resonances, with singlets at 0.54(C-18) and 0.79ppm (C-19). There were doublet methyls due to C-21, C-26, C-27, and C-28 at 0.81(*J*=7), 0.82(*J*=7), 0.90(*J*=6), and 1.00ppm (*J*=6Hz). In particular, the nonidentity of the C-26 and C-27 methyl groups required the presence of an asymmetric center at C-24. Significant parts of the spectrum were similar to that of ergosterol. In particular, ergosterol showed its 22-23 olefinic hydrogens as a broad ended triplet at 5.15ppm and four secondary C-methyl groups as doublets centered at 0.82, 0.83, 0.90, and 1.01ppm. This clearly defined the nature of the side chain as being identical with that of ergosterol.

Mass Data:

Benzoate: Found: C, 83.6%; H, 10.1; calcd for $C_{35}H_{52}O_2$; C, 83.6; H, 10.0. The mass spectrum of the free alcohol showed a molecular ion at 398*m/e* ($C_{28}H_{46}O$) with significant fragment peaks at 383*m/e* (M^+ - CH_3), 380(M^+ - H_2O), 365(M^+ - CH_3 and -H_2O), 273(M^+ - side chain), and 255(M^+ - side chain and - H_2O).

References

D. H. R. Barton, J. E. T. Corrie, D. A. Widdowson, M. Bard, and R. A. Woods; Biosynthesis of Terpenes and Steroids. Part IX. The Sterols of Some Mutant Yeasts and Their Relationship to the Biosynthesis of Ergosterol; J. Chem. Soc., Trans Perkin I, pp. 1326-1333(1974).

D. H. R. Barton, A. A. L. Gunatilaka, T. R. Jarman, D. A. Widdowson, M. Bard, and R. A. Woods; Biosynthesis of Terpenes and Steroids. Part X. The Sterols of Some Yeast Mutants Doubly Defective in Ergosterol Biosynthesis; J. Chem. Soc., Trans Perkin I, pp. 88-92(1975).

L. W. Parks, F. T. Bond, E. D. Thompson, and P. R. Starr; $\Delta^{8(9),22}$-Ergostadiene-3β-ol, an Ergosterol Precursor Accumulated in Wild-type and Mutants of Yeast; J. Lipid Res., Vol. 13, pp. 311-316(1972).

Common/Systematic Name

(24*S*)24-Methylcholesta-7,22,24(28)-trien-3β-ol; Ergosta-7,22,24-trien-3β-ol

Molecular Formula/Molecular Weight

$C_{28}H_{44}O$; MW = 396.33922

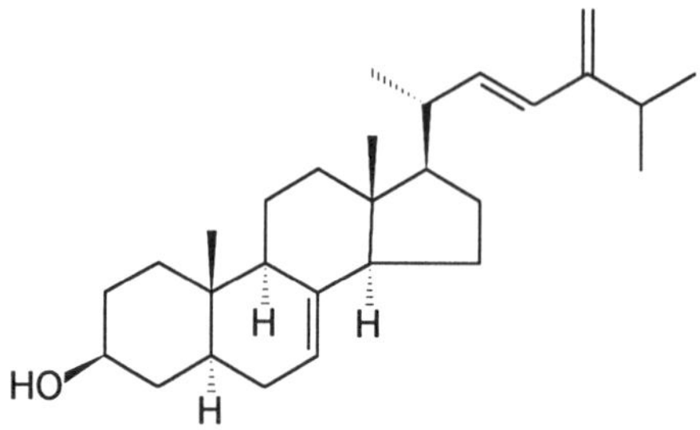

General Characteristics

Crystals from acetone-methanol; 119-121°C; $[\alpha]_D^{25}$ + 36.5° (*c*=0.85, in $CHCl_3$). Benzoate, crystals from chloroform-methanol; mp., 139-141°C; $[\alpha]_D^{26}$ + 13.5° (*c*=1.56, in $CHCl_3$). Acetate, plates from chloroform-methanol; 133-136°C; $[\alpha]_D^{24}$ + 6.4° (*c*=0.85, in $CHCl_3$).

Fungal Source

Saccharomyces cerevisiae, S. carlsbergensis, Candida albicans, C. utilis, and *Leptosphaeria typhae.*

Isolation/Purification

Yeast cells were refluxed in 10% methanolic KOH, methanol was removed *in vacuo*, and ether soluble material was chromatographed by alumina column chromatography, eluting with various combinations of benzene, acetone, light petroleum ether or methanol. Alternatively, they were converted into benzoate derivatives and separated with preparative silica gel GF_{254} TLC followed by silver nitrate-impregnated TLC.

Spectral Data

UV:

λ_{max} 232(ϵ = 18,600), 226sh (18,300), and 240nm sh (12,800); benzoate, λ_{max} 232nm (ϵ = 33,800); acetate, λ_{max} 227sh (ϵ = 22,100), 232(23,200), and 240nm sh (15,700).

IR:

(KBr) 3400, 1640, 1600, and 890cm^{-1}; benzoate, 1710, 1640, 1600, and 890cm^{-1}; acetate, 1710, 1640, 1600, and 890cm^{-1}.

^{1}H NMR:
(CDCl$_3$) 4.12(1H, d, *J*=16Hz, H-23); 4.50(1H, dd, *J*=8.0, 16Hz, H-22); 4.90(1H, m, H-7); 5.25(2H, m, H-28); 6.46(1H, m, H-3α); 7.52(1H, septet, *J*=7.0Hz, H-25); 8.98(6H, d, *J*=7.0Hz, 26- and 27- CH$_3$); 9.23(3H, s, 10-Me); and 9.46τ (3H, s, 13-Me).

Mass Data:
Found: C, 84.5%; H, 11.0; calcd for $C_{28}H_{44}O$; C, 84.8; H, 11.2; benzoate, found: C, 83.7; H, 9.4; calcd for $C_{35}H_{48}O_2$; C, 83.9; H, 9.7; acetate, LREIMS: 438(M^+), 423, 395, 378, 363, 354, 315, and 313*m/e* (100%); found: C, 82.0%; H, 10.6; calcd for $C_{30}H_{46}O_2$; C, 82.1; H, 10.6.

References

D. H. R. Barton, J. E. T. Corrie, D. A. Widdowson, M. Bard, and R. A. Woods; Biosynthesis of Terpenes and Steroids. Part IX. The Sterols of Some Mutant Yeasts and Their Relationship to the Biosynthesis of Ergosterol; J. Chem. Soc., Perkin Trans I, pp. 1326-1333(1974).

D. H. R. Barton, U. M. Kempe, and D. A. Widdowson; Investigations on the Biosynthesis of Steroids and Terpenes. Part VI. The Sterols of Yeast; J. Chem. Soc., Perkin Trans I, pp. 513-522(1972).

Common/Systematic Name

(24*S*)24-Methylcholesta-8,14,24(28)-trien-3β-ol; Ergosta-8,14,24-trien-3β-ol

Molecular Formula/Molecular Weight

$C_{28}H_{44}O$; MW = 396.33922

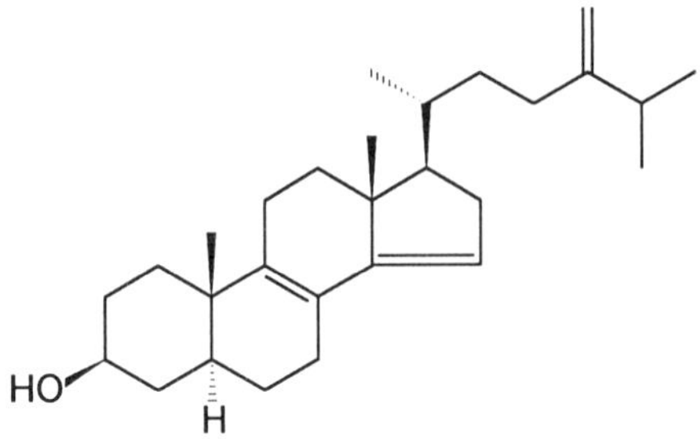

General Characteristics

Benzoate, plates from chloroform-methanol; mp., 148-151°C; $[\alpha]_D$ - 4.1° (*c*=0.38).

Fungal Source

Saccharomyces cerevisiae.

Isolation/Purification

Yeast cells were refluxed in 10% methanolic KOH, methanol was removed *in vacuo*, and ether-soluble material was converted into benzoate derivatives. The benzoates were separated with preparative silica gel GF_{254} TLC followed by silver nitrate-impregnated TLC.

Spectral Data (Benzoate)

UV:

λ_{max} 230(ϵ = 25,800) and 251nm sh (21,400).

Mass Data:

LREIMS: 500(M^+, 100%), 485, 457, 441, 378, 375, 373, 363, 335, 279, 271, 253, 251, 239, 238, and 237*m/e*; found: C, 83.8; H, 9.5; calcd for $C_{35}H_{48}O_2$; C, 83.9%; H, 9.7.

References

D. H. R. Barton, J. E. T. Corrie, D. A. Widdowson, M. Bard, and R. A. Woods; Biosynthesis of Terpenes and Steroids. Part IX. The Sterols of Some Mutant Yeasts and Their Relationship to the Biosynthesis of Ergosterol; J.Chem. Soc., Trans Perkin I, pp. 1326-1333 (1974).

D. H. R. Barton, A. A. L. Gunatilaka, T. R. Jarman, D. A. Widdowson, M. Bard, and R. A. Woods; Biosynthesis of Terpenes and Steroids. Part X. The Sterols of Some Yeast Mutants Doubly Defective in Ergosterol Biosynthesis; J. Chem. Soc., Trans Perkin I, pp. 88-92 (1975).

Common/Systematic Name

Ergosta-8,22,24(28)-trien-3β-ol

Molecular Formula/Molecular Weight

$C_{28}H_{44}O$; MW = 396.33922

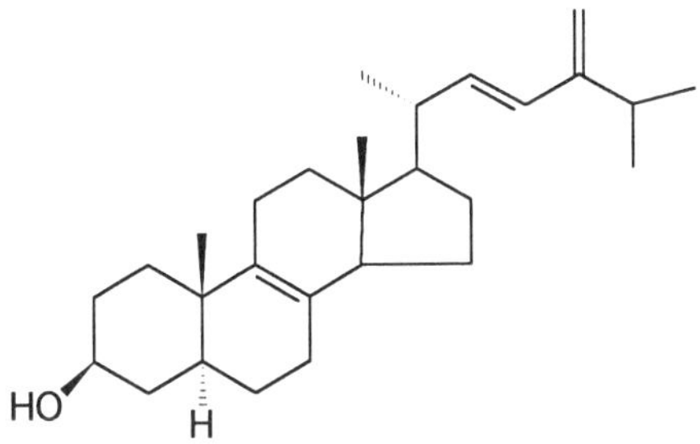

Fungal Source

Saccharomyces cerevisiae (SM36 mutant strain).

Isolation/Purification

Yeast cells were refluxed in 10% methanolic KOH, methanol was removed *in vacuo*, and ether-soluble material was converted into benzoate derivatives. The benzoates were separated with preparative silica gel GF_{254} TLC followed by silver nitrate-impregnated TLC.

Spectral Data

^{1}H NMR:

($CDCl_3$) 0.65(s, 13β-methyl); 1.02(s, 10β-methyl); and 1.04ppm (9H, d, 20-methyl and 25-methyls).

Mass Spectrum:

Benzoate, LREIMS: 500*m/e* (M^+).

References

D. H. R. Barton, J. E. T. Corrie, D. A. Widdowson, M. Bard, and R. A. Woods; Biosynthesis of Terpenes and Steroids. Part IX. The Sterols of Some Mutant Yeasts and Their Relationship to the Biosynthesis of Ergosterol; J. Chem. Soc., Trans Perkin I, pp. 1326-1333(1974).

D. H. R. Barton, A. A. L. Gunatilaka, T. R. Jarman, D. A. Widdowson, M. Bard, and R. A. Woods; Biosynthesis of Terpenes and Steroids. Part X. The Sterols of Some Yeast Mutants Doubly Defective in Ergosterol Biosynthesis; J. Chem. Soc., Trans Perkin I, pp. 88-92(1975).

<u>Common/Systematic Name</u>

Ergosta-5,7,9(11),22-tetraen-3β-ol

<u>Molecular Formula/Molecular Weight</u>

$C_{28}H_{42}O$; MW = 394.32357

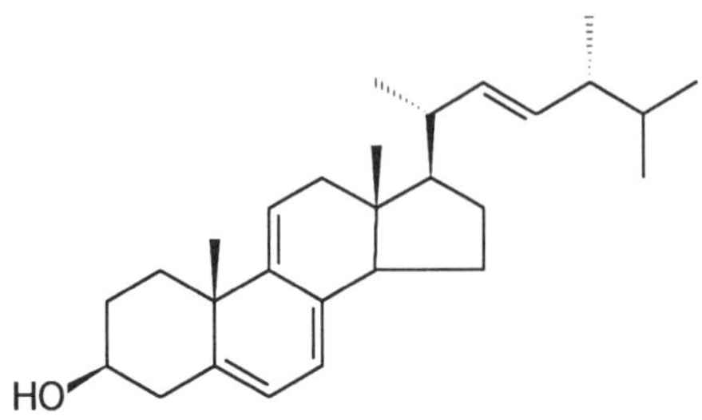

<u>Fungal Source</u>

Mucor rouxii, Saccharomyces cerevisiae, and *Gibberella fujikuroi.*

<u>Isolation/Purification</u>

Purification was by TLC (silica gel GF), followed by repeated crystallization to give pure ergosta-5,7,9,22-tetraen-3β-ol.

<u>Spectral Data</u>

UV:

λ_{max} 231(ϵ = 21,000), 271(11,700), 282(12,000), and 293nm (6,600).

IR:

(Nujol mull) 3350, 1610, 1060, 1040, 980, and 900cm^{-1}.

^{1}H NMR:

9.37(3H, s, 13-Me); 9.08(3H, s, 10-Me); 8.95(9H, d, *J*=6.5Hz, 21-, 26-, 25-Me); 6.42(1H, m, 3α-H); and 5.23-5.21ppm (2H, m, 28-H).

Mass Data:

Benzoate, Anal. calcd for ($C_{35}H_{46}O_2$); C, 84.29%; H, 93.0; found: C, 83.26; H, 9.15.

<u>TLC Data</u>

R_f 0.33 with silica gel plates developed with benzene-diethyl ether (9:1, v/v).

<u>HPLC Data</u>

Relative retention time was 0.56 (relative to cholesterol) performed with C_{18} reversed phase column eluted with 4% aqueous methanol at ambient temperature with a flow rate of 1.6ml/min.

GC Data

Relative retention time was 1.11 relative to cholesterol using 3% SE-30 at 245°C.

References

L. Atherton, J. M. Duncan, and S. Safe; Isolation and Biosynthesis of Ergosta-5,7,9(11),22-tetraen-3β-ol from *Mucor rouxii*; J. Chem. Soc., Chem. Commun., pp. 882-883(1972).

D. H. R. Barton, U. M. Kempe, and D. A. Widdowson; Investigations on the Biosynthesis of Steroids and Terpenes. Part VI. The Sterols of Yeast; J. Chem. Soc., Perkin Trans I, pp. 513-522(1972).

W. D. Nes, S. Xu, and W. F. Haddon; Evidence for Similarities and Differences in the Biosynthesis of Fungal Sterols; Steroids, Vol. 53, pp. 533-558(1989).

W. B. Turner and D. C. Aldridge; Fungal Metabolites II; Academic Press, New York, 310 pp. (1983).

Common/Systematic Name

Ergosta-5,8,22,24(28)-tetraen-3β-ol

Molecular Formula/Molecular Weight

$C_{28}H_{42}O$; MW = 394.32357

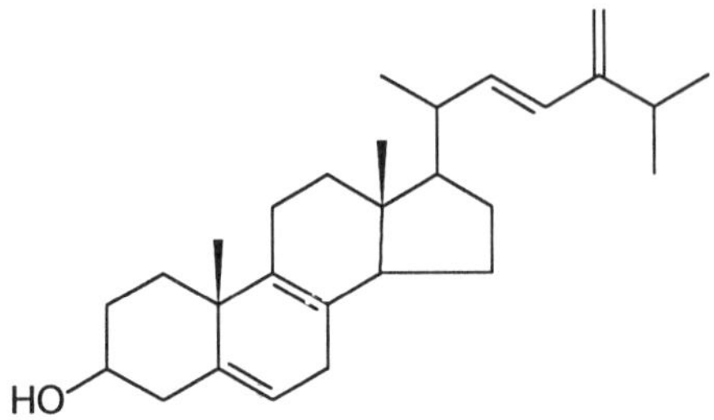

General Characteristics

Acetate, crystals from chloroform-methanol; mp., 104-106°C.

Fungal Source

Saccharomyces cerevisiae.

Isolation/Purification

The sterol was purified by preparative argentation TLC (silica gel coated with $AgNO_3$) of the acetate derivative using benzene as mobile phase. 3β-Acetoxyergosta-5,8,22,24(28)-tetraene was eluted from a band at R_f 0.33 and recrystallized from chloroform-methanol.

Spectral Data

UV:

λ_{max}^{Hexane} (acetate) 225sh(ε=25,000), 231(26,200), and 239.5nm sh (16,700).

^{1}H NMR:

($CDCl_3$) (acetate) 0.70(3H, s, H_3-18); 1.09(6H, d, J=7.0Hz, H_3-26 and H_3-27); 1.09(3H, d, H-21); 1.21(3H, s, H-19); 2.08(3H, s, CH_3CO-); 4.64(1H, m, H-3); 4.82(2H, m, H-28); 5.41(1H, m, H-6); 5.51(1H, dd, $J_{20,22}$=8.5Hz); and 5.92ppm (1H, d, $J_{22,23}$=15.3Hz).

Mass Spectrum:

LREIMS: (acetate) 436(M^+, 5.9%), 376(70.5), 361(100), 253(25.3), 251(23.2), 211(12.2), 157(31.2), 143(24.8), 123(17.2), and 81m/e (23.0).

Reference

A. M. Pierce, A. M. Unrau, A. C. Oehlschlager; and R. A. Woods; Azasterol Inhibitors in Yeast. Inhibition of the Δ^{24}-Sterol Methyltransferase and the 24-Methylene Sterol $\Delta^{24(28)}$-Reductase in Sterol Mutants of *Saccharomyces cerevisiae*; Can. J. Biochem., Vol. 57, 112-116(1979).

Common/Systematic Name
Ergosterone
Ergosta-4,7,22-trien-3-one

Common/Systematic Name
$C_{28}H_{42}O$; MW = 394.32357

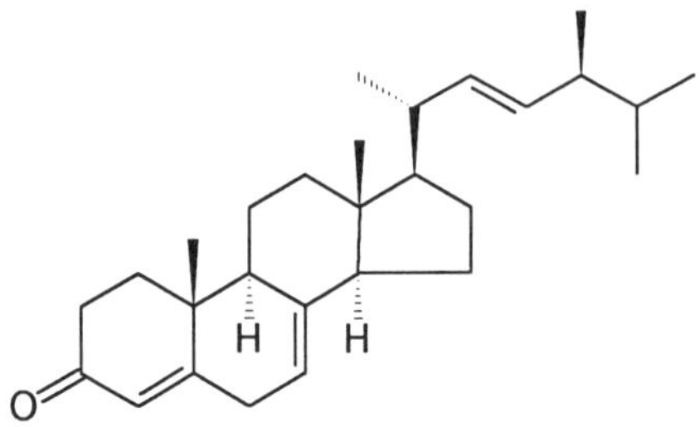

General Characteristics
Crystals from acetone-light petroleum; 134°C; $[\alpha]_D$ -12° (*c*=4.5, in $CHCl_3$).

Fungal Source
Fomes annosus.

Isolation/Purification
Sporophores were extracted with light petroleum and evaporated to dryness to give a yellow wax. The extract was chromatographed on a column of neutral alumina with increasing concentrations of chloroform in benzene. The benzene-chloroform (1:1, v/v) eluant contained ergosterol and ergosta-4,7,22-trien-3-one which were separated by preparative TLC using chloroform as developing solvent.

Spectral Data

UV:
λ_{max} 240nm (log ϵ = 4.11).

IR:
(KBr) 1680cm^{-1}.

Mass Data:
Found: C, 84.8; H, 10.5; calcd. for $C_{28}H_{42}O$: C, 85.2; H, 10.7%.

References
F. Johnson, G. T. Newbold, and F. S. Spring; Steroids. Part XIII. The Conversion of Ergosterol into Progesterone; J. Chem. Soc. (C), pp. 1302-1306(1954).

H. D. Munro and O. C. Musgrave; Extractives from Sporophores of Some *Fomes* Species; J. Chem. Soc. (C), pp. 685-688(1971).

Common/Systematic Name

Ergosta-7,22-diene-3β,5α-diol

Common/Systematic Name

$C_{28}H_{46}O_2$; MW = 414.34978

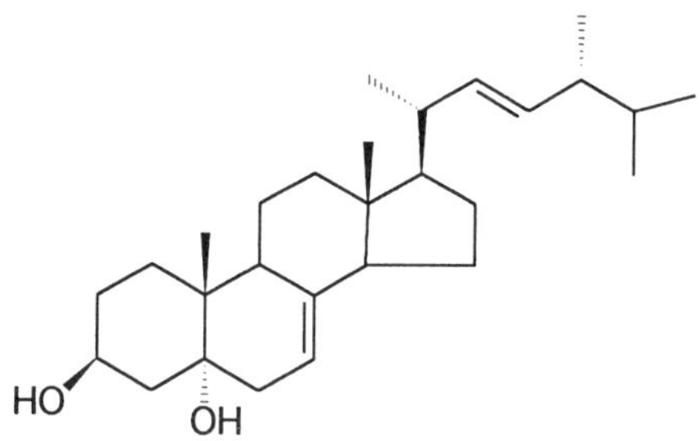

Fungal Source

Saccharomyces cerevisiae.

Biological Activity

Shown to be an intermediate in the biosynthesis of ergosterol.

Reference

M. Fryberg, A. C. Oehlschlager, and A. M. Unrau; Formation of A 5α-Hydroxy Sterol by *Saccharomyces cerevisiae;* Biochem. Biophys. Res. Commun., Vol. 51, pp. 219-222(1973).

Common/Systematic Name

Ergosta-8,24(28)-diene-3β,6α-diol

Common/Systematic Name

$C_{28}H_{46}O_2$; MW = 414.34978

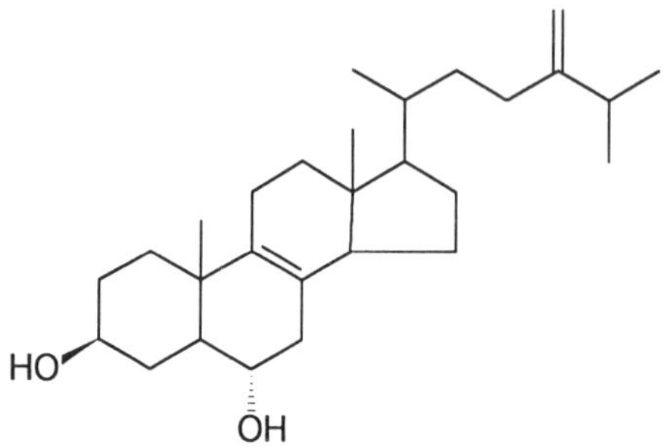

Fungal Source

Saccharomyces cerevisiae (mutant erg 2).

Isolation/Purification

The non-saponifiable fraction was acetylated and separated by preparative TLC on silica gel 254, $AgNO_3$ plates developed in benzene. Ergosta-8,24(28)-diene-3β,6α-diol was finally purified from preparative TLC by eluting the band at R_f 0.14.

Biological Activity

Shown to be an intermediate in the biosynthesis of ergosterol.

Spectral Data

^{1}H NMR:

Acetate: ($CDCl_3$) 0.61(H-18); 1.08(H-19); 0.97(*J*=7.5Hz, H-21); 1.05(*J*=7.0Hz, H-26, 27); 2.05, 2.07($COOCH_3$); ca. 4.65(H-3α); ca. 3.90(H-6β); and 4.65ppm (*J*=5.0Hz, H-28).

^{13}C NMR:

Acetate ($CDCl_3$) 34.8(C-1); 27.0(C-2); 72.7(C-3); 28.1(C-4); 44.3(C-5); 70.2(C-6); 33.9(C-7); 125.9(C-8); 134.5(C-9); 37.4(C-10); 22.4(C-11); 36.4(C-12); 41.8(C-13); 51.2(C-14); 23.4(C-15); 28.4(C-16); 54.4(C-17); 11.0(C-18); 18.4(C-19); 35.9(C-20); 18.4(C-21); 34.4(C-22); 30.8(C-23); 156.5(C-24); 33.5(C-25); 21.6(C-26); 21.6(C-27); 105.3(C-28); 20.9, 21.2($COOCH_3$); 170.2; and 170.6ppm ($COOCH_3$).

Mass Spectrum:

HREIMS: Acetate 498.3716*m/e*; $C_{28}H_{46}O_2$, requires 498.3709. CIMS: 498*m/e*.

Reference

A. M. Pierce, H. D. Pierce, Jr., A. M. Unrau, and A. C. Oehlschlager; Identification of Ergosta-8,24-diene-3β,6α-diol in a Δ^8 to Δ^7 Sterol Isomerase-Blocked Mutant; Lipids, Vol. 14, pp. 876-879(1979).

Common/Systematic Name

Portensterol; Hydroxyisoergosterol

Molecular Formula/Molecular Weight

$C_{28}H_{44}O_2$; MW = 412.33413

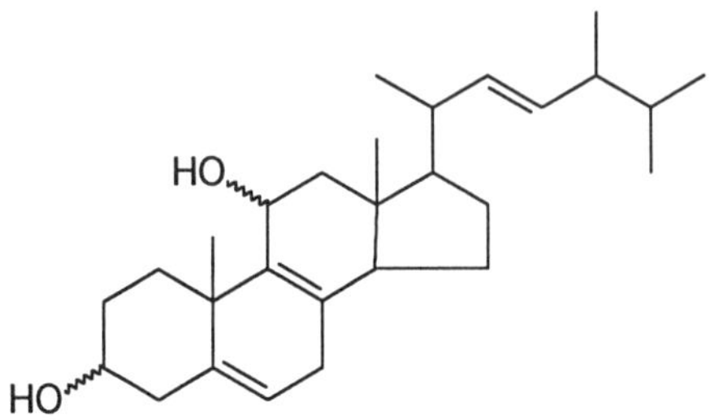

General Characteristics

Colorless irregular prisms from ethanol; 248-250°C; 255°C in capillary tube; $[\alpha]_D^{20}$ - 53° (*c*=0.010, in $CHCl_3$/MeOH). Acetate: $C_{32}H_{48}O_4$; mp., 160°C in capillary tube; $[\alpha]_D^{20}$ - 119° (*c*=0.010, in $CHCl_3$). Portensterol was very soluble in methanol and ethanol; soluble in chloroform, acetone, and benzene; sparingly soluble in ethyl ether; and insoluble in petroleum ether.

Fungal Source

Clitocybe nebularis, Rhodopaxillus nudus, and *Tricholoma portentosum.*

Isolation/Purification

Fruit bodies of *Clitocybe nebularis, Rhodopaxillus nudus,* and *Tricholoma portentosum* were extracted with benzene, evaporated to dryness, redissolved in minimal benzene and chromatographed on a neutral alumina (15% water) column. The column was eluted with a gradient from benzene to ethyl ether. Rechromatography of the impure metabolite with neutral alumina (25% water) eluted with a gradient from ethyl ether to 10% ethanol/ethyl ether followed by recrystallization from ethanol yielded pure portensterol.

Spectral Data

Mass Data:

Found: C, 77.42; H, 10.61; O, 12.22%; calcd. for $C_{28}H_{44}O_2 \cdot H_2O$; C, 78.09; H, 10.77; O, 11.15%.

GC Data

Column: 10% UCW 98 coated onto Chromosorb GAW; flame-ionization detector at 290°C; column temperature 260°C isothermal; injector temperature 240°C; retention time acetate, 7.2 min, TMS 10.3 min.

Reference

F. Regerat and H. Pourrat; Isolement d'un nouveau stérol, le portenstérol, des carpophores de *Tricholoma portenosum, Rhodopaxillus nudus, et Clitocybe nebularis*; Annales Pharmaceutiques Françaises, Vol. 34, pp. 323-328(1976).

Common/Systematic Name

Pyrocalciferol

Molecular Formula/Molecular Weight

$C_{28}H_{44}O$; MW = 396.33922

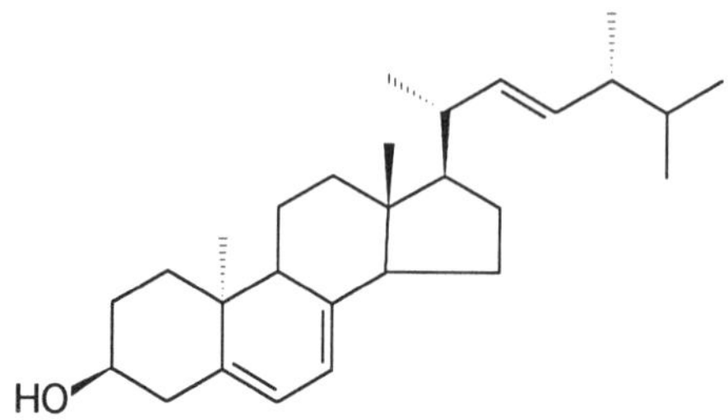

General Characteristics

Mycelium extracted with petroleum ether, removal of extracting solvent followed by crystallization from ethanol yielded pure pyrocalciferol. Crystals from ethanol; 93-95°C; $[\alpha]_D^{25}$ + 502°; positive Liebermann-Burchard test.

Fungal Source

Penicillium notatum and *Trichoderma koningii.*

Isolation/Purification

Mycelium extracted with petroleum ether; removal of extracting solvent followed by crystallization from ethanol yielded pure pyrocalciferol.

Spectral Data

UV:

λ_{max}^{MeOH} 295(ϵ = 3111), 285(4375), 271(4733), and 263nm (4368).

IR:

(Nujol) 3436(OH), 3322, 2873, 1736, 1724, 1643, 1618, 1471, 1390, 1370, 1242, 1163, 1073, 1059, 1042, 975, 883, 806, and 727cm^{-1}.

^{1}H NMR:

($CDCl_3$) 9.35 and 9.22(two singlets for angular methyl groups at carbons 10 and 13); 9.1(12H, d, *gem*-dimethyls and two C-methyls at C-25); 5.75(OH at C-3, disappeared on deuteration); 4.78(2H, s, two protons in ring B at C-6 and -7); and 7.5-8.8τ (23H, broad multiplet for remaining protons of the ring and side chain).

Mass Data:

LREIMS: 396(M^+, intense), 378(- H_2O), 363(-CH_3), 337, 231, 211, 199, 197, and 147*m/e*; found: C, 85.04; H, 10.95; O, 4.11%; calcd. for $C_{28}H_{44}O$; C, 84.79; H, 11.18; O, 4.03%

Reference

A. Kamal, R. Akhtar, and A. A. Qureshi; Studies in The Biochemistry of Microorganisms; Pakistan J. Sci., Ind. Res., Vol. 14, pp. 71-78(1971).

Common/Systematic Name

Ergosta-5,7,22,24(28)-tetraen-3ß-ol

Molecular Formula/Molecular Weight

$C_{28}H_{42}O$; MW = 394.32357

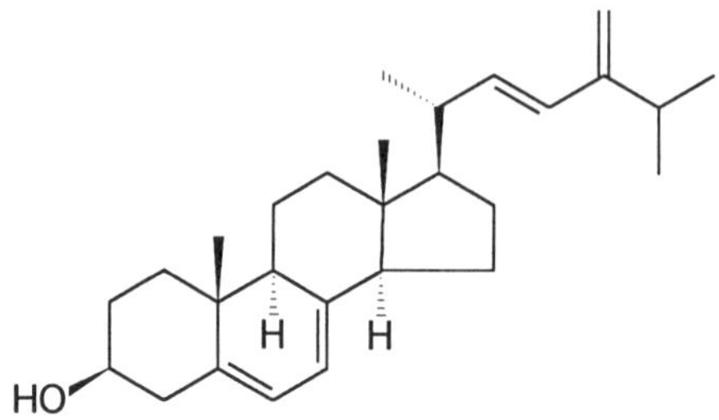

Fungal Source

Aspergillus oryzae, Candida albicans, C. utilis, Saccharomyces cerevisiae, and *Gibberella fujikuroi.*

Spectral Data

UV:

λ_{max} 232 and 282nm.

TLC Data

R_f 0.33 with silica gel plates developed with benzene-diethyl ether (9:1, v/v).

HPLC Data

Relative retention time was 0.67 (relative to cholesterol) performed with C_{18} reversed phase column eluted with 4% aqueous methanol at ambient temperature with a flow rate of 1.6ml/min.

GC Data

Relative retention time was 1.30 relative to cholesterol using 3% SE-30 at 245°C.

References

D. H. R. Barton, U. M. Kempe, and D. A. Widdowson; Investigations on the Biosynthesis of Steroids and Terpenes. Part VI. The Sterols of Yeast; J. Chem. Soc., Perkin Trans I, pp. 513-522(1972).

M. Fryberg, A. C. Oehlschlager, and A. M. Urau; Sterol Biosynthesis in Antibiotic Sensitive and Resistant *Candida*; Archiv. Biochem. and Biophys., Vol. 173, pp. 171-177(1975).

W. D. Nes, S. Xu, and W. F. Haddon; Evidence for Similarities and Differences in The Biosynthesis of Fungal Sterols; Steroids, Vol. 53, pp. 533-558(1989).

W. B. Turner and D. C. Aldridge; Fungal Metabolites II; Academic Press, New York, p. 316(1983).

Common/Systematic Name

(24*S*)24-Methylcholesta-7,22-dien-3-one; Ergosta-7,22-dien-3-one

Molecular Formula/Molecular Weight

$C_{28}H_{44}O$; MW = 396.33922

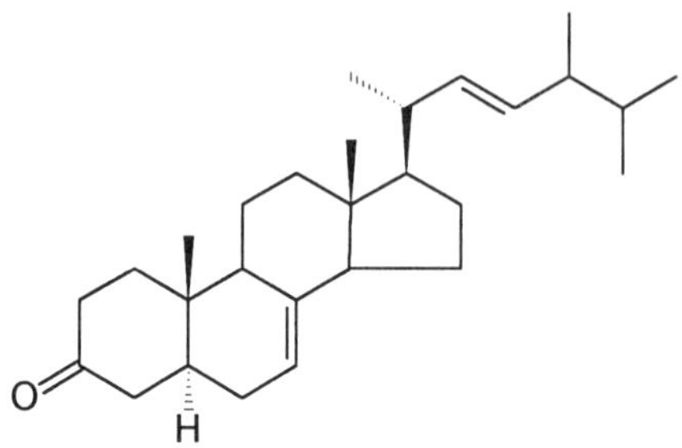

General Characteristics

Needles from acetone; mp., 182-183.5°C; $[\alpha]_D^{22}$ + 1.63° (*c*=4.90, in $CHCl_3$) (mp., 183-185°C; $[\alpha]_D$ + 1.0°).

Fungal Source

Ganoderma applanatum = *Fomes applanatus, Fomes fomentarius, F. annosus,* and *F. pini.*

Isolation/Purification

Sporophores were collected from the trunks of dead hairy birch trees in the Forest of Birse, near Aboyne, and chopped into small pieces. The air-dried fungus was exhaustively extracted with light petroleum followed by evaporation to yield a yellow wax. The light petroleum extract was evaporated to dryness, redissolved in ethyl ether, washed with aqueous sodium hydrogen carbonate, followed by acidification. The ethereal solution was evaporated to dryness, redissolved in benzene and chromatographed through a column of neutral alumina which was eluted with light petroleum-benzene (1:1, v/v), (1:2, v/v), benzene, benzene-chloroform (3:4, v/v), chloroform, chloroform-methanol (4:1, v/v), and methanol. The benzene fraction was crystallized from acetone to yield ergosta-7,22-dien-3-one as needles.

Spectral Data

IR:

(KBr) 1717cm^{-1} (six membered ring C=O).

Mass Data:

HREIMS: 396.3392(M^+, $C_{28}H_{44}O$) and 244.1828*m/e* ($C_{17}H_{24}O$); found: C, 84.75; H, 11.3; calcd for $C_{28}H_{44}O$: C, 84.8; H, 11.2%.

TLC Data

R_f 0.41 using Kieselgel plates developed with petroleum ether-benzene-ethyl acetate, 85:10:5 (v/v/v).

References

H. R. Arthur, T. G. Halsall, and R. D. Smith; The Chemistry of Triterpenes and Related Compounds. Part XXXIII. The Isolation of Ergosta-7,22-dien-3-one from *Fomes fomentarius*; Vol. 29, pp. 2603-2605(1958).

H. D. Munro and O. C. Musgrave; Extractives from Sporophores of Some *Fomes* Species; J. Chem. Soc. (C), pp. 685-688(1971).

H. Ripperger and H. Budikieqicz; Steroide aus *Ganoderma applanatum*; Phytochem., Vol. 14, pp. 2297-2298(1975).

Common/Systematic Name

Cerevisterol

Ergosta-7,22-diene-5α,6β,7β-triol

Molecular Formula/Molecular Weight

$C_{28}H_{46}O_3$; MW = 430.34470

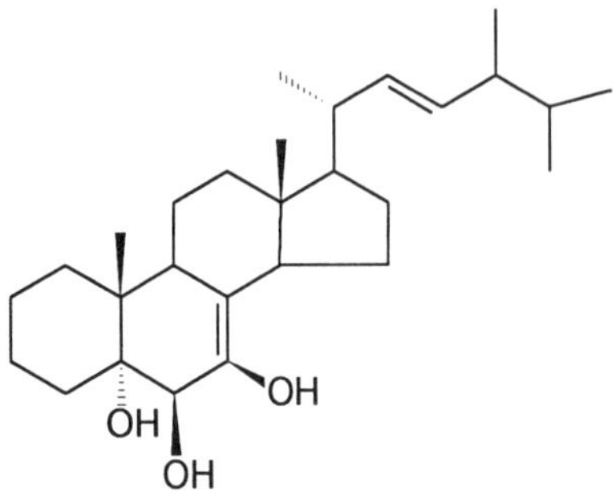

General Characteristics

Crystals; mp., 252-255°C; $[\alpha]_D$ - 80.5° (pyridine); diacetate; mp., 167-171°C; $[\alpha]_D$ - 147°.

Fungal Source

Acremonium luzulae, Boletaceae (family), *Cantharellus cibarius, Fomes allardii, Fusarium moniliforme, Polyporus dryadeus,* and *Russula sardonia.*

Isolation/Purification

Cultures extracted with butanol and the residue was purified by silica gel column chromatography using benzene-ethyl acetate as eluant.

Spectral Data

Mass Spectrum:

LREIMS: 412(M^+ -18), 394, 376, 361, 269, 251, and 69*m/e*.

References

D. Andina, M. De Bernardi, A. Del Vecchio, G. Fronza, G. Mellerio, G. Vidari, and P. Vita-Finzi; Sesquiterpenes from *Russula sardonia*; Phytochemistry, Vol. 19, pp. 93-97(1980).

P. Ceccherelli, R. Fringuelli, and G. F. Madruzza; Cerevisterol and Ergosterol Peroxide from *Acremonium luzulae*; Phytochemistry, Vol. 14, p. 1434(1975).

Yu. P. Cherotchenko and A. N. Shivrina; Cerevisterol in Fungi of the Family *Boletaceae;* Dokl. Akad. Naukssr., Vol. 212, pp. 486-487(1973).

E. P. Serebryakov, A. V. Simolin, V. F. Kuchrov, and B. V. Roskkynov; New Metabolites of *Fusarium moniliforme* Sheld.; Tetrahedron, Vol. 26, pp. 5215-5223(1970).

W. B. Turner and D. C. Aldridge; Fungal Metabolites II; Academic Press, New York, 316 pp. (1983).

Common/Systematic Name

14α-Methylcholest-7-en-3β-ol

Molecular Formula/Molecular Weight

$C_{28}H_{48}O$; MW = 400.37052

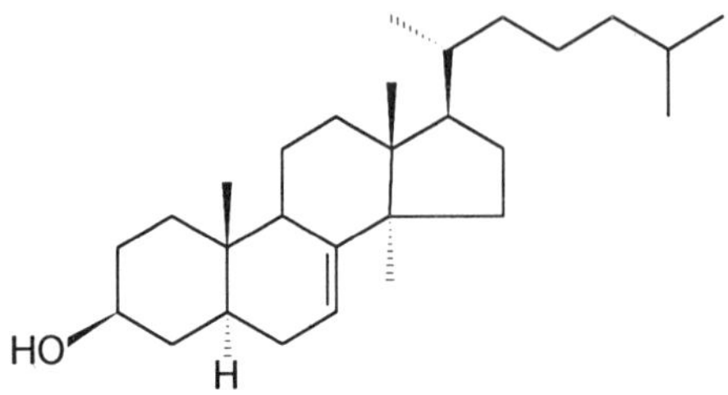

Fungal Source

Blastocladia ramosa.

Isolation/Purification

The neutral lipid was fractionated on Woelm neutral grade II alumina, the sterols were isolated by digitonin precipitation, and final purification was by argentation chromatography.

Spectral Data

Mass Spectrum:

14α-Methylcholest-7-en-3β-yl acetate; LREIMS: 442(M^+), 427, 381, 329, 287, 273, 255, 245, 227, 201, 161, 147, 133, 107, 95, 81, 69, 55, and 43*m/e* (100%).

GLC Data

Analysis of free sterols was with 3% SE-30, 1% QF-1 and 3% OV-17; sterol acetates were with SE-30 and QF-1.

Reference

M. A. Southall, J. J. Motta, and G. W. Patterson; Identification and Phylogenetic Implications of Fatty Acids and Sterols in Three Genera of Aquatic *Phycomycetes*; Amer. J. Botany, Vol. 62, pp. 246-252(1972),

Common/Systematic Name

14α-Methylcholest-8-en-3β-ol

Molecular Formula/Molecular Weight

$C_{28}H_{48}O$; MW = 400.37052

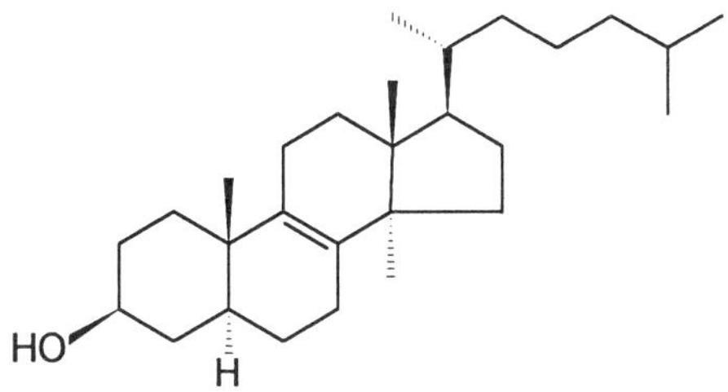

Fungal Source

Blastocladia ramosa.

Isolation/Purification

The neutral lipid was fractionated on Woelm neutral grade II alumina, the sterols were isolated by digitonin precipitation, and final purification was by argentation chromatography.

Spectral Data

Mass Spectrum:

14α-Methylcholest-8-en-3β-yl acetate; LREIMS: 442(M^+), 427, 382, 371, 368, 353, 329, 287, 273, 255, 245, 229, 227, 213, 201, 187, 173, 161, 145, 119, 107, 95, 69, and 43*m/e* (100%).

GLC Data

Analysis of free sterols was with 3% SE-30, 1% QF-1, and 3% OV-17; sterol acetates were with SE-30 and QF-1.

Reference

M. A. Southall, J. J. Motta, and G. W. Patterson; Identification and Phylogenetic Implications of Fatty Acids and Sterols in Three Genera of Aquatic *Phycomycetes*; Amer. J. Botany, Vol. 62, pp. 246-252(1972).

Common/Systematic Name

14α-Methylzymosterol

Molecular Formula/Molecular Weight

$C_{28}H_{46}O$; MW = 398.35487

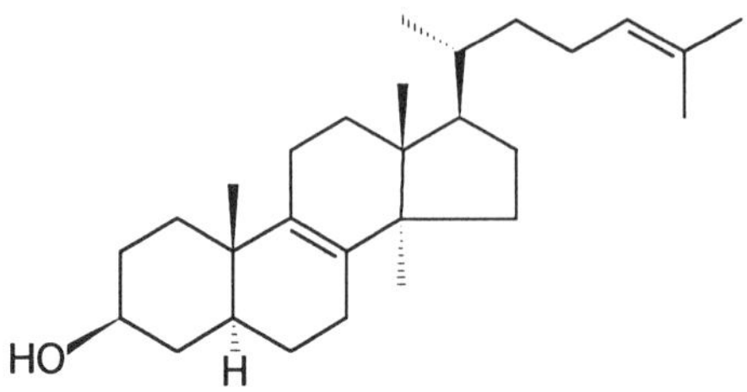

General Characteristics

Crystals from methanol; mp., 103-104.5°C.

Fungal Source

Saccharomyces cerevisiae (25-Aza-inhibited cultures) and *Candida albicans* (mutant).

Isolation/Purification

The yeast cells were saponified, free sterols separated by TLC followed by recrystallization from methanol. Sterol acetates separated by argentation TLC chromatography.

Spectral Data

^{1}H NMR:

($CDCl_3$) 0.72(3H, s, C-18 H); 0.89(3H, s, C-32 H); 0.90(3H, d, $J_{20,21}$=6.0Hz, C-21 H); 0.95(3H, s, C-19 H); 1.60 and 1.68(2 X 3H, 2 X s, C-26 and C-27 H); 3.57(1H, m, C-3 H); and 5.08ppm (1H, m, C-24 H).

Mass Spectrum:

LREIMS: 398(M^+, 71.9%), 383(100), 365, 313, 271, 245, 231, 219, 213, 201, 187, 161, 159, 149, 147, 145, 137, 135, 133, 131, 123, 121, 119, 109, 107, 105, 95, 93, 91, 83, 81, 79, 69, 67, 55, 43, and 41*m/e*.

TLC Data

TLC was conducted with silica gel HF-254 + 366 (Merck type 60) containing rhodamine 6G developed in methylcyclohexane-ethyl acetate (3:1, v/v). Sterol acetates were separated with argentation TLC chromatography (25% $AgNO_3$) developed with diethyl ether.

GC Data

Analysis of TMS derivatives was with OV-101 at 245°C; acetates with OV-101 at 245°C and 3% SilicAR-10C on Gas Chrom -Q at 220°C.

Reference

A. M. Pierce, R. B. Mueller, A. M. Unrau, and A. C. Oehlschlager; Metabolism of Δ24-Sterols by Yeast Mutants Blocked in Removal of the C-14 Methyl Group; Can. J. Biochem., Vol. 56, pp. 794-800(1978).

Common/Systematic Name

4α-Methylzymosterol

Formula/Molecular Weight

$C_{28}H_{46}O$; MW = 398.35487

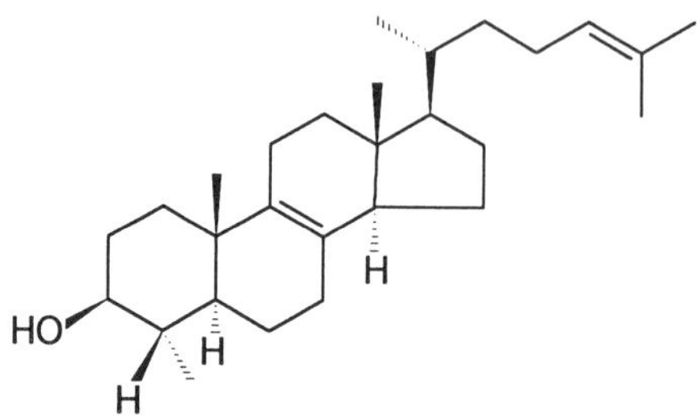

General Characteristics

(Acetate) Plates from acetone; mp., 143-145.5°C; $[\alpha]_D^{25}$ + 65.4° (mp., 138-139°C); $[\alpha]_D$ + 61°.

Fungal Source

Saccharomyces cerevisiae, S. carlsbergensis, Candida albicans, Candida sp., *C. utilis, Hemispora stellata, Pichia* sp., and *Torulopsis glabrata.*

Isolation/Purification

Yeast sterol residues were absorbed on an alumina grade V column and eluted with different ratios of benzene-petroleum ether; 4α-methylzymosterol eluted in the benzene-petroleum ether, 2:3 (v/v), fraction. The partially purified fraction was chromatographed again on alumina grade V. The 4α-methylzymosterol was acetylated and final purification was accomplished by fractional crystallization; triangulation from acetone-methanol.

References

D. H. R. Barton, D. M. Harrison, G. P. Moss, and D. A. Widdowson; Investigations on the Biosynthesis of Steroids and Terpenoids. Part II. Role of 24-Methylene Derivatives in the Biosynthesis of Steroids and Terpenoids; J. Chem. Soc. (C), pp. 775-785(1970).

D. H. R. Barton, U. M. Kempe, and D. A. Widdowson; Investigations on the Biosynthesis of Steroids and Terpenoids. Part VI. The Sterols of Yeast; J. Chem. Soc., Perkin Trans I, pp. 513-522(1972).

3

C_{29} Sterols

(24*R*) 24-Ethylcholesta-4,6,8(14),22-tetraen-3-one
(24*S*) 24-Ethylcholest-5-en-3β-ol (Sitosterol)
(24*R*) 24-Ethylcholes-5-en-3β-ol (Clionasterol)
(24*R*) 24-Ethylcholesta-5,22-dien-3β-ol (Stigmasterol)
(24*R*) 24-Ethylcholesta-5,24(28)-dien-3β-ol (Fucosterol)
(24*R*) 24-Ethylcholesta-5,7,24-trien-3β-ol (7-Dehydrofucosterol)
(24*R*) 24-Ethylcholest-22-en-3β-ol
Stigmasta-7(*Z*),24(28)-dien-3β-ol
(24*S*) 24-Ethylcholest-7-en-3β-ol (Stigmast-7-en-3β-ol)
(24*S*) 24-Ethylcholesta-5,7-dien-3β-ol (Stigmasta-5,7-dien-3β-ol)
(24*S*) 24-Ethylcholesta-5,22-dien-3β-ol (Poriferasterol)
(24*S*) 24-Ethylcholestan-3β-ol (Stigmastanol)
4α,14α-Dimethylcholesta-8,24-dien-3β-ol
14-Demethyllanosterol
14-Methylfecosterol
4,4-Dimethylcholesta-8,14,24-trien-3β-ol
Lanosta-7,9(11),24-trien-3β-ol
Senexonol
Senexadione
Oxidosenexone

Common/Systematic Name

(24*R*)24-Ethylcholesta-4,6,8(14),22-tetraen-3-one

Molecular Formula/Molecular Weight

$C_{29}H_{42}O$; MW = 406.32357

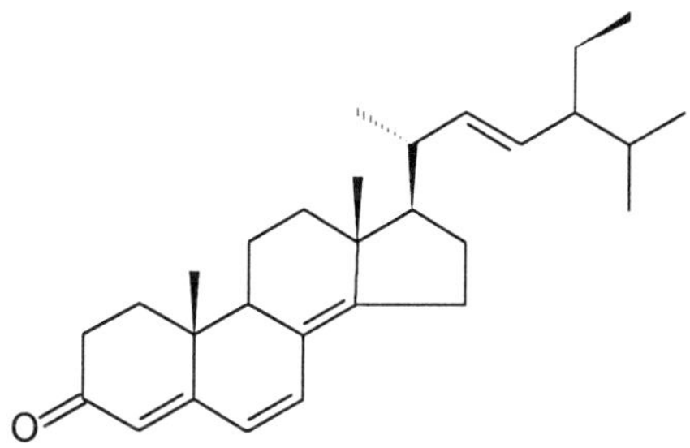

General Characteristics

Crystals from petroleum ether; mp., 107-109.5°C (mixed with C_{28} analog); highly fluorescent.

Fungal Source

Moldy flour; most abundant molds were *Aspergillus niger, A. candidus, A. versicolor,* and *A. flavus.*

Isolation/Purification

Fungal cultures were extracted twice with methanol, saponified with ethanolic KOH, and extracted with *n*-hexane. The *n*-hexane extracts were reduced in volume, applied to a silica gel 60 column eluted in succession with hexane, ethyl acetate-hexane (5:95, v/v), and ethyl acetate-hexane(10:90). (24*R*)24-Ethylcholesta-4,6,8(14),22-tetraen-3-one purification was monitored with long-wave UV light and was eluted in the 5% ethyl acetate-hexane fraction; crystallization was from petroleum ether.

Spectral Data

UV:

λ_{max}^{EtOH} 351(log ϵ = 25,800), 282(5,800), and 239nm (4,100).

IR:

(KBr) 2940, 2860, 1650, 1562, 1445, 1347, 1320, 1265, 1217, 1191, 964, 946, 872, and 758cm^{-1}.

Mass Spectrum:

LREIMS: 406(M^+, 6%), 268(100), 253(17), 240(7), and 214*m/e* (22).

References

R. G. Cooks, R. D. Daftary, and Y. Pomeranz; Changes in Mold-Damaged Wheat Flours Stored at Various Temperatures; J. Agric. Food Chem., Vol. 18, pp. 620-623(1970).

W. D. Nes and E. J. Parish (eds.); Analysis of Sterols and Other Biologically Significant Steroids; Academic Press, Inc. New York, NY, pp. 341 (1989).

W. B. Turner and D. C. Aldridge; Fungal Metabolites II; Academic Press, Inc., New York, NY, p. 631 (1983).

J. D. Weete; Structure and Function of Sterols in Fungi; Advances in Lipid Research, Vol. 23, pp. 115-167(1989).

Common/Systematic Name

Sitosterol; β-Sitosterol; Stigmast-5-en-3β-ol; Cupreol; Rhamnol; α-Dihydrofucosterol; α-Phytosterol; Cinchol; Quebrachol; Sitosterin; 22,23-Dihydrostigmasterol (24*S*)24-Ethylcholest-5-en-3β-ol

Molecular Formula/Molecular Weight

$C_{29}H_{50}O$; MW = 414.38617

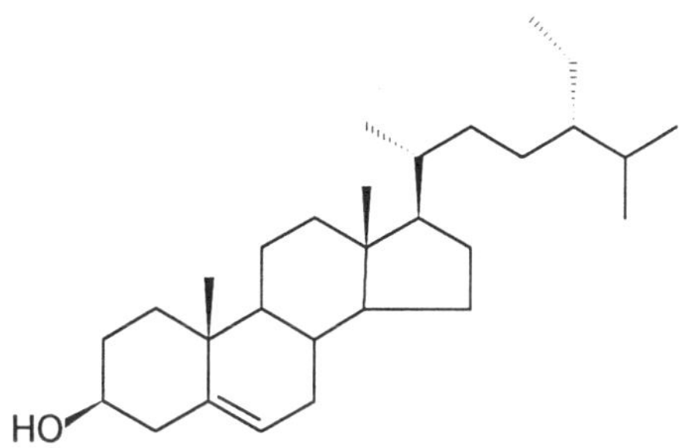

General Characteristics

Plates from alcohol; mp., 140°C; $[\alpha]_D^{25}$ -37° (*c*=2.0, in $CHCl_3$); precipitated by digitonin. Acetate, $C_{31}H_{52}O_2$; mp., 127-128°C; $[\alpha]_D^{25}$ -41° (*c*= 2.0, in $CHCl_3$).

Fungal Source

Allomyces macrogynus, Aspergillus oryzae, Atta sexdens rubropilosa, Candida sp., *C. utilis, Coriolus versicolor, Fusarium sporotrichioides, Glomus mosseae, Hypochytrium catenoides, Physarum polycephalum, Rhizophlyctis rosea,* and *Stachybotrys alternans.*

Isolation/Purification

See R. C. Heupel; Isolation and Primary Characterization of Sterols, In Analysis of Sterols and Other Biologically Significant Steroids; W. D. Nes and E. J. Parish (eds.), Academic Press Inc., New York, NY, 341 pp. (1989).

Spectral Data

UV:

End absorption.

^{1}H NMR:

H-1, 1.15, 1.89; H-2, 1.85, 1.56; H-3, 3.53; H-4, 2.28; H-6, 5.36 (br d, *J*=5Hz); H-7, 1.53; H-8, 1.93; H-9, 0.98; H-11, 1.51; H-12, 1.19, 2.03; H-14, 1.00; H-15, 1.13, 1.61; H-16, 1.35, 1.86; H-17, 1.10; H-18, 0.68 (s) ; H-19, 1.009 (s) ; H-20, 1.37; H-21, 0.922 (d, *J*=7Hz); H-22, 1.04, 1.35; H-23, 1.20; H-24, 0.97; H-25, 1.70; H-26(*Pro-S*, 0.836 (d, *J*=7Hz); H-27(*Pro-R*), 0.814 (*J*=7Hz); H-28, 1.27 and H-29, 0.846ppm (t, *J*=7Hz).

^{13}C NMR:

C-1, 37.2; C-2, 31.6; C-3, 71.8; C-4, 42.5; C-5, 14; C-6, 121.7; C-7, 31.8; C-8, 31.9; C-9, 50.1; C-10, 36.5; C-11, 21.1; C-12, 39.7; C-13, 42.3; C-14, 56.7; C-15, 24.3; C-16, 28.2; C-17, 56.0; C-18, 11.9; C-19, 19.4; C-20, 36.1; C-21, 18.8; C-22, 33.9; C-23, 26.0; C-24, 45.8; C-25, 29.1; C-26(*S*), 19.8; C-27(*R*),19.1; C-28, 23.0 and C-29, 12.0ppm.

Mass Data:

414(M^+, 100%), 399(45), 396(54), 381(42), 329(50), 285, 273(54), 255(60), 231(58), and 213*m/e* (74). Acetate derivative; 456(M^+), 441, 396(100%), 381, 329, 327, 315, 313, 273, 255(45), and 213*m/e* (29); C 83.99%, H 12.15%, O 3.86%.

References

W. D. Nes and E. J. Parish (eds.); Analysis of Sterols and Other Biologically Significant Steroids; Academic Press Inc., New York, NY, 341 pp. (1989).

W. B. Turner and D. C. Aldridge; Fungal Metabolites II; Academic Press, Inc., New York, NY, 631 pp. (1983).

J. D. Weete; Structure and Function of Sterols in Fungi, Advances in Lipid Research, Vol. 23, pp. 115-167(1989).

Common/Systematic Name

Clionasterol; 22-Dihydroporiferasterol
(24*R*)24-Ethylcholest-5-en-3β-ol

Molecular Formula/Molecular Weight

$C_{29}H_{50}O$; MW = 414.38617

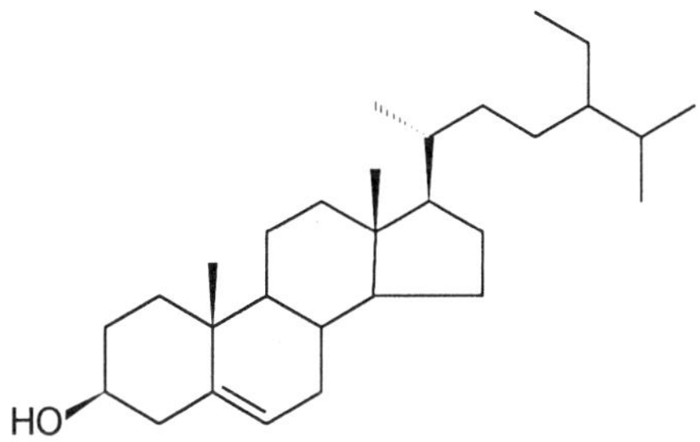

General Characteristics

Monohydrate plates from alcohol when dry; mp., 147-148°C; $[\alpha]_D^{20}$ -43° (*c*=1.9, in $CHCl_3$). Acetate, $C_{31}H_{52}O_2$; mp., 143-144°C; $[\alpha]_D^{20}$ -45.3° (*c*=1.9, in $CHCl_3$).

Fungal Source

Physarum polycephalum and *P. flavicomum.*

Isolation/Purification

See R. C. Heupel; Isolation and Primary Characterization of Sterols, In Analysis of Sterols and Other Biologically Significant Steroids; W. D. Nes and E. J. Parish (eds.), Academic Press Inc., New York, NY, 341 pp. (1989).

Spectral Data

UV:

End absorption.

^{13}C NMR:

C-15, 24.3; C-16, 28.2; C-17, 56.0; C-18, 11.9; C-20, 36.3; C-21, 18.8; C-22, 33.9; C-23, 26.4; C-24, 46.1; C-25, 29.0; C-26(*S*), 19.6; C-27(*R*), 19.0; C-28, 23.0; and C-29, 12.3ppm.

Mass Spectrum:

414(M^+, 100%), 399(45), 396(54), 381(42), 329(50), 285, 273(54), 255(60), 231(58), and 213*m/e* (74%). Acetate derivative: 456(M^+), 441, 396(100%), 381, 329, 327, 315, 313, 273, 255(45), and 213*m/e* (29).

References

W. D. Nes and E. J. Parish (eds.); Analysis of Sterols and Other Biologically Significant Steroids; Academic Press Inc., New York, NY, 341 pp. (1989).

W. B. Turner and D. C. Aldridge; Fungal Metabolites II; Academic Press, Inc., New York, NY, 631 pp. (1983).

J. D. Weete; Structure and Function of Sterols in Fungi, Advances in Lipid Research, Vol. 23, pp. 115-167(1989).

Common/Systematic Name

Stigmasterol; Stigmasta-5,22-dien-3β-ol
(24*R*)24-Ethylcholesta-5,22-dien-3β-ol

Molecular Formula/Molecular Weight

$C_{29}H_{48}O$; MW = 412.37052

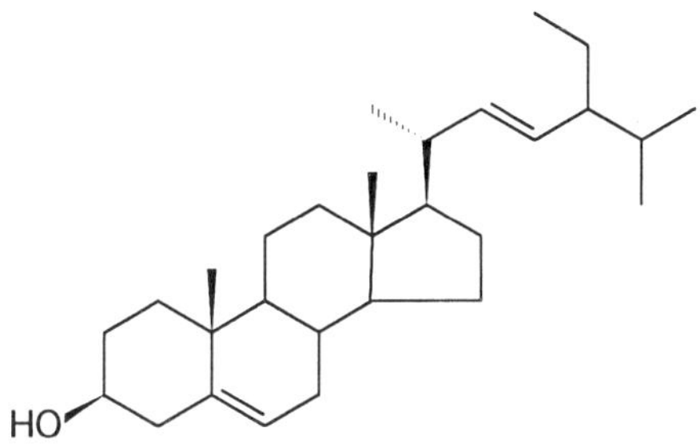

General Characteristics

Monohydrate; crystals from alcohol, when dry; mp., 170°C; $[\alpha]_D^{22}$ -51° (*c*=2, in $CHCl_3$). Insoluble in water; soluble in the usual organic solvents. Acetate derivative, $C_{31}H_{50}O_2$; mp., 144°; $[\alpha]_D^{20}$ -55.6° (*c*=2, in $CHCl_3$).

Isolation/Purification

See R. C. Heupel; Isolation and Primary Characterization of Sterols, In Analysis of Sterols and Other Biologically Significant Steroids; W. D. Nes and E. J. Parish (eds.), Academic Press Inc., New York, NY, 341 pp. (1989).

Fungal Source

Physarum polycephalum, Debaromyces hansenii, Fusarium sporotrichioides, Glomus mosseae, Hypochytrium catenoides, Pullularia pullulans, and *Stachybotrys alternans.*

Spectral Data

UV:

End absorption.

^{13}C NMR:

C-15, 24.2; C-16, 29.0; C-17, 56.1; C-18, 12.1; C-20, 40.5; C-21, 21.1; C-22, 138.4; C-23, 129.3; C-24, 51.3; C-25, 31.9; C-26, 21.3; C-27, 19.0; C-28, 25.4; and C-29, 12.3ppm.

(Assignments for skeletal carbon signals were almost consistent with cholesterol)

Mass Spectrum:

(Acetate derivative) 454(M^+), 411, 439, 394(100), 379, 351, 315, 313, 273, 255(95), 253, and 213*m/e*.

References

W. D. Nes and E. J. Parish (eds.); Analysis of Sterols and Other Biologically Significant Steroids; Academic Press, Inc., New York, NY, 341 pp. (1989).

W. B. Turner and D. C. Aldridge; Fungal Metabolites II; Academic Press, Inc., New York, NY, 631 pp. (1983).

J. D. Weete; Structure and Function of Sterols in Fungi, Advances in Lipid Research, Vol. 23, pp. 115-167(1989).

Common/Systematic Name

Fucosterol

(24*R*)24-Ethylcholesta-5,24(28)-dien-3β-ol

Molecular Formula/Molecular Weight

$C_{29}H_{48}O$; MW = 412.37052

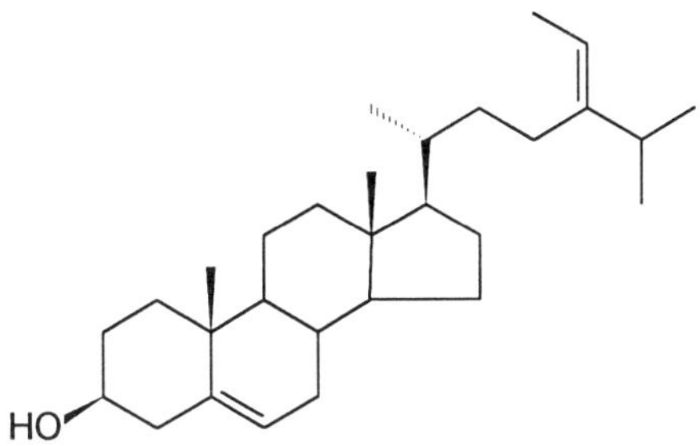

General Characteristics

Crystals from methanol; mp., 124°C; $[\alpha]_D^{20}$ -38.4° ($CHCl_3$). Soluble in most organic solvents. Acetate derivative: crystals; mp., 120-121°C; $[\alpha]_D^{20}$ -45° ($CHCl_3$).

Fungal Source

Fucus vesiculosus (predominant sterol in the brown marine algae, Phaeophyceae), *Achyla bisexualis, Aplanopsis terrestris, Apodachlya brachynema, A. minima, Apodachlyella completa, Leptolegnia caudata, Pythiopsis cymosa, Saprolegnia ferax*, and *S. megasperma*.

Isolation/Purification

The mycelium from *Achlya bisexualis* was air dried, saponified with ethanolic KOH, and the lipids were extracted with hexane. The non-saponifiable material was chromatographed on alumina (activity III), and squalane and other hydrocarbons were eluted with light petrol. Further elution with diethyl ether gave a sterol fraction which was subsequently acetylated and separated by preparative TLC on silica gel impregnated with silver nitrate.

Also, see R. C. Heupel; Isolation and Primary Characterization of Sterols, In Analysis of Sterols and Other Biologically Significant Steroids; W. D. Nes and E. J. Parish (eds.), Academic Press Inc., New York, NY, 341 pp. (1989).

Spectral Data

UV:

End absorption.

^{13}C NMR:

($CDCl_3$) C-15, 24.3; C-16, 28.2; C-17, 55.7; C-18, 11.9; C-20, 36.4; C-21, 18.7; C-22, 35.2; C-23, 25.7; C-24, 146.7; C-25, 34.8; C-26, 22.2; C-27, 115.4; and C-28, 13.1.

Mass Data:

Acetate derivative; 454(M^+), 439, 394,379, 356, 341, 315, 313, 296(100%), 281, 273, 255, 253, and 213*m/e*. C, 84.40%, H, 11.72%. O, 3.88%.

TLC Data

TLC was carried out on silica gel G plates impregnated with $AgNO_3$ and rhodamine 6G. Plates were developed with benzene. R_f fucosteryl acetate, 0.60-0.70.

GC Data

GC was performed using 1% QF-1; Fucosteryl acetate had relative retention time of 1.51 (relative to cholesteryl acetate).

References

W. D. Nes and E. J. Parish (eds.); Analysis of Sterols and Other Biologically Significant Steroids; Academic Press Inc., New York, NY, 341 pp. (1989).

C. R. Popplestone and A. M. Unrau; Major Sterols of *Achlya bisexualis*; Phytochemistry, Vol. 12, pp. 1131-1133(1973).

W. B. Turner and D. C. Aldridge; Fungal Metabolites II; Academic Press, Inc., New York, NY, 631 pp. (1983).

J. D. Weete; Structure and Function of Sterols in Fungi, Advances in Lipid Research, Vol. 23, pp. 115-167(1989).

Common/Systematic Name

7-Dehydrofucosterol
(24*R*)24-Ethylcholesta-5,7,24(28)-trien-3β-ol

Molecular Formula/Molecular Weight

$C_{29}H_{46}O$; MW = 410.35487

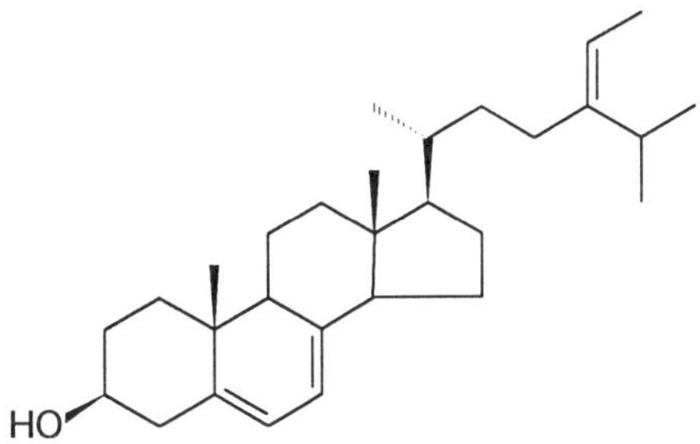

General Characteristics

Colorless crystals from methanol; mp., 127-130°C (129-131°C).

Fungal Source

Achlya bisexualis

Isolation/Purification

The mycelium from *Achlya bisexualis* was air dried, saponified with ethanolic KOH, and the lipids were extracted with hexane to give a yellow oil. The non-saponifiable material was chromatographed on alumina (activity III), and squalene and other hydrocarbons were eluted with light petrol. Further elution with diethyl ether gave a sterol fraction which was subsequently acetylated and separated by preparative TLC on silica gel impregnated with silver nitrate.

Also, See R. C. Heupel; Isolation and Primary Characterization of Sterols, In Analysis of Sterols and Other Biologically Significant Steroids; W. D. Nes and E. J. Parish (eds.), Academic Press Inc., New York, New York, 341pp (1989)

Spectral Data

UV:

Typical 5,7-diene system (see ergosterol).

Mass Spectrum:

Acetate derivative: 392(M^+ - $MeCO_2H$), corresponding to a molecular weight of 452, 297, 253, 143, 157, and 158*m/e*.

TLC Data

TLC was carried out on silica gel G plates impregnated with $AgNO_3$ and rhodamine 6G. Plates were developed with 10% ethyl acetate-benzene. R_f (7-dehydrofucosteryl acetate) 0.19-0.33.

GC Data

GC was performed using 1% QF-1. Fucosteryl acetate had relative retention time of 1.67 (relative to cholesteryl acetate).

References

W. D. Nes and E. J. Parish (eds.); Analysis of Sterols and Other Biologically Significant Steroids; Academic Press, Inc., New York, New York, 341pp. (1989).

C. R. Popplestone and A. M. Unrau; Major Sterols of *Achlya bisexualis*; Phytochemistry, Vol. 12, pp. 1131-1133(1973).

W. B. Turner and D.C. Aldridge; Fungal Metabolites II; Academic Press, Inc., New York, New York, 631 pp. (1983).

J. D. Weete; Structure and Function of Sterols in Fungi, Advances in Lipid Research; Vol. 23, pp. 115-167(1989).

Common/Systematic Name

24-Ethylcholesta-22-enol; Δ^{22}-Stigmasta-3β-ol
(24*R*)24-Ethylcholest-22-en-3β-ol

Molecular Formula/Molecular Weight

$C_{29}H_{50}O$; MW = 414.38617

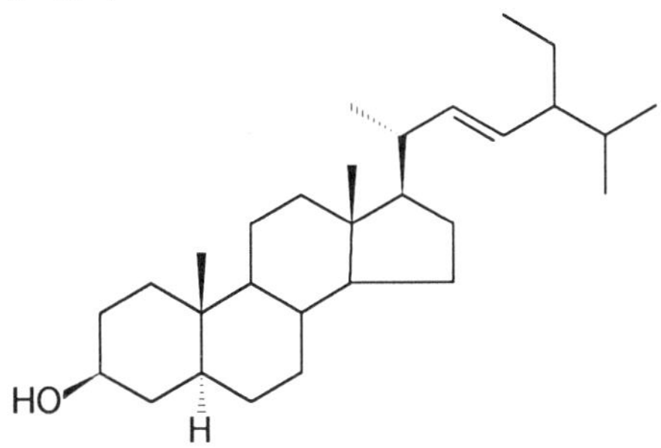

General Characteristics

Anisotropic needles from ethanol; mp., 156-157°C; $[\alpha]_D$ +5.2°. Acetate, glistening plates from ether-methanol; mp., 156°C.

Fungal Source

Dictyostelium discoideum.

Isolation/Purification

The neutral alcoholic lipids were chromatographed on neutral alumina (activity grade III); eluted increasing amounts of benzene in petroleum ether. 24-Ethylcholest-22-enol was isolated and crystallized (from ethanol solution) from the 25% and 50% benzene in petroleum ether fractions.

Also see R. C. Heupel; Isolation and Primary Characterization of Sterols, In Analysis of Sterols and Other Biologically Significant Steroids; W. D. Nes and E. J. Parish (eds.), Academic Press Inc., New York, NY, 341 pp. (1989).

Biological Activity

Demonstrated weak acrasin-type activity (aggregation of individual slime mold cells into multicellular units) as measured by the Shaffer bioassay.

Spectral Data

UV:

End absorption.

IR:

(CS_2) strong absorption maxima at 3.42 and 3.50μ (C-H stretching), 9.64μ (C-O stretching of OH) and 10.29μ (*trans*-disubstituted double bond); weak absorption maxima at 8.56, 8.82, 9.31, 10.10, 10.44, and 10.71μ which was correlated with a 3ß-hydroxy A/B *trans*-sterol.

Mass Data:

LREIMS: (acetate derivative) 456(M^+, 27%), 413(M^+ - C_3H_7), 441(M^+ - CH_3), 396(M^+ - ROH, 13), 381(M^+ - CH_3 - ROH), 353(M^+ - C_3H_7- ROH), 317(23), 315(32), 275, 257(100), 255(50), and 215*m/e* (29); found: C, 83.98%; H, 12.09%.

References

E. Heftmann, B. E. Wright, and G. U. Liddel; The Isolation of Stigmast-22-en-3ß-ol from *Dictyostelium discoideum*; Archives of Biochemistry and Biophysics, Vol. 91, pp. 266-270(1960).

W. D. Nes and E. J. Parish (eds.); Analysis of Sterols and Other Biologically Significant Steroids; Academic Press Inc., New York, NY, 341 pp. (1989).

W. B. Turner and D. C. Aldridge; Fungal Metabolites II; Academic Press, Inc., New York, NY, 631 pp. (1983).

J. D. Weete; Structure and Function of Sterols in Fungi, Advances in Lipid Research, Vol. 23, pp. 115-167(1989).

Common/Systematic Name

Stigmasta-7(*Z*),24(28)-dien-3β-ol

Molecular Formula/Molecular Weight

$C_{29}H_{48}O$; MW = 410.35487

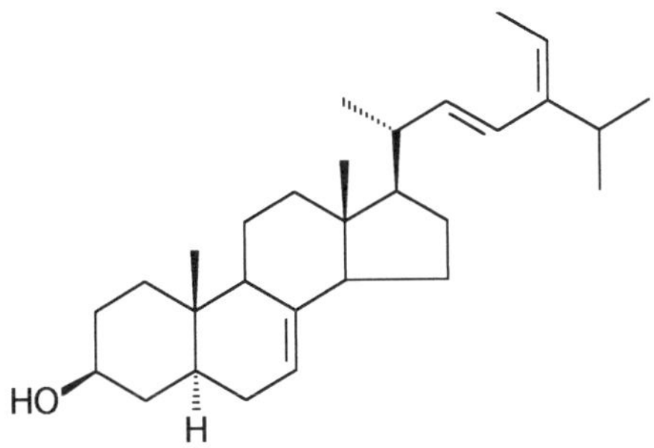

Fungal Source

Aspergillus oryzae, Cronartium fusiforme, Dictyuchus monosporus, Melampsora lini, and *Uromyces phaseoli* (bean rust spores).

Isolation/Purification

Spores were broken and extracted with chloroform-methanol, 2:1, v/v. The concentrated extract was chromatographed by preparative TLC using silica gel with a solvent of ethyl ether-heptane-acetic acid (70:30:1, v/v/v). The sterols were eluted from the TLC adsorbent with chloroform and rechromatographed (preparative TLC) with a solvent of 0.5% acetic acid in chloroform. The sterols were acetylated and chromatographed by argentation TLC using 0.5% acetic acid in chloroform as developing solvent. Final purification was by recrystallization from acetone solution.

See R. C. Heupel; Isolation and Primary Characterization of Sterols, In Analysis of Sterols and Other Biologically Significant Steroids; W. D. Nes and E. J. Parish (eds.), Academic Press Inc., New York, NY, 341 pp. (1989).

Spectral Data

UV:

End absorption.

^{1}H NMR:

The resonances were 0.535(s, CH_3-18); 0.81(s, CH_3-19); 0.93(broad doublet, *J*=7Hz, CH3-21); 0.97(doublet, *J*=7Hz, CH_3-26 and CH_3-27); 1.58(doublet, *J*=7Hz, CH_3-29); 2.01(singlet, CH_3-CO); 2.82(septet, *J*=7Hz, CH-25); 4.72(broad, CH-3); and 5.18ppm (cal)(broad,CH-7 and CH-28).

Mass Spectrum:

The mass spectrum of the sterol acetate showed a molecular ion at 454 *m/e* (36%); strong peaks were observed at 439(3.6, M^+ - CH_3), 394(1.4, M^+- acetate), 379(3.1, M^+- CH_3 and acetate), 356(46.7, M^+- C_7H_{14} of side chain), 341(5.6, M^+- C_7H_{14} and CH_3), 313(100, M^+-side chain-2H), 296(5.7, 356 - acetate), 288(7.9, M^+- side chain and Cls, 17), 281(4.9, 296-CH_3), 255 (12.4, M^+- side chain and acetate), and 213*m/e* (18.6%, M^+ - side chain – 42, and -acetate).

References

H. K Lin and H. W. Knoche; Origin of Sterols in Uredospores of *Uromyces phaseoli*; Phytochemistry, Vol. 13, pp. 1795-1799(1974).

H. K Lin, R. J. Langenbach, and H. W. Knoche; Sterols of *Uramyces phaseoli* Uredospores; Phytochemistry, Vol. 11, pp. 2319-2322(1972).

W. D. Nes and E. J. Parish (eds.); Analysis of Sterols and Other Biologically Significant Steroids; Academic Press,Inc., New York, NY, 341 pp. (1989).

W. B. Turner and D. C. Aldridge; Fungal Metabolites II; Academic Press, Inc., New York, NY, 631 pp. (1983).

J. D. Weete; Structure and Function of Sterols in Fungi, Advances in Lipid Research, Vol. 23, pp. 115-167(1989).

Common/Systematic Name

Stigmast-7-en-3β-ol

(24*S*)24-Ethylcholest-7-en-3β-ol

Molecular Formula/Molecular Weight

$C_{29}H_{50}O$; MW = 414.38617

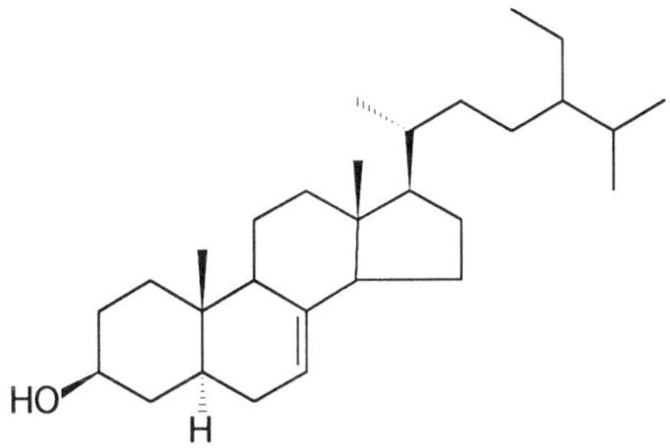

Fungal Source

Aspergillus oryzae, Cronartium fusiforme, Linderina pennispora, Melampsora lini, Puccinia graminis var. *tritici, P. striiformis, Uromyces phaseoli,* and *Ustilago nuda.*

Isolation/Purification

Spores were broken and extracted with chloroform-methanol, 2:1, v/v. The concentrated extract was chromatographed by preparative TLC using silica gel with a solvent of ethyl ether-heptane-acetic acid (70:30:1, v/v/v). The sterols were eluted from the TLC adsorbent with chloroform and rechromatographed (preparative TLC) with a solvent of 0.5% acetic acid in chloroform. The sterols were acetylated and chromatographed by argentation TLC using 0.5% acetic acid in chloroform as developing solvent. Final purification was by recrystallization from acetone solution.

See R. C. Heupel; Isolation and Primary Characterization of Sterols, In Analysis of Sterols and Other Biologically Significant Steroids; W. D. Nes and E. J. Parish (eds.), Academic Press Inc., New York, NY, 341 pp. (1989).

Spectral Data

UV:

End absorption.

Mass Spectrum:

(Acetate derivative) LREIMS: 456(M^+, 81%), 441(18), 396(21), 381(13), 315(11), 273(13), 255(84), 229(27), 213(40), and 288*m/e* (9); 456(M^+, 26.7), peaks at 441(26, M^+ - CH_3), 396(52, M^+ - acetate), 381(27.4, M^+ - acetate and CH_3), 315(17, M^+ - side chain and 2H), 288(22.3, M^+ - side chain and $C_{16,17}$), 273(14, M^+ - side chain and 42), 255(100, M^+ - side chain and acetate), and 213*m/e* (65.6%, M^+ - side chain -42 and -acetate).

References

H-K Lin, R. J. Langenbach, and H. W. Knoche; Sterols of *Uromyces phaseoli* Uredospores; Phytochemistry, Vol. 11, pp. 2319-2322(1972).

W. D. Nes and E. J. Parish (eds.); Analysis of Sterols and Other Biologically Significant Steroids; Academic Press Inc., New York, NY, 341 pp. (1989).

W. B. Turner and D. C. Aldridge; Fungal Metabolites II; Academic Press, Inc., New York, NY, 631 pp. (1983).

J. D. Weete; Structure and Function of Sterols in Fungi, Advances in Lipid Research, Vol. 23, pp. 115-167(1989).

Common/Systematic Name

Stigmasta-5,7-dien-3β-ol
(24*S*)24-Ethylcholesta-5,7-dien-3β-ol

Molecular Formula/Molecular Weight

$C_{29}H_{48}O$; MW = 412.37052

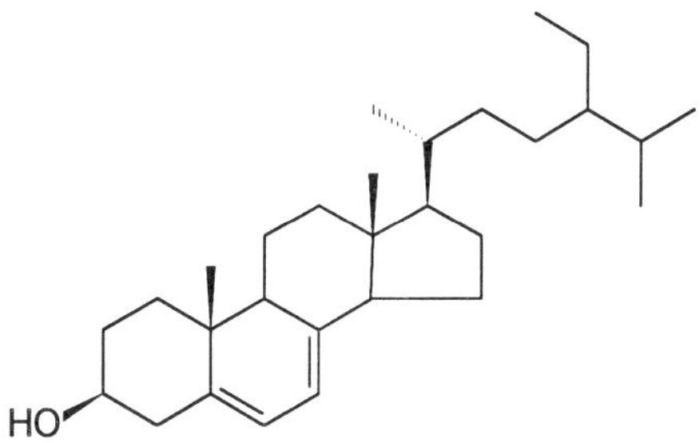

Fungal Source

Cronartium fusiforme and *Melampsora lini.*

Isolation/Purification

See R. C. Heupel; Isolation and Primary Characterization of Sterols, In Analysis of Sterols and Other Biologically Significant Steroids; W. D. Nes and E. J. Parish (eds.), Academic Press Inc., New York, NY, 341 pp. (1989).

Spectral Data

UV:

Typical 5,7-diene chromophore (i.e. 7-dehydrocholesterol or ergosterol).

Mass Spectrum:

(Acetate derivative) LREIMS: 454(M^+, 4%), 394(100), 379(40), 339, 313, 271, 253(37), 227(43), 211(17), 158(40), and 143*m/e* (55).

References

W. D. Nes and E. J. Parish (eds.); Analysis of Sterols and Other Biologically Significant Steroids; Academic Press Inc., New York, NY, 341 pp. (1989).

W. B. Turner and D. C. Aldridge; Fungal Metabolites II; Academic Press, Inc., New York, NY, 631 pp. (1983).

J. D. Weete; Structure and Function of Sterols in Fungi, Advances in Lipid Research, Vol. 23, pp. 115-167(1989).

Common/Systematic Name

Poriferasterol; Stigmasta-5,22-dien-3β-ol
(24*S*)24-Ethylcholesta-5,22-dien-3β-ol

Molecular Formula/Molecular Weight

$C_{29}H_{48}O$; MW = 412.37052

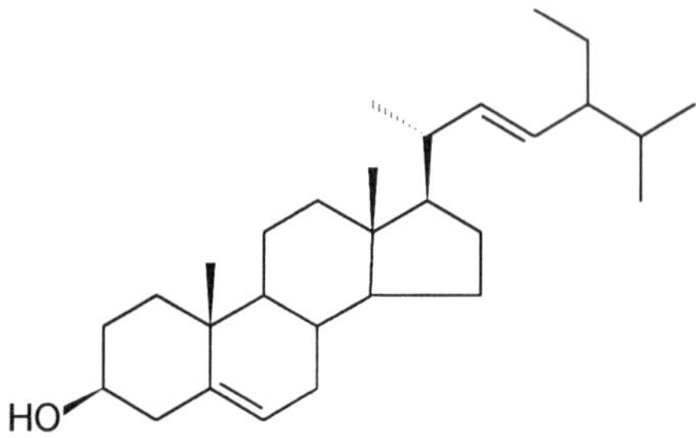

Fungal Source

Physarum flavicomum and *P. polycephalum.*

Isolation/Purification

See R. C. Heupel; Isolation and Primary Characterization of Sterols, In Analysis of Sterols and Other Biologically Significant Steroids; W. D. Nes and E. J. Parish (eds.), Academic Press Inc., New York, NY, 341 pp. (1989).

Spectral Data

UV:

End absorption.

^{13}C NMR:

($CDCl_3$) C-15, 24.4; C-16, 28.8; C-17, 56.1; C-18, 12.1; C-20, 40.5; C-21, 21.0; C-22, 138.3; C-23, 129.4; C-24, 51.3; C-25, 31.9; C-26(*S*), 19.0; C-27, 21.2; C-28, 25.4; and C-29, 12.4ppm.

References

W. D. Nes and E. J. Parish (eds.); Analysis of Sterols and Other Biologically Significant Steroids; Academic Press Inc., New York, NY, 341 pp. (1989).

W. B. Turner and D. C. Aldridge; Fungal Metabolites II; Academic Press, Inc., New York, NY, 631 pp. (1983).

J. D. Weete; Structure and Function of Sterols in Fungi, Advances in Lipid Research, Vol. 23, pp. 115-167(1989).

Common/Systematic Name

Stigmastanol; Dihydro-β-sitosterol; β-Sitostanol; Fucostanol
(24*S*)24-Ethylcholestan-3β-ol

Molecular Formula/Molecular Weight

$C_{29}H_{52}O$; MW = 416.40182

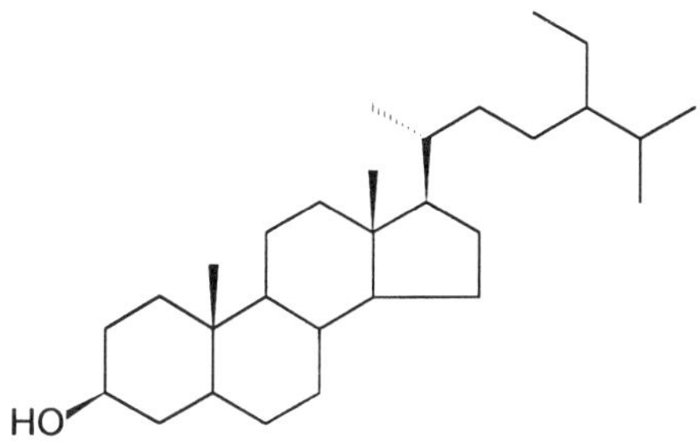

General Characteristics

Monohydrate: crystals; mp., 138-139°C. When dry mp. 144-145°C; $[\alpha]_D^{20}$ +25° (*c*=1.1, in $CHCl_3$); Acetate: $C_{31}H_{54}O_2$; mp., 137-138°; $[\alpha]_D^{20}$ + 14° (*c*=1.8, in $CHCl_3$).

Fungal Source

Dictyostelium discoideum and *Physarum polycephalum.*

Isolation/Purification

See R. C. Heupel: Isolation and Primary Characterization of Sterols, In Analysis of Sterols and Other Biologically Significant Steroids; W. D. Nes and E. J. Parish (eds.), Academic Press Inc., New York, NY, pp. 1-27(1989).

Spectral Data

Mass Data:
C 83.58%, H 12.58%, and O 3.84%.

Reference

W. B. Turner and D. C. Aldridge; Fungal Metabolites II; Academic Press Inc., New York, New York, p. 337(1983).

Common/Systematic Name

4α,14α-Dimethylcholesta-8,24-dien-3β-ol; 4α,14α-Dimethylzymosterol

Molecular Formula/Molecular Weight

$C_{29}H_{48}O$; MW = 412.37052

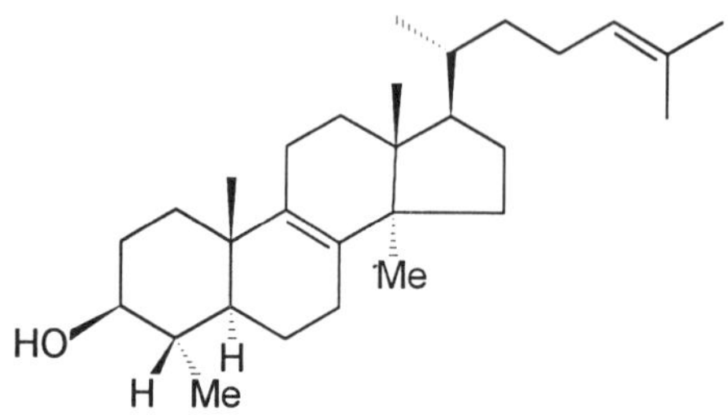

Fungal Source

Saccharomyces cerevisiae (strains D-587-4B, Nys 1, and Erg2) and *Candida albicans.*

Isolation/Purification

The yeast cells were saponified and sterols (as acetate derivatives) separated by TLC on $AgNO_3$-impregnated silica gel. Further purification of free sterols was achieved by liquid chromatography.

Spectral Data

^{1}H NMR:

($CDCl_3$) TMS derivative: 0.71(s, 13-CH_3); 0.88(s, 14-CH_3); 0.93(20-CH_3); 0.96(s, 10-CH_3); 0.99(d, J=6Hz, 4α-CH_3); 1.60 and 1.68ppm (geminal vinyl methyl groups at C-26 and C-27); 5.12(a vinyl proton at C-24); and 3.42ppm (a methyl group at C-4, d, J=6Hz which collapsed to a singlet at 3.42ppm when proton at C-4 was irradiated at 4.85ppm). With Eu(dpm)$_3$, had 2.08(s, 10-CH_3, -1.12); and 3.42(d, J=6Hz, 4α-methyl, -2.43); and 4.84ppm (H-4).

Mass Spectrum:

LREIMS: 412($M^{+\cdot}$), 397, 379 and 245*m/e*; trimethylsilyl ether showed M^+ at nominal mass of 484*m/e*.

TLC Data

R_f 0.56 using silica gel G TLC plates developed with benzene-ethyl acetate (4:1, v/v).

HPLC Data

R_f 2.07 minutes using μBondapak C_{18} column with a solvent system consisting of tetrahydrofuran-acetonitrile-water, 5:5:2 v/v/v, and a flow rate of 0.5ml/min.

GLC Data

Relative retention time of trimethylsilyl ether relative to squalene was 4.78 minutes using 1% neopentyl glycol succinate on Chromosorb W.

References

P. J. Trocha, S. J. Jasne, and D. B Sprinson; Novel Sterols in Ergosterol Deficient Yeast Mutants; Biochemical and Biophysical Research Communications, Vol.59, pp. 666-671(1974).

P. J. Trocha, S. J. Jasne, and D. B Sprinson; Yeast Mutants Blocked in Removing the Methyl Group of Lanosterol at C-14. Separation of Sterols by High-Pressure Liquid Chromatography; Biochemistry, Vol. 16, pp. 4721-4722(1977).

Common/Systematic Name

14-Demethyllanosterol; 4,4-Dimethylzymosterol

Molecular Formula/Molecular Weight

$C_{29}H_{48}O$; MW = 412.37052

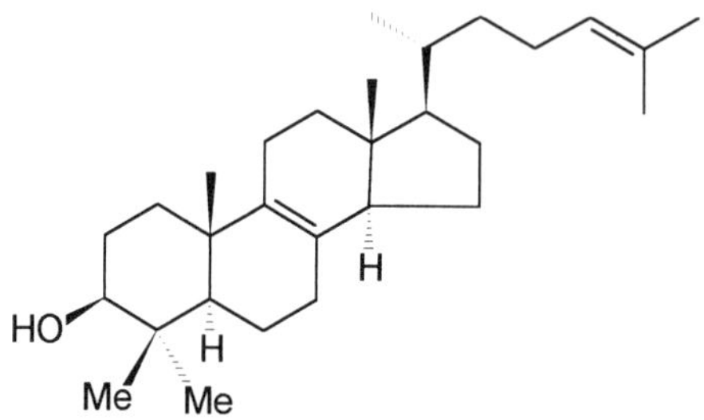

General Characteristics

Colorless needles from methanol; mp., 136-138.5°C; $[\alpha]_D^{23}$ + 40.4° (*c*=1.0, in $CHCl_3$); positive Tortelli-Jaffe and Liebermann-Burchard tests; acetate; mp., 128-132°C; $[\alpha]_D^{23}$ + 41.8°.

Fungal Source

Saccharomyces cerevisiae, Candida albicans, C. sp., *C. utilis,* and *Torulopsis glabrata.*

Isolation/Purification

Isolated from the lipid fraction by preparative TLC on Kieselgel G using hexane-benzene (1:4, v/v) as developing solvent followed by crystallization from methanol.

Spectral Data

IR:

3470(OH), 1680(C=C), and 1630cm^{-1} (CH=C).

^{1}H NMR:

(CCl_4) 8.45(3H); 8.40(26-, 27-CH_3); 8.05(CH_2-7); 6.92(CH-3); and 5.05τ (CH-24).

Mass Data:

HREIMS: $C_{29}H_{48}O \bullet H_2O$ (421.7); found: C, 82.97; H, 11.40; calcd for $C_{29}H_{48}O$; C, 82.6%; H, 11.71.

References

D. H. R. Barton, U. M. Kempe, and D. A. Widdowson; Investigations on the Biosynthesis of Steroids and Terpenoids. Part VI. The Sterols of Yeast; J. Chem. Soc., Perkin Trans I, pp. 513-522(1972).

H. Morimoto, I. Imada, T. Murata, and N. Matsumoto; Uber die Sterine von *Candida utilis;* Liebigs. Ann. Chem., Vol. 78, pp. 230-240(1967).

Common/Systematic Name

14α-Methylergosta-8,24-dien-3β-ol; 14-Methylfecosterol

Molecular Formula/Molecular Weight

$C_{29}H_{48}O$; MW = 412.37052

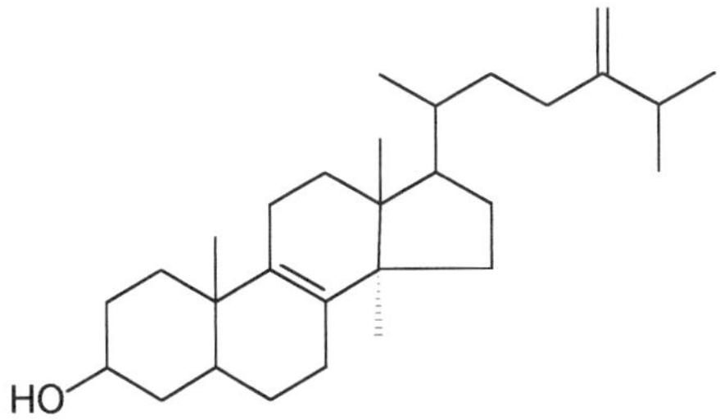

Fungal Source

Saccharomyces cerevisiae (strains D-587-4B, Nys 1, and Erg 2) and *Ustilago maydis*.

Isolation/Purification

The yeast cells were saponified and sterols (as acetate derivatives) separated by $AgNO_3$-impregnated silica gel TLC. Further purification of free sterols was achieved by liquid chromatography (see below).

Spectral Data

^{1}H NMR:

($CDCl_3$) TMS derivative, 0.72(3H, s, H-13); 0.92(9H, d, *J*= 5-6Hz, H-20 and C-$C(CH_3)_2$); 0.99(3H, s, H-14; 1.06(3H, s, H-10); 4.68 and 4.73ppm (br s, C=CH_2). With Eu(dpm)^3, 2.57(H-10, -1.65); other methyl group protons were shifted only about 0.2ppm.

Mass Spectrum:

LREIMS: 412(M^+), 397(base peak), 379, 369, 313, 299, 245, and 231*m/e*; trimethylsilyl ether showed M^+ at nominal mass of 484*m/e*.

TLC Data

R_f 0.47 using silica gel G TLC plates developed with benzene-ethyl acetate (4:1, v/v).

HPLC Data

Retention time 1.94 minutes using μBondapak C_{18} column with a solvent system consisting of tetrahydrofuran-acetonitrile-water, 5:5:2 (v/v/v), and a flow rate of 0.5ml/min.

GLC Data

Relative retention time of trimethylsilyl ether relative to squalene was 4.00 minutes using 1% neopentyl glycol succinate on Chromosorb W.

References

N. N. Ragsdale; Specific Effects of Triarimol on Sterol Biosynthesis in *Ustilago maydis*; Biochimica et Biophysica Acta, Vol. 380, pp. 81-96(1975).

P. J. Trocha, S. J. Jasne, and D. B Sprinson; Novel Sterols in Ergosterol Deficient Yeast Mutants; Biochemical and Biophysical Research Communications, Vol.59, pp. 666-671(1974).

P. J. Trocha, S. J. Jasne, and D. B Sprinson; Yeast Mutants Blocked in Removing the Methyl Group of Lanosterol at C-14. Separation of Sterols by High-Pressure Liquid Chromatography, Biochemistry, Vol. 16, pp. 4721-4722(1977).

Common/Systematic Name

4,4-Dimethylcholesta-8,14,24-trien-3β-ol

Molecular Formula/Molecular Weight

$C_{29}H_{46}O$; MW = 410.35487

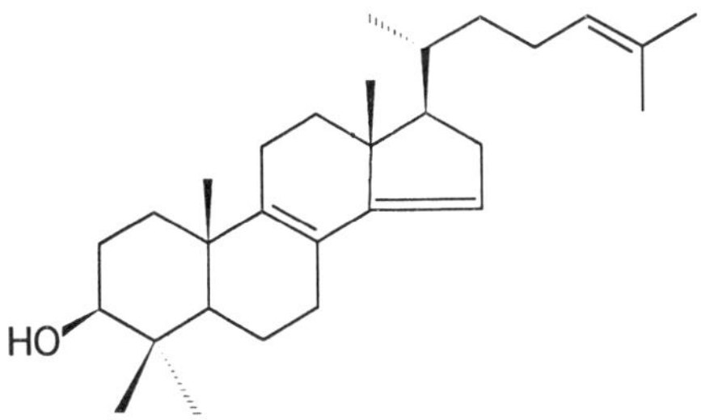

General Characteristics

Purified as 3β-acetoxy derivative; crystals from methanol; mp., 126-128.5°C.

Fungal Source

Saccharomyces cerevisiae cultures treated with 23-azacholesterol.

Isolation/Purification

The sterol mixture from the non-saponifiable fraction of yeast cultures grown in the presence of 23-azacholesterol was separated by preparative TLC using silica gel HF_{254} containing 0.2% rhodamine 6G developed with methylcyclohexane-ethyl acetate, 3:1. The 4,4-dimethylcholesta-8,14,24-trien-3β-ol containing fraction was acetylated and further purified by preparative TLC using silica gel HF_{254} impregnated with 25% silver nitrate and 0.2% rhodamine 6G developed with benzene. The 4,4-dimethylcholesta-8,14,24-trien-3β-ol containing fraction was chromatographed on silica gel HF_{254} developed with benzene followed by recrystallization from methanol.

Spectral Data

UV:

3β-acetoxy derivative: λ_{max}^{Hexane} 249nm (ε=19,600).

^{1}H NMR:

1.04(10β-methyl); 0.81(13β-methyl) and (acetate) 1.06 and 0.81ppm (10β-methyl and 13β-methyl).

Mass Spectrum:

LREIMS: 410*m/e* (M^+); acetate: 452*m/e* (M^+).

GLC Data

3% Silar-10C on Gas Chrom Q; retention time relative to cholestanol acetate; 4,4-dimethylcholesta-8,14,24-trien-3β-yl acetate relative retention time 1.92; OV-101 capillary relative retention time 1.11.

References

H. D. Pierce, Jr., A. M. Pierce, R. Srinivasan, A. M. Unrau, and A. C. Oehlschlager; Azasterol Inhibitors in Yeast; Biochimica et Biophysica Acta, Vol. 529, pp. 429-437(1978).

W. B. Turner and D. C. Aldridge; Fungal Metabolites II; Academic Press, New York, pp. 327(1983).

Common/Systematic Name

Lanosta-7,9(11),24-triene-3β,21-diol

Molecular Formula/Molecular Weight

$C_{29}H_{46}O_2$; MW = 426.34978

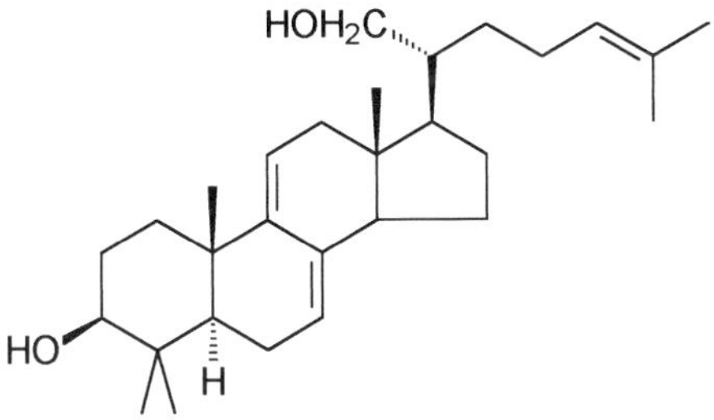

General Characteristics

Colorless needles from ethanol; 194-197°C; $[\alpha]_D^{25}$ + 72° (c=3%, in $CHCl_3$).

Fungal Source

Polyporus pinicola and *Trichoderma koningii*.

Isolation/Purification

Mycelium was extracted with petroleum ether, extraction solvent removed, followed by crystallization from ethanol to yield pure pyrocalciferol. The mother liquor was subjected to preparative TLC (solvent system petroleum ether-ethyl ether, 4:1v/v; R_f 0.50) followed by crystallization of lanosta-7,9(11),24-triene-3β,21-diol from ethanol.

Spectral Data

UV:

λ_{max}^{MeOH} 325(E = 31,152) and 275nm (20,157).

IR:

(Nujol) 3448(OH), 2941, 2857, 1645, 1604, 1471(CH=CH), and 1371cm^{-1}.

^{1}H NMR:

($CDCl_3$) 8.40(15H, s, 5 x CH_3 groups); 6.68(8H, m, 4 x CH_2 groups); 5.15(3H, m, 3 x CH groups); and 4.581τ.

Mass Data:

LREIMS: 440(M^+, intense), 425, 422, 408, 394, 368, 346, 339, 338, 313, 312, 311, 285, 280, 264, 255, 231, 222, 213, 207, 199, 185, 171, 129, and 110*m/e*; found: C, 81.59; H, 11.35; O, 7.45%; calcd. for $C_{30}H_{48}O_2$; C, 81.76; H, 10.98; O, 7.26%

Reference

A. Kamal, R. Akhxar, and A. A. Qureshi; Studies in The Biochemistry of Microorganisms; Pakistan J. Sci., Ind. Res., Vol. 14, pp. 71-78(1971).

Common/Systematic Name

Senexonol

6α,22-Dihydroxy-4α,14α-dimethylcholesta-8,24-dien-3-one

Molecular Formula/Molecular Weight

$C_{29}H_{46}O_3$; MW = 442.34470

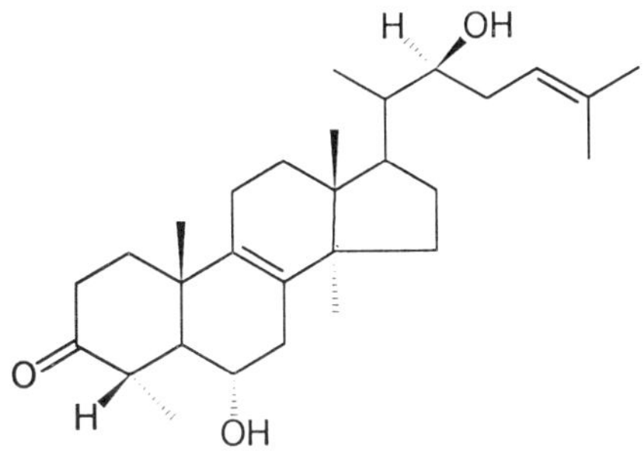

General Characteristics

Crystals; mp., 118-119°C; $[\alpha]_D$ + 78° (in $CHCl_3$); positive Liebermann-Burchard and tetranitromethane color reactions.

Fungal Source

The wood-rotting fungus *Fomes senex*.

Spectral Data

Mass Spectrum:

LREIMS: 426(M^+) and 69*m/e* (base peak).

References

A. K. Batta and S. Rangaswami; Structure of Senexonol, a Novel Tetracyclic Nortriterpene from *Fomes senex*; Indian J. Chem., Vol. 7, pp. 1063-1064(1969).

A. K. Batta and S. Rangaswami; New Tetracyclic Triterpenes from *Fomes senex*: Senexonol, Senexdione, Oxidosenexone and Senexdiolic Acid; Current Science, Vol. 39, pp. 416-417(1970).

S. Rangaswami; New Tetracyclic Triterpenes of *Fomes senex*; J. Indian Chem. Soc., Vol. L-11, pp. 179-184(1975).

Common/Systematic Name

Senexadione

4α,14α-Dimethylcholesta-8,24-diene-3,22-dione

Molecular Formula/Molecular Weight

$C_{29}H_{44}O_2$; MW = 424.33413

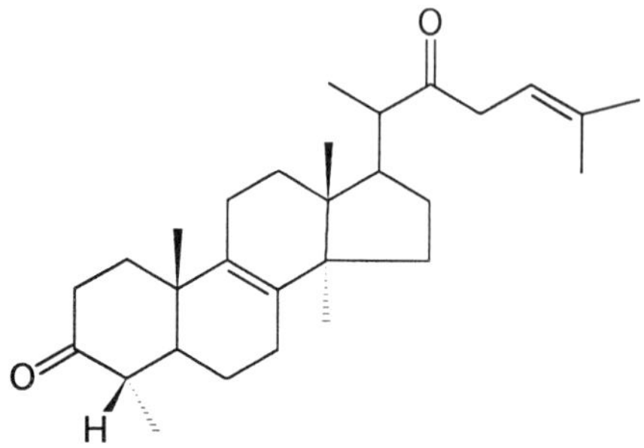

General Characteristics

Crystals; mp., 108-110°C; $[\alpha]_D$ + 52° (in $CHCl_3$); positive Liebermann-Burchard and tetranitromethane color reactions.

Fungal Source

The wood-rotting fungus *Fomes senex*.

Spectral Data

Mass Spectrum:

LREIMS: 424*m/e* (M^+).

References

A. K. Batta and S. Rangaswami; Structure of Senexonol, a Novel Tetracyclic Nortriterpene from *Fomes senex*; Indian J. Chem., Vol. 7, pp. 1063-1064(1969).

A. K. Batta and S. Rangaswami; New Tetracyclic Triterpenes from *Fomes senex*: Senexonol, Senexdione, Oxidosenexone and Senexdiolic Acid; Current Science, Vol. 39, pp. 416-417(1970).

S. Rangaswami; New Tetracyclic Triterpenes of *Fomes senex*; J. Indian Chem. Soc., Vol. L-11, pp. 179-184(1975).

Common/Systematic Name

Oxidosenexone

Molecular Formula/Molecular Weight

$C_{29}H_{46}O_2$; MW = 426.34978

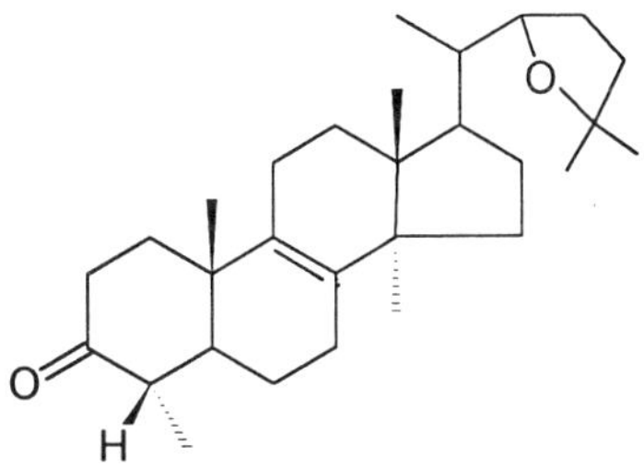

Fungal Source

The wood-rotting fungus *Fomes senex*.

References

A. K. Batta and S. Rangaswami; Structure of Senexonol, a Novel Tetracyclic Nortriterpene from *Fomes senex*; Indian J. Chem., Vol. 7, pp. 1063-1064(1969).

A. K. Batta and S. Rangaswami; New Tetracyclic Triterpenes from *Fomes senex*: Senexonol, Senexdione, Oxidosenexone, and Senexdiolic Acid; Current Science, Vol. 39, pp. 416-417(1970).

S. Rangaswami; New Tetracyclic Triterpenes of *Fomes senex*; J. Indian Chem. Soc., Vol. L-11, pp. 179-184(1975).

4

C_{30} Sterols

Lanosterol
Lanoast-8-en-3β-ol
Lanoasta-8,23-dien-3β,25-diol
Lanoasta-8,24-dien-3β,23-diol
Lanoasta-8,23-dien-3α,22,25-triol
3β-Hydroxylanosta-7,9(11),24-trien-21-oic acid
3β,15α-Dihydroxylanosta-8,24-diene-21,26-dioic acid
Trametenolic acid
15α-Hydroxytrametenolic acid
Lanost-8-ene-3β,22-diol
Obtusifoliol
Parkeol
Obliquol (inotodiol)
Deacetylechinodone
Deacetyl-3-epiechinodol
Deacetoxy-3-epiechinodol
Deacetoxyechinodol
Deacetoxyechinodone
Pinicolic acid A
Zeorin
Glochidone
Glochidonol
Betulin
Cyathic acid
Friedelin
Alnusenone
Officinalic acid
Senexdiolic acid

Common/Systematic Name

Lanosterol; Lanosta-8,24-dien-3β-ol

Molecular Formula/Molecular Weight

$C_{30}H_{50}O$; MW = 426.38617

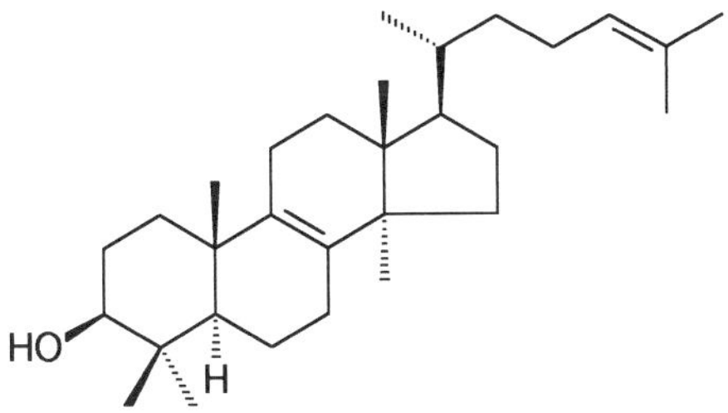

General Characteristics

Crystals; mp., 138-140°C; $[\alpha]_D^{20}$ + 62.0° ($CHCl_3$). Acetate; mp., 131.5-133°C; $[\alpha]_D^{20}$ + 65.5° (*c*=1.12, in $CHCl_3$).

Fungal Source

Agaricus campestris, Allomyces macrogynus, Alternaria kikuchiana, Aspergillus oryzae, Blastocladiella emersonii, Blastocladia ramosa, Candida albicans, C utilis, Catenaria anguillulae, Chytridium confervae, Coriolus heteromorphus, C. pargamenus, Cryptoderma citrinum, Fomitopsis pinicola, Grifola frondosa, Microporus labelliformis, Mucor pusillus, Neurospora crassa, Phycomyces blakesleeanus, Physarum flavicomum, P. polycephalum, Pichia sp., *Saccharomyces cerevisiae, Torulopsis glabrata, Hyphochytrium catenoides, Hymenomycetes* spp., *Rhizidiomyces apophysatus, Rhizophlyctis rosea, Spizellomyces punctatum*, and *Uromyces phaseoli*.

Isolation/Purification

See R. C. Heupel; Isolation and Primary Characterization of Sterols, In Analysis of Sterols and Other Biologically Significant Steroids; W. D. Nes and E. J. Parish (eds.), Academic Press Inc., New York, NY, 341 pp. (1989).

Spectral Data

UV:

End absorption.

Mass Spectrum:

LREIMS: (acetate derivative) 468(M^+, 65%), 454(100), 393(74), 383(5), 341(4), 325(5), 311(5), 301(6), 289(5), 271(5), 255(6), 243(8), 241(9), 229(9), 227(10), and 215*m/e* (10).

References

W. D. Nes and E. J. Parish (eds.); Analysis of Sterols and Other Biologically Significant Steroids; Academic Press Inc., New York, NY, 341 pp. (1989).

W. B. Turner and D. C. Aldridge; Fungal Metabolites II; Academic Press, Inc., New York, NY, 631 pp. (1983).

J. D. Weete; Structure and Function of Sterols in Fungi, Advances in Lipid Research, Vol 23, pp. 115-167(1989).

J. D. Weete, M. S. Fuller, M. Q. Huang, and S. Gandhi; Fatty Acids and Sterols of Selected *Hyphochytriomycetes* and *Chytridiomycetes*; Experimental Mycology, Vol. 13, pp. 183-195(1989).

Common/Systematic Name

Lanost-8-en-3β-ol; 24-Dihydrolanosterol

Molecular Formula/Molecular Weight

$C_{30}H_{52}O$; MW = 428.40182

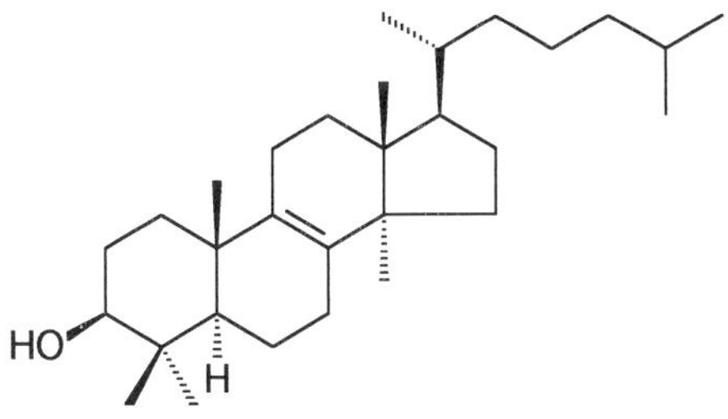

General Characteristics

Crystals from methanol; mp., 144.5-145.5°C; $[\alpha]_D$ + 65.7° (*c*=1.6, in $CHCl_3$).

Fungal Source

Saccharomyces cerevisiae, *Fomes fomentarius*, *F. annosus*, and *F. pini*.

Isolation/Purification

Lanost-8-en-3β-ol was purified from yeast sterol residues using column chromatography [alumina-Brockmann activity grade III, TLC using silica gel GF_{254} (Merck), and silica gel impregnated with silver nitrate (20%) with a solvent system consisting of benzene-light petroleum or ethyl acetate-benzene (20:80, v/v].

Spectral Data

IR:

(KBr) 3380 and 1630cm^{-1}.

Reference

H. D. Munro and O. C. Musgrave; Extractives from Sporophores of Some *Fomes* Species; J. Chem. Soc. (C), pp. 685-688(1971).

Common/Systematic Name

Lanosta-8,23-diene-3β,25-diol

Molecular Formula/Molecular Weight

$C_{30}H_{50}O_2$; MW = 442.38108

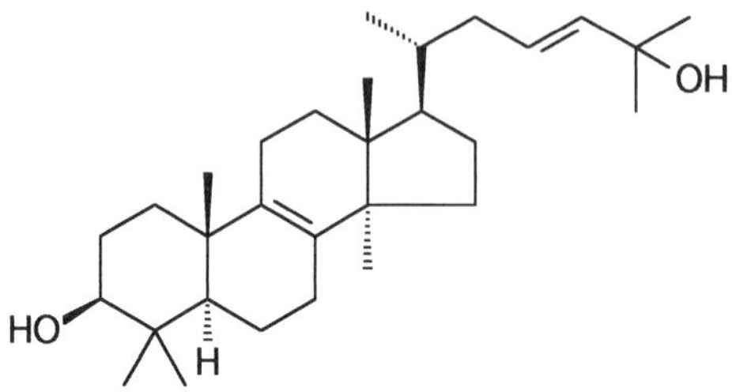

General Characteristics

Crystals; mp., 189-190°C; $[\alpha]_D^{21}$ + 60.3.

Fungal Source

Peridia of *Scleroderma aurantium*.

Isolation/Purification

Fresh peridia were extracted with acetone, concentrated, and partitioned between 60% aqueous ethanol and benzene. Purification was achieved by repeated silicic acid column chromatography.

Also, see R. C. Heupel; Isolation and Primary Characterization of Sterols, In Analysis of Sterols and Other Biologically Significant Steroids; W. D. Nes and E. J. Parish (eds.), Academic Press Inc., New York, NY, 341 pp. (1989).

Spectral Data

IR:

(KBr) 3380, 1635, and 1090cm^{-1}.

^{1}H NMR:

($CDCl_3$) 1.00(s), H-30; 0.81(s), H-30; 0.98(s), H-19; 0.70(s), H-18; 0.87(s), H-32; 0.89(d, $J_{21,20}$= 6Hz, H-21; 1.31(s), H-26; 1.31(s), H-27; 3.23 (m, J= 15Hz, H-3); 5.59(complex unresolved multiplet, H-23); 5.59(complex unresolved multiplet, H-24); and 1.57ppm, OH.

Mass Spectrum:

LREIMS: 442(M^+), 427($C_{29}H_{47}O_2$), 424($C_{30}H_{48}O$), 409($C_{29}H_{45}O$), 391($C_{29}H_{45}O$), 327($C_{23}H_{35}O$), 309($C_{23}H_{33}$), and 255*m/e* ($C_{19}H_{27}$).

References

W. D. Nes and E. J. Parish (eds.); Analysis of Sterols and Other Biologically Significant Steroids; Academic Press Inc., New York, NY, 341 pp. (1989).

W. B. Turner and D. C. Aldridge; Fungal Metabolites II; Academic Press, Inc., New York, NY, 319 pp. (1983).

J. Vrkoč, M. Buděšínský and L. Dolejš; Constituents of the Basidiomycete *Scleroderma aurantium*; Phytochemistry, Vol. 15, pp. 1782-1784(1976).

J. D. Weete; Structure and Function of Sterols in Fungi, Advances in Lipid Research, Vol. 23, pp. 115-167(1989).

J. D. Weete, M. S. Fuller, M. Q. Huang, and S. Gandhi; Fatty Acids and Sterols of Selected *Hyphochytriomycetes* and *Chytridiomycetes*; Experimental Mycology, Vol. 13, pp. 183-195(1989).

Common/Systematic Name

Lanosta-8,24-diene-3β,23-diol

Molecular Formula/Molecular Weight

$C_{30}H_{50}O_2$; MW = 442.38108

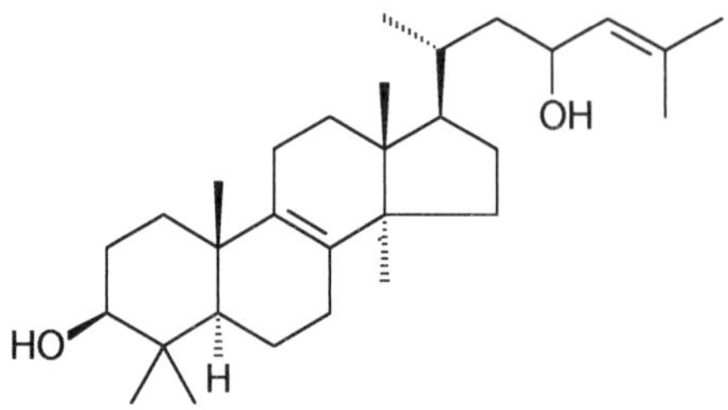

General Characteristics

Crystals; mp., 189-190°C; $[\alpha]_D^{21}$ + 37.4° (in $CHCl_3$).

Fungal Source

Scleroderma aurantium.

Isolation/Purification

Fresh peridia of *Scleroderma aurantium* were homogenized in acetone and the concentrated extract was distributed between 60% aqueous ethanol and benzene. The benzene layer was concentrated and the free acids were converted to methyl esters. The methyl ester mixture was repeatedly chromatographed on silica gel (15% water) columns using a gradient from benzene to benzene-acetone (9:1, v/v) to obtain purified metabolites.

Spectral Data

IR:

γ_{max}^{kbR} 3400, 1635, and 1090cm^{-1}.

^{1}H NMR:

($CDCl_3$) 1.00(s, H-30); 0.81(s, H-31); 0.98(s, H-19); 0.72(s, H-18); 0.87(s, H-32); 0.98(d, $J_{21,20}$= 6Hz, H-21); 1.67(broadened signals with $J_{26,24}$=$J_{27,24}$=1.3Hz, H-26); 1.68(broadened signals with $J_{26,24}$=$J_{27,24}$=1.3Hz, H-27); 3.23 (broadened signals with $J_{26,24}$=$J_{27,24}$=13Hz, H-3); 4.46(dt, $J_{23,22}$= 3.0 and 9.0Hz, H-23); 5.20(broad doublet, $J_{24,23}$= 9.0Hz, $J_{24,26}$=$J_{24,27}$=13Hz); and 1.55ppm (OH).

Mass Spectrum:

LREIMS: 442(M$^+$), 427($C_{29}H_{47}O_2$), 424($C_{30}H_{48}O$), 409($C_{29}H_{45}O$), 391($C_{29}H_{43}$), 327($C_{23}H_{35}O$), 309($C_{23}H_{33}$), and 255*m/e* ($C_{19}H_{27}$).

Reference

J. Vrkoč, M. Buděšínský and L. Dolejš; Constituents of the Basidiomycete *Scleroderma aurantium*; Phytochemistry, Vol. 15, pp. 1782-1784(1976).

Common/Systematic Name

(22*R*)-Lanosta-8,23-diene-3α,22,25-triol; (22*R*)-3α,22,25-Trihydroxylanosta-8,23-diene (Isolated and identified as diacetate, (22*R*)-25-Hydroxylanosta-8,23-diene-3α,22-diacetate)

Molecular Formula/Molecular Weight

$C_{30}H_{50}O_3$; MW = 458.37600
(Diacetate: $C_{34}H_{54}O_5$; MW = 542.39713)

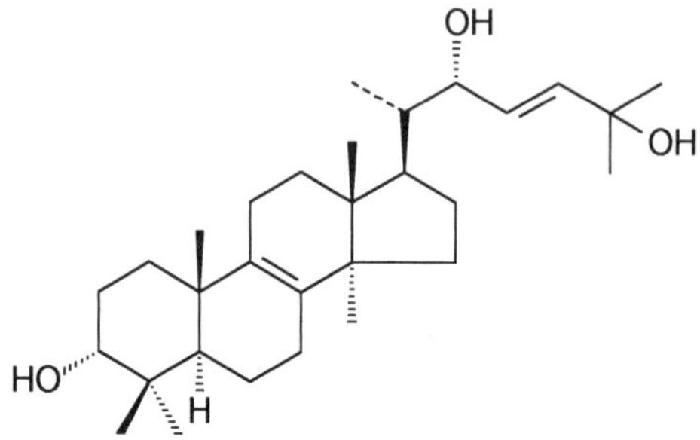

General Characteristics

Diacetate: crystals from methanol; mp., 190-192°C; $[\alpha]_D$ + 6° (*c*=1.5, in $CHCl_3$).

Fungal Source

Fruiting bodies of *Pisolithus tinctorius*.

Isolation/Purification

Dried fruit bodies were extracted with methanol, which was divided into benzene soluble and insoluble fractions. The benzene soluble fraction was chromatographed on a silica gel flash column to give five fractions. Fraction 3 was acetylated and further fractionated with silica gel column chromatography using hexane-ethyl acetate (7:3, v/v) to yield purified (22*R*)-25-hydroxylanosta-8,23-diene-3α,22-diacetate).

Spectral Data

^{1}H NMR:

($CDCl_3$) (Diacetate) 0.72 (3H, s); 0.87(6H, s); 0.92 (3H, s); 0.94 (3H, d, *J*= 6Hz); 1.00 (3H, s); 1.32(6H, s); 2.05 (3H, s); 2.06 (3H, s); 4.57(1H, d, *J*= 7Hz); 5.25 (1H, dd, J_1= 7.5Hz, J_2= 3.7Hz); 5.56(1H, d, J_1= 15Hz, J_2= 7.5Hz); and 5.79ppm (1H, d, *J*= 15Hz).

Mass Spectrum:

542[M^+, $C_{34}H_{54}O_5$]$^+$ (44%), 482[M-AcOH]$^+$ (5), 467[M-AcOH-Me]$^+$ (32), 449[M-AcOH-Me-H_2O]$^+$ (42), 407[M-2AcOH-Me]$^+$ (26), 389[M-2AcOH-Me-H_2O]$^+$ (26), 356[M-$C_{10}H_{18}O_3$]$^+$ (11), 341[M-$C_{10}H_{18}O_3$]$^+$ (11), and 325*m/e* [M-$C_{10}H_{18}O_3$-AcOH]$^+$ (14).

Reference

A. M. Lobo, M. de Abreu, S. Prabhakar, R. Jones, H. S. Rzepa, and D. J. Williams; Triterpenoids of the Fungus *Pisolithus tinctorius*; Phytochem., Vol. 27, pp. 3569-3572 (1988).

Common/Systematic Name

3β-Hydroxylanosta-7,9(11),24-trien-21-oic acid

Molecular Formula/Molecular Weight

$C_{30}H_{46}O_3$; MW = 454.34470

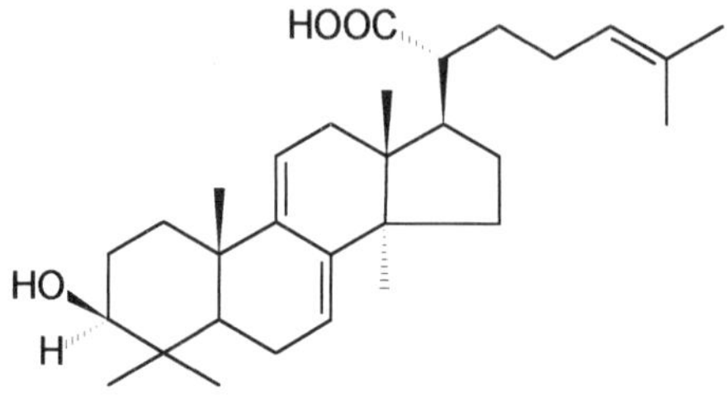

General Characteristics

Methyl ester; colorless needles from methanol; mp., 142-142.5°C; $[\alpha]_D^{24}$ + 42° (*c*=1.0, in $CHCl_3$).

Fungal Source

Sclerotia of *Poria cocos.*

Isolation/Purification

The powdered sclerotia were extracted with ethyl ether, the extract was concentrated, and crude pachymic acid was separated out. The ethyl ether mother liquor was evaporated to dryness, washed with hexane, methylated with diazomethane, and the methyl esters were chromatographed on a alumina column (Wako) eluted with benzene-ether (95:5, v/v). The acid was recrystallized from methanol as colorless needles.

Biological Activity

"Fuling" is the naturally occurring sclerotium of *Poria cocos* and has been frequently prescribed in Chinese medicine as a diuretic and used for palpitations.

Spectral Data

UV:

λ_{max}^{EtOH} (Methyl Ester) 237(log ϵ = 4.08), 244 (4.15), and 252nm (3.97).

IR:

(KBr) 3310(OH), 1725(C=O), and 810cm^{-1} (>C=C-).

^{1}H NMR:

($CDCl_3$) 9.42(3H, s); 9.13(6H, s); 9.05(3H, s); 9.02(3H, s); 8.43(3H, s); 8.33(3H, s); 6.78(1H, m); 6.36(3H, s); 4.94(1H, m); 4.72(1H, m); and 4.53τ (1H, m).

Mass Data:
Anal. calcd. for $C_{31}H_{48}O_3$: C, 79.43; H, 10.32; found: C, 79.26%; H, 10.11%.

Reference

A. Kanematsu and S. Natori; Triterpenoids of Hoelen (fuling), Sclerotia of *Poria cocos* Wolf. II. 3β-Hydroxylanosta-7,9(11),24-trien-21-oic Acid; Chem. Pharm.Bull., Vol. 18, pp. 779-783(1970).

Common/Systematic Name

3β,15α-Dihydroxylanosta-8,24-diene-21,26-dioic acid

Molecular Formula/Molecular Weight

$C_{30}H_{46}O_6$; MW= 502.32944

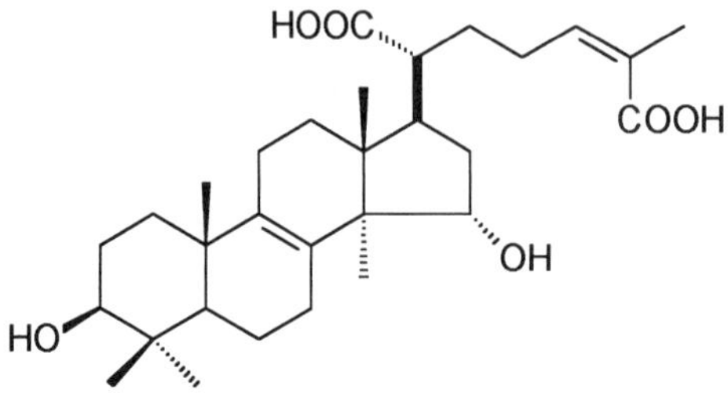

General Characteristics

Dimethyl derivative, needles from methanol, mp 149-153°C; $[\alpha]_D$ + 59° (*c*=1.0, in $CHCl_3$); diacetate derivative, needles from methanol, mp 172-178°C.

Fungal Source

Trameetes odorata.

Isolation/Purification

The finely ground fungus was extracted with cold methanol for three weeks at room temperature. The extract when concentrated under reduced pressure gave a waxy gel which was extracted with ethyl ether, filtered, and separated into acidic and neutral fractions. The acidic fraction was methylated and chromatographed on deactivated alumina eluted with benzene. The dimethyl 3β,15α-dihydroxylanosta-8,24-diene-21,26-dioate containing fraction was further purified by preparative TLC using benzene-methanol, 9:1(v/v) followed by crystallization from methanol to yield needles of dimethyl 3β,15α-dihydroxylanosta-8,24-diene-21,26-dioate .

Spectral Data (Dimethyl ester)

UV:

λ_{max}^{MeOH} 222 nm (ε = 17,000).

IR:

(KBr) 3550, 3385 (OH), 1735 (CO_2Me), 1702 (α,β-unsaturated CO_2Me), 1653 (C=C), 1060 (OH) and 755cm^{-1} (trisubstituted C=C).

^{1}H NMR:

($CDCl_3$) 0.77, 0.80, 0.93, 0.96, and 0.99ppm (tertiary methyls); 1.63 (2OH's exchanged with D_2O), 1.80 (Me_2C=); 3.20 (1H, m, H-3 proton); 3.65, 3.72 (2 X 3H, 2 x CO_2Me); 4.20 (1H, m, H-15 proton); and 6.65ppm (1H, m, H-24 proton). Diacetate derivative; 0.80, 0.87, 0.87, 0.98, and 0.98ppm (tertiary methyls); 1.8(MeC=); 2.04(6H, 2 x OAc); 3.66, 3.83(2 x CO_2Me); 4.45(1H, m, H-3 proton); 4.96(1H, m, H-15 proton); and 6.65 ppm (1H, m, H-24 proton).

Mass Data:

HREIMS 530.3601*m/e* (M^+) $C_{32}H_{50}O_6$ requires 530.3607. Found: C, 71.8; H, 9.25; Calcd. for $C_{32}H_{50}O_6$: C, 72.4; H, 9.5%.

References

W. B. Turner and D. C. Aldridge; Fungal Metabolites II; Academic Press, New York, p.320 (1983).

R. C. Cambie, R. N. Duve and J. C. Parnell, Constituents of *Trametes odorata*, New Zealand Journal of Science, Vol. 15, pp. 200-208 (1972).

Common/Systematic Name

Trametenolic acid

3β-Hydroxylanosta-8,24-dien-21-oic acid

Molecular Formula/Molecular Weight

$C_{30}H_{48}O_3$; MW = 456.36035

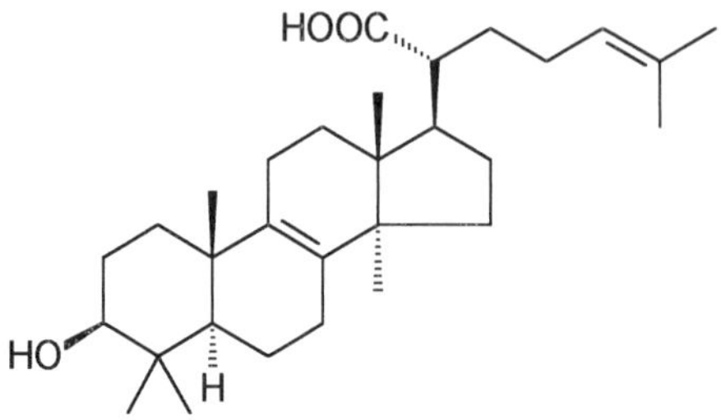

General Characteristics

Colorless needles from methanol; mp., 252-258°C; $[\alpha]_D$ + 45° (in $CHCl_3$); acetate derivative, needles from methanol; mp., 145-146 ° C; $[\alpha]_D^{20}$ + 68 ° (*c*=0.75, in $CHCl_3$).

Fungal Source

Daedalea trabea, Fomes senex, Gloephyllum abietinum, G. odoratum, G. sepiarium, G. striatum, G. trabeum, Heterobasidion tasmanica, Laetiporus sulphureus, Lentinus lepideus, Lenzites trabea, Phellinus gilvus, Spongiporus appendiculatus, Trametes odorata, and *Veluticeps angularis.*

Isolation/Purification

The fruiting bodies of *Phellinus gilvus* collected from the Changa Manga Forest near Lahore (Specimen No. 6376 deposited at herbarium PCSIR Peshawar) were extracted with EtOH at room temperature for 24 days. The brown extract was concentrated under reduced pressure and chromatographed on silica gel eluted with C_6H_6 followed by crystallization from methanol solution to give a crystalline mixture. Further elution with $CHCl_3$ and crystallization with MeOH gave colorless needles of pinicolic acid. Elution with Me_2CO-$CHCl_3$ (5:95, v/v) and crystallization from MeOH gave colorless needles of trametanolic acid.

Spectral Data

UV:

λ_{max} 236, 243, and 252nm.

IR:

methyl ester ($CHCl_3$) 3600, 3425, and 1725cm^{-1}.

^{1}H NMR:

(CDCl$_3$) 0.72, 0.83, 0.88, 0.98, and 0.99ppm (tertiary methyls); 1.58, 1.68(Me_2C=); 3.65(CO_2Me); and 5.08ppm (m, C-24 proton). Acetate derivative; 0.72, 0.87, 0.87, 0.87, and 0.98ppm (tertiary methyls); 1.58, 1.68(Me_2C=); 2.03(OAc); 3.65(CO2Me); and 5.l0ppm.

Mass Data:

Peaks at 441*m/e* = M^+-15(M+-CH_3), 423*m/e* = M^+-15,-18(M^+-CH_3-H_2O); Acetate derivative; found: C, 77.0%; H, 10.2; calcd. for $C_{33}H_{52}O_4$: C, 77.3%; H, 10.2.

<u>TLC Data</u>

Silica gel plates with sodium carbonate and a solvent of benzene-acetone, 9:11, v/v) gave an R_f of .0.42.

<u>References</u>

S. Ahmad, G. Hussain, and S. Razaq; Triterpenoids of *Phellinus gilvus*; Phytochemistry, Vol. 15, p. 2000(1976).

R. C. Cambie and J. C. Parnell; Chemistry of Fungi. Part VII Constituents of *Heterobasidion tasmanica;* New Zealand J. Sci., Vol. 14, pp. 292-298(1991).

W. B. Turner and D. C. Aldridge; <u>Fungal Metabolites II</u>; Academic Press, New York, p. 321(1983).

V. R. Villanueva; Triterpenes et Sterols de *Lenzites trabea*, Phytochemistry, Vol. 10, pp. 427-4300970).

Common/Systematic Name

15α-Hydroxytrametenolic acid

3β,15α-Dihydroxylanosta-8,24-dien-21-oic acid

Molecular Formula/Molecular Weight

$C_{30}H_{48}O_4$; MW= 472.35526

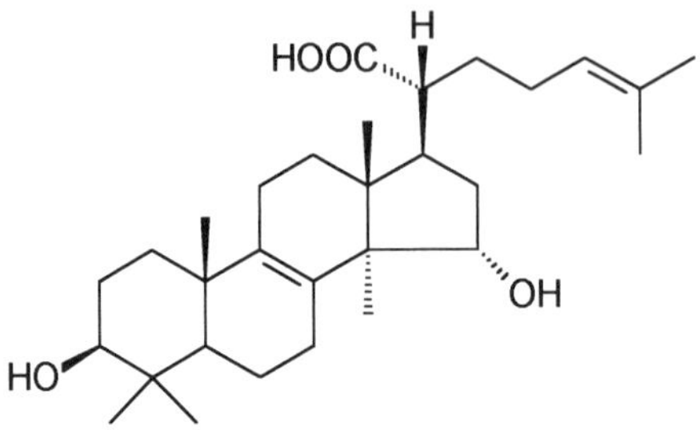

General Characteristics

Methyl ester, needles from methanol, mp. 180-182°C; $[\alpha]_D$ + 64° (*c*=1.2, in $CHCl_3$)
Diacetate, needles from methanoi, mp. 104-106°C;

Fungal Source

Daedalea trabea, Lenzites trabea, and *Trameetes odorata* .

Isolation/Purification

The mycelium was defatted with light petroleum and extracted with ethyl ether. The acidic fraction was separated from the neutral fraction and methylated with diazomethane. The methylated fraction was eluted on an alumina column with benzene-ether (19:1,v/v). 15α-Hydroxytrametenolic acid was finally purified as the methyl ester on a silica gel column impregnated with 12.5% silver nitrate eluted with benzene-ether (9:1,v/v) followed by crystallization from methanol.

Spectral Data

IR:

Methyl ester, (KCl) 3448, 1724 and 1259cm^{-1}.

Mass Spectrum:

LREIMS peaks at M^+-15(M+-CH_3), M^+-15,-18(M^+-CH_3-H_2O), M^+-15,-36(M^+-CH_3-$2H_2O$). (Methyl ester) 486.3703*m/e* (M^+) calc. for $C_{31}H_{50}O_4$ 486.3709. Found: C, 76.5; H, 10.6%; $C_{31}H_{50}O_4$ requires C, 76.5; H, 10.4%;

TLC Data

Silica gel with sodium carbonate (solvent, benzene-acetone, 9:11,v/v); $R_f = 0.54$.

Reference

V. R. Villanueva, Triterpenes et Sterols de *Lenzites Trabea*; Phytochemistry, vol. 10, pp. 427-430 (1970).

W. Lawrie, J. McLean and J. Watson, A New Triterpenoid Acid from *Lenzites trabea*, J. Chem. Soc. (C), pp. 1776-1779 (1967).

Common/Systematic Name
Obtusifoliol

Molecular Formula/Molecular Weight
$C_{30}H_{50}O$; MW = 426.38617

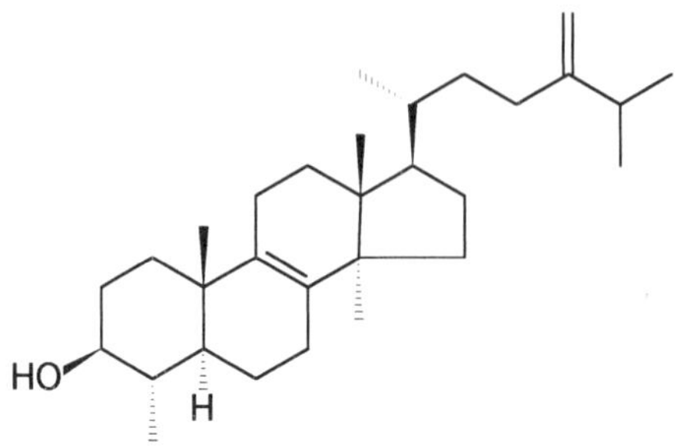

Fungal Source
Candida albicans, Monilinia fructigena, Mucor pusillus, Saccharomyces cerevisiae (strains D-587-4B, Nys 1, and Erg 2), and *Ustilago maydis.*

Isolation/Purification
The yeast cells were saponified and sterols (as acetate derivatives) separated by $AgNO_3$-impregnated silica gel TLC. Further purification of free sterols was achieved by liquid chromatography.

See R. C. Heupel; Isolation and Primary Characterization of Sterols, In Analysis of Sterols and Other Biologically Significant Steroids; W. D. Nes and E. J. Parish (eds.), Academic Press Inc., New York, NY, 341 pp.(1989).

Spectral Data

UV:
End absorption.

^{1}H NMR:
($CDCl_3$) TMS derivative: 0.72(s, 13-CH_3); 0.93(d, *J*=7Hz, 20-CH_3 and C-C(CH_3)$_2$); 0.99(s, 14-CH_3); 0.99(d, *J*=6Hz, 4α-methyl); 1.06(s, 10-CH_3); 4.67 and 4.72ppm (br s, C=CH_2). With Eu(dpm)$_3$, 1.98(s, 10-CH_3, -0.92); 3.23(d, *J*=6.0Hz, 4α-methyl, - 2.24); and 4.60ppm (H-4). Upon irradiation at 4.60, the doublet of 4α-methyl collapsed to a singlet at 3.23ppm.

^{13}C NMR:

($CDCl_3$) C-1, 34.9; C-2, 31.1; C-3, 76.4; C-4, 39.2; C-5, 47.0; C-6, 20.7; C-7, 28.2; C-8, 135.5*; C-9, 134.5*; C-10, 36.3; C-11, 21.7; C-12, 25.5; C-13, 44.5; C-14, 49.8; C-15, 31.2; C-16, 31.0; C-17, 50.3; C-18, 15.7; C-19, 18.7**; C-20, 36.4; C-21, 18.2**; C-22, 34.9; C-23, 30.7; C-24, 156.8; C-25, 33.8; C-26, 21.8***; C-27, 22.0***; C-28, 105.9; C-30, 15.0; and C-32, 24.4ppm.

*, **, *** Assignments may be interchangeable.

Mass Spectrum:

426*m/e* (M^+); (trimethylsilyl ether) 498*m/e*; acetate derivative, 468(M^+, 6%) , 453(M^+ - 15, 4), 408(M^+ - ROH, 100), 393(81), 385, 369, 343, 341, 324(6), 309, 301, 283(31), 281, 269, and 241*m/e* (15).

TLC Data

R_f 0.57 using Silica gel G TLC plates developed with benzene-ethyl acetate (4:1, v/v).

HPLC Data

Retention time 1.96 minutes using μBondapak C_{18} column with a solvent system consisting of tetrahydrofuran-acetonitrile-water, 5:5:2 (v/v/v), and a flow rate of 0.5ml/min.

GC Data

Relative retention time of trimethylsilyl ether relative to squalene was 3.13 minutes using 1% neopentyl glycol succinate on Chromosorb W.

References

W. D. Nes and E. J. Parish (eds.); Analysis of Sterols and Other Biologically Significant Steroids; Academic Press Inc., New York, NY, pp. 341 (1989).

N. N. Ragsdale; Specific Effects of Triarimol on Sterol Biosynthesis in *Ustilago maydis*; Biochimica et Biophysica Acta, Vol. 380, pp. 81-96(1975).

P. J. Trocha, S. J. Jasne and D. B Sprinson, Novel Sterols in Ergosterol Deficient Yeast Mutants, Biochem.Biophys.Res.Commun., Vol. 59, pp. 666-671 (1974).

P. J. Trocha, S. J. Jasne, and D. B Sprinson; Yeast Mutants Blocked in Removing the Methyl Group of Lanosterol at C-14. Separation of Sterols by High-Pressure Liquid Chromatography; Biochemistry, Vol. 16, pp. 4721-4722(1977).

W. B. Turner and D. C. Aldridge; Fungal Metabolites II; Academic Press, New York, New York, 327 pp. (1983).

J. D. Weete; Structure and Function of Sterols in Fungi; Advances in Lipid Research, Vol. 23, pp. 115-167(1989).

Common/Systematic Name
Parkeol

Molecular Formula/Molecular Weight
$C_{30}H_{50}O$; MW = 426.38617

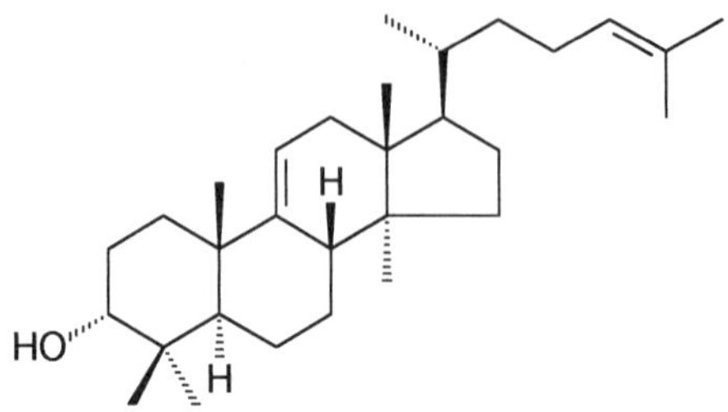

General Characteristics
Parkeol was obtained as needles from chloroform-methanol; mp., 159.5-160°C; $[\alpha]_D$ + 78.4° (*c*=1.49, in $CHCl_3$). Parkeyl benzoate gave needles from chloroform-methanol; mp., 200-203°C (200.5-201°C); $[\alpha]_D$ + 93 .4° (*c*=0.57, in $CHCl_3$)($[\alpha]_D$ + 96.4°). Parkeyl acetate was obtained as plates; mp., 159.5-160°C; $[\alpha]_D$ + 90° (*c*=1.82, in $CHCl_3$).

Fungal Source
Saccharomyces cerevisiae.

Isolation/Purification
Parkeol was purified from sterol residues of yeast using column chromatography (alumina Brockmann activity grade III), TLC using silica gel GF_{254} (Merck), and silica gel impregnated with 20% silver nitrate with a solvent system consisting of benzene-light petroleum or ethyl acetate-benzene (20:80, v/v).

Spectral Data

^{1}H NMR:
($CDCl_3$)(benzoate) 4.80(1H, m, H-11); 4.96(1H, t, *J*=6Hz, H-24); 5.36(1H, m, H-3α); 8.36 (3H, s, Me-26 or 27); 8.44 (3H, s, Me-27 or 26); 8.91(3H, s, Me-10); 8.98 (3H, s, Me-4α); 9.09 (3H, s, Me-4β); 9.27(3H, s, Me-14); and 9.35τ (3H, s, Me-13).

Reference
D. H. R. Barton, U. M. Kempe, and D. A. Widdowson; Investigations on the Biosynthesis of Steroids and Terpenoids. Part VI. Sterols of Yeast; J. Chem. Soc., Perkin Trans I, pp. 513-522(1972).

Common/Systematic Name

Obliquol; Inotodiol

Molecular Formula/Molecular Weight

$C_{30}H_{50}O_2$; MW = 442.38108

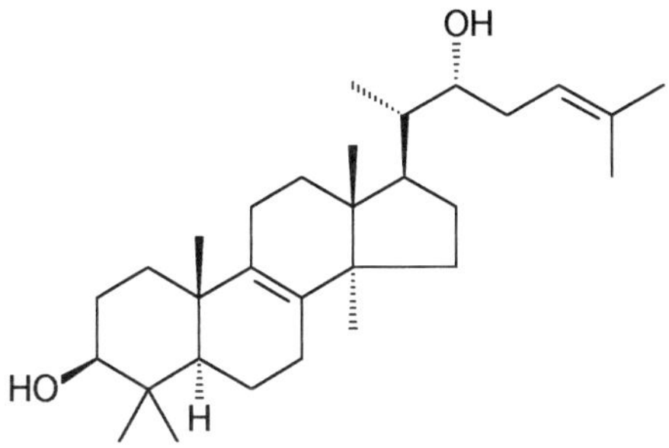

General Characteristics

Needles from acetone; mp., 188-190°C; $[\alpha]_D^{20}$ + 61° (*c*=1.19, in $CHCl_3$); diacetate, lamina from ethanol-water; mp., 162-164°C; $[\alpha]_D^{20}$ + 39° (*c*=1.39, in $CHCl_3$); dibenzoate, needles from acetone-ether; mp., 179-181°C; $[\alpha]_D^{20}$ + 75.3° (*c*=1.13, in $CHCl_3$). The Liebermann-Burchard test gave a pale yellow color turning green, then orange. The Salkowski test gave a rust-colored ring, tetranitromethane gave a yellow color and the Rosenheim test was negative.

Fungal Source

Poria obliqua = *Inonotus obliquus*.

Isolation/Purification

The dried, powdered, sterile conks were extracted with hexane and the hexane extract concentrated to a residue. The residue was heated with ethyl ether and filtered. The ethereal filtrate was extracted with 1% sodium hydroxide which produced a flocculent insoluble material. The flocculent material was suspended in water, acidified, and extracted with ether. The residue was recrystallized from acetone and chromatographed on alumina to yield obliquol.

Spectral Data

UV:

λ_{max} 244(ε = 270) and 252nm (250).

IR:

3350, 1410, 1048, and 1030cm^{-1}.

Mass Data:
Anal. calcd for $C_{30}H_{50}O_2$: C, 81.39; H, 11.38 found: C, 81.72; H, 11.48%.

References

L. B. Kier; Triterpenes of *Poria obliqua*; J. Pharmaceutical Sciences, Vol. 50, pp. 471474(1961).

L. B. Kier and W. S. Brey, Jr.; Structural Studies on The Triterpene Obliquol; J. Pharmaceutical Sciences, Vol. 52, pp. 465-468(1963).

Common/Systematic Name
Deacetylechinodone

Molecular Formula/Molecular Weight
$C_{30}H_{46}O_3$; MW = 454.34470

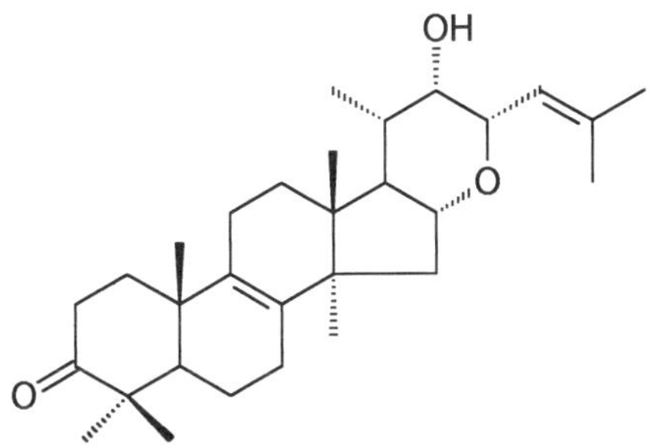

General Characteristics
Colorless needles from methanol; mp., 186-187°C; $[\alpha]_D^{20.5}$ + 54.2° (*c*=1.0, in $CHCl_3$).

Fungal Source
Fruit bodies of *Echinodontium tsugicola*. The fungus is indigenous to Japan and saprophytic to *Tsuga diversifolia*.

Isolation/Purification
Fruit bodies of *Echinodontium tsugicola*, collected at Marunuma, Gumma Prefecture, Japan, in October 1969, were air-dried, crushed, and extracted twice with ethyl ether at room temperature. The combined extracts were treated with hot benzene to remove an orange-yellow amorphous powder. The benzene extract was chromatographed on a column of alumina (Woelm neutral) and eluted successively with (1) benzene, (2) benzene-ether (99:1, v/v), (3) benzene-ether (98.5:1.5, v/v), (4) ether, and (5) MeOH. Fraction (3) was rechromatographed under similar conditions and each fraction was examined by TLC and GLC. The fractions eluted with benzene that contained deacetoxyechinodone and echinodone were combined and further separated by preparative thin-layer chromatography (silica gel HF_{254}). Echinodone and deacetoxyechinodone were recrystallized from MeOH to give colorless needles. The next fraction containing a diterpene ester was recrystallized from MeOH to colorless prisms. The fractions eluted with benzene-ether (98:2, v/v) containing deacetoxy-3-echinodol and deacetoxyechinodol were combined and further purified by preparative thin-layer chromatography. Deacetoxy-3-epiechinodol was recrystallized from MeOH, while deacetoxyechinodol was obtained as colorless needles from MeOH. The benzene-ether (95:5, v/v) fractions were collected and again separated by preparative thin-layer chromatography into three components: 3-epiechinodol, deacetylechinodol, and deacetyl-3-epiechinodol.

Spectral Data

UV:
λ_{max}^{EtOH} 278nm (log ε 1.8).

IR:
(KBr) 3440 and 1710cm^{-1}.

^{1}H NMR:
($CDCl_3$) 3.75 (m, H-16); 2.96 (2.96 (t, *J*=9.0Hz, H-22); 3.95 (t, *J*=9.0Hz, H-23); 5.18 (bd, *J*=9.0Hz, H-24); 0.72; (3H, H-18); 1.12 (H-19); H-21, (3H) obscured by other peaks; 1.77(bs, H-26); 1.05 (H-30, -31); and 1.07ppm (H-32).

Mass Data:
HREIMS: 454.342*m/e* (M^+); calcd for $C_{30}H_{46}O_3$ 454.345; LREIMS: 71(8%), 297(47), 313(100), 439(8), and 454*m/e* (M^+, 17).

Reference

A. Kanematsu and S. Natori; Triterpenoids of *Echinodontium tsugicola*; Chem. Pharm. Bull., Vol. 20, pp. 1993-1999(1972).

Common/Systematic Name

Deacetyl-3-epiechinodol

Molecular Formula/Molecular Weight

$C_{30}H_{48}O_3$; MW = 456.36035

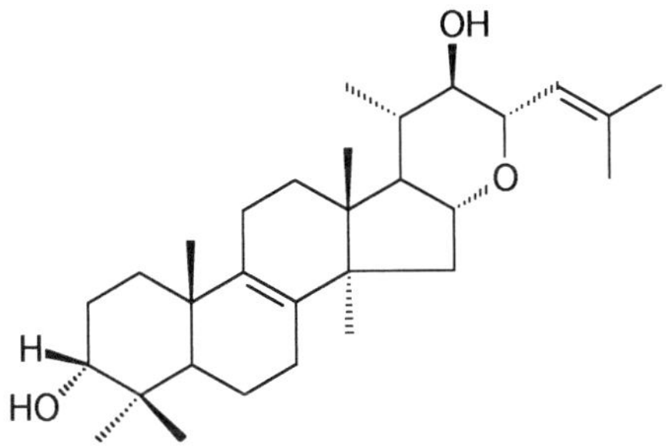

General Characteristics

Colorless needles from methanol; mp., 232-233 °C; $[\alpha]_D^{17}$ + 45 ° (*c*=1.0, in $CHCl_3$).

Fungal Source

Fruit bodies of *Echinodontium tsugicola*. The fungus is indigenous to Japan and saprophytic to *Tsuga diversifolia*.

Isolation/Purification

(See deacetylechinodone).

Spectral Data

IR:

(KBr) 3480cm^{-1}.

^{1}H NMR:

($CDCl_3$) 3.42 (m, broadened triplet of about 4Hz, H-3); 3.75 (m, H-16); 2.92 (t, *J*=9.0Hz, H-22); 3.90 (t, *J*=9.0Hz, H-23); 5.14 (bd, *J*=8.0Hz, H-24); 0.69 (3H, H-18); 0.97 (3H, H-19); obscured by other peaks (3H, H-26, -27); 1.75 (bs); 0.97 (3H, H-30) 0.88 (H-31); and 1.03ppm (3H, H-32).

Mass Data:

Anal calcd for $C_{30}H_{48}O_3$: C, 78.9; H, 10.6; found: C, 78.6%; H, 10.5.

Reference

A. Kanematsu and S. Natori; Triterpenoids of *Echinodontium tsugicola*; Chem. Pharm. Bull., Vol. 20, pp. 1993-1999(1972).

Common/Svstematic Name

Deacetoxy-3-epiechinodol

Molecular Formula/Molecular Weight

$C_{30}H_{48}O_2$; MW = 440.36543

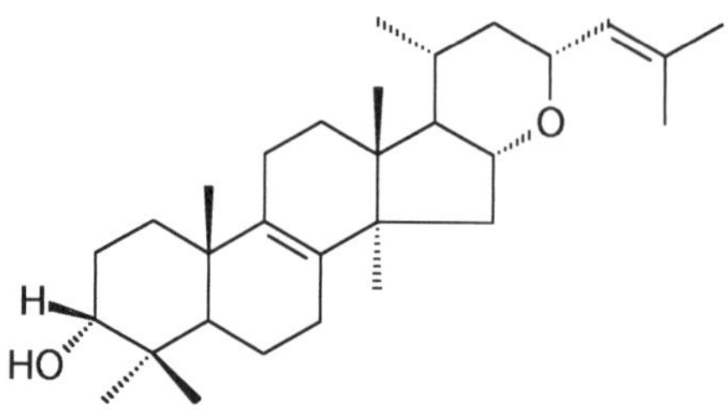

General Characteristics

Colorless needles from methanol; mp., 194-195 °C; $[\alpha]_D^{19}$ +32.5 ° (*c*=1.0, in $CHCl_3$).

Fungal Source

Fruit bodies of *Echinodontium tsugicola*. The fungus is indigenous to Japan and saprophytic to *Tsuga diversifolia*.

Isolation/Purification

(See deacetylechinodone).

Spectral Data

IR:

(KBr) 3450cm^{-1}.

^{1}H NMR:

($CDCl_3$) 3.39 (m, broadened triplet of about 4Hz, H-3); 3.63(m, H-16); 4.12(bt, *J*=9.0Hz, H-23); 5.12(bd, *J*=9.0Hz, H-24); 0.68 (3H, H-18); 1.00 (3H, H-19); 0.95(3H, d, *J*=6.0Hz, H-21); 1.73(3H each, bs, H-26, -27); 1.02 (3h,H, H-30); 0.91 (3H, H-31); and 1.07ppm (3H, H-32).

Mass Data:

HREIMS: 440.366*m/e* (M^+), calcd for $C_{30}H_{48}O_2$ 440.365; LREIMS: 109(75%), 273(4), 299(7), 315(4), 425(100), and 440*m/e* (M^+, 75); anal calcd. for $C_{30}H_{48}O_2$: C, 81.8; H, 11.0; found: C, 81.3%; H, 10.9.

Reference

A. Kanematsu and S. Natori; Triterpenoids of *Echinodontium tsugicola*; Chem. Pharm. Bull., Vol. 20, pp. 1993-1999(1972).

Common/Systematic Name

Deacetoxyechinodol

Molecular Formula/Molecular Weight

$C_{30}H_{48}O_2$; MW = 440.36543

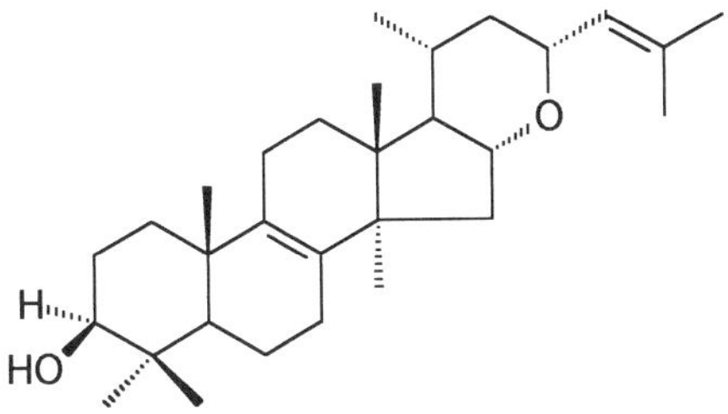

General Characteristics

Colorless needles from methanol; mp., 194-195°C; $[\alpha]_D^{18}$ +58.3° (c=0.96, in $CHCl_3$).

Fungal Source

Fruit bodies of *Echinodontium tsugicola*. The fungus is indigenous to Japan and saprophytic to *Tsuga diversifolia*.

Isolation/Purification

(See deacetylechinodone)..

Spectral Data

IR:

(KBr) 3450cm^{-1}.

^{1}H NMR:

($CDCl_3$) 3.20(m, H-3); 3.60(m, H-16); 4.10(bt, *J*=9.0Hz, H-23); 5.12(bd, *J*=9.0Hz, H-24); 0.68 (3H, H-18); 1.03 (3H, H-19); 0.91 (3H, d, *J*=6.0Hz, H-21); 1.72(3H each, bs, H 26, 27); 1.03 (3H, H-30); 0.82(3H, H-31); and 1.06ppm (3H, H-32).

Mass Data:

HREIMS: 440.366*m/e* (M$^+$), calcd for $C_{30}H_{48}O_2$ 440.365; LREIMS: 109(33%), 273(4), 299(7), 315(2), 425(100), and 440*m/e* (M$^+$, 73); anal calcd. for $C_{30}H_{48}O_2$: C, 81.8; H, 11.0; found: C, 81.3; H, 10.9%.

Reference

A. Kanematsu and S. Natori; Triterpenoids of *Echinodontium tsugicola*; Chem. Pharm. Bull., Vol. 20, pp. 1993-1999(1972).

Common/Systematic Name

Deacetoxyechinodone

Molecular Formula/Molecular Weight

$C_{30}H_{46}O_2$; MW = 438.34978

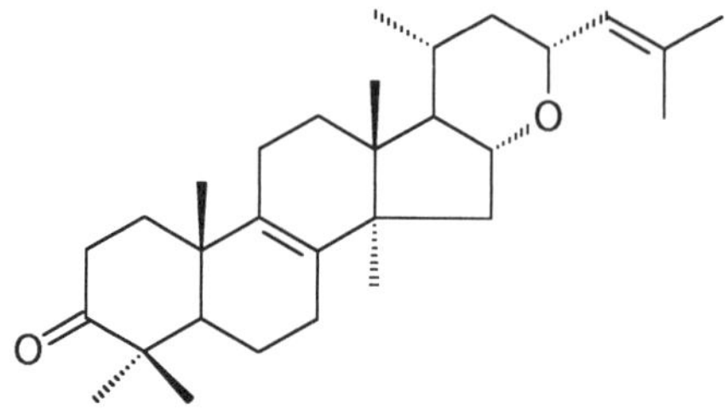

General Characteristics

Colorless needles from methanol; mp., 182-183°C; $[\alpha]_D^{24}$ + 84.1° (*c*=0.82. in $CHCl_3$).

Fungal Source

Fruit bodies of *Echinodontium tsugicola*. The fungus is indigenous to Japan and saprophytic to *Tsuga diversifolia*.

Isolation/Purification

(See deacetylechinodone).

Spectral Data

IR:

(KBr) 1710cm^{-1}.

^{1}H NMR:

($CDCl_3$) 3.60(m, H-16); 4.12(bt, *J*=9.0Hz, H-23); 5.15(bd, *J*=9.0Hz, H-24); 0.70 (3H, H-18); 1.18 (3H, H-19); 0.92(3H, d, *J*=6.0Hz, H-21); 1.72(3H each, bs H-26, -27); 1.11 (3H each, H-30, -31); and 1.13ppm (3H, H-32).

Mass Data:

HREIMS: 438.346*m/e* (M^+), calcd for $C_{30}H_{46}O_2$ 438.350; LREIMS: 109(100%), 271(5), 297(12), 313(3), 423(49), and 438*m/e* (M^+, 40); anal calcd. for $C_{30}H_{46}O_2$: C, 81.8; H, 11.0; found: C, 81.3; H, 10.9%.

Reference

A. Kanematsu and S. Natori; Triterpenoids of *Echinodontium tsugicola*; Chem. Pharm. Bull., Vol. 20, pp. 1993-1999(1972).

Common/Systematic Name
Pinicolic acid A

Molecular Formula/Molecular Weight
$C_{30}H_{46}O_3$; MW = 454.34470

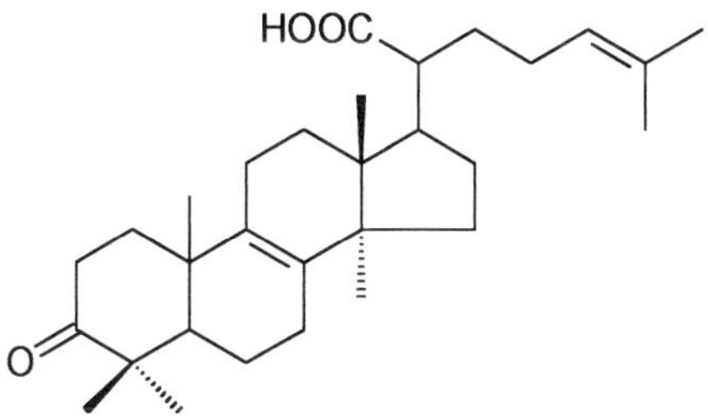

General Characteristics
Colorless needles of pinicolic acid from methanol; mp., 198-202°C; $[\alpha]_D$ + 65° (in $CHCl_3$); methyl ester; mp., 120-121°C; $[\alpha]_D^{20}$ + 69° (*c*=0.27, in $CHCl_3$).

Fungal Source
Phellinus gilvus, *Fomes senex*, and *Heterobasidion tasmanica*.

Isolation/Purification
The fruiting bodies of *Phellinus gilvus* collected from the Changa Manga Forest near Lahore (Specimen No. 6376 deposited at herbarium PCSIR Peshawar) were extracted with EtOH at room temperature for 24 days. The brown extract when concentrated under reduced pressure was chromatographed on silica gel eluted with benzene and crystallized from methanol solution to give a crystalline mixture. Further elution with $CHCl_3$ and crystallization with MeOH gave colorless needles of pinicolic acid.

Spectral Data

IR:
γ_{max} 1700 and 1730cm^{-1}.

^{1}H NMR:
($CDCl_3$) methyl ester, 0.76; 0.89; 1.07; 1.09; 1.09(tertiary methyls); 1.58; 1.67(Me_2C=); 3.66(CO_2Me); and 5.66ppm (m, C-24 proton). ,

Mass Data:
LREIMS: 454*m/e* (M). Methyl ester: found; C, 79.1%; H, 10.3; O, 10.6; calcd for $C_{31}H_{48}O_3$; C, 79.4%; H, 10.3; O, 10.2.

References

S. Ahmad, G. Hussain, and S. Razaq; Triterpenoids of *Phellinus gilvus*; Phytochemistry, Vol. 15, p. 2000(1976).

R. C. Cambie and J. C. Parnell; Chemistry of Fungi. Part VII. Constituents of *Heterobasidion tasmanica*; New Zealand J. Sci., Vol. 14, pp. 292-298(1991).

W. B. Turner and D. C. Aldridge; Fungal Metabolites II; Academic Press, New York, p.321 (1983).

Common/Systematic Name

Zeorin

Molecular Formula/Molecular Weight

$C_{30}H_{52}O_2$; MW = 444.39673

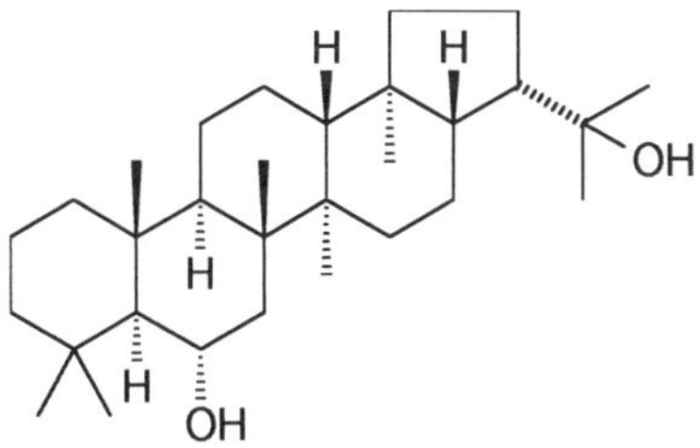

General Characteristics

Crystals from chloroform; mp., 222-224°C; $[\alpha]_D^{24}$ + 73.41° (*c*=2-3%, in $CHCl_3$). Acetate; crystals from chloroform-methanol; mp., 225-230°C; $[\alpha]_D$ + 78° (*c*=0.5, in $CHCl_3$). Benzoate; crystals from benzene-acetone; mp., 236-246°C; $[\alpha]_D$ + 73° (*c*=0.5; in $CHCl_3$).

Fungal Source

Found in *Cyathus helenae*, plants from the genus *Glochidion*, and the lichen, *Nephroma arcticum*.

Isolation/Purification

Mycelium was extracted with ethyl acetate; the ethyl acetate was washed with water, dried, and concentrated to give a brown oil. The crude extract was chromatographed on a column containing silica gel eluted with methylene chloride. Zeorin was crystallized from chloroform solution.

Spectral Data

UV:

λ_{max} 228nm (log ϵ=4.00).

IR:

($CHCl_3$) 1660 and 885cm^{-1}.

Mass Data:

Acetate: Found: C, 79; H, 11.05. $C_{32}H_{54}O_3$ requires C, 78.95; H, 11.2%; benzoate: Found: C, 80.7; H, 9.95. $C_{37}H_{56}O_3$ requires C, 80.95; H, 10.3%.

References

W. A. Ayer, L. M. Browne, J. R. Mercer, D. R. Taylor, and D. Edward; Metabolites of Bird's Nest Fungi. Part 8. Some Minor Metabolites of *Cyathus helenae* and Some Correlations Among the Cyanthins; Can. J. Chem., Vol. 56, pp. 717-721(1978).

D. H. R. Barton and T. Bruun; Triterpenoids. Part VI. Some Observations on the Constitution of Zeorin; J. Chem. Soc. (London), pp. 1683-1690(1952).

D. H. R. Barton, P. De Mayo, and J. C. Orr; Triterpenoids. Part XXIV. Further Investigations on the Constitution of Zeorin; J. Chem. Soc. (London), pp. 2239-2248(1958).

A. K. Ganguly, T. R. Govindachari, P. A. Modhamed, A. D. Rahimtulla, and N. Viswanathan; Chemical Constituents of *Glochidion hohenackeri*; Tetrahedron, Vol. 22, pp. 1513-1519(1966).

I. Yosioka, T. Nakanishi, and I. Kitagawa; The Chemical Proof of Hoptane Skeleton of Zeorin; Chem. Pharm. Bull., Vol. 15, pp. 353-355(1967).

Common/Systematic Name

Glochidone

Molecular Formula/Molecular Weight

$C_{30}H_{46}O$; MW = 422.35487

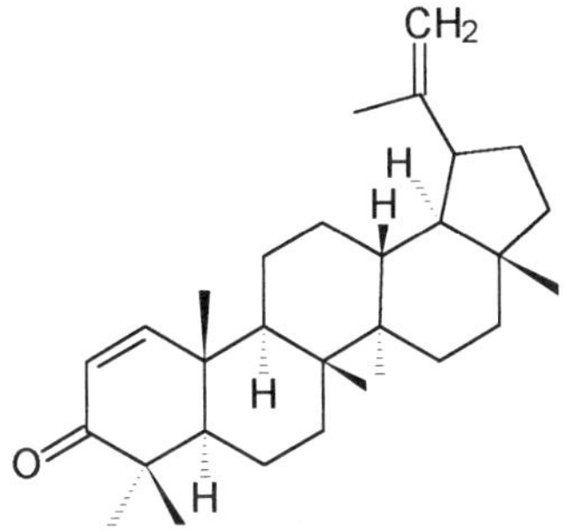

General Characteristics

Crystals; mp., 158-160°C; needles from chloroform-methanol; mp., 164-165°C; $[\alpha]_D^{24}$ + 73.41° (*c*=2-3%, in $CHCl_3$).

Fungal Source

Cyathus helenae and in plants from the genus *Glochidion*.

Isolation/Purification

The mycelium was extracted with ethyl acetate; the ethyl acetate was washed with water, dried, and concentrated to give a brown oil. The crude extract was chromatographed on a column containing silica gel eluted with methylene chloride to yield purified glochidone.

Spectral Data

UV:

λ_{max} 228nm (log ε=4.00).

IR:

($CHCl_3$) 1660 and 885cm^{-1}.

Mass Data:

Found: C, 84.72; H, 10.65. $C_{30}H_{46}O$ requires C, 85.24; H, 10.97%.

References

W. A. Ayer, L. M. Browne, J. R. Mercer, D. R. Taylor, and D. E. Ward; Metabolites of Bird's Nest Fungi. Part 8. Some Minor Metabolites of *Cyathus helenae* and Some Correlations Among the Cyanthins; Can. J. Chem., Vol. 56, pp. 717-721(1978).

A. K. Ganguly, T. R. Govindachari, P. A. Modhamed, A. D. Rahimtulla, and N. Viswanathan; Chemical Constituents of *Glochidion hohenackeri*, Tetrahedron, Vol. 22, pp. 1513-1519(1966).

Common/Systematic Name
Glochidonol

Molecular Formula/Molecular Weight
$C_{30}H_{48}O_2$; MW = 440.36543

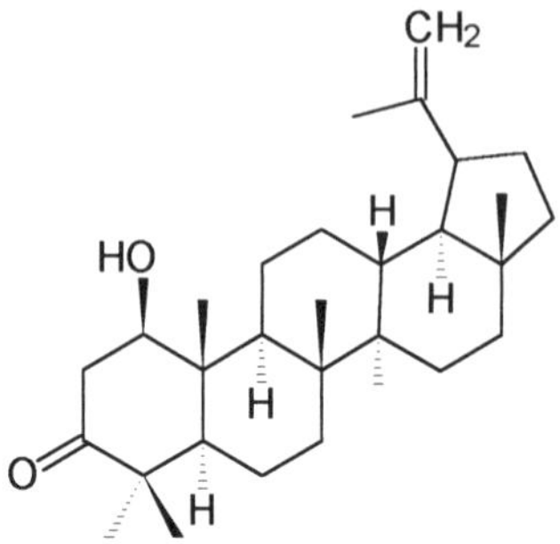

General Characteristics
Amorphous solid from light petroleum; mp., 228°C; acetate derivative, needles from methanol; mp., 194-195°C; $[\alpha]_D$ + 49.4° (in $CHCl_3$). Readily converts to glochidone.

Fungal Source
Cyathus helenae, also in plants from the genus *Glochidion*.

Isolation/Purification
The mycelium was extracted with ethyl acetate; the ethyl acetate was washed with water, dried, and concentrated to give a brown oil. The crude extract was chromatographed on a column containing silica gel eluted with methylene chloride. An amorphous solid was obtained from light petroleum.

Spectral Data

UV:
λ_{max} End absorption.

IR:
(Nujol mull) 3430(OH), 3080, 1725(C=O), 1650, and 880cm^{-1}.

Mass Data:
Acetate derivative: Found: C, 80.00; H, 10.2. $C_{32}H_{50}O_3$ requires C, 79.6; H, 10.40%.

TLC Data
Silica gel G developed with chloroform, R_f 0.20.

References

W. A. Ayer, L. M. Browne, J. R. Mercer, D. R. Taylor, and D. E. Ward, Metabolites of Bird's Nest Fungi. Part 8. Some Minor Metabolites of *Cyathus helenae* and Some Correlations Among the Cyanthins; Can. J. Chem., Vol. 56, pp. 717-721(1978).

W. H. Hui and M. L. Fung; An Examination of the *Euphorbiaceae* of Hong Kong. Part VI. Isolation and Structure of Glochidonol, a New Triterpene Ketol from *Glochidion wrightii*; J. Chem. Soc. (C), pp. 1710-1712(1969).

Common/Systematic Name

Betulin (lup-20(29)-ene-3β,28-diol)

Molecular Formula/Molecular Weight

$C_{30}H_{50}O_2$; MW = 442.38108

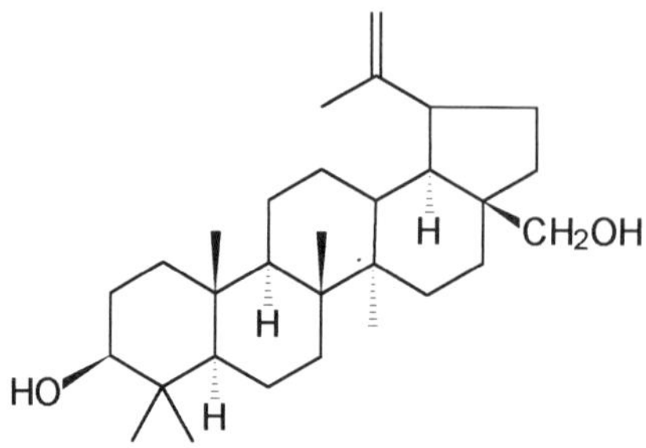

General Characteristics

Needles from methanol; mp., 250-251°C; $[\alpha]_D^{20}$ + 16° (*c*=0.3, in EtOH); diacetate derivative; mp., 216-217°C.

Fungal Source

The wood-rotting fungus, *Polyporus pinicola*. Betulin is an important constituent of the bark of birch trees and it has been suggested that the compound may be translocated from the bark directly into the fungus (in this case *Fomes fomentarius*). Although *Polyporus pinicola* is usually found on pines, betulin may arise in a similar manner or from small pieces of bark adhering to the fungus.

Spectral Data

IR:

(Nujol) 3279, 1658, 1265, and 915cm^{-1}.

^{1}H NMR:

($CDCl_3$) 0.78(3H, s, 24-CH_3); 0.85(3H, s, 25-CH_3); 0.99(6H, s, 23-, 27-CH_3); 1.04(3H, s, 26-CH_3); 1.40(2H, s, D_2O exchangeable); 1.70(3H, s, 29-CH_3); 3.25(1H, m, 3-H); 3.43; 3.75(2H, 28-CH_2); and 4.63, 4.73ppm (2H, 30-CH_3).

Mass Data:

LREIMS: 442(M^+, 91%), 411(82), 424(14), 234(44), 220(30), 203(62), 205(10), 207(85), and 189*m/e* (94.5). Diacetate, found: C, 81.0; H, 11.35%; calcd for $C_{30}H_{50}O_2$: C, 81.4; H, 11.4%; found: C, 77.4; H, 10.3%; calcd for $C_{34}H_{54}O_4$: C, 77.5; H, 10.3%.

References

R. C. Cambie; Chemistry of Fungi, 12. Betulin from *Polyporus pinicola* Fr.; New Zealand J. Science, Vol. 21, pp. 565-567(1978).

W. B. Turner and D. C. Aldridge; Fungal Metabolites II; Academic Press, New York, p. 330 (1983).

Common/Systematic Name

Cyathic acid

Molecular Formula/Molecular Weight

$C_{30}H_{48}O_4$; MW = 472.35526

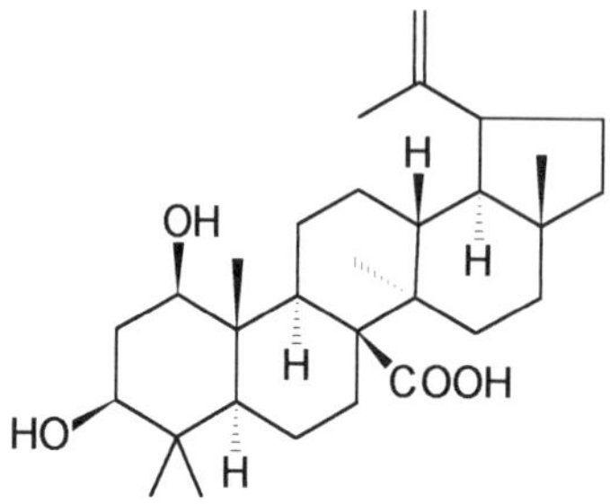

General Characteristics

Crystals from chloroform; mp., 307-309°C (after sublimation at 280°C); *O,O*-diacetylcyathic acid: clear white needles from methanol; mp., 311-312°C; sublimed at 280°C.

Fungal Source

Cyathus helenae

Isolation/Purification

Cyathic acid was purified as *O,O*-diacetylcyathic acid. The mycelium was extracted with methanol, concentrated to dryness, tritiated with methanol, concentrated, and chromatographed on Sephadex LH_{20} eluted with methanol. The first fractions contained glochidonol followed by glochidone and finally *O,O*-diacetylcyathic acid.

Spectral Data

UV:

λ_{max}^{MeOH} 210nm (ϵ = 100).

IR:

(Thin film) 3400, 2960, and 1685cm^{-1}; *O,O*-diacetylcyathic acid: ($CHCl_3$) 3500-3100, 3080, 2900, 1741, 1685, 1250, and 1240cm^{-1}.

Mass Data:

LREIMS: 472(M^+, 65%), 457(21), 445(19), 429(34), 219(14), 218(67), 203(77), 189(74), 175(50), 135(83), 121(100), 107(96), and 95*m/e* (94). *O,O*-Diacetylcyathic acid: 556(M^+, 36%), 496(9), 436(30), 219(34), 218(100), 203(79), 189(75), 175(43), 161(33), 149(35), 147(47), 135(91), and 121*m/e* (86); found: C, 72.39; H, 9.46. $C_{32}H_{52}O_6$ requires C, 72.35; H, 9.41%.

References

W. A. Ayer, L. M. Browne, J. R. Mercer, D. R. Taylor, and D. Edward, Metabolites of Bird's Nest Fungi. Part 8. Some Minor Metabolites of *Cyathus helenae* and Some Correlations Among the Cyanthins; Can. J. Chem., Vol. 56, pp. 717-721(1978).

R. J. Flanagan, Ph. D. Thesis, University of Alberta, Edmonton, Canada, (1978).

Common/Systematic Name
Friedelin

Molecular Formula/Molecular Weight
$C_{30}H_{50}O$; MW = 426.38617

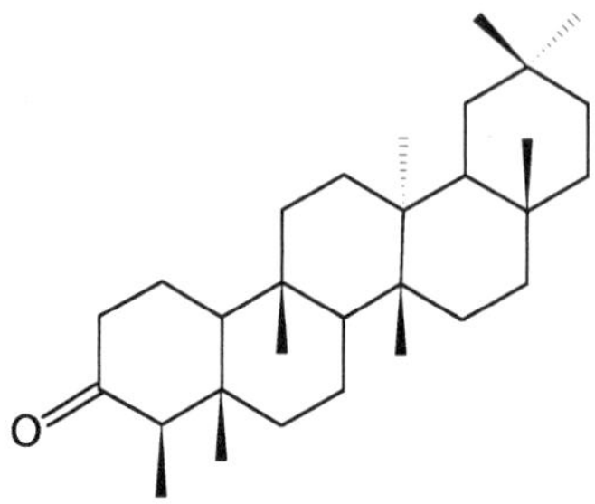

General Characteristics
Crystals from light petroleum; mp., 255-257°C (262-263°C); $[\alpha]_D$ - 25° (c=1.2, in $CHCl_3$) ($[\alpha]_D$ - 21°).

Fungal Source
Ganoderma applanatum.

Isolation/Purification
Crushed, dried pieces of the fungus were extracted with ethanol. The extract was evaporated to dryness and the residue extracted with boiling light petroleum, filtered, and evaporated to dryness to give an oil. The oil was chromatographed on a silica gel column eluted with light petroleum followed by light petroleum-benzene, 9:1 (v/v).

Spectral Data

IR:
($CHCl_3$) 1712cm^{-1} (C=O).

Mass Spectrum:
LREIMS: 426(M^+, 4%, $C_{30}H_{50}O$), 411(3, $C_{29}H_{47}O$), 341(2, $C_{25}H_{41}$), 302(5, $C_{21}H_{34}O$), 273(12, $C_{19}H_{29}O$), 205(13), and 69*m/e* (100).

References
J. Protiva, H. Skorkovská, J. Urban, and A. Vystrčil; Triterpenes and Steroids from *Ganoderma applanatum*; Collection Czechoslov. Commun., Vol. 45, pp. 2710-2713(1980)

W. B. Turner and D. C. Aldridge; Fungal Metabolites II; Academic Press, New York, p. 331 (1983).

Common/Systematic Name
Alnusenone; Friedoolean-5-en-3-one

Molecular Formula/Molecular Weight
$C_{30}H_{48}O$; MW = 424.37052

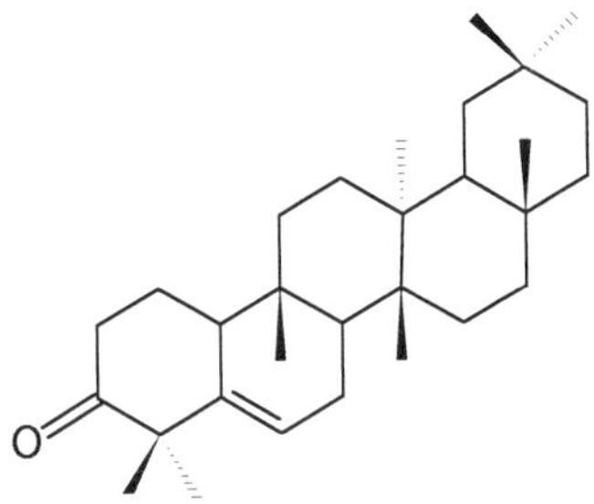

General Characteristics
Crystals from light petroleum; mp., 230-233°C (247°C); $[\alpha]_D$ + 24° (*c*=0.05, in $CHCl_3$).

Fungal Source
Ganoderma applanatum.

Isolation/Purification
Crushed, dried pieces of the fungus were extracted with ethanol. The extract was evaporated to dryness and the residue extracted with boiling light petroleum, filtered, and evaporated to dryness to give an oil. The oil was chromatographed on a silica gel column eluted with light petroleum followed by light petroleum-benzene, 9:1 (v/v).

Spectral Data

IR:
($CHCl_3$) 1712cm^{-1} (C=O).

Mass Spectrum:
LREIMS: 424(M$^+$, 12%, $C_{30}H_{48}O$), 409(6, $C_{29}H_{45}O$), 274(55, $C_{20}H_{34}$), 259(37, $C_{19}H_{31}$), 205(36), and 69*m/e* (100).

References
J. Protiva, H. Skorkovská, J. Urban, and A. Vystrčil; Triterpenes and Steroids from *Ganoderma applanatum*; Collection Czechoslov. Commun., Vol. 45, pp. 2710-2713(1980).

W. B. Turner and D. C. Aldridge; Fungal Metabolites II; Academic Press, New York, p. 331 (1983).

Common/Systematic Name

Officinalic acid

Molecular Formula/Molecular Weight

$C_{30}H_{44}O_6$; MW = 500.31379

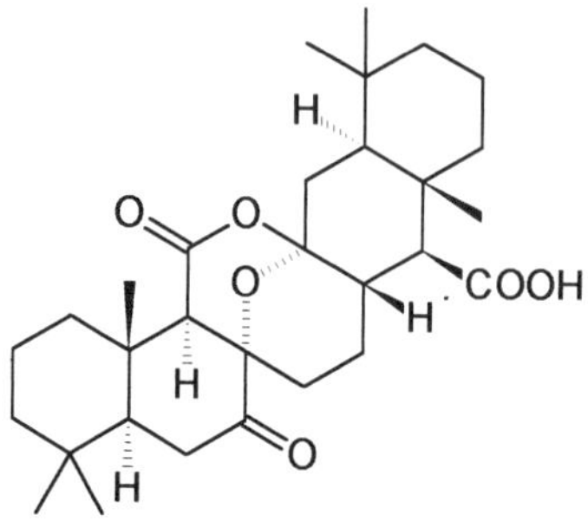

General Characteristics

Crystals; mp., 272°C; $[\alpha]_D$ -60° (*c*=0.5, in dioxane). Monomethyl ester; mp., 236°C; $[\alpha]_D$ -54° (*c*=0.15, in dioxane).

Fungal Source

Fomes officinalis, a wood-rotting fungus found on the trunks of living or dead coniferous trees in the Pacific Northwest United States, Canada and in Europe.

Spectral Data

IR:

3210, 1730, and 1709 cm^{-1}; monomethyl ester, no OH and 1736cm^{-1}.

^1H NMR:

($CDCl_3$) 2.05.3; 172.6, and 168.5ppm (3 carbonyls). Monomethyl ester: $C_{31}H_{46}O_6$, 105.3ppm (ketal carbon); six methyl resonances at 0.82, 0.84, 0.85, 0.89, 1.04, and 1.29ppm (all singlets); 2.84(1H, d, *J*=13Hz); 3.24(2H, s); and 2.49ppm (2H, s).

References

W. W. Epstein, F. W. Sweat, G. VanLear, F. M. Lovell, and E. J. Gabe; Structure and Stereochemistry of Officinalic Acid, A Novel Triterpene from *Fomes officinalis*; J. Am. Chem. Soc., Vol. 101, pp. 2748-2750(1979).

W. B. Turner and D. C. Aldridge; Fungal Metabolites II; Academic Press, New York, p. 331 (1983).

Common/Systematic Name
Senexdiolic acid

Molecular Formula/Molecular Weight
$C_{30}H_{48}O_4$; MW = 472.35526

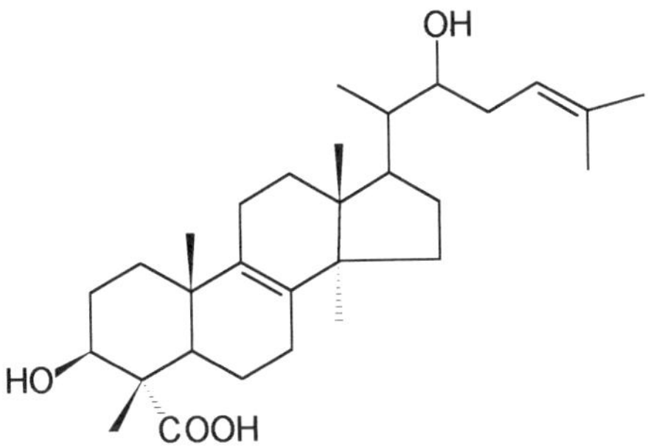

General Characteristics
Positive Liebermann-Burchard and tetranitromethane tests.

Fungal Source
The wood-rotting fungus *Fomes senex*.

Spectral Data

Mass Spectrum:
LREIMS: 440(M^+ - 32) and 69*m/e* (base peak).

References
A. K. Batta and S. Rangaswami; Structure of Senexonol, a Novel Tetracyclic Nortriterpene from *Fomes senex*; Indian J. Chem., Vol. 7, pp. 1063-1064(1969).

A. K. Batta and S. Rangaswami; New Tetracyclic Triterpenes from *Fomes senex*: Senexonol, Senexdione, Oxidosenexone, and Senexdiolic Acid; Current Science, Vol. 39, pp. 416-417(1970).

S. Rangaswami; New Tetracyclic Triterpenes of *Fomes senex*; J. Indian Chem. Soc., Vol. L-11, pp. 179-184(1975).

5

C_{31} Sterols

Dehydroeburicoic acid ketone
3-Ketodehydrosulfurenic acid
Eburicodiol
Eburical
24-Methyl-24,25-dihydrolanosterol
Stowardolic acid
Stowardonic acid
Eburicoic acid
Tumulosic acid
Polyporenic acid C
6α-Hydroxypolyporenic acid C
Polyporenic acid D
Dehydroeburicoic acid
Dehydrotumulosic acid
24-Methylene-24,25-dihydrolanosterol (Eburicol)
Pisolactone
3-Oxopisolactone
(22*S*, 24*R*) 24-Methyllanost-8-en-22,28-epoxy-3β,28α-diol
(22*S*, 24*S*) 24-Methyllanost-8-en-22,28-epoxy-3β,28β-diol
24-Methylenelanosterol
(22*S*) Lanosta-8,24 (28)-diene-3β,22-diol

Common/Systematic Name

Dehydroeburicoic acid ketone

Molecular Formula/Molecular Weight

$C_{31}H_{46}O_3$, MW = 466.34470

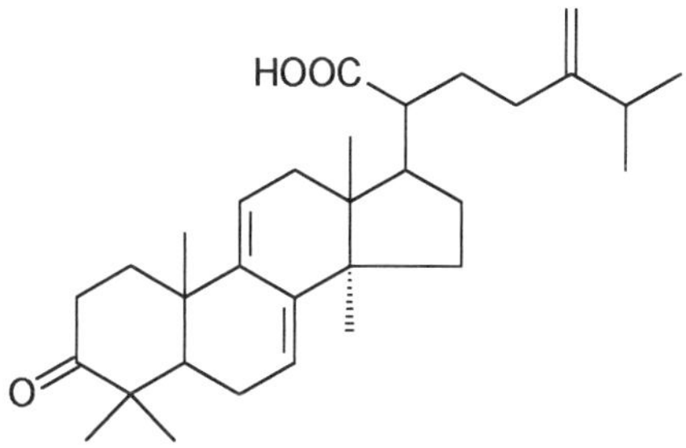

General Characteristics

Crystals of methyl ester from methanol; mp., 157-159°C; $[\alpha]_D^{25}$ + 30.2° (*c*=0.86, in $CHCl_3$). Ethyl ester; mp., 240-242°C; $[\alpha]_D^{24}$ + 29° (*c*=1.2, in pyridine); 2,4-dinitrophenylhydrazone; mp., 144-145°C.

Fungal Source

Fomes officinalis.

Spectral Data

UV:
252(log ε = 3.79), 243(3.93), and 236nm (3.90).

IR:
(KBr) 3450, 1718, 1640, and 893cm^{1}.

Mass Spectrum:
LREIMS: 466*m/e* (M^+).

Reference

K. E. Schulte, G. Wicker, and H. Fachmann; Zur Struktur Der "Agaricolsaure", Tet. Lett., pp. 4823-4825(1967).

Common/Systematic Name

3-Ketodehydrosulfurenic acid
(Isolated and identified as methyl-3-ketodehydrosulfurenate)

Molecular Formula/Molecular Weight

$C_{31}H_{46}O_4$, MW = 482.33961

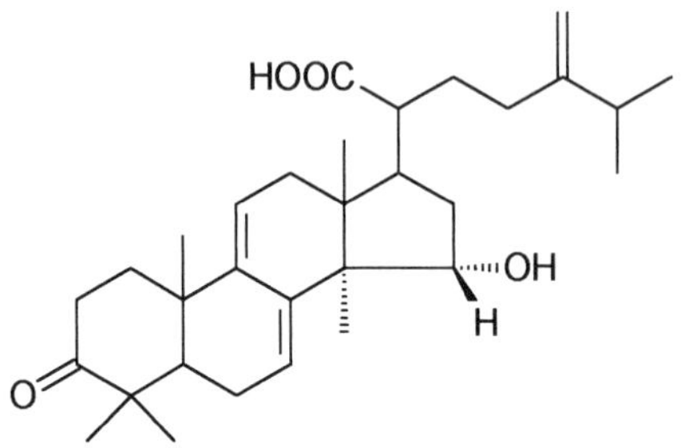

General Characteristics

Methyl-3-ketoacetyldehydrosulfurenate: Long slender needles from ethanolic KOH; mp., 173-175°C; $[\alpha]_D^{24}$ + 67.5° (*c*=2.0, in $CHCl_3$).

Fungal Source

Fomes officinalis.

Isolation/Purification

The chloroform extract from the fungus was separated into neutral, strong and weak acids. The weak acid fraction was reduced *in vacuo*, taken up in methanol, and allowed to stand for several days. The precipitate was removed and the methanol mother liquor removed, reduced *in vacuo,* and acetylated. The resulting weak acids were methylated, and chromatographed on Florisil to yield methyl-3-ketoacetyldehydrosulfurenate which was allowed to stand overnight in 0.1N ethanolic KOH to yield long slender needles.

Spectral Data (Methyl ester)

UV:

λ_{max} 236(log ε = 4.10), 2.43(4.23), and 252nm (4.00).

IR:

(KBr) 3540, 1725, 1700, 1650, 1060, and 890cm^{-1}.

^{1}H NMR:

($CDCl_3$) several methyl resonances in the region of 8.8-9.5(21H); 6.32(3H); 5.32(1H); 5.25(1H); 4.48(1H); and 4.101(1H).

Mass Data:
LREIMS: 496*m/e* (M^+).

Reference

C. G. Anderson, W. W. Epstein, and G. Van Lear; Minor Triterpenoids of *Fomes officinalis*; Phytochemistry, Vol. 11, pp. 2847-2852(1972).

Common/Systematic Name

Eburicodiol

24-Methylenelanost-8-ene-3β,21-diol

Molecular Formula/Molecular Weight

$C_{31}H_{52}O_2$; MW = 456.39673

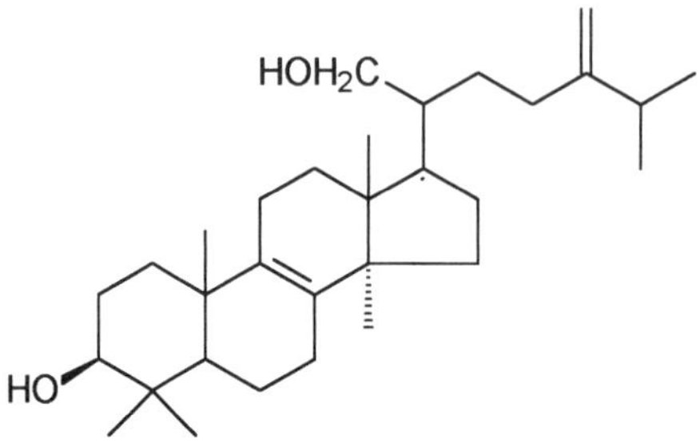

General Characteristics

Needles from methanol; mp., 211-212°C; $[\alpha]_D^{28}$ + 57°(*c*=1.2, in $CHCl_3$); diacetate; mp., 132-133°C; $[\alpha]_D^{28}$ + 46°(*c*=1.0, in $CHCl_3$).

Fungal Source

Fomes officinalis.

Isolation/Purification

The non-saponifiable lipids were chromatographed on Florex. The benzene fractions were combined, concentrated, and rechromatographed on Florex. The benzene-ether (1:1, v/v) eluents were concentrated and rechromatographed on Florex; the ether fraction was crystallized from methanol to yield purified eburicodiol as needles.

Spectral Data

Mass Spectrum:

LREIMS: 456*m/e* (M^+).

Reference

C. G. Anderson and W. W. Epstein; Metabolic Intermediates in the Biological Oxidation of Lanosterol to Eburicoic Acid; Phytochemistry, Vol. 10, pp. 2713-2717(1971).

Common/Systematic Name

Eburical

24-Methylenelanost-8-en-3β-ol-21-al

Molecular Formula/Molecular Weight

$C_{31}H_{50}O_2$; MW = 454.38108

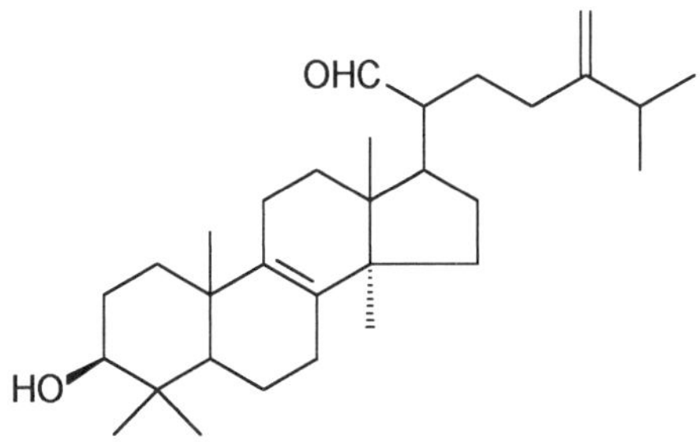

General Characteristics

Needles from methanol; mp., 177-179°C; $[\alpha]_D^{26}$ + 57° (*c*=1.25, in $CHCl_3$); acetate, small plates from methanol; mp., 154-155°C; $[\alpha]_D^{24}$ + 57° (*c*=1.4, in $CHCl_3$).

Fungal Source

Fomes officinalis.

Isolation/Purification

The neutral extract was defatted and chromatographed on Florex. The benzene fraction was rechromatographed on Florisil; the eburical eluted with benzene-ethyl acetate (95:5, v/v) from the benzene-ethyl acetate (8:2, v/v) fraction chromatographed on a second Florisil column.

Spectral Data

IR:

(CCl_4) 2850sh, 2741(w), and 1725 cm^{-1}; acetate: 2849sh (w), 2740(w), and 1725cm^{-1} (s).

^1H NMR:

($CDCl_3$) 1 proton singlet at 9.55ppm (CHO); (CCl_4) acetate, broad 1 proton singlet at 9.58ppm (CHO).

Mass Spectrum:

LREIMS: 454*m/e* (M^+); acetate 496*m/e* (M^+).

Reference

C. G. Anderson and W. W. Epstein; Metabolic Intermediates in the Biological Oxidation of Lanosterol to Eburicoic Acid; Phytochemistry, Vol. 10, pp. 2713-2717(1971).

Common/Systematic Name

24-Methyl-24,25-dihydrolanosterol

Molecular Formula/Molecular Weight

$C_{31}H_{54}O$; MW = 442.41747

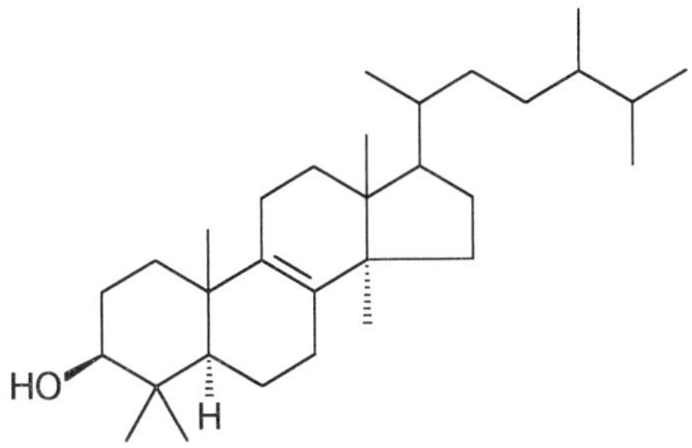

Fungal Source

Candida albicans, *C. utilis*, and *Neurospora crassa*.

Isolation/Purification

The lyophilized mycelium was extracted with acetone; benzene-acetone (1:1, v/v); ethyl acetate-absolute methanol (1:1, v/v) and methanol. The fatty acids were removed by partition and the organic phase was concentrated and chromatographed on silica gel G TLC plates using hexane-diethyl ether-methanol (80:20:2, v/v/v) as solvent to separate free sterols from sterol esters. The sterols and sterol esters were eluted from TLC plates with ethyl ether, the sterol esters saponified, and both free sterols and free sterols from saponification rechromatographed as above. The sterols were analyzed by GLC using OV-101 and Silar 10C on Gas Chrom Q at 220°C.

References

M. Fryberg, A. C. Oehlschlager, and A. M. Unrau; Sterol Biosynthesis in Antibiotic Sensitive and Resistant *Candida*; Archives of Biochem. and Biophys., Vol. 173, pp. 171177(1975).

A. M. Pierce, H. D. Pierce, Jr., A. M. Unrau, and A. C. Oehlschiager; Lipid Composition and Polyene Antibiotic Resistance of *Candida albicans* Mutants; Can. J. Biochem., Vol. 56, pp. 135-142(1978).

A. M. Pierce, H. D. Pierce, Jr., A. M. Unrau, A. C. Oehlschlager, R. E. Subden, and R. L. Renaud; The Biosynthesis of the Free Sterols and Sterol Esters of *Neurospora crassa*; Can. J. Biochem. Vol. 57, pp. 112-116(1979).

R. L. Renaud, R. E. Subden, A. M. Pierce, and A. C. Oehlschlager; Sterol Composition of *Neurospora crassa*, Lipids, Vol. 13, pp. 56-58(1978).

Common/Systematic Name

Stowardolic acid

3α-Hydroxy-24ξ-methyl-23-oxolanosta-8,25-dienoic acid

Molecular Formula/Molecular Weight

$C_{31}H_{48}O_4$; MW = 484.35526

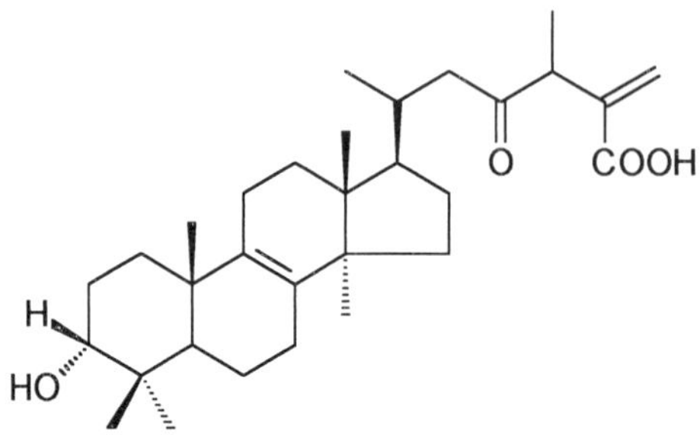

General Characteristics

Needles from acetone-light petroleum; mp., 172-181°C.

Fungal Source

Trametes stowardii.

Isolation/Purification

Fungal cultures were extracted with 95% ethanol, evaporated to dryness, and redissolved in ethyl ether. The ether soluble fraction was partitioned between 2N HCl and 5% $NaHCO_3$. Acidification of the bicarbonate solution and extraction with ether gave a crude product that was chromatographed on a silicic acid column followed by crystallization from acetone-light petroleum to yield stowardolic acid.

Spectral Data

UV:

λ_{max} 210nm (ε = 10,400).

IR:

(Nujol null) 2500-3600, 1670-1740, 1065, and 960cm^{-1}.

^{1}H NMR:

($CDCl_3$) 3.45(*J*=6.0Hz, H-3); 3.7(*J*=7.0Hz, H-24); 5.90(*J*=3.0Hz, H-26); 6.2(2.5Hz, H-26); 1.35(3H, d, *J*=7.0Hz, H-31); and 5.1ppm (2H, OH, CO_2H).

Mass Spectrum:

HREIMS: 484.355*m/e* (M^+); C31H4804 requires 484.355.

Reference

H. T. Cheung, J. C. F. Seeto, and T. R. Watson; Triterpene Metabolite of *Trametes stawardii*; Aust. J. Chem., Vol. 26, pp. 609-618(1973).

Common/Systematic Name
Stowardonic acid

Molecular Formula/Molecular Weight
$C_{31}H_{46}O_4$; MW = 482.33961

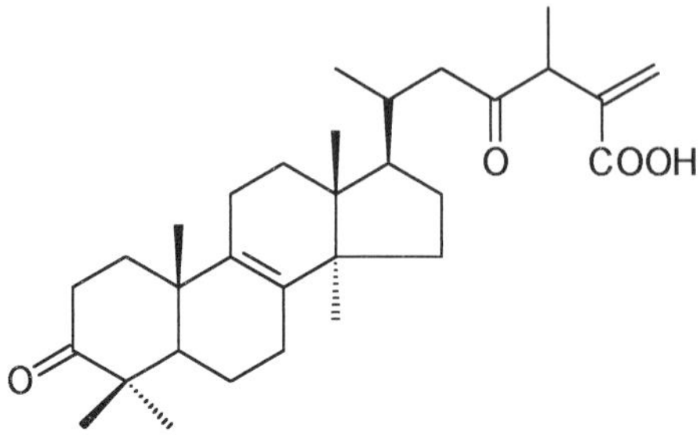

General Characteristics
Crystals from ethyl acetate; mp., 180-184°C.

Fungal Source
Trametes stowardii.

Isolation/Purification
Fungal cultures were extracted with 95% ethanol, evaporated to dryness, and redissolved in ethyl ether. The ether soluble fraction was partitioned between 2N HCl and 5% $NaHCO_3$. Acidification of the bicarbonate solution and extraction with ether gave a crude product that was chromatographed on a silicic acid column followed by crystallization from ethyl acetate to yield stowardonic acid.

Spectral Data

UV:
λ_{max} 210nm (ε = 11,100).

IR:
(Nujol null) 2500-3600, 1730, 1680, and 940cm^{-1}.

^{1}H NMR:
($CDCl_3$) 3.7(q, *J*=7.0Hz, H-24); 5.95(*J*=2.5Hz, H-26); 6.2(1.5Hz, H-26); 1.35(3H, d, *J*=7.0Hz, H-31); and 9.0ppm (1H, CO_2H).

Mass Spectrum:
HREIMS: 482.341*m/e* (M^+); $C_{31}H_{46}0_4$ requires 482.340.

Reference

H. T. Cheung, J. C. F. Seeto, and T. R. Watson; Triterpene Metabolite of *Trametes stowardii*; Aust. J. Chem., Vol. 26, pp. 609-618(1973).

Common/Systematic Name

Eburicoic acid

Molecular Formula/Molecular Weight

$C_{31}H_{50}O_3$; MW = 470.37600

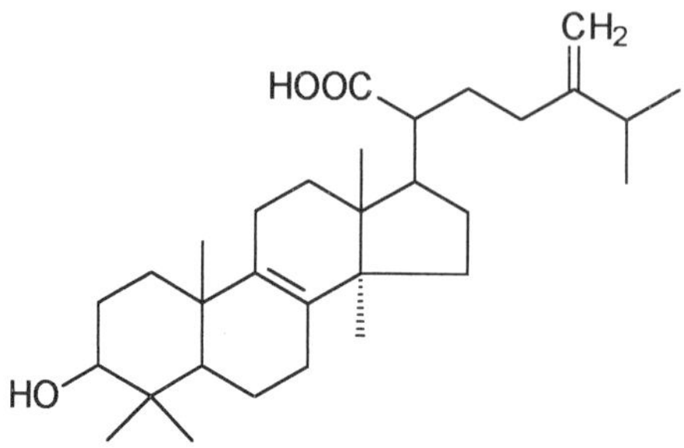

General Characteristics

Needles from ethanol; mp., 289-290°C (sublime); $[\alpha]_D^{20}$ + 35° (*c*=1.0, in pyridine); acetate; mp., 256-258°C.

Fungal Source

Daedalea trahea, Fomes officinalis, Gloephyllum ahietinum, G. sepiarium, G. striatum, G. trabeum, Laetiporus sulphureus, Lentinus lepideus, Lenzites trabea, Polyporus cretaceous, Trametes lilacino-gilva, and *Spongiporus appendiculatus.*

Spectral Data

UV:

λ_{max} 236, 243, and 252nm.

IR:

(KBr) 3538, 3422(OH), 2940, 2870, 2838, 1720, and 1702cm^{-1}.

Mass Spectrum:

LREIMS: 470(M^+), 456, 455, 441, 437, 423, 409, and 395*m/e*.

TLC Data

Silica gel with sodium carbonate (solvent, benzene-acetone, 9:11)gave R_f= 0.52.

References

L. H. Briggs, R. C. Cambie, I. C. Dean, S. H. Dromgoole, B. J. Fergus, W. B Ingram, K. G. Lewis, C. W. Small, R. Thomas and D. A. Walker; 10. Metabolites of Some Fungi, New Zealand Journal of Science, Vol. 18, pp. 565-576(1975)

K. E. Schulte, G. Rucker, and H. Fachmann, Zur Struktur Der "Agaricolsaure", Tet. Lett., pp. 4823-4825(1967).

V. R. Villanueva; Triterpenes et Sterols de *Lenzites trabea*; Phytochemistry, Vol. 10, pp. 427-430(1970).

Common/Systematic Name
Tumulosic acid

Molecular Formula/Molecular Weight
$C_{31}H_{50}O_4$; MW = 486.37091

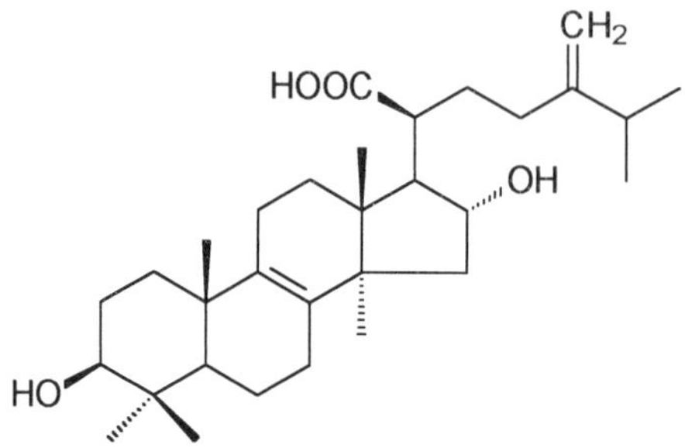

General Characteristics
Methyl ester, needles from aqueous methanol; mp., 163-164°C; $[\alpha]_D$ + 25°.

Fungal Source
Daedalea dickinsii, Melanoporia cajanderi, M. juniperina, M. nigra, M. purpuracea, M. rosea, Spongiporus appendiculatus, Trametes feel, T. lilacino gilva, and *T. dickinsii.*

Isolation/Purification (purified as methyl ester)
The fungus was extracted with ethyl ether, methylated, chromatographed on alumina, and eluted with benzene (3 different fractions) and benzene-chloroform, 1:1 (v/v). The later benzene fraction yielded a 1:1, v/v, mixture of methyl tumulosate and methyl dehydrotumulosate. Methyl tumulosate was obtained by hydrolysis of methyl-3-*O*-methyltumulosate followed by precipitation with water and recrystallization from aqueous methanol.

Biological Activity
Antibiotic.

Spectral Data (Methyl ester)

UV:
λ_{max} 243nm (broad, ε = 500).

IR:
($CHCl_3$) 3540, 3400, 725, 1645, and 890cm^{-1} ($C{=}CH_2$).

^{1}H NMR:
($CDCl_3$) 4.7(2H, m, vinylic protons); 4.10(1H, m, H-16); 3.7(3H, s, CO_2Me); and

3.2ppm(1H, q after D_2O, *J*=4.0, 10Hz, H-3); and 1.07ppm (OH).

Mass Data:
Found: C, 74.2; H, 10.3; calcd for $C_{32}H_{52}O_4 \bullet H_2O$; C, 74.1; H, 10.5.

References

J. T. Pinhey, B. J. Ralph. J. J. H. Simes, and M. Wootton; Extractives of Fungi 1. The Constituents of *Trametes lilacino-gilva*; Aust. J. Chem., Vol. 23, pp. 2141-2146(1970).

J. T. Pinhey, B. J. Ralph, J. J. H. Simes, and M. Wootton; Extractives of Fungi II. The Constituents of *Trametes feei*, 6α-Hydroxypolyporenic Acid C; Aust. J. Chem., Vol. 24, pp. 609-619(1971).

Common/Systematic Name

Polyporenic acid C

Molecular Formula/Molecular Weight

$C_{31}H_{46}O_4$; MW = 482.33961

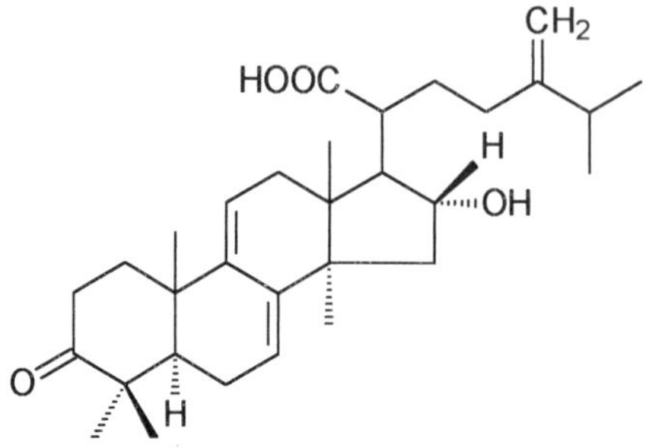

General Characteristics

Methyl ester derivative: crystals from aqueous methanol; mp., 198-199°C; $[\alpha]_D$ + 10°.

Fungal Source

Daedalea dickinsii, D. tanakae, Fomitopsis pinicola, Melanoporia cajanderi, M. juniperina, M. nigra, M. purpuracea, M. rosea, Piptoporus betulinus, Poria cocos, Trametes feel, and *T. dickinsii*.

Isolation/Purification

The fungus was extracted with ethyl ether, methylated, chromatographed on alumina, and eluted with benzene (3 different fractions) and benzene-chloroform, 1:1 (v/v). The first benzene fraction was crystallized from methanol solution, saponified, remethylated, and purified by alumina column chromatography and recrystallization from aqueous methanol to give the purified methyl ester of polyporenic acid C.

Biological Activity

Antibiotic.

Spectral Data (Methyl ester)

UV:

λ_{max} 236(ε = 14,500), 243(16,600), and 252nm (11,100).

IR:

($CHCl_3$) 3600, 3480(OH), 1725(CO_2Me), 1705(cyclohexanone), 1645, and 890cm^{-1} ($C=CH_2$); (Nujol) 3400, 1730, 1685, 1645, and 890cm^{-1}.

^{1}H NMR:
(CDCl$_3$) 5.5(2H, m, vinylic protons at C-7 and C-11); 4.77 and 4.72(2H, m, C=CH$_2$); 4.1(1H, m, CHOH); 3.7(3H, s, CO$_2$Me); and 1.9ppm (OH).

Mass Data:
LREIMS: 496(M$^+$), calcd. for $C_{32}H_{46}O_4$ 496; Found: C, 77.3; H, 9.7; calcd for $C_{32}H_{46}O_4$; C, 77.4; H, 9. 7.

Reference
J. T. Pinhey, B. J. Ralph, J. J. H. Simes, and M. Wootton; Extractives of Fungi II. The Constituents of *Trametes feei*, 6α-Hydroxypolyporenic Acid C; Aust. J. Chem., Vol. 24, pp. 609-619(1971).

Common/Systematic Name
6α-Hydroxypolyporenic acid C

Molecular Formula/Molecular Weight
$C_{31}H_{48}O_5$; MW = 500.35017

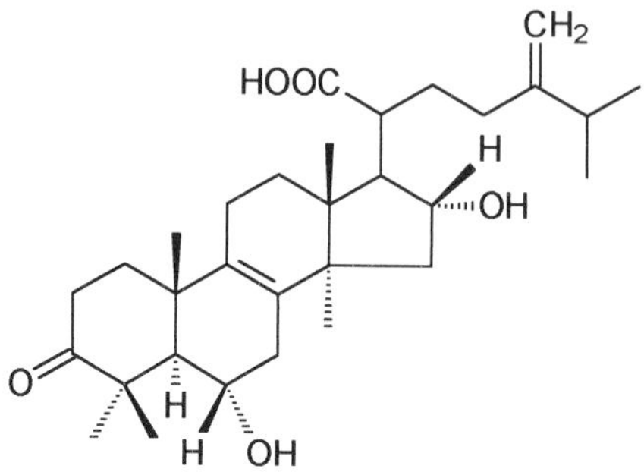

General Characteristics
Methyl ester derivative: crystals from aqueous acetone; mp., 222-223°C; $[\alpha]_D$ + 99°.

Fungal Source
Trametes feei.

Isolation/Purification
The fungus was extracted with ethyl ether, methylated, chromatographed on alumina, eluted with benzene (3 different fractions) and benzene-chloroform, 1:1 (v/v). The benzene-chloroform fraction was purified by preparative TLC (silica gel HF_{254}; chloroform-methanol, 50:1, v/v). The more non-polar component was saponified and purified by preparative TLC (silica gel HF_{254}; chloroform), recrystallized from aqueous acetone to give purified the methyl ester of 6a-hydroxypolyporenic acid C

Biological Activity
Antibiotic.

Spectral Data (Methyl ester)

UV:
λ_{max} 237(E = 12,600), 243(16,600), and 252nm (9,900).

IR:
($CHCl_3$) 3600, 3440(OH), 1730(CO_2Me), 1705(cyclohexanone), 1645, and 890cm^{-1} ($C{=}CH_2$); (Nujol) 3540, 3450, 1735, 1675, 1645, and 890cm^{-1}.

^{1}H NMR:

($CDCl_3$) 5.5(2H, m, vinylic protons at C-7 and C-11); 4.78 and 4.70(2H, m, $C=CH_2$); 4.4(1H, d after deuteration, *J*=l0Hz, C-6OH); 4.1(1H, m, C-16HOH); 3.7(3H, s, CO_2Me); and 1.3-0.63ppm (7 X Me).

Mass Data:
HREIMS: 512.3509*m/e* (M^+); calcd. for $C_{32}H_{48}O_5$ 512.3502; found: C, 74.7; H, 9.5; calcd for $C_{32}H_{48}O_5$, C, 75.0%; H, 9.4%.

Reference

J. T. Pinhey, B. J. Ralph, J. J. H. Simes, and M. Wootton; Extractives of Fungi II. The Constituents of *Trametes feei*, 6α-Hydroxypolyporenic Acid C; Aust. J. Chem., Vol. 24, pp. 609-619(1971).

Common/Systematic Name

Polyporenic acid D

3α-Hydroxy-4,4,14-trimethyl-5α-ergosta-8,24(28)-dien-26-oic acid

Molecular Formula/Molecular Weight

$C_{31}H_{50}O_3$; MW = 470.37600

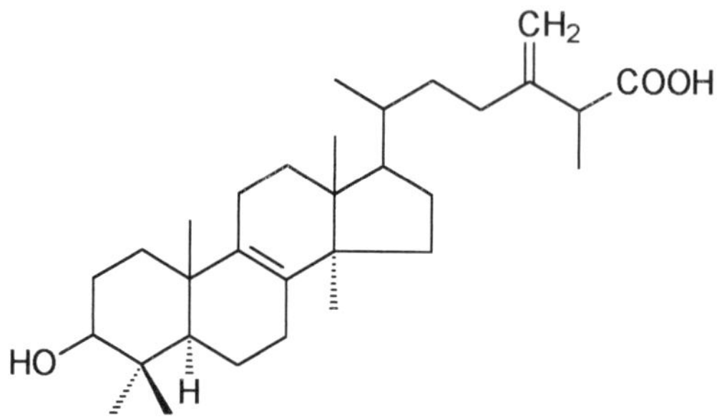

General Characteristics

A white crystalline solid from methanol; mp., 268-270°C; $[\alpha]_D$ + 40° (in pyridine); acetate, a white solid; mp., 235°C; $[\alpha]_D$ + 23.2°.

Fungal Source

Polyporus officinalis.

Isolation/Purification

Soxhlet extraction of the powdered fungus with n-hexane gave an extract which deposited a white crystalline material on chilling; recrystallization from methanol gave a white crystalline solid.

Spectral Data

IR:

3525(OH), 2700-2500, 1710, 1698 (C=O), 1640(-C=CH_2), and 898cm^{-1}.

^{1}H NMR:

(C_5D_5N) 0.90-1.02(6 methyls); 1.23(d, *J*=2Hz, H-27); 4.81(2H, d, H-28); and 3.27ppm (1H, t, H-3).

Mass Spectrum:

LREIMS: 470(M^+), 468, 455, 437, 409, 354, 279, 273, 247, 234, 220, 219, 211, 210, 203, 202, 201, 186, 185, 144, 116 and 115*m/e*.

Reference

R. K. Thappa. S. G. Agarwal, K. L. Dhar, and C. K. Atal; A New Triterpenic Acid from the Wood Rotting Fungi; Phytochemistry, Vol. 20, pp. 1746-1747(1981).

Common/Systematic Name

Dehydroeburicoic acid

Molecular Formula/Molecular Weight

$C_{31}H_{48}O_3$; MW= 468.36035

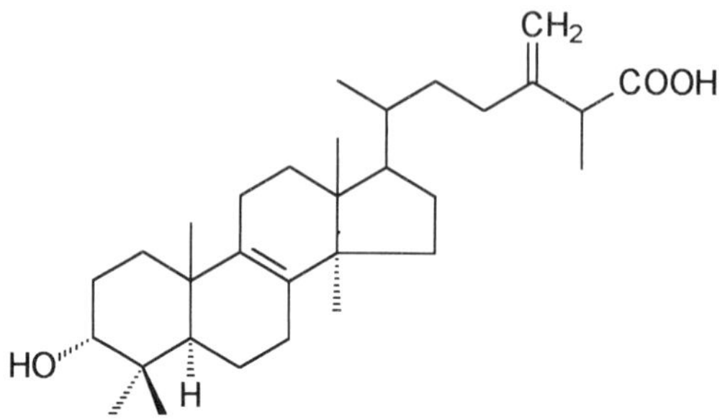

General Characteristics

Crystals; mp., 240-242°C; $[\alpha]_D^{21}$ + 28° (*c*=0.8). Methyl ester; 159-160°C; $[\alpha]_D^{21}$ + 29.5° (*c*=2.5).

Fungal Source

Daedalea trabea and *Fomes officinalis*.

Spectral Data

Mass Spectrum:

LREIMS: 468*m/e* (M^+).

References

L. H. Briggs, R. C. Cambie, I. C. Dean, S. H. Dromgoole, B. J. Fergus, W. B Ingram, K. G. Lewis, C. W. Small, R. Thomas, and D. A. Walker; 10. Metabolites of Some Fungi; New Zealand Journal of Science, Vol. 18, pp. 565-576(1975).

K. E. Schulte, G. Rocker, and H. Fachmann; Zur Struktur Der "Agaricolsaure", Tet. Lett. pp. 4823-4825 (1967).

Common/Systematic Name
Dehydrotumulosic acid

Molecular Formula/Molecular Weight
$C_{31}H_{48}O_4$; MW = 484.35526

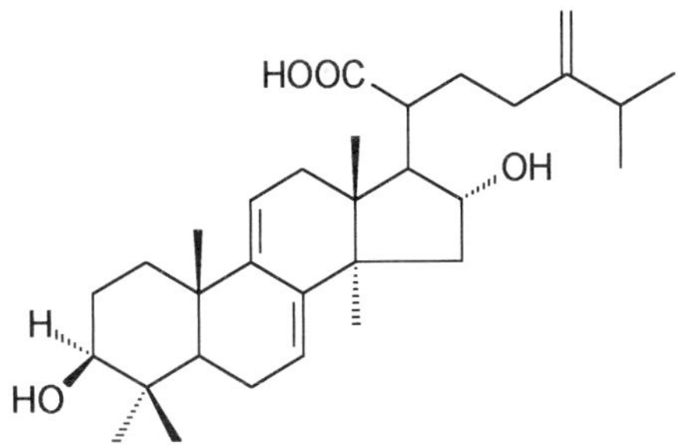

General Characteristics
Crystals from methanol; mp., 135-137°C.

Fungal Source
Daedalea dickinsii, *Melanoporia juniperina*, *M. rosea*, *Trametes dickinsii*, *T. feei*, and *T. lilacino gilva*.

Spectral Data (All spectral data on methylate)

UV:
λ_{max}^{EtOH} 236(log ε = 3.91), 244(3.92), and 252nm (3.73).

IR:
(KBr) 3440, 1715, 1648, and 890cm^{-1}.

^{1}H NMR:
($CDCl_3$) 3.66(3H, s); 3.22(1H, m); 4.06(1H, m); 4.65(1H, m); and ca 5.30ppm (1H, m).

Reference
A. Yokoyama and S. Natori; Triterpenoids of Lanostane Group from Fruit Bodies of Nine *Basidiomycetous* Species; Chem. Pharm. Bull., Vol. 22, pp. 877-883(1974).

Common/Systematic Name

Eburicol

24-Methylene-24,25-dihydrolanosterol

Molecular Formula/Molecular Weight

$C_{31}H_{52}O$; MW = 440.40182

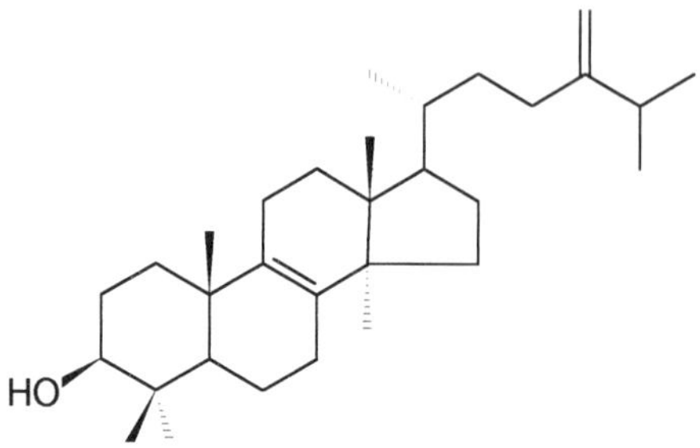

General Characteristics

A clear yellow color with Liebermann-Burchard reagent, indicating the presence of a 14α-methyl group; crystals; mp., 150-151 °C.

Fungal Source

Phycomyces blakesleeanus, Candida albicans, C. utilis, Neurospora crassa, and *Agaricus campestris.*

Isolation/Purification

The fungi were extracted twice with ethyl ether-ethanol (1:1), filtered, diluted with water and extracted with ethyl ether-petrol and saponified. The crude extract was chromatographed on alumina (Brockmann Grade III). The 4,4-dimethyl sterol fraction was acetylated, chromatographed by preparative TLC on 10% $AgNO_3$-silica plates developed with 40% benzene-hexane to give purified eburicol.

Spectral Data

UV:

End absorption.

IR:

(KBr) 887 and 1640cm^{-1}.

^{1}H NMR:

($CDCl_3$) (acetate derivative) Signals at 9.31 (C-18 protons); 9.12(4α, 4ß and 14α-methyl protons; 9.02(C-19 protons); 9.00(C-26 protons); 8.95(C-27 protons); 7.98(3ß-*O*-acetyl protons); 5.35(C-28 protons); and 5.50τ (3α- proton).

Mass Spectrum:

(acetate derivative) 482(M^+), 467(M^+ - Me), 422(M^+ - AcOH), and 407*m/e* (M^+ - [AcOH + Me]).

Reference

G. Goulston, E. Ian Mercer, and L. J. Goad; The Identification of 24-Methylene-24,25-dihydrolanosterol and Other Possible Ergosterol Precursors in *Phycomyces blakesleeanus* and *Agaricus campestris*; Phytochemistry, Vol. 14, pp. 457-462(1975).

Common/Systematic Name

Pisolactone

(22*S*)24-Methyllanost-8-en-22,28-epoxy-3β-ol-28-one

Molecular Formula/Molecular Weight

$C_{31}H_{50}O_3$; MW = 470.37600

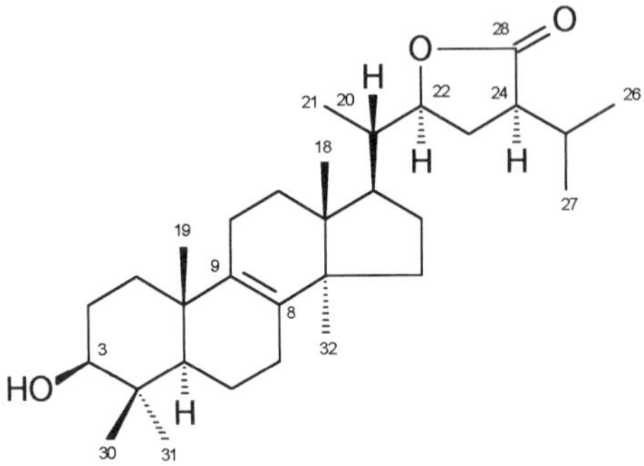

General Characteristics

Colorless prisms from methanol; mp., 279-280°C; $[\alpha]_D^{21}$ + 60° (*c*=1.0, in $CHCl_3$).

Monoacetate, crystals from methanol; mp., 280°C; $[\alpha]_D^{28}$ + 53°(*c*=1.0, in $CHCl_3$).

Fungal Source

Spores of *Pisolithus tinctorius*.

Isolation/Purification

Spores were extracted with ethyl ether and purified by chromatography.

Also see Heupel, R.C.; Isolation and Primary Characterization of Sterols, In Analysis of Sterols and Other Biologically Significant Steroids; W. D. Nes and E. J. Parish (eds.); Academic Press Inc., New York, NY, 341 pp, (1989).

Spectral Data

IR:

(KBr) 3410, and 1745cm^{-1}; acetate derivative: 1746, and 1725cm^{-1}.

^{1}H NMR:

Four singlets; 0.72(3H), 0.97(6H), 0.81(3H), and 0.88ppm (3H) were attributed to methyl groups at C-18, C-19, C-30, C-31, and C-32, respectively.

References

A. M. Lobo, M. de Abreu, S. Prabhakar, and L. S. Godinho; Pisolactone, A Novel Triterpenoid Isolated From The Fungus *Pisolithus tinctorius*; Tetrahedron Letters, Vol.

24, pp. 2205-2208(1983).

W. D. Nes and E. J. Parish (eds.); Analysis of Sterols and Other Biologically Significant Steroids; Academic Press Inc., New York, NY, 341 pp.(1989).

W. B. Turner and D. C. Aldridge; Fungal Metabolites II; Academic Press, Inc., New York,; NY, p. 319 (1983).

Common/Systematic Name
3-Oxopisolactone

Molecular Formula/Molecular Weight
$C_{31}H_{48}O_3$; MW = 468.36035

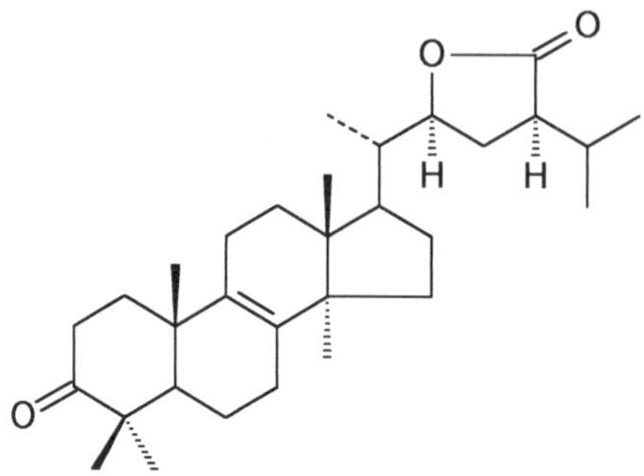

General Characteristics
White crystalline solid; mp., 248-250°C; $[\alpha]_D$ +79° (*c*=1.4, in $CHCl_3$).

Fungal Source
Fruiting bodies of *Pisolithus tinctorius*.

Isolation/Purification
Powdered, dried fruiting bodies were extracted by percolation with Et_2O followed by MeOH. Removal of the Et_2O and of the MeOH gave brown oils which were chromatographed separately on a column of silica gel. The Et_2O extract was purified on a chromatographic column eluted with hexane-EtOAc (6:4, v/v). Three fractions were collected (F-1, F-2 and F-3) by order of the compounds eluted and analyzed by TLC (silica gel GF_{254}, hexane-EtOAc, 7:3, v/v). Purification of Fraction F-1 by flash column chromatography (silica gel, hexane-EtOAc, 9:1, v/v) gave a white crystalline solid of 3-oxopisolactone.

Spectral Data

IR:
(KBr) 1705, and 1740cm^{-1} (C=O).

^{1}H NMR:
($CDCl_3$) 0.71(3H, s); 0.87(3H, s); 0.89(3H, d, *J*=7Hz); 0.91(3H, d, *J*=7Hz); 0.98(3H, d, *J*=7Hz); 1.02(3H, s); 1.07(3H, s); 1.08(3H, s); 2.47(1H, m); 2.50(2H, m); and 4.35ppm (1H, m).

Mass Spectrum:

LREIMS: 468*m/e* $[M^+, C_{31}H_{48}O_3]^+$ (20%), 453 $[M-Me]^+$ (35), 435$[M-Me-H_2O]^+$ (10), 313$[M-C_9H_{15}O_2]^+$ (5), 298$[M-C_9H_{15}O_2-Me]^+$ (15).

<u>Reference</u>

A. M. Lobo, M. de Abreu, S. Prabhakar, R. Jones, H. S. Rzepa, and D. J. Williams; Trıterpenoids of The Fungus *Pisolithus tinctorius*; Phytochem., Vol. 27, pp. 3569-3574(1988).

Common/Systematic Name

(22*S*,24*R*)-24-Methyllanost-8-en-22,28-epoxy-3β,28α-diol

Molecular Formula/Molecular Weight

$C_{31}H_{52}O_3$, MW = 472.39165

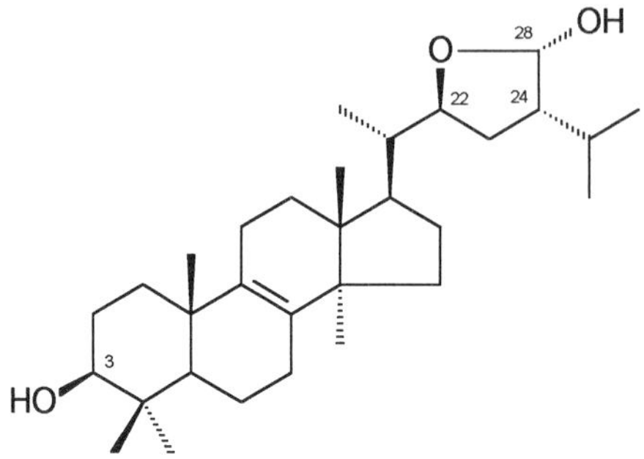

General Characteristics

Colorless amorphous powder.

Fungal Source

Fruiting bodies of *Pisolithus tinctorius* (proposed to be non-enzymatically formed from pisolactone through its enolic form at position 28).

Isolation/Purification

Dried fruit bodies were extracted with methanol, which was divided into benzene soluble and insoluble fractions. The benzene soluble fraction was chromatographed on a silica gel flash column to give five fractions. Fraction 4 was further fractionated with silica gel flash column chromatography using benzene-ethyl acetate (8:1, v/v) followed by a medium pressure liquid chromatography (silica gel) eluting with benzene-ethyl acetate (7:1, v/v) to afford 24-methyllanosta-8,24(28)-diene-3β,22ζ-diol and pisolactone. Fraction five was fractionated using silica gel flash column chromatography and medium pressure liquid chromatography using ODS with methanol as eluting solvent to give pisolactone, (22*S*,24*R*)-24-methyllanost-8-en-22,28-epoxy-3β,28α-diol, (22*S*,24*R*)-24-Methyllanost-8-en-22,28-epoxy-3β,28β-diol, and 24-methyllanosta-8,24(28)-diene-3β,22ζ-diol.

Biological Activity

(22*S*,24*R*)-24-Methyllanost-8-en-22,28-epoxy-3β,28α-diol suppressed proliferation of mouse spleen lymphocytes stimulated with the mitogens, concanavalin A and lipopolysaccharide. ID_{50} calculated to be 2.3μg/ml against the Con A-induced proliferation and 1.9μg/ml against the LPS-induced proliferation of mouse spleen lymphocytes.

Spectral Data

^{1}H NMR:
1.23(m, H-1); 1.74(m); 1.55(m, H-2); 1.68(m); 3.24(dd, *J*=11.6, 4.5Hz, H-3, H-3); 1.05(dd, *J*=12.5, 2.0Hz, H-5); 1.52(m, H-6); 1.68(m); 2.04(2H, m, H-7); 2.02(2H, m, H-11); 1.21(m, H-12); 1.75(m); 130(m, H-15), 1.80(m); 138(m, H-16); 2.03(m); 1.82(m, H-17); 0.70(3H, s, H-18); 0.98(3H, s, H-19); 1.46(m, H-20); 0.91(3H, d, *J*=6.4Hz, H-21); 4.28(ddd, *J*=10.9, 5.1, 2.0Hz, H-22); 1.42(m, H-23); 1.84(m); 1.78(m, H-24); 1.60(m, H-25); 0.91(3H, d, *J*= 6.4Hz, H-26); 0.99(3H, d, *J*=6.8Hz, H-27); 5.17(d, *J*=3.4Hz, H-28); 0.81(3H, s, H-29); 1.00(3H, s, H-30); and 0.90ppm (3H, s, H-31).

^{13}C NMR:
C-1, 35.58(t); C-2, 27.85(t); C-3, 78.99(d); C-4, 38.89(s); C-5, 50.40(d); C-6, 18.27(t); C-7, 26.49(t); C-8, 34.35(s); C-9, 34.47(s); C-10, 37.02(s); C-11, 21.02(t); C-12, 30.92(t); C-13, 44.63(s); C-14, 49.72(s); C-15, 30.97(t); C-16, 27.87(t); C-17, 47.88(d); C-18, 15.55(q); C-19, 19.16(q); C-20, 39.05(d); C-21, 12.98(q); C-22, 80.52(d); C-23, 33.40(t); C-24, 54.63 (d); C-25, 30.58(d); C-26, 20.76(q); C-27, 21.41(q); C-28, 102.21(d); C-29, 15.43(q); C-30, 27.97(q); and C-31, 24.40ppm (q).

Mass Spectrum:
HRFAB-MS: 472.3918*m/e* (M^+); calcd for $C_{31}H_{52}O_3$, 472.3916.

Reference

H. Fujimoto, M. Nakayama, Y. Nakayama, and M. Yamazaki; Isolation and Characterization of Immunosuppressive Components of Three Mushrooms, *Pisolithus tinctorius*, *Microporus flabelliformis* and *Lenzites betulina*; Chem. Pharm. Bull., Vol. 42, pp. 694-697(1994).

Common/Systematic Name

(22*S*,24*S*)-24-Methyllanost-8-en-22,28-epoxy-3β,28β-diol

Molecular Formula/Molecular Weight

$C_{31}H_{52}O_3$; MW = 472.39165

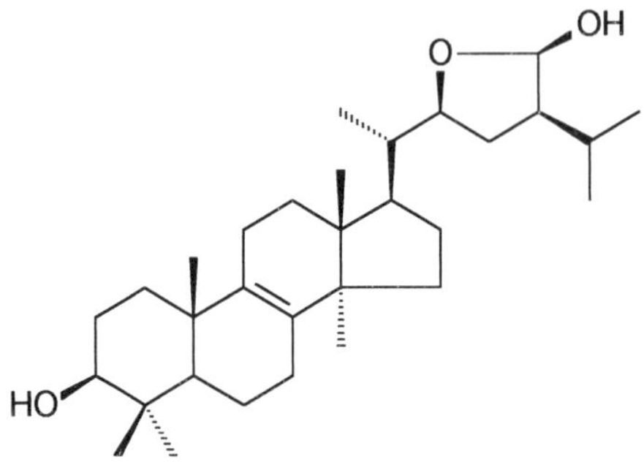

General Characteristics

Colorless amorphous powder.

Fungal Source

Fruiting bodies of *Pisolithus tinctorius* (proposed to be non-enzymatically formed from pisolactone through its enolic form at position 28).

Isolation/Purification

See (22*S*,24*R*)-24-Methyllanost-8-en-22,28-epoxy-3β,28α-diol isolation.

Biological Activity

Suppressed proliferation of mouse spleen lymphocytes stimulated with the mitogens, concanavalin A and lipopolysaccharide. ID_{50} calculated to be 2.3μg/ml against the Con A-induced proliferation and 1.9μg/ml against the LPS-induced proliferation of mouse spleen lymphocytes.

Spectral Data

^{1}H NMR:

H-1, 1.23(m, H-1); 1.74(m); 1.55(m, H-2); 1.68(m); 3.24(dd, *J*=11.6, 4.5Hz, H-3); 1.05(dd, *J*=12.5, 2.0Hz, H-5); 1.52(m, H-6); 1.68(m); 1.04(21, m, H-7); 2.02(21, m, H-11); 1.21(m, H-12); 1.75(m); 1.70(m, H-15); 1.80(m); 1.54(m, H-16); 2.01(m); 1.86(m, H-17); 0.70(31, s,); H-18); 0.98(31, s, H-19); 1.57(m, H-20); 0.93(31, d, *J*=6.4Hz, H-21); 4.13(ddd, *J*=10.9, 5.1, 2.0Hz, H-22); 1.22(m, H-23); 1.76(m); 1.72(m, H-24); 1.67(m, H-25); 0.99(31, d, *J*=6.8Hz, H-26); 0.95(d, *J*=6.8Hz, H-27); 5.27(d, *J*=3.9Hz, H-28); 0.81(s, H-29); 1.00(s, H-30); and 0.90ppm (3H, s, H-31).

^{13}C NMR:

C-1, 35.58(t); C-2, 27.85(t); C-3, 78.99(d); C-4, 38.89(s); C-5, 50.40(d); C-6, 18.27(t); C-7, 26.49(t); C-8, 34.35(s); C-9, 34.47(s); C-10, 37.02(s); C-11, 21.02(t); C-12, 30.93(t); C-13, 44.61(s); C-14, 49.77(s); C-15, 30.89(t); C-16, 27.92(t); C-17, 48.00(d); C-18, 15.58(q); C-19, 19.16(q); C-20, 39.35(d); C-21, 13.01(q); C-22, 83.01(d); C-23, 31.29(t); C-24, 52.94(d); C-25, 27.81(d); C-26, 21.48(q); C-27, 22.06(q); C-28, 97.77(d); C-29, 15.43(q); C-30, 27.97(q); and C-31, 24.36ppm (q).

Mass Spectrum:

HRFAB-MS: 472.3918*m/e* (M^+); calcd for $C_{31}H_{52}O_3$, 472.3916.

Reference

H. Fujimoto, M. Nakayama, Y. Nakayama, and M. Yamazaki; Isolation and Characterization of Immunosuppressive Components of Three Mushrooms, *Pisolithus tinctorius, Microporus flabelliformis* and *Lenzites betulina*; Chem. Pharm. Bull., Vol. 42, pp. 694-697(1994).

Common/Systematic Name

24-Methylenelanosterol

24-Methylene-5α-lanost-8-en-3β-ol

Molecular Formula/Molecular Weight

$C_{31}H_{52}O$; MW = 440.40182

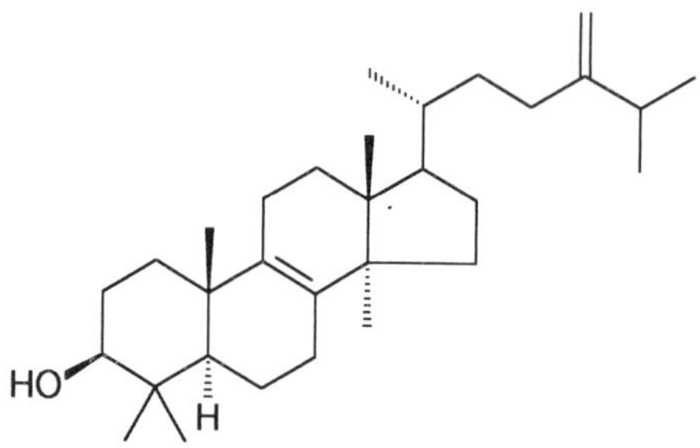

Isolation/Purification

The unsaponifiable material from *Pichia* sp. was separated into five fractions by preparative TLC (Wakogel B-10 coated plates developed with hexane-ether, 7:3, v/v). Preparative argentation (10% silver nitrate coated onto Wakogel B-10 plates, developed with hexane-benzene, 3:2, v/v) TLC of the acetylated 4,4-dimethyl fraction produced lanosterol and 24-methylenelanosterol.

Fungal Source

Pichia sp. (an alkane-utilizing yeast).

Spectral Data

Mass Spectrum:

LREIMS: 440(M^+, 6%), 425(M^+ - CH_3, 17), 407(M^+ - CH_3 - H_2O, 11), 356(M^+ C_6H_{12}, 2), 341(M^+ - C_6H_{12} - CH_3, 4), 323(M^+ - C_6H_{12} - CH_3 - H_2O, 6), 315(M^+ - C_9H_{17}, 3), 297(M^+ - C_9H_{17} - H_2O, 3), 273(M^+ - C_9H_{17} - C_3H_6, 5), 259[M^+ - C_9H_{17} - C3H6- CH_2 (-14α-methyl - H), 11], and 241*m/e* (M^+ - C_9H_{17} - C3H6 - CH_2 - H_2O, 11); weak ions at 438(M^+ - H) and 423*m/e* (M^+ - CH_3) indicative of a Δ^8 bond.

GC Data

OV-17 coated onto Gas Chrom-Z; column temperature 256°C; detector 280°C; carrier gas N_2 at 50ml/min. Relative retention time relative to β-sitosterol for 24-methylene-lanosterol was1.18 and the acetate derivative was 1.43.

Reference

T. M. Jeong, T. Itoh, T. Tamura, and T. Matsumoto; Sterols from a Species of *Pichia*, A *n*-Alkane-utilizing Yeast; Steroids, Vol. 25, pp. 741-751(1975).

Common/Systematic Name

Lanosta-8,24(28)-diene-3β,22-diol; (22*S*)-Lanosta-8,24(28)-diene-3β,22-diol

Molecular Formula/Molecular Weight

$C_{31}H_{52}O_2$; MW = 456.39673

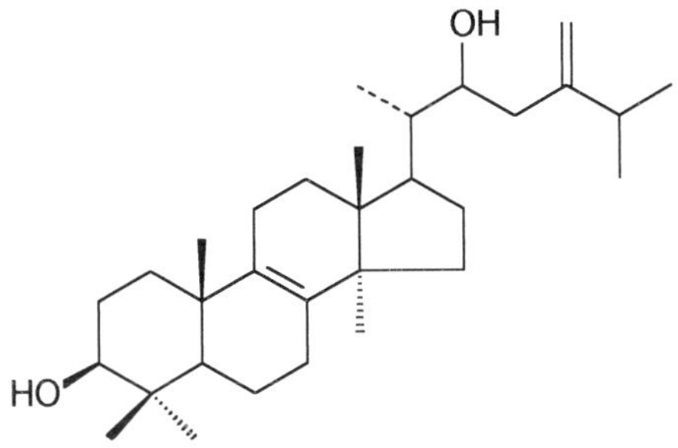

General Characteristics

Crystals from methanol; mp., 161-165°C; $[\alpha]_D$ + 29° (*c*=0.16, in $CDCl_3$).

Fungal Source

Fruiting bodies of *Pisolithus tinctorius*.

Isolation/Purification

Powdered, dried fruiting bodies were extracted by percolation with Et_2O followed by MeOH. Removal of the Et_2O and MeOH gave brown oils which were chromatographed separately on a column of silica gel. The Et_2O extract was purified on a chromatographic column eluted with hexane-EtOAc (6:4,v/v). Three fractions were collected (F-1, F-2 and F-3 by order of the compounds eluted) and analyzed by TLC (silica gel GF_{254}, hexane-EtOAC, 7:3, v/v). A portion of fraction two was acetylated and purified by silica gel column chromatography (hexane-ethyl acetate, 9:1, v/v) to yield the diacetate of lanosta-8,24(28)diene-3β,22-diol, which was recovered after saponification and recrystallization from methanol.

Spectral Data

IR:

(KBr) 3450 and 890cm^{-1}.

^{1}H NMR:

($CDCl_3$) 0.70(3H,s); 0.81(3H, s); 0.91(3H, s); 0.92(3H, d, *J*=6 Hz); 0.98(3H, s); 1.01(3H, d, *J*=6 Hz); 1.04(3H, s); 1.06(3H, d, *J*=6Hz); 3.23(1H, m); 3.80(1H, m); 4.78(1H, s): and 4.88ppm (1H, s).

^{13}C NMR:
($CDCl_3$) 153.9; 135.0; 134.8; 128.6; 78.3; and 71.0ppm.

Mass Spectrum:
LREIMS: 456[M^+, $C_{31}H_{52}O_2$]+ (26%), 441$[M\text{-}Me]^+$ (7), 438 $[Me\text{-}H_2O]^+$ (1), 423$[M\text{-}Me\text{-}H_2O]^+$ (9), 372$[M\text{-}C_6H_{12}]^+$ (16), 357$[M\text{-}C_6H_{12}\text{-}Me]^+$ (100), 339$[M\text{-}C_6H_{12}\text{-}Me\text{-}H_2O]^+$ (45), 314$[M\text{-}C_9H_{18}O]^+$ (10), and 281*m/e* $[M\text{-}C_9H_{18}O\text{-}H_2O]^+$ (18).

Reference

A. M. Lobo, M. de Abreu, S. Prabhakar, R. Jones, H. S. Rzepa, and D. J. Williams; Triterpenoids of The Fungus *Pisolithus tinctorius*; Phytochem., Vol. 27, pp. 3569-3574(1988).

6

C_{32} Sterols

Echinodol
3-Epiechinodol
Echinodone
(22*R*) 3α,25-Dihydroxylanosta-8,23-diene-22-acetate
12β-Acetoxy-4,4-dimethyl-24-methylene-5α-cholesta-8,14-diene-3β,11α-diol

Common/Systematic Name

Echinodol

Molecular Formula/Molecular Weight

$C_{32}H_{50}O_4$; MW = 498.37091

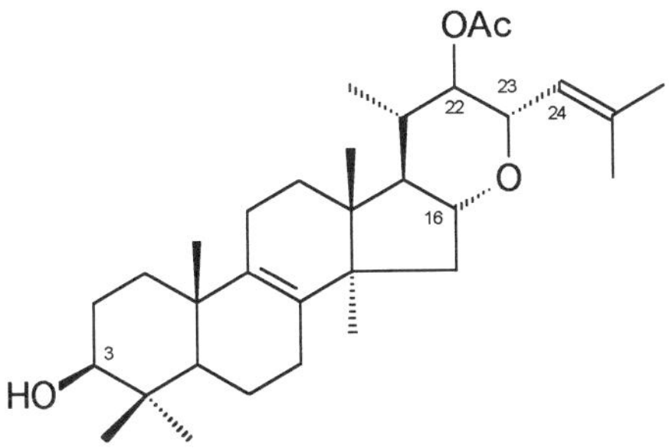

General Characteristics

Crystals; mp., 236-238°C; $[\alpha]_D^{25}$ +48°; diacetate; mp., 270-271°C; $[\alpha]_D^{25}$ +51°.

Fungal Source

Echinodontium tinctorium, a wood-rotting fungus commonly found on conifers, particularly in the Northwest.

Isolation/Purification

The dried fungus was extracted with petroleum ether, followed by column chromatography on alumina and repeated recrystallizations.

Biological Activity

The fungus was used as a bactericidal agent by the Wasco Indians, and tumor-inhibiting activity of crude material was indicated in preliminary tests.

Spectral Data

^{1}H NMR:

($CDCl_3$) H-3, 3.20(m); H-16, 3.74(m); H-22, 4.51(t, J 9Hz); H-23, 4.10(t, *J*=9Hz); H-24, 5.16(allylically broadened doublet); 18-CH_3, 0.69; 19-CH_3, 1.00; 21-CH_3, 0.90(d, *J*=7.0Hz); 26, 27-CH_3, 1.69(d, *J*=1.5Hz); 30-CH_3, 0.82; 31-CH_3, 1.00; and 32CH_3, 1.05ppm.

Reference

F. H. Bond, D. S. Fullerton, L. A. Sciuchetti, and P. Catalfomo; The Isolation and Structure of Echinodol, a Triterpene Acetate; J. Am. Chem. Soc., Vol 88, pp. 3882-3883 (1966).

Common/Systematic Name

3-Epiechinodol

Molecular Formula/Molecular Weight

$C_{32}H_{50}O_4$; MW = 498.37091

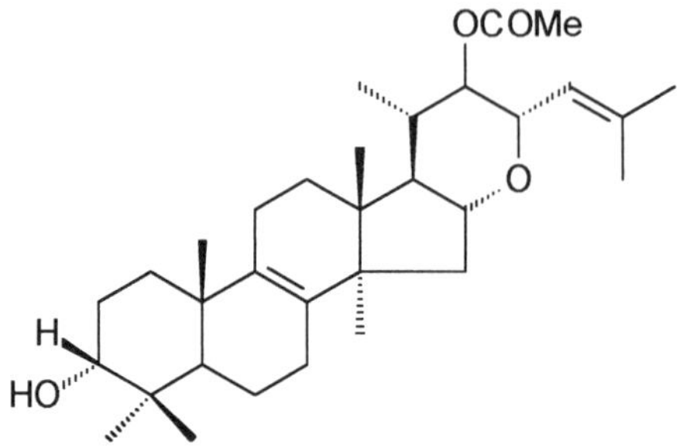

General Characteristics

Colorless needles from methanol; mp., 230-231°C; $[\alpha]_D^{19}$ + 45° (*c*=1.0, in $CHCl_3$).

Fungal Source

Fruit bodies of *Echinodontium tsugicola*. The fungus is indigenous to Japan and saprophytic to *Tsuga diversifolia*.

Isolation/Purification

Fruit bodies of *Echinodontium tsugicola*, collected at Marunuma, Gumma Prefecture, Japan, in October 1969, were air-dried, crushed, and extracted twice with ether at room temperature. The combined extracts were treated with hot benzene to remove an orange-yellow amorphous powder. The benzene extract was chromatographed on a column of alumina (Woelm neutral) and eluted successively with (1) benzene, (2) benzene-ether (99:1, v/v), (3) benzene-ether (98.5:1.5, v/v), (4) ethyl ether, and (5) MeOH. Fraction (3) was rechromatographed under similar conditions and each fraction was examined by TLC and GLC. The fractions eluted with benzene that contained deacetoxyechinodone and echinodone were combined and further separated by preparative thin-layer chromatography (silica gel HF_{254}). Echinodone and deacetoxyechinodone were recrystallized from MeOH to give colorless needles. The next fraction containing a diterpene ester was recrystallized from MeOH to colorless prisms. The fractions eluted with benzene-ether (98:2, v/v) containing deacetoxy-3-echinodol and deacetoxyechinodol were combined and further purified by preparative thin-layer chromatography. Deacetoxy-3-epiechinodol was recrystallized from MeOH, while deacetoxyechinodol was obtained as colorless needles from MeOH. The benzene-ether (95:5, v/v) fractions were collected and again separated by preparative thin-layer chromatography into three components 3-epiechinodol, deacetylechinodol and deacetyl-3-epiechinodol.

Spectral Data

IR:
(KBr) 3480, 1750, and 1228cm^{-1}.

^{1}H NMR:
($CDCl_3$) 3.42(m, broadened triplet of about 4Hz, H-3); 3.79(m, H-16); 4.52(t, *J*=9.0Hz, H-22); 4.03(t, *J*=9.0Hz, H-23); 5.12(bd, *J*=9.0Hz, H-24); 0.70 (3H, H-18); 0.98 (3H, H-19); 0.91(3H, d, *J*=6.0Hz, H-21); 1.70(3H each, bs, H-26, -27); 0.98 (3H, H-30); 0.87; 1.05 (H-32); and 1.98ppm (3H, $-OCOCH_3$.)

Mass Data:
Anal calcd for $C_{32}H_{50}O_4$: C, 77.1; H, 10.1; found: C, 76.8; H, 10.0%.

Reference

A. Kanematsu and S. Natori; Triterpenoids of *Echinodontium tsugicola*; Chem. Pharm. Bull., Vol. 20, pp. 1993-1999(1972).

Common/Systematic Name

Echinodone

Molecular Formula/Molecular Weight

$C_{32}H_{48}O_4$ = 496.35526

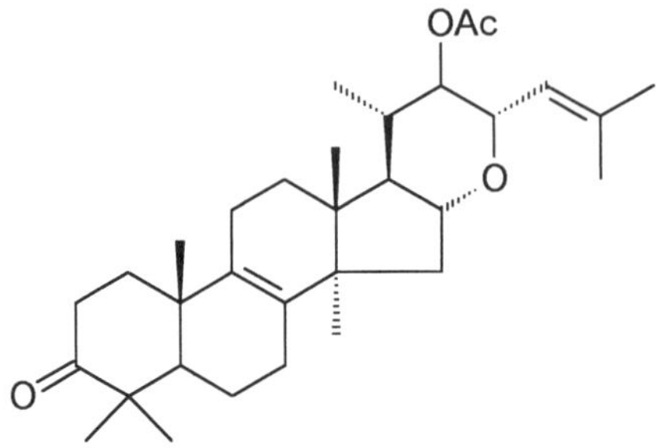

General Characteristics

Colorless needles from methanol; mp., 225-227°C (227-229°C); $[\alpha]_D^{24}$ + 67° (*c*=1.0, in $CHCl_3$).

Fungal Source

Fruit bodies of *Echinodontium tsugicola*. The fungus is indigenous to Japan and saprophytic to *Tsuga diversifolia*.

Isolation/Purification

See 3-epiechinodol for isolation.

Spectral Data

UV:

λ_{max}^{EtOH} 278nm (log ε = 1.8).

IR:

(KBr) 1748, 1705, and 1225cm^{-1}.

^{1}H NMR:

($CDCl_3$ 3.20(m, H-3); 3.72(m, H-16); 4.52(t, *J*=9.0Hz, H-22); 4.03(t, *J*=9.0Hz, H-23); 5.12(bd, *J*=9.0Hz, H-24); 0.68 (CH_3-18); 0.98 (CH_3-19); 0.90 (d, *J*=6.0Hz, CH_3-21); 1.69(6H, bs, CH_3-26, -27); 0.80 (6H, CH_3-30, -31); 0.98; 1.03(CH_3-32); and 1.98ppm (3H, $OCOCH_3$, H-22).

Mass Spectrum:

HREIMS: 496.346*m/e* (M), calcd for $C_{32}H_{48}O_4$ 496.352; LREIMS: 71(10%), 297(55), 313(8), 421(16), 436(100), 481(2), and 496*m/e* (M^+, 3).

Reference

A. Kanematsu and S. Natori; Triterpenoids of *Echinodontium tsugicola*; Chem. Pharm. Bull., Vol. 20, pp. 1993-1999(1972).

Common/Systematic Name

(22*R*)25-Hydroxylanosta-8,23-dien-3α-ol-22-monoacetate;
(22*R*)3α,25-Dihydroxylanosta-8,23-diene-22-acetate

Molecular Formula/Molecular Weight

$C_{32}H_{52}O_4$; MW = 500.38656

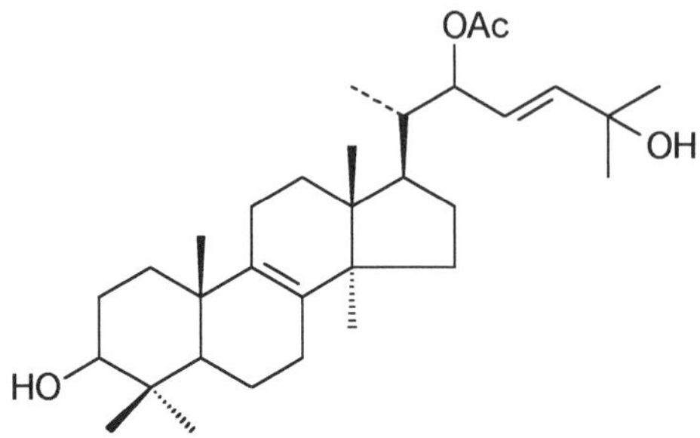

General Characteristics

Crystals from methanol; mp., 187-190°C.

Fungal Source

Fruiting bodies of *Pisolithus tinctorius*.

Isolation/Purification

Dried fruit bodies were extracted with methanol, which was divided into benzene soluble and insoluble fractions. The benzene soluble fraction was chromatographed on a silica gel flash column using hexane-ethyl acetate (7:3, v/v) to yield purified (22*R*)-3α,25-dihydroxylanosta-8,23-diene-22-acetate.

Spectral Data

IR:

(KBr) 3370(OH) and 1720cm^{-1} (C=O).

^{1}H NMR:

($CDCl_3$) 0.70(3H, s); 0.82(3H, s); 0.87(3H, s); 0.94(6H, d, J=6Hz); 0.97(3H, s); 0.98(3H, s); 1.32(6H, s); 2.05(3H, s); 3.42(1H, $J_{1,2}$=7Hz); 5.30(1H, dd, J_1=7.5Hz, J_2=3.7Hz); 5.65(1H, dd, J_1=14.7Hz, J_2=7.5Hz); and 5.86ppm (1H, d, J=14.7Hz).

Mass Spectrum:

500[M^+, $C_{32}H_{52}O_4$]$^+$ (10%), 485[M-Me]$^+$, 467[M-H_2O-Me]$^+$ (2), 440[M-AcOH]$^+$ (26), 425[M-AcOH-Me]$^+$ (30), 407[M-AcOH-Me-H_2O]$^+$ (40), 389[M-AcOH-Me-2H_2O]$^+$ (4), 325[M-$C_8H_{13}O_3$-H_2O]$^+$ (14), and 314*m/e* [M-$C_8H_{13}O_3$]$^+$ (8).

Reference
A. M. Lobo, M. de Abreu, S. Prabhakar, R. Jones, H. S. Rzepa, and D. J. Williams; Triterpenoids of The Fungus *Pisolithus tinctorius*; Phytochem., Vol. 27, pp. 3569-3574 (1988).

Common/Systematic Name

12β-Acetoxy-4,4-dimethyl-24-methylene-5α-cholesta-8,14-diene-3β,11α-diol

Molecular Formula/Molecular Weight

$C_{32}H_{50}O_4$; MW = 498.37091

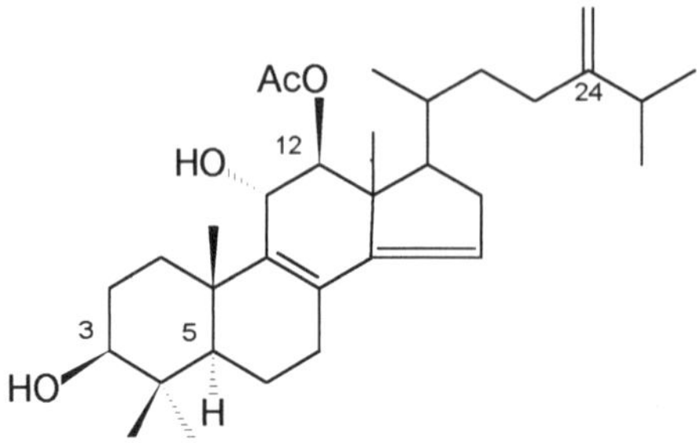

General Characteristics

Crystallized from acetone-hexane as needles; mp., 180-181°C; $[\alpha]^{25}$ + 32.7° (*c*=0.44, in $CHCl_3$).

Fungal Source

Fusarium sporotrichioides 921

Isolation/Purification

Fungal cultures were extracted with ethyl alcohol, evaporated to dryness, diluted with water, and partitioned with light petroleum, diethyl ether, ethyl acetate, chloroform, and methylene chloride. The light petroleum and diethyl ether fractions were combined and treated with diazomethane. The mixture was concentrated *in vacuo* and the residue diluted with light petroleum ether. White needles of T-2 toxin formed and were removed by filtration. 3β-Acetoxy-4,4-dimethyl-24-methylene-5α-cholesta-8,14-diene-3β,11α-diol was isolated from the mother liquor by chromatography on a Florisil column eluted with light petroleum and light petroleum-ethyl acetate followed by crystallization from acetone-hexane.

Spectral Data

IR:

(KBr) 3450, 1712, 1260, 1060, and 880cm^{-1}.

^{1}H NMR:

($CDCl_3$) 2.07(s, 12-OAc); 3.33(m, H-3); 4.3(s, H-11); 4.75(2H, d, *J*=5Hz, H-28); 5.03(s, H-12); and 5.67ppm (s, H-15).

^{13}C NMR:

(CDCl$_3$) C-1, 34.36; C-2, 27.38; C-3, 79.16; C-4, 39.04; C-5, 50.15; C-6, 18.07; C-7, 26.94; C-8, 125.85; C-9 138.39; C-10, 37.17; C-11, 68.55; C-12, 79.02; C-13, 46.45; C-14, 146.69; C-15, 128.82; C-16, 35.17; C-17, 48.97; C-18, 16.68; C-19, 22.05; C-20, 33.25; C-21, 18.07; C-22, 33.25; C-23, 30.89; C-24, 156.38; C-25, 33.81; C-26, 21.85; C-27, 21.99; C-28, 105.99; C-29, 15.51; C-30, 28.19; CH_3CO, 170.73; and CH_3CO, 21.26ppm.

Mass Spectrum:

LREIMS: 498(M^+), 438(M^+ - 60), and 423*m/e* (M^+ - 60 - 15).

Reference

B. Yagen, P. Horn, A. Z. Joffe, and R. H. Cox; Isolation and Structural Elucidation of a Novel Sterol Metabolite of *Fusarium sporotrichioides* 921; J. Chem. Soc., Perkin Trans I, pp. 2914-2917(1980).

7

Miscellaneous Sterols

Pachymic acid (3β-Acetoxy-16α-hydroxylanosta-8,24(28)-dien-21-oic acid)
3-*O*-Acetyltumulosic acid
Carbomethoxyacetylquercinic acid
Carboxyacetylquercinic acid
12β-Hydroxycarboxyacetylquercinic acid
12β-Hydroxycarbomethoxyacetylquercinic acid
Carboxyacetylstowardolic acid
23,24,25,26,27-Pentanorlanost-8-ene-3β,22-diol
24*E*-Ethylidene-5α-lanosta-8-ene-3β-acetate
22ξ-Acetoxy-3β,23ξ-dihydroxy-24(28)*Z*-ethylidenelanost-8-ene
22ξ-Acetoxy-3β,23ξ-dihydroxy-24(28)-methylenelanost-8-ene
Terretonin
3β,6β-Dihydroxy-4,4,14α-trimethyl-5α-pregn-8-en-20-one
3α-Hydroxy-4,4,14α-trimethyl-5α-pregn-8-en-20-one
3α-Hydroxy-5β-cholest-11-en-24-oic acid

Common/Systematic Name

Pachymic acid

3β-Acetoxy-16α-hydroxylanosta-8,24(28)-dien-21-oic acid

Molecular Formula/Molecular Weight

$C_{33}H_{52}O_5$; MW = 528.38148

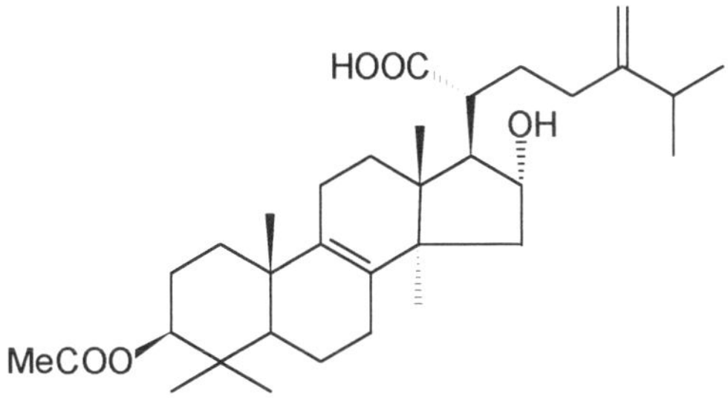

General Characteristics

Methyl ester: crystals; mp., 185-187°C; $[\alpha]_D^{14}$ + 41.5°.

Fungal Source

Sclerotia of *Poria cocos.*

Isolation/Purification

The powdered sclerotia were extracted with ethyl ether and the extract was concentrated, from which crude pachymic acid separated out. The crude acid was methylated with diazomethane followed by final purification on an alumina column.

Biological Activity

"Fuling" is the naturally occurring sclerotium of *Poria cocos* and has been frequently prescribed in Chinese medicine as a diuretic and for palpitations.

Spectral Data (Methyl ester)

UV:

λ_{max}^{MeOH} 234(ϵ= 2,500), 243(2,950), and 251nm (2,000).

^{1}H NMR:

($CDCl_3$) 0.58; 0.72; 0.87; 0.96; 1.06; and 1.13ppm.

References

A. Kanematsu and S. Natori; Triterpenoids of Hoelen (fuling), Sclerotia of *Poria cocos* Wolf. II. 3β-Hydroxylanosta-7,9(11),24-trien-21-oic Acid; Chem. Pharm Bull., Vol. 18, pp. 779-783(1970).

J. Valisolalao, L. Bang. J-P. Beck, and G. Ourisson; Etude Chimique et Biochimique de Drogues Chinoises. V. Cytotoxicite de Triterpenes de *Poria cocos* (Polyporacee et de Substances Analogues); Bull. Soc. Chim. Fr., pp. 473-477(1980).

Common/Systematic Name
3-*O*-Acetyltumulosic acid

Molecular Formula/Molecular Weight
$C_{33}H_{52}O_5$; MW = 528.38148

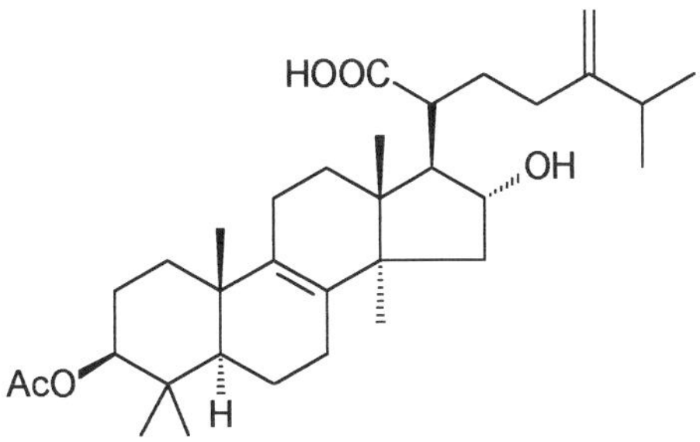

General Characteristics
Methyl ester, crystals from aqueous methanol; mp., 184-186°C; $[\alpha]_D$ + 41°.

Fungal Source
Trametes feei, *T. lilacino gilva*, and *Poria cocos*.

Isolation/Purification
The fungus was extracted with ethyl ether, methylated, chromatographed on alumina, eluted with benzene (3 different fractions), and benzene-chloroform, 1:1(v/v). The first benzene fraction in methanol and dioxane together with sodium borohydride was stirred at room temperature for 2 hours. Water was added and the resultant precipitate was chromatographed on alumina using benzene and recrystallized from aqueous methanol to give the purified methyl ester of 3-*O*-acetyltumulosic.

Biological Activity
Antibiotic

Spectral Data (Methyl ester)

UV:
λ_{max} 243nm (broad, ε = 500).

IR:
($CHCl_3$) 3480(OH), 1715(unresolved CO_2Me, OAc), 1260(OAc), 1645, and 887cm^{-1} ($C{=}CH_2$).

^{1}H NMR:

($CDCl_3$) 4.7(2H, m, vinylic protons); 4.50(1H, q, *J*=4.0, 10Hz, C<u>H</u>OAc); 4.1(1H, m, CHOH); 3.7(3H, s, CO_2Me); 2.04(3H, s, OCOMe); and 1.82ppm (OH).

Mass Data:

LREIMS: 542*m/e* (M); calcd for $C_{34}H_{54}O_5$, 542; found: C, 75.2; H, 9.9%; calcd for $C_{34}H_{54}O_5$; C, 75.2; H, 10.0%.

References

J. T. Pinhey, B. J. Ralph, J. J. H. Simes, and M. Wootton; A New Source of Tumulosic Acid and 3-O-acetate tumulosic Acid; Aust. J. Chem., Vol. 23, pp. 2141-2146(1970).

J. T. Pinhey, B. J. Ralph, J. J. H. Simes, and M. Wootton; Extractives of Fungi II. The Constituents of *Trametes feei*, 6α-Hydroxypolyporenic Acid C; Aust. J. Chem., Vol. 24, pp. 609-619(1971).

Common/Systematic Name

Carbomethoxyacetylquercinic acid

Molecular Formula/Molecular Weight

$C_{35}H_{54}O_7$; MW = 586.38696

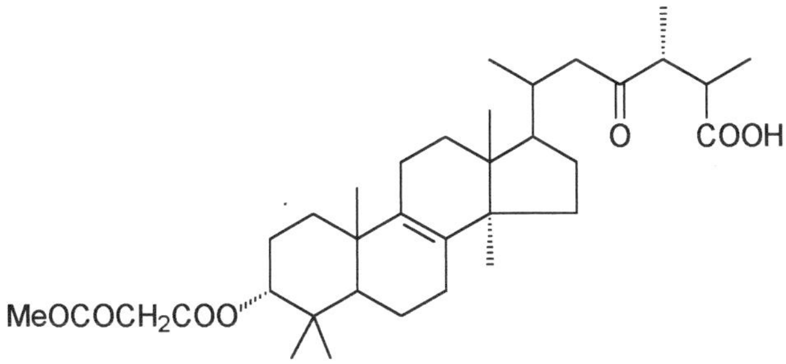

General Characteristics

Crystals; mp., 165.5-167°C; $[\alpha]_D$ + 0.7°.

Fungal Source

Trametes dickinsii (Japanese name Hoorokutake).

Spectral Data

^{1}H NMR:

(Methylated derivative) 6.59(s, 2H); 6.72(s, 3H); and 6.33τ (s, 3H).

Reference

H. Inouye and K. Tokura; Absolute Configuration of Carboxyacetylquercinic Acid, Chemical Correlation of Carboxyacetylquercinic Acid with Lanosterol; Z. Naturforsch., Teil B, Vol. 25, pp. 1194-1195(1970).

Common/Systematic Name
Carboxyacetylquercinic acid

Molecular Formula/Molecular Weight
$C_{34}H_{52}O_7$; MW = 572.37130

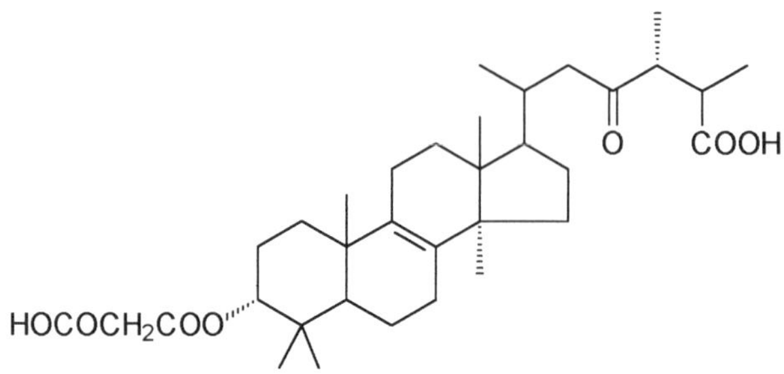

General Characteristics
Crystals; mp., 221-224°C; $[\alpha]_D$ + 8.8°.

Fungal Source
Trametes dickinsii (Japanese name Hoorokutake) and *Daedalea quercina*.

Spectral Data

^{1}H NMR:
(Methylated derivative) 6.59(s, 2H); 6.72(s, 3H); and 6.33τ (s, 3H).

References
H. Inouye and K. Tokura; Absolute Configuration of Carboxyacetylquercinic Acid, Chemical Correlation of Carboxyacetylquercinic Acid with Lanosterol; Z. Naturforsch., Teil B, Vol. 25, pp. 1194-1195(1970).

S. Natori, A. Yokoyama (nee Kanematsu), and K. Aoshima; Distribution of Tetracyclic Triterpenoids of Lanostane Group in Pore Fungi (*Basidiomycetes*). In Chemistry in Botanical Classifications, G. Bendz, J. Santesson, eds., Nobel Foundation, Stockholm, Academic Press, New York, pp. 173-179(1974).

Common/Systematic Name
12β-Hydroxycarboxyacetylquercinic acid

Molecular Formula/Molecular Weight
$C_{34}H_{52}O_8$; MW = 588.36622

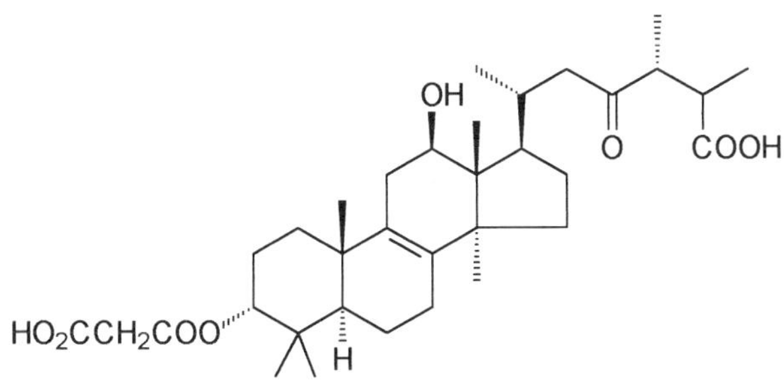

Fungal Source
Melanoporia rosea and *M. nigra*.

Isolation/Purification
Fruit bodies were extracted with ethyl ether, methylated, and purified by preparative TLC using silica gel H or HF_{254} and developing solvents of hexane-ethyl acetate (7:3, v/v) and chloroform-methanol (9:1, v/v); detection was by spraying 20% vanillin-phosphoric acid and heating.

Spectral Data

^{1}H NMR:
($CDCl_3$) 0.74(3H, s, H-18); 1.00(3H, s, H-19); 0.92(6H, H-30, -18); 0.88(3H, s, H-32); 1.04(3H, d, *J*=6Hz, H-27); 1.12(3H, d, *J*=6Hz, H-28); 4.66(1H, m); 3.61(3H, s, OCH_3); and 3.38ppm (2H, s, OCH_2-). C-21 methyl appeared at 0.9-1.0ppm overlapping with other methyl signals.

References
S. Natori, A. Yokoyama (*nee* Kanematsu), and K. Aoshima; Distribution of Tetracyclic Triterpenoids of Lanostane Group in Pore Fungi (*Basidiomycetes*). In Chemistry in Botanical Classifications, G. Bendz, J. Santesson, eds., Nobel Foundation, Stockholm, Academic Press, New York, pp. 173-179(1974).

W. B. Turner and D. C. Aldridge; Fungal Metabolites II; Academic Press, New York, p. 323 (1983).

A. Yokoyama and S. Natori; Triterpenoids of Lanostane Group from Fruit Bodies of Nine *Basidiomycetous* Species; Chem. Pharm. Bull., Vol. 22, pp. 877-883(1974).

A. Yokoyama, S. Natori, and K. Aoshima; Distribution of Tetracyclic Triterpenoids of Lanostane Group and Sterols in The Higher Fungi Especially of the *Polyporaceae* and Related Families; Phytochemistry, Vol. 14, pp. 487-497(1975).

Common/Systematic Name
12β-Hydroxycarbomethoxyacetylquercinic acid

Molecular Formula/Molecular Weight
$C_{35}H_{54}O_8$; MW = 602.38187

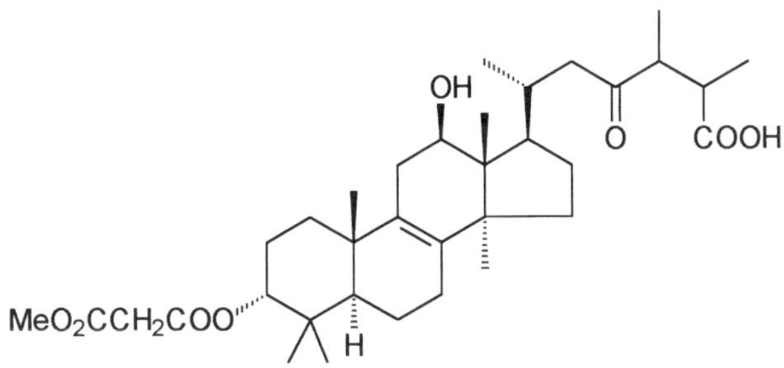

General Characteristics
(Methyl ester) Crystals from methanol; mp., 116-117 °C; $[\alpha]_D^{21}$ + 4.66 ° (*c*=0.97, in $CHCl_3$); oily monoacetate; $[\alpha]_D^{21}$ + 5.26°; diol monomethyl ester; $[\alpha]_D^{21}$ + 19.0°.

Fungal Source
Melanoporia rosea.

Isolation/Purification
Fruit bodies were extracted with ethyl ether, methylated, and purified by preparative TLC using silica gel H or HF_{254} and developing solvents of hexane-ethyl acetate (7:3, v/v) and chloroform-methanol (9:1, v/v); detection was by spraying 20% vanillin-phosphoric acid and heating.

Spectral Data

IR:
(methyl ester) 3420, 1750, 1735, 1708, and 1700cm^{-1}.

^{1}H NMR:
($CDCl_3$) 0.72(3H, s, H-18); 1.01(3H, s, H-19); 0.92(6H, H-30, -18); 0.87(3H, s, H-32); 1.03(3H, d, *J*=6Hz, H-27); 1.12(3H, d, *J*=6Hz, H-28); 4.68(1H, m); 4.10(1H, m); 3.62(3H, s, OCH_3); and 3.38ppm (2H, s, OCH_2-). C-21 methyl appeared at 0.9-1.0ppm overlapping with other methyl signals.

Mass Spectrum:

(methyl ester) HREIMS: 616.409(M^+, calcd for $C_{36}H_{56}O_8$: 616.398), 598(M^+ - H_2O), 583(M^+ - CH_3 - H_2O), 498(M^+ - $C_4H_6O_4$), 483(M^+ - CH_3 - $C_4H_6O_4$), 465(M^+ - CH_3 H_2O - $C_4H_6O_4$), 325(M^+ - CH_3- $C_4H_6O_4$ - $C_8H_{14}O_3$), and 307*m/e* (M^+ - CH_3- H_2O - $C_4H_6O_4$ - $C_8H_{14}O_3$).

References

W. B. Turner and D. C. Aldridge; Fungal Metabolites II; Academic Press, New York, p.323 (1983).

A. Yokoyama and S. Natori; Triterpenoids of Lanostane Group from Fruit Bodies of Nine *Basidiomycetous* Species; Chem. Pharm. Bull., Vol. 22, pp. 877-883(1974).

Common/Systematic Name

Carboxyacetylstowardolic acid

3α-Carboxyacetoxy-24ξ-methyl-23-oxolanosta-8,25(26)-dien-27-oic acid

Molecular Formula/Molecular Weight

$C_{34}H_{50}O_7$; MW = 570.35565

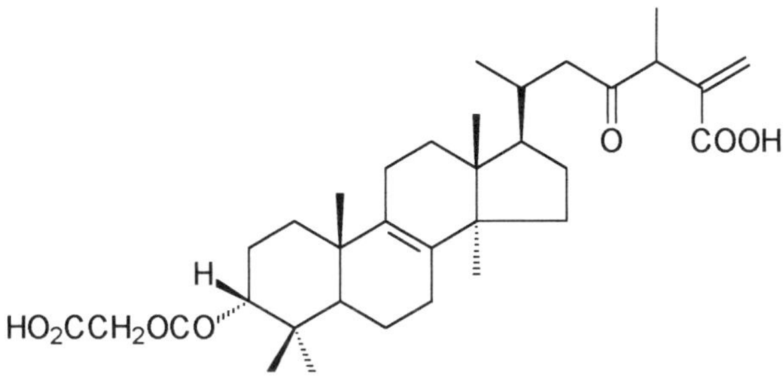

General Characteristics

Crystals from aqueous methanol; mp., 184-186°C.

Fungal Source

Trametes stowardii.

Isolation/Purification

Fungal cultures were extracted with 95% ethanol, evaporated to dryness, and redissolved in ethyl ether. The ether soluble fraction was partitioned between 2N HCl and 5% $NaHCO_3$. Acidification of the bicarbonate solution and extraction with ether gave a crude product that was chromatographed on a silicic acid column.

Spectral Data

UV:

λ_{max} 210nm (ε = 11,800)

IR:

(Nujol mull) 1765, 1740, 1725, 1715, and 1695cm^{-1}.

^{1}H NMR:

($CDCl_3$) 4.75(*J*=6.0Hz, H-3); 3.7(q, *J*=7.0Hz, H-24); 5.95(3.0Hz, H-26); 6.2(2.5Hz, H-26); 1.35(3H, d, *J*=7.0Hz, H-31); 10.0(2H, OH, CO_2H); and 3.45ppm (2H, $COCH_2CO$).

Reference

H. T. Cheung, J. C. F. Seeto, and T. R. Watson; Triterpene Metabolite of *Trametes stowardii*; Aust. J. Chem., Vol. 26, pp. 609-618(1973).

Common/Systematic Name

23,24,25,26,27-Pentanorlanost-8-ene-3β,22-diol

Molecular Formula/Molecular Weight

$C_{25}H_{42}O_2$; MW = 374.31848

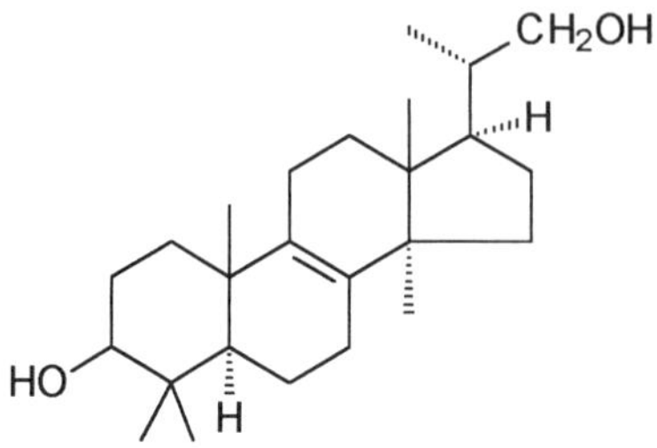

General Characteristics

Prisms or needles from methanol; mp., 206-207°C after a transition at 90°C for loss of solvent; diacetate; mp., 175-177°C.

Fungal Source

Verticillium lecanii (entomopathogenic fungus).

Isolation/Purification

The dried mycelium was extracted with petrol followed by ethyl acetate. The extracts in benzene were chromatographed on a silica gel column eluted from benzene to ethyl ether. Crystallization of the benzene-ethyl ether fraction from methanol yielded the purified 23,24,25,26,27-pentanorlanost-8-ene-3β,22-diol.

Spectral Data

UV:

End absorption.

IR:

(KBr) 3380 and 1622(w) cm^{-1}.

^{1}H NMR:

0.70; 0.79; 0.86; 0.96; 0.98(3H, all s); 1.01(3H, d, *J*=6.0Hz); 1.1-2.0(19H); 2.0(20H, br); 3.21(dd, *J*=11.5Hz, H-3); 3.34; and 3.63ppm (AB, *J*=10.5, 6.5, 2.0Hz, H-22).

^{13}C NMR:

C-1, 35.6(2); C-2, 27.6*(2); C-3, 79.0(1); C-4, 38.9(0); C-5, 50.4(1); C-6, 18.3(2); C-7, 27.9*(2); C-8, 134.3** (0); C-9, 134.6**(0); C-10, 37.0(0); C-11, 21.0(2); C-12, 26.5(2); C-13, 44.7(0); C-14, 49.6(0); C-15, 30.9*** (2); C-16, 31.0*** (2); C-17, 46.81; C-18, 15.9 3 ; C-19, 19.1 3 ; C-20, 39.5 1 ; C-21, 16.7 3 ; C-22, 68.2 2; C-28, 24.3(3); C-29, 28.0(3); and C-30, 15.4ppm*** (3).

*, **, *** Assignments may be reversed.

Mass Data:

HREIMS: 374.3185*m/e* (M^+, $C_{25}H_{42}O_2$ requires 374.3185); found: C, 79.6; H, 11.4; calcd C, 80.1; H, 11.3%.

Reference

J. F. Grove; 23,24,25,26,27-Pentanorlanost-8-ene-3β,22-diol from *Verticillium lecanii*; Phytochemistry, Vol. 23, pp. 1721-1723(1984).

Common/Systematic Name

E-24-Ethylidene-5α-lanost-8-ene-3β-acetate

Molecular Formula/Molecular Weight

$C_{34}H_{56}O_2$; MW = 496.42803

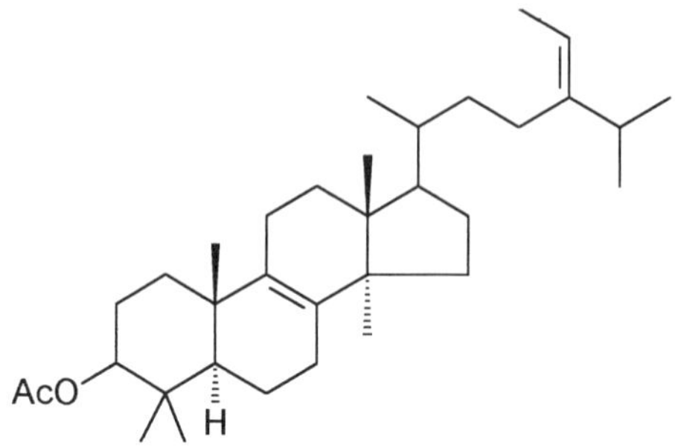

General Characteristics

Crystals; mp., 135-138°C.

Fungal Source

Mucor rouxii isolated from sheep rumen contents.

Isolation/Purification

The freeze-dried mycelium of *Mucor rouxii* was extracted with chloroform-methanol (2:1, v/v) and purified by argentation TLC chromatography.

Also, see Heupel, R.C.; Isolation and Primary Characterization of Sterols, In Analysis of Sterols and Other Biologically Significant Steroids; W. D. Nes and E. J. Parish (eds.); Academic Press Inc., New York, New York, 341 pp. (1989).

Spectral Data

UV:

End absorption.

^{1}H NMR:

($CDCl_3$) 0.69(3H, H-18); 0.87(9H, H-4α, -4β, -14); 0.95(3H, H-21); 1.00(3H, H-19); 0.97(3H, H-26,-27); 1.58(3H, H-29); 2.05(3H, 3β-O-$COCH_3$); 2.19(1H, H-25); 4.50(1H, H-3α); and 5.19ppm (1H, H-28); free sterol 0.60(3H, s, H-18); 0.81 and 0.87(6H, s, H-4α and -4β); 0.96-1.02(15H, m, H-19, -14, -21, -26, and -27); 1.57(3H, d, *J*=6.0Hz, H-29); 2.29(1H, heptet, *J*=6.0Hz, H-25); 3.24(1H, q, *J*=5.0Hz, H-3α); and 5.29ppm (1H, q, *J*=6.0Hz, H-28).

Mass Spectrum:

LREIMS: 496, 481, and 421*m/e*; free sterol, HREIMS: 454.4169*m/e* (M^+); calcd for $C_{32}H_{54}O$, 454.4174.

References

W. D. Nes and E. J. Parish (eds.); Analysis of Sterols and Other Biologically Significant Steroids; Academic Press Inc., New York, New York, 341 pp. (1989).

S. Safe and L. M. Safe; The Isolation and Synthesis of *E*-24-Ethylidene-5α-lanost-8-en-3β Acetate; Canadian J. Chem., Vol. 53, pp. 3247-3249(1975).

W. B. Turner and D. C. Aldridge; Fungal Metabolites II; Academic Press, New York, New York, 631 pp. (1983).

Common/Systematic Name

22ξ-Acetoxy-3β,23ξ-dihydroxy-24(28)*Z*-ethylidenelanost-8-ene

Molecular Formula/Molecular Weight

$C_{34}H_{56}O_4$; MW = 528.41786

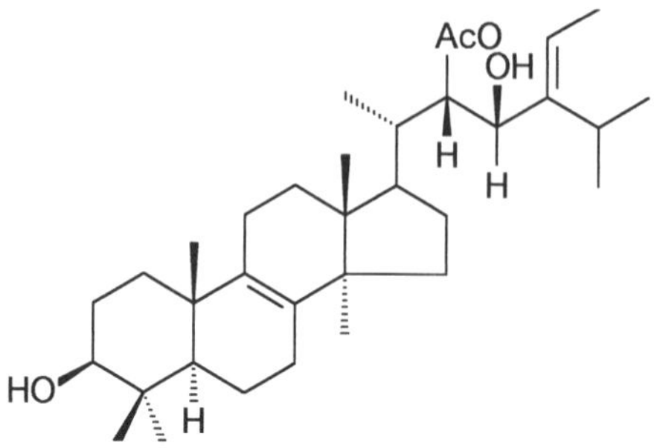

General Characteristics

Free alcohol ($C_{32}H_{54}O_3$): mp., 192-196°C; $[\alpha]_D^{20}$ + 39° (*c*=1.1, in $CHCl_3$); triacetate: mp., 160-165°C; $[\alpha]_D^{20}$ + 30° (*c*=0.9 in $CHCl_3$); triol: mp., 193-195°C; $[\alpha]_D^{20}$ + 38° (*c*=1.1, in $CHCl_3$).

Fungal Source

Pisolithus tinctorius.

Isolation/Purification

Purified by extraction with dichloromethane and chromatographic purification of the crude extract.

Spectral Data

IR:

(KBr) 1735cm^{-1}.

^{1}H NMR:

($CDCl_3$) Free alcohol: 0.745; 0.819; 0.921; 0.990; and 1.005 attributed to C-18, C-30, C-31, C-32, and C-19 tertiary methyl groups (each 3H, s); 0.936(3H, d, *J*=6.0Hz) assigned to secondary methyl group at C-21; 1.078(3H, d, *J*=6.0Hz), 1.152(3H, d, *J*=6.0Hz); and 2.833(1H, d, *J*=6.0Hz) assigned to the isopropyl group attached to C-24; 1.470(3H, d, *J*=6.0Hz); and 5.636(1H, q, *J*= 6.0Hz) suggested for CH_3-CH=CR_2 group; 3.225(1H, m, H-3α); 3.984(1H, d, *J*=9.0Hz); and 3.627(1H, d, *J*=9.0Hz) attributed to carbons bearing secondary OH groups.

Reference

A. M. Lobo, P. Macedo de Abreu, S. Prabhakar, and L. S. Godinho; Novel Triterpenoids from The Fungus *Pisolithus tinctorius*; Tetrahedron Letters, Vol. 26, pp. 2589-2592(1985).

Common/Systematic Name

22ξ-Acetoxy-3β,23ξ-dihydroxy-24-methylenelanost-8-ene

Molecular Formula/Molecular Weight

$C_{33}H_{54}O_4$; MW = 514.40221

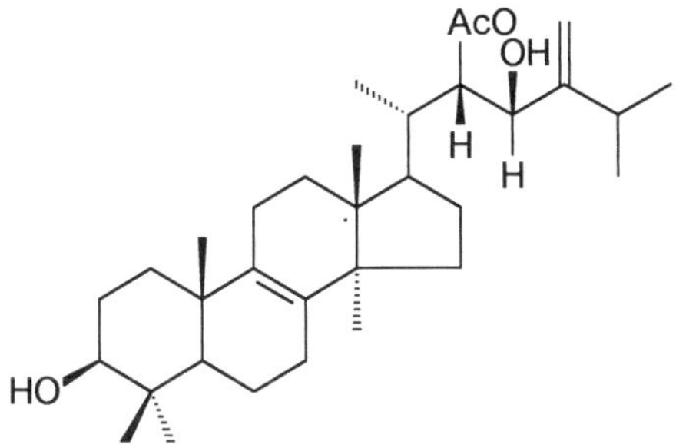

General Characteristics

Triol; mp., 193-195°C; $[\alpha]_D^{20}$ + 38° (*c*=1.1, in $CHCl_3$).

Fungal Source

Pisolithus tinctorius.

Isolation/Purification

Purified by extraction with dichloromethane and chromatographic purification (preparative TLC) of the crude extract.

Spectral Data

IR:

(KBr) 1735cm^{-1}.

^{1}H NMR:

($CDCl_3$) Spectrum similar to 22ξ-acetoxy-3β,23ξ-dihydroxy-24(28)*Z*-ethylidene-lanost-8-ene except for substitution on side chain.

Reference

A. M. Lobo, P. Macedo de Abreu, S. Prabhakar, and L. S. Godinho; Novel Triterpenoids from The Fungus *Pisolithus tinctorius*; Tetrahedron Letters, Vol. 26, pp. 2589-2592 (1985).

Common/Systematic Name

Terretonin

Molecular Formula/Molecular Weight

$C_{26}H_{32}O_9$; MW = 488.20463

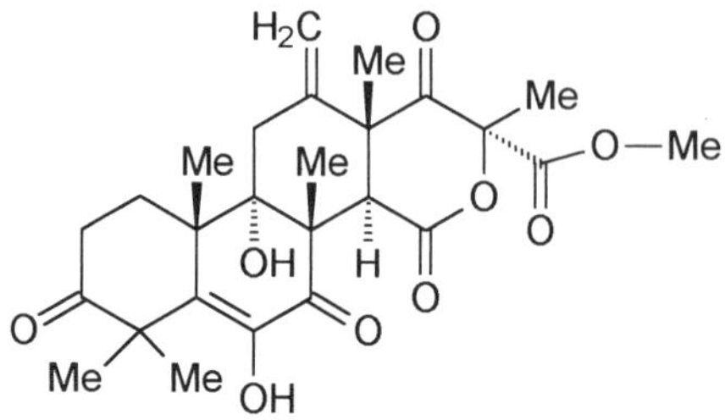

Fungal Source

Aspergillus terreus.

Isolation/Purification

Fungal cultures were extracted with chloroform, filtered, dried over anhydrous sodium sulfate, and reduced in volume. The crude extract was chromatographed on a Florisil column eluted with a gradient from benzene to ethyl acetate. The biologically active fractions were combined, reduced in volume, and chromatographed on a Woelm neutral grade IV alumina column eluted with a gradient from benzene to ethyl acetate. The active fractions were combined, reduced in volume, and allowed to sit at 5°C until crystals formed.

Biological Activity

The metabolite had an LD_{50} in day-old cockerels of between 250-375mg/kg body weight.

Spectral Data

UV:

λ_{max}^{MeOH} 276nm (ε = 7,960).

IR:

(KBr) 3470, 3340, 2940, 1764, 1748, 1729, 1708, 1685, 1641, 1452, 1372, 1307, 1265, 1252, 1171, 1150, 1118, 1018, 999, and 919cm^{-1}.

CD:

(*c*=0.54, in MeOH) $[\theta]_{221}$ - 9200°, $[\theta]_{275}$ -20,900°, $[\theta]_{305}$ -3400°, $[\theta]_{317}$ - 3,700°, and $[\theta]_{329}$ - 2,700.

^{1}H NMR:

(CDCl$_3$) 1.22; 1.45; 1.48 x 2; 1.73; 1.94(s, 3H x 6, H-20 ABC, H-21 ABC, H-24 ABC, H-26 ABC, H-28 ABC, H-30 ABC); 1.80(cm, 1H, H-1A eq); 1.80(s, ex, 1H, H-25); 2.28(bd, *J*=14Hz, 1H, H-11 A or B); 2.38(ddd, *J*=14, 11,8Hz, 1H, H-1Bax); 2.54(ddd, *J*=19, 11, 8Hz, 1H, H-2Bax); 2.73(bdd, *J*=19, 8Hz, 1H, H-2Aeq); 2.99(bd, *J*=14Hz, 1H, H-11 A or B); 3.55(s, 1H, H-14); 3.81(s, 3H, H-34 ABC); 5.11, 5.49(bs, 1H x 2, H-27 AB); and 6.16ppm (s, ex, 1H, H-22).

^{13}C NMR:

(CDCl$_3$) C-1, 32.7[a]; C-2, 34.4[a]; C-3, 214.5; C-4, 52.2; C-5, 130.0[b]; C-6, 139.8[b]; C-7, 197.3; C-8, 43.2; C-9, 77.5; C-10, 47.7[c]; C-11, 28.1; C-12, 139.1; C-13, 49.1[c]; C-14, 43.9; C-15, 168.2; C-17, 85.2; C-18, 201.8; C-20, 21.3[d]; C-21, 24.0[d]; C-24, 21.3[d]; C-26, 19.4[d]; C-27, 115.2; C-28, 18.6[d]; C-30, 23.5[d]; C-31, 168.2; and C-34, 53.4ppm.

[a], [b], [c], [d] Assignments may be reversed.

Mass Data:

HREIMS: 488.2066(M$^+$, calcd for $C_{26}H_{32}O_9$, 488.2046), 473.1844(- CH_3), 470.1959 (-H_2O), 357.1702(- $C_5H_7O_4$), and 330.1850*m/e* (- $C_6H_6O_5$); Found C, 64.07%; H, 6.80; and O by difference 29.13. Required for $C_{26}H_{32}O_9$: C, 63.9%; H, 6.6; 0,29.5.

<u>TLC Data</u>

Appeared as a green spot at R_f 0.3 using silica gel GH-R plates developed with chloroform-acetone, 93:7, v/v and detected with 50% ethanolic sulfuric acid spray followed by heating at 125° for 5 min.

<u>Reference</u>

J. P. Springer, J. W. Dorner, R. J. Cole, and R. H. Cox; Terretonin, A Toxic Compound from *Aspergillus terreus*; J. Organic Chemistry, Vol. 44, pp. 4852-4854(1979).

Common/Systematic Name

3β,6β-Dihydroxy-4,4,14α-trimethyl-5α-pregn-8-en-20-one

Molecular Formula/Molecular Weight

$C_{24}H_{38}O_3$; MW = 374.28210

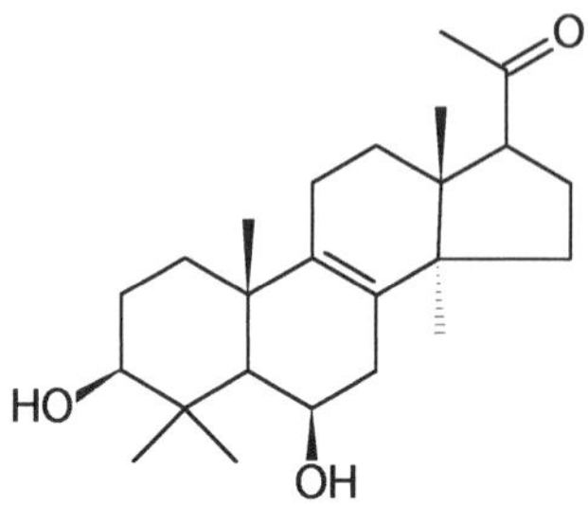

General Characteristics

Prisms from isopropyl ether; mp., 188-189°C; $[\alpha]_D^{24}$ + 20° (*c*=1.0, in $CHCl_3$); diacetate: crystals from methanol; mp., 258-260°C; $[\alpha]_D$ + 24° (*c*=0.24, in $CHCl_3$).

Fungal Source

Fomes officinalis

Isolation/Purification

See Epstein and G. van Lear; J. Org. Chem., Vol. 31 p. 3434(1966).

Spectral Data

IR:

(CCl_4) 3350 and 1685cm^{-1}; diacetate: 1739, 1718, and 1247cm^{-1}.

ORD:

$[\alpha]_{600}$ + 54°, $[\alpha]_{589}$ + 54°, $[\alpha]_{313}$ + 895°, $[\alpha]_{268}$ - 574°, and $[\alpha]^{247}$ 0° (*c*=0.14, in dioxane).

^{1}H NMR:

($CDCl_3$) 9.28, 9.14, 9.05, 9.01, and 7.80τ (methyl resonances in the ratio of 1:1:2:1:1); diacetate; 7.99τ (6H) methyl resonances.

Mass Data:

LREIMS: 374*m/e* (M^+); anal calcd for $C_{24}H_{38}O_3$: C, 76.96; H, 10.23%; Found: C, 76.66; H, 10.24%; diacetate, 398*m/e* (M^+ - 60).

Reference

C. G. Anderson, W. W. Epstein, and G. Van Lear; Minor Triterpenoids of *Fomes officinalis*; Phytochemistry, Vol. 11, pp. 2847-2852(1972).

Common/Systematic Name

3α-Hydroxy-4,4,14α-trimethyl-5α-pregn-8-en-20-one

Molecular Formula/Molecular Weight

$C_{24}H_{38}O_2$; MW = 358.28718

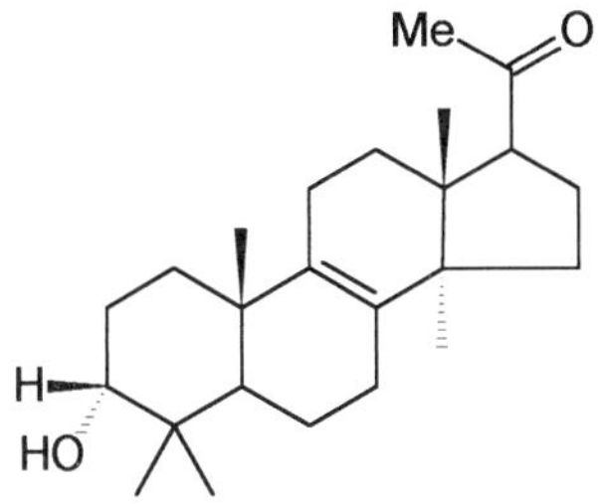

General Characteristics

Crystals from methanol; mp., 230-232°C; $[\alpha]_D$ + 109° (*c*=0.18, in hexane); acetate, crystals from methanol; mp., 250-252°C; $[\alpha]_D$ + 48° (*c*=0.05, in $CHCl_3$).

Fungal Source

Fomes officinalis

Isolation/Purification

Finely ground fungus was extracted with petroleum ether, concentrated, and redissolved in methanol. The methanol solution was allowed to sit for several days; the precipitate was removed and the mother liquor was separated into neutral, basic and acidic components. The neutral materials were recrystallized from methanol to give ergosterol. The mother liquor was concentrated and treated with complexing material (urea) to remove fatty substances, which were discarded. The neutral material was isolated, saponified, and non-complexing materials chromatographed on Florisil. The less polar fractions were rechromatographed on Florisil; 3α-hydroxy-4,4,14α-trimethyl-5α-pregn-8-en-20-one eluted from the column with ether-benzene. Final purification was by recrystallization from methanol.

Spectral Data

UV:

λ_{max}^{EtOH} 235(log ε=3.08), 243(3.11), and 250nm (2.90).

IR:

3350 and 1695cm^{-1}; diacetate, 1739, 1718 and 1247cm^{-1}.

ORD:

$[\alpha]_{625}$ + 90°, $[\alpha]_{589}$ + 109°, $[\alpha]_{312}$ + 2370°, $[\alpha]_{267}$ - 2170°, and $[\alpha]^{247}$ - 1950° (*c*=0.18, in dioxane).

^{1}H NMR:
(CDCl$_3$) 9.27, 9.12, 9.02, and 7.92τ (methyl resonances in the ratio of 1:1:3:1).

Mass Data:
LREIMS: 358*m/e* (M^+).

TLC Data
Silica gel G plates developed with acetone-chloroform (1:9, v/v).

Reference
W. W. Epstein and G. Van Lear; Metabolites of *Fomes officinalis*; J. Org. Chem., Vol. 31, pp. 3434-3435(1966).

Common/Systematic Name

3α-Hydroxy-5β-cholest-11-en-24-oic acid

Molecular Formula/Molecular Weight

$C_{24}H_{38}O_3$; MW = 374.28210

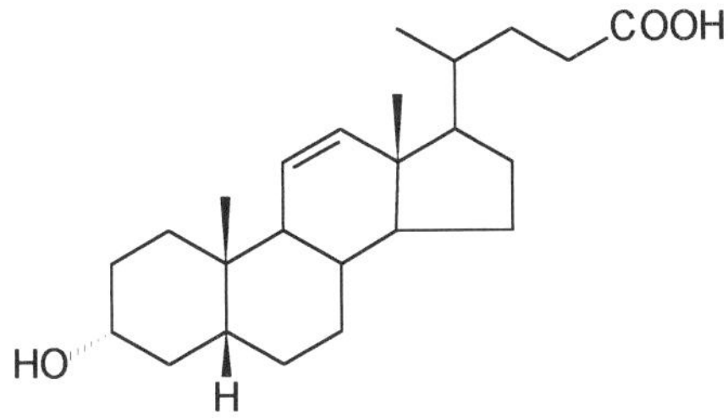

General Characteristics

Crystallized from benzene solution to give plates; mp., 166°C; $[\alpha]_D^{19}$ + 25.9° (*c*=2.2, in $CHCl_3$). Others 164-166°C; 165-165.5°C; and 168-169°C. Acetoxy derivative: crystals from methanol; mp., 126-127°C.

Fungal Source

Curvularia spp.

Spectral Data

IR:

(KBr) 3420(OH), 2660 (H-bonded OH), and 1729cm^{-1} (carboxylic C=O).

^{1}H NMR:

($CDCl_3$) 3.91, 4.57, and *ca.* 7.72τ (AMXm, J_{AM}=10, J_{AX}=2, J_{MX}=3Hz, -CH=CH-CH=); 4.26br (1H, s, =CHOH); *ca.* 6.35(1H, $J_{1,2}$ *ca.* 20Hz, axial =CHOH); and 9.12 and 9.28τ (each 3H, s, CH_3 at C-10 and C-13); acetoxy derivative: 3.90, 4.58 and *ca.* 7.70τ (AMXm, J_{AM}=10, J_{AX}=2, J_{MX}=3Hz, -CH=CH-CH=); *ca.* 5.26(1H, $J_{1,2}$ *ca.* 20Hz, axial =CHOAc); 7.99(3H, s, CH_3CO_2R); and 9.12 and 9.28τ (each 3H, s, CH_3 at C-10 and C-13).

Mass Data:

LREIMS: 374(M$^+$, 9%), 356(100), 341(35), 255(45), 228(400), and 213*m/e* (25); acetoxy derivative: 416(M$^+$, 0.2%), 356(66), 341(29), 255(71), 228(63), and 213*m/e* (39). Found: C, 76.8; H, 10.3%; $C_{24}H_{38}O_3$ requires C, 76.95; H, 10.25%; acetoxy derivative HREIMS: 356.2717*m/e*, $C_{24}H_{36}O_2$ requires 356.2715.

Reference

H. D. Munro, O. C.Musgrave, and A. B. Turner; 3α-Hydroxy-5β-chol-11-en-24-oic Acid, a New Fungal Metabolite; J. Chem. Soc., Perkin Trans. I, pp. 1597-1598(1974).

8

Antheridiols and Oogoniols

Antheridiol
23-Deoxyantheridiol
Oogoniol
Oogoniol-1
Oogoniol-2
Oogoniol-3
15-Keto-oogoniol-2
24(28)-Dehydro-oogoniol
24(28)-Dehydro-oogoniol-1
24(28)-Dehydro-oogoniol-2
24(28)-Dehydro-oogoniol-3

Common/Systematic Name
Antheridiol

Molecular Formula/Molecular Weight
$C_{29}H_{42}O_5$; MW = 470.30322

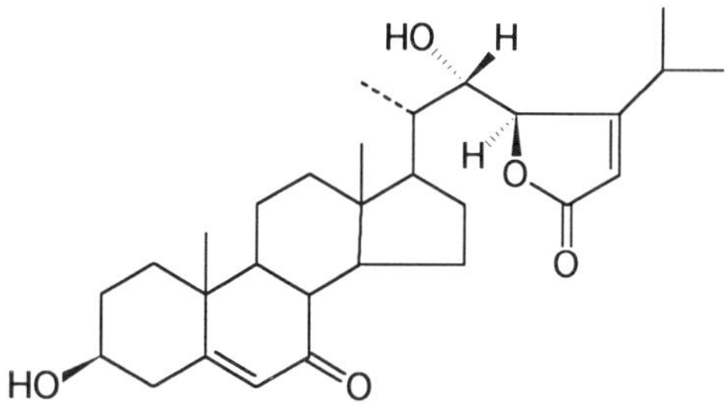

General Characteristics
Crystals; mp., 250-255°C (dec.)

Fungal Source
Achlya bisexualis.

Isolation/Purification
The fungal mycelium was removed by filtration. The filtrate was adjusted to pH 4.00 and the hormone activity was extracted with methylene chloride. Concentration of the extract produced a brown gum that was chromatographed on silica gel with ethyl acetate as eluant. Biological activity eluted late in the column elution. Final purification was achieved by preparative TLC of the biologically active fraction on silica gel H with chloroform-methanol (10:1, v/v) as solvent. Antheridiol formed crystals in the yellow gum eluted from preparative TLC which were removed by dissolving the impurities away from the crystals with methanol-ethyl ether.

Biological Activity
Sex hormone secreted by the female that induces formation of antheridial hyphae on the male.

Spectral Data

UV:
λ_{max} 222nm (log ϵ 4.22).

IR:
3390, 1742, and 1672cm^{-1}.

^{1}H NMR:
(CDCl$_3$-MeOD, 3:1, v/v) 0.70(18-H); 1.17 and 1.22(pair of d, *J*=7Hz, isopropyl CH$_3$); 1.20(19-H); 4.95(d, *J*=8Hz, 23-H); 5.69(6-H); and 5.77ppm (24'-H).

Mass Spectrum:
LREIMS: 470(M$^+$), 344, 326, 298, 287, 269, and 251*m/e*.

References

G. P. Arsenault, K. Biemann, A. W. Barksdale, and T. C. McMorris; The structure of Antheridiol, a Sex Hormone in *Achlya bisexualis*; J. Am. Chem. Soc., Vol. 90, pp. 5635-5636 (1968).

J. A. Edwards, J. S. Mills, J. Sundeen, and J. H. Fried; The Synthesis of the Fungal Sex Hormone Antheridiol; J. Am. Chem. Soc., Vol. 91, pp. 1248-1249(1969).

J. A. Edwards, J. Sundeen, W. Salmond, T. Iwadare, and J. H. Fried; A New Synthetic Route to the Fungal Sex Hormone Antheridiol and the Determination of its Absolute Stereochemistry; Tetrahedron Letters, pp. 791-794(1972).

T. C. McMorris and A. W. Barksdale; Isolation of a Sex Hormone from the Water Mold *Achlya bisexualis*; Nature, Vol. 215, pp. 320-321(1967).

Common/Systematic Name

23-Deoxyantheridiol

Molecular Formula/Molecular Weight

$C_{29}H_{42}O_4$; MW = 454.30831

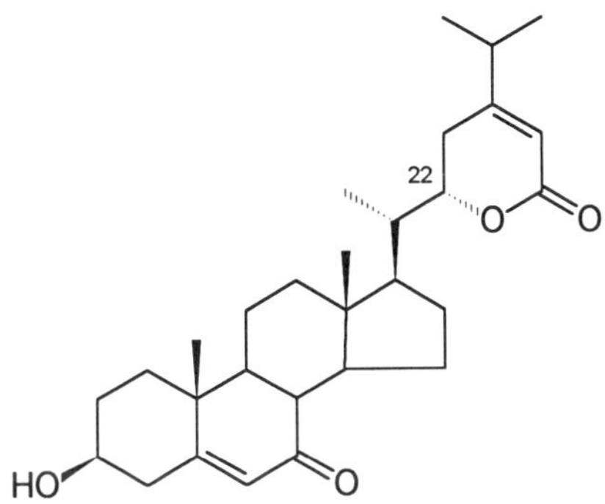

General Characteristics

Crystals; mp., 265-270°C.

Fungal Source

Achlya ambisexualis (No. 734) and *A. bisexualis* (No. 369).

Isolation/Purification

Liquid cultures of the mold were extracted with methylene chloride and the concentrated extract shaken briefly with dilute sodium hydroxide solution to remove fatty acid material. The solvent was removed and the residue chromatographed on silica gel eluted with ethyl acetate-petroleum (1:1, v/v).

Biological Activity

Sexual reproduction in *Achlya* is mediated by the steroid hormones antheridiol and the oogoniols. Sexual reproduction is initiated by the vegetative female mycelium which secretes the first hormone, antheridiol (hormone A). In response to antheridiol, male hyphae react by putting out numerous thin lateral branches which secrete another hormone, the oogoniol (hormone B). These in turn induce the development of spherical oogonial initials on female hyphae. Oogonial initials are believed to secrete relatively large amounts of antheridiol (hormone A), which directs the growth of the antheridial hyphae to the developing oogonium, leading to conjugation of the sex organs. Deoxyantheridiol possessed an activity of about 1/100th that of antheridiol but may have been due to the contamination with small amounts of antheridiol.

Spectral Data

UV:

λ_{max}^{EtOH} 230nm (ϵ = 17,900); treatment with acid changed the spectrum to λ_{max} 212 and 278nm.

IR:

ν_{max}^{KBr} 3470, 1706, 1672, and 1633cm^{-1}.

CD:

$[\theta]_{400}$ +0°, $[\theta]_{384}$ -30°, $[\theta]_{380}$ -65°, $[\theta]_{376}$ -77°, $[\theta]_{370}$ +0°, $[\theta]_{360}$ +330°, $[\theta]_{350}$ +1160°, $[\theta]_{340}$ +2170°, $[\theta]_{330}$ +3030°, $[\theta]_{320}$ +2870°, $[\theta]_{310}$ +2110°, $[\theta]_{300}$ +1040°, $[\theta]_{289}$ +0°, $[\theta]_{280}$ -2180°,$[\theta]_{270}$ -6300°, $[\theta]_{260}$ -14,710°, $[\theta]_{250}$ -24,740°, $[\theta]_{240}$ -22,400° and $[\theta]_{230}$ -14,140°.

^{1}H NMR:

($CDCl_3$) 0.71(18-H); 1.02(1H, d, *J*=6.0Hz, 21-H); 1.10(3H, d, *J*=7.0Hz, isopropyl CH_3); 1.19(1H, 19-H); 3.3-3.8(1H, m, 3α-H); 4.3, 4.4(1H, pair of broad triplets, 22-H); 5.69(6H); and 5.75ppm (1H, lactone olefinic H).

Mass Spectrum:

LREIMS: 454(M^+, 28%), 436(23), 316(41), 245(100), 192(33), 161(30), and 139*m/e* (45).

Reference

D. M. Green, J. A. Edwards, A. W. Barksdale, and T. C. McMorris; The Isolation and Structure of 23-Deoxyantheridiol and the Synthesis of its C-22 Epimer; Tetrahedron, Vol. 27, pp. 1199-1203(1971).

Common/Systematic Name

Oogoniol

Molecular Formula/Molecular Weight

$C_{29}H_{48}O_5$; MW = 476.35017

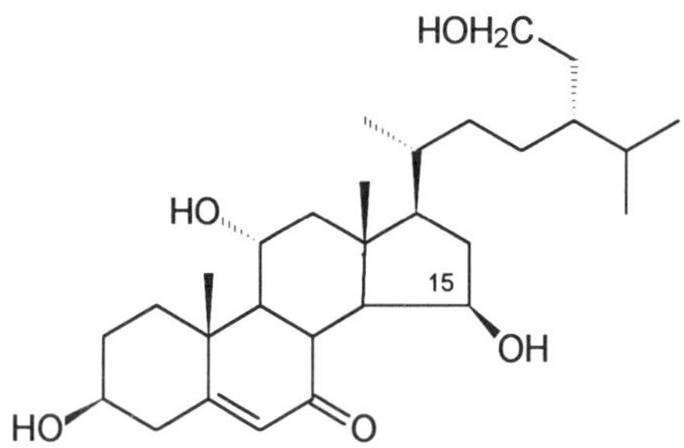

Fungal Source

Achlya heterosexualis and *A. bisexualis*.

Biological Activity

Sexual reproduction in *Achlya* is mediated by the steroid hormones antheridiol and the oogoniols. Sexual reproduction is initiated by the vegetative female mycelium which secretes the first hormone, antheridiol (hormone A). In response to antheridiol, male hyphae react by putting out numerous thin lateral branches which secrete another hormone, the oogoniol (hormone B). These in turn induce the development of spherical oogonial initials on female hyphae. Oogonial initials are believed to secrete relatively large amounts of antheridiol (hormone A), which directs the growth of the antheridial hyphae to the developing oogonium, leading to conjugation of the sex organs.

References

T. C. McMorris, R. Seshadri, G. R. Weihe, G. P. Arsenault, and A. W. Barksdale; Structures of Oogoniol-1, -2, and -3, Steroidal Sex Hormones of the Water Mold, *Achlya*; J. Amer. Chem. Soc., Vol. 97, pp. 2544-2545(1975).

M. W. Preus and T. C. McMorris; The Configuration at C-24 in Oogoniol (24*R*-3β,11α,15β,29-Tetrahydroxystigmast-5-en-7-one and Identification of 24(28)-Dehydrooogoniols as Hormones in *Achlya*; J. Amer. Chem. Soc., Vol. 101, pp. 3066-3071(1979).

Common/Systematic Name

Oogoniol-1

Molecular Formula/Molecular Weight

$C_{33}H_{54}O_6$; MW = 546.39204

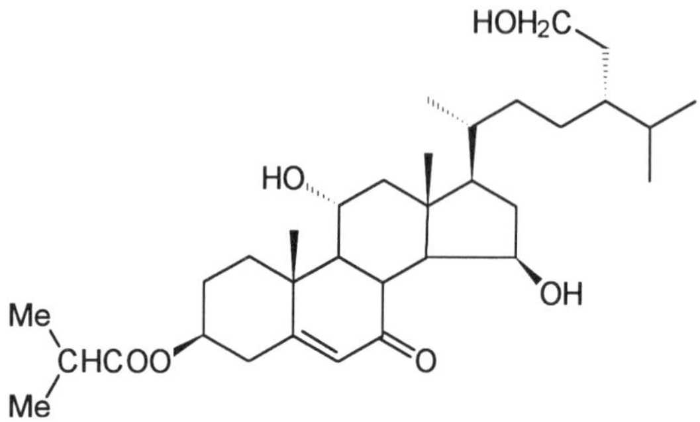

Fungal Source

Achlya heterosexualis.

Isolation/Purification

Crude active crystalline fractions from chromatography of methylene chloride extracts of the culture liquid were further purified by multiple development preparative thin-layer chromatography with ethyl acetate-petroleum ether (1:1, v/v) to give oogoniol-1.

Biological Activity

Sexual reproduction in *Achlya* is mediated by the steroid hormones antheridiol and the oogoniols. Sexual reproduction is initiated by the vegetative female mycelium which secretes the first hormone, antheridiol (hormone A). In response to antheridiol, male hyphae react by putting out numerous thin lateral branches which secrete another hormone, the oogoniol (hormone B). These in turn induce the development of spherical oogonial initials on female hyphae. Oogonial initials are believed to secrete relatively large amounts of antheridiol (hormone A), which directs the growth of the antheridial hyphae to the developing oogonium, leading to conjugation of the sex organs.

Spectral Data

UV:

λ_{max}^{EtOH} 235nm (ϵ = 13,500).

IR:

$\nu_{max}^{Chloroform}$ 3625, 3525, 1732, and 1662cm^{-1}.

^{1}H NMR:

($CDCl_3$) 0.84(t, *J*=7Hz, 29-H); 0.95(d, *J*=6Hz, 27-H); 0.98(18-H); 1.01(d, *J*=7.0Hz, 21-H, probably due to the presence of the dehydro derivatives also present in the sample); 1.16(d, *J*=7Hz, isobutyrate-CH_3); 1.36(19-H); 3.65(m, 26-H); 4.16(m, 11β-H, OH); 4.71(m, 3αH, 15α-H); and 5.82ppm (s, 6-H).

Mass Spectrum:

LREIMS: 458(100), 301(5), 299(21), 283(30), and 161*m/e* (62). HREIMS: 546.3908*m/e* (calcd for $C_{33}H_{54}O_6$, 546.3920).

References

T. C. McMorris, R. Seshadri, G. R. Weihe, G. P. Arsenault, and A. W. Barksdale; Structures of Oogoniol-1, -2, and -3, Steroidal Sex Hormones of the Water Mold, *Achlya*; J. Amer. Chem. Soc., Vol. 97, pp. 2544-2545(1975).

M. W. Preus and T. C. McMorris; The Configuration at C-24 in Oogoniol (24R-3β,11α,15β,29-Tetrahydroxystigmast-5-en-7-one) and Identification of 24(28)-Dehydrooogoniols as Hormones in *Achlya*; J. Amer. Chem. Soc., Vol. 101, pp. 3066-3071(1979).

Common/Systematic Name

Oogoniol-2

Molecular Formula/Molecular Weight

$C_{32}H_{52}O_6$; MW = 532.37639

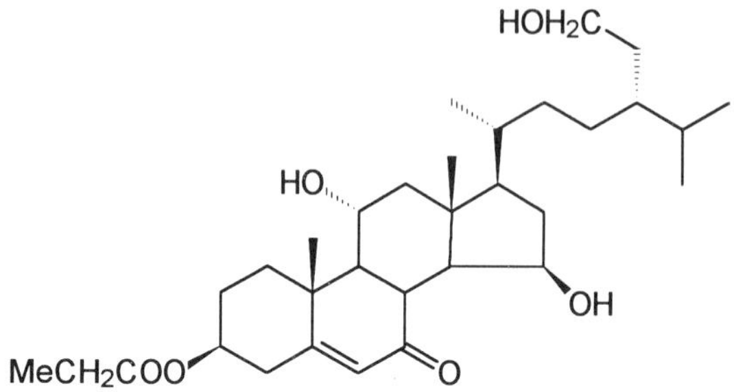

General Characteristics

Crystals from ethyl acetate-petroleum ether; mp., 161-163°C.

Fungal Source

Achlya heterosexualis.

Isolation/Purification

Crude active crystalline fractions from chromatography of methylene chloride extracts of the culture liquid were further purified by multiple development preparative thin-layer chromatography with ethyl acetate-petroleum ether (1:1, v/v), to give oogoniol-2.

Biological Activity

Sexual reproduction in *Achlya* is mediated by the steroid hormones antheridiol and the oogoniols. Sexual reproduction is initiated by the vegetative female mycelium which secretes the first hormone, antheridiol (hormone A). In response to antheridiol, male hyphae react by putting out numerous thin lateral branches which secrete another hormone, the oogoniol (hormone B). These in turn induce the development of spherical oogonial initials on female hyphae. Oogonial initials are believed to secrete relatively large amounts of antheridiol (hormone A), which directs the growth of the antheridial hyphae to the developing oogonium, leading to conjugation of the sex organs.

Spectral Data

UV:

λ_{max}^{EtOH} 235nm (ϵ = 14,300).

IR:
$\nu_{max}^{Chloroform}$ 3625, 3540, 1735, and 1660cm^{-1}.

^{1}H NMR:
($CDCl_3$) 0.84(t, *J*=7Hz, 29-H); 0.95(d, *J*=7 Hz, 27-H); 0.98(18-H); 1.05(d, *J*=7.0Hz, 21-H); 1.10(t, *J*=6 Hz, propionate-CH_3); 1.36(19-H); 2.32(q, *J*=6.0Hz, propionate-CH_2); 3.65(m, 26-H); 4.16(m, 11β-H, OH); 4.74(m, 3α-H, 15α-H); and 5.82ppm (s, 6H).

Mass Spectrum:
LREIMS: 532(M^+, 9), 458(100), 440(27), 301(7), 299(20), 283(28) and 161*m/e* (57);
HREIMS: 532.3762*m/e* (calcd for $C_{32}H_{52}O_6$, 532.3764).

References
T. C. McMorris, R. Seshadri, G. R. Weihe, G. P. Arsenault, and A. W. Barksdale; Structures of Oogoniol-1, -2, and -3, Steroidal Sex Hormones of the Water Mold, *Achlya*; J. Amer. Chem. Soc., Vol. 97, pp. 2544-2545(1975).

M. W. Preus and T. C. McMorris; The Configuration at C-24 in Oogoniol (24R-3β,11α,15β,29-Tetrahydroxystigmast-5-en-7-one) and Identification of 24(28)-Dehydrooogoniols as Hormones in *Achlya*; J. Amer. Chem. Soc., Vol. 101, pp. 3066-3071(1979).

Common/Systematic Name

Oogoniol-3

Molecular Formula/Molecular Weight

$C_{31}H_{50}O_6$; MW = 518.36074

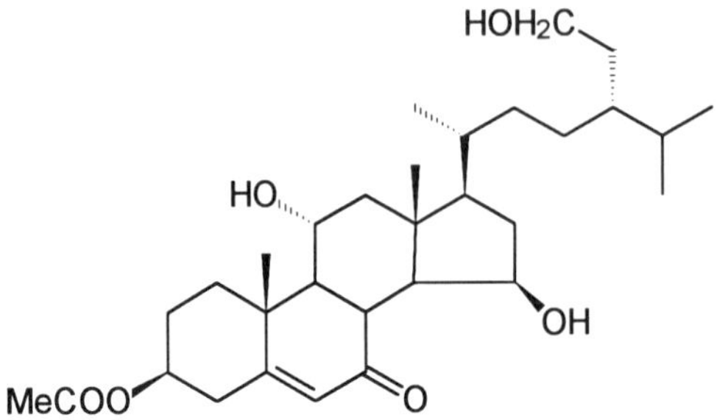

Fungal Source

Achlya heterosexualis.

Isolation/Purification

Crude active crystalline fractions from chromatography of methylene chloride extracts of the culture liquid were further purified by multiple development preparative thin-layer chromatography with ethyl acetate-petroleum ether (1:1, v/v), to give oogoniol-3.

Biological Activity

Sexual reproduction in *Achlya* is mediated by the steroid hormones antheridiol and the oogoniols. Sexual reproduction is initiated by the vegetative female mycelium which secretes the first hormone, antheridiol (hormone A). In response to antheridiol, male hyphae react by putting out numerous thin lateral branches which secrete another hormone, the oogoniol (hormone B). These in turn induce the development of spherical oogonial initials on female hyphae. Oogonial initials are believed to secrete relatively large amounts of antheridiol (hormone A), which directs the growth of the antheridial hyphae to the developing oogonium, leading to conjugation of the sex organs.

Spectral Data

IR:

$\nu_{max}^{Chloroform}$ 3620, 1735, and 1660cm^{-1}.

^{1}H NMR:

($CDCl_3$) 0.84(t, *J*=7Hz, 29-H); 0.95(d, *J*= 6Hz, 27-H); 0.98(18-H); 1.03(d, *J*=7.0Hz, 21-H); 1.36(19-H); 2.04(acetate); 3.66(m, 26-H); 4.14(m, 11β-H, OH); 4.72(m, 3α-H, 15α-H); and 5.82ppm (s, 6-H).

Mass Spectrum:
HREIMS: 518.3588*m/e* (M^+).

References

T. C. McMorris, R. Seshadri, G. R. Weihe, G. P. Arsenault, and A. W. Barksdale; Structures of Oogoniol-1, -2, and -3, Steroidal Sex Hormones of the Water Mold, *Achlya*; J. Amer. Chem. Soc., Vol. 97, pp. 2544-2545(1975).

M. W. Preus and T. C. McMorris; The Configuration at C-24 in Oogoniol (24R-3β,11α,15β,29-Tetrahydroxystigmast-5-en-7-one) and Identification of 24(28)-Dehydrooogoniols as Hormones in *Achlya*; J. Amer. Chem. Soc., Vol. 101, pp. 3066-3071(1979).

Common/Systematic Name

15-Keto-oogoniol-2

Molecular Formula/Molecular Weight

$C_{32}H_{50}O_6$; MW = 530.36074

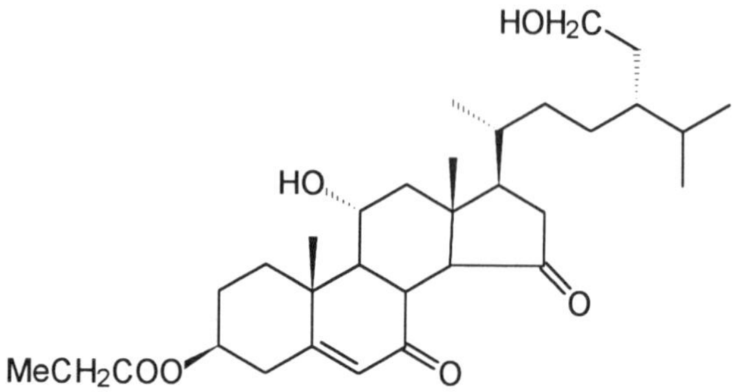

Fungal Source

Achlya heterosexualis.

Isolation/Purification

Crude active crystalline fractions from chromatography of methylene chloride extracts of the culture liquid were further purified by multiple development preparative thin-layer chromatography with ethyl acetate-petroleum ether (1:1, v/v) followed by crystallization from methanol. The dehydro and 15-keto derivatives occurred as minor metabolites together with parent metabolites and final purification could only be achieved with HPLC (μ-Bondapak C_{18} column eluted with a mobile phase of 70% methanol in water and a flow rate of 2.6ml/min at 2000psi).

Biological Activity

Sexual reproduction in *Achlya* is mediated by the steroid hormones antheridiol and the oogoniols. Sexual reproduction is initiated by the vegetative female mycelium which secretes the first hormone, antheridiol (hormone A). In response to antheridiol, male hyphae react by putting out numerous thin lateral branches which secrete another hormone, the oogoniol (hormone B). These in turn induce the development of spherical oogonial initials on female hyphae. Oogonial initials are believed to secrete relatively large amounts of antheridiol (hormone A), which directs the growth of the antheridial hyphae to the developing oogonium, leading to conjugation of the sex organs.

References

T. C. McMorris; Sex Hormones of the Aquatic Fungus *Achlya*; Lipids, Vol. 13, pp. 716-722(1978).

M. W. Preus and T. C. McMorris; The Configuration at C-24 in Oogoniol (24R-3β,11α, 15β,29-Tetrahydroxystigmast-5-en-7-one) and Identification of 24(28)-Dehydro-oogoniols as Hormones in *Achlya*; J. Amer. Chem. Soc., Vol. 101, pp. 3066-3071(1979).

Common/Systematic Name

24(28)-Dehydro-oogoniol

Molecular Formula/Molecular Weight

$C_{29}H_{46}O_5$; MW = 474.33452

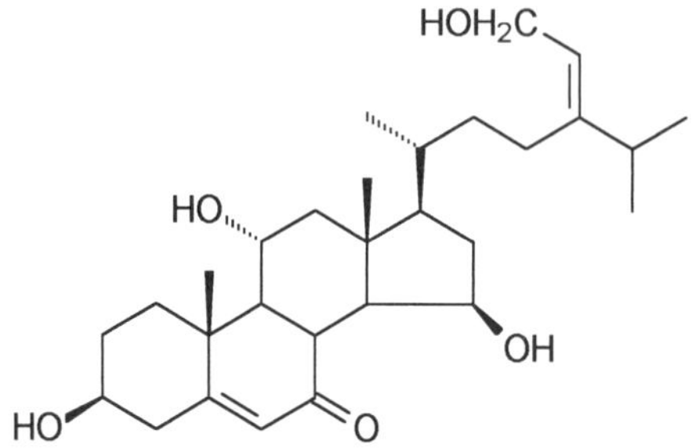

Fungal Source

Achlya heterosexualis.

Isolation/Purification

Crude active crystalline fractions from chromatography of methylene chloride extracts of the culture liquid were further purified by multiple development preparative thin-layer chromatography with ethyl acetate-petroleum ether (1:1, v/v) followed by crystallization from methanol. The dehydro derivatives occurred as minor metabolites together with parent metabolites and final purification could only be achieved with HPLC (μ-Bondapak C_{18} column eluted with a mobile phase of 70% methanol in water and a flow rate of 2.6ml/min at 2000psi).

Biological Activity

Sexual reproduction in *Achlya* is mediated by the steroid hormones antheridiol and the oogoniols. Sexual reproduction is initiated by the vegetative female mycelium which secretes the first hormone, antheridiol (hormone A). In response to antheridiol, male hyphae react by putting out numerous thin lateral branches which secrete another hormone, the oogoniol (hormone B). These in turn induce the development of spherical oogonial initials on female hyphae. Oogonial initials are believed to secrete relatively large amounts of antheridiol (hormone A), which directs the growth of the antheridial hyphae to the developing oogonium, leading to conjugation of the sex organs. 24(28)Dehydro-oogoniol was active at 50μg/ml (induced oogonial initials).

References

T. C. McMorris; Sex Hormones of the Aquatic Fungus *Achlya*; Lipids, Vol. 13, pp. 716-722(1978).

M. W. Preus and T. C. McMorris; The Configuration at C-24 in Oogoniol (24R-3β,11α, 15β,29-Tetrahydroxystigmast-5-en-7-one) and Identification of 24(28)-Dehydrooogoniols as Hormones in *Achlya*; J. Amer. Chem. Soc., Vol. 101, pp. 3066-3071(1979).

Common/Systematic Name

24(28)-Dehydro-oogoniol-1

11α,15β,29-Trihydroxy-5,24(28)(*E*)-dien-7-one-3β-isobutyrate

Molecular Formula/Molecular Weight

$C_{33}H_{52}O_6$; MW = 544.37639

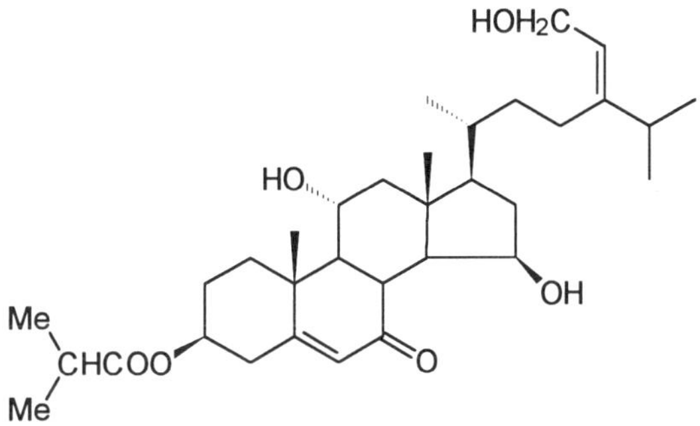

Fungal Source

Achlya heterosexualis.

Isolation/Purification

Crude active crystalline fractions from chromatography of methylene chloride extracts of the culture liquid were further purified by multiple development preparative thin-layer chromatography with ethyl acetate-petroleum ether (1:1, v/v) followed by crystallization from methanol. The dehydro derivatives occurred as minor metabolites together with parent metabolites,and final purification could only be achieved with HPLC (μ-Bondapak C_{18} column eluted with a mobile phase of 70% methanol in water and a flow rate of 2.6ml/min at 2000psi). 24(28)-Dehydro-oogoniol-1 eluted at 11.9 minutes.

Biological Activity

Sexual reproduction in *Achlya* is mediated by the steroid hormones antheridiol and the oogoniols. Sexual reproduction is initiated by the vegetative female mycelium which secretes the first hormone, antheridiol (hormone A). In response to antheridiol, male hyphae react by putting out numerous thin lateral branches which secrete another hormone, the oogoniol (hormone B). These in turn induce the development of spherical oogonial initials on female hyphae. Oogonial initials are believed to secrete relatively large amounts of antheridiol (hormone A), which directs the growth of the antheridial hyphae to the developing oogonium, leading to conjugation of the sex organs. 24(28)-Dehydro-oogoniol-1 was active at 50μg/ml (induced oogonial initials).

Spectral Data

^{1}H NMR:

($CDCl_3$) 0.84(t, *J*=7Hz, 29-H); 0.94(d, *J*=7Hz, 27-H); 0.98(3-H, s); 1.01(6-H, d, *J*=7.0Hz, isobutyrate CH_3, C-26, and C-27 methyls); 1.36(3-H, s); 3.65(broad peak, H-29, the methylene protons on C-29 sometimes appear together with the C-11β protons at about 4.1ppm); 4.13(m, 11β-H); 4.69(m, 3α-H, 15α-H, 15α-H); and 5.83ppm (s).

Mass Spectrum:

LREIMS: 456(M^+ -88, 100), 440(31), 425(9), 422(11), 407(6), 301(6), 299(3), 283(37), 265(14), 245(9), 227(11), and 161*m/e* (54).

Reference

M. W. Preus and T. C. McMorris; The Configuration at C-24 in Oogoniol (24R-3β,11α, 15β,29-Tetrahydroxystigmast-5-en-7-one) and Identification of 24(28)-Dehydro-oogoniols as Hormones in *Achlya*; J. Amer. Chem. Soc., Vol. 101, pp. 3066-3071(1979).

Common/Systematic Name

24(28)-Dehydro-oogoniol-2

Molecular Formula/Molecular Weight

$C_{32}H_{50}O_6$; MW = 530.36074

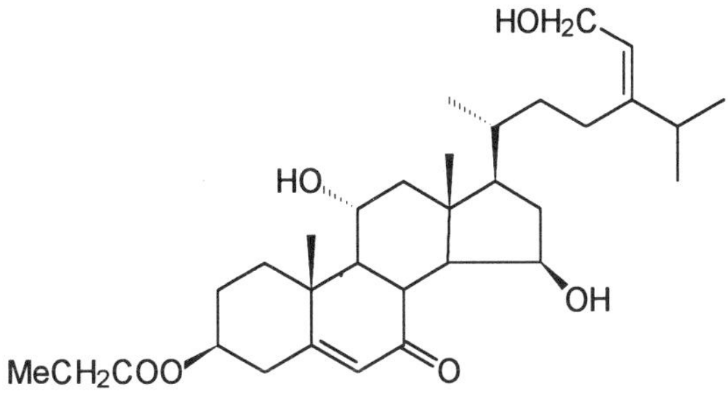

Fungal Source

Achlya heterosexualis.

Isolation/Purification

Crude active crystalline fractions from chromatography of methylene chloride extracts of the culture liquid were further purified by multiple development preparative thin-layer chromatography with ethyl acetate-petroleum ether (1:1) followed by crystallization from methanol. The dehydro derivatives occurred as minor metabolites together with parent metabolites, and final purification could only be achieved with HPLC (μ-Bondapak C_{18} column eluted with a mobile phase of 70% methanol in water and a flow rate of 2.6ml/min at 2000psi).

Biological Activity

Sexual reproduction in *Achlya* is mediated by the steroid hormones antheridiol and the oogoniols. Sexual reproduction is initiated by the vegetative female mycelium which secretes the first hormone, antheridiol (hormone A). In response to antheridiol, male hyphae react by putting out numerous thin lateral branches which secrete another hormone, the oogoniol (hormone B). These in turn induce the development of spherical oogonial initials on female hyphae. Oogonial initials are believed to secrete relatively large amounts of antheridiol (hormone A), which directs the growth of the antheridial hyphae to the developing oogonium, leading to conjugation of the sex organs. 24(28)-Dehydro-oogoniol-2 was active at 50μg/ml (induced oogonial initials).

Spectral Data

^{1}H NMR:

($CDCl_3$) 0.84(t, J=7Hz, 29-H); 0.94(d, J=7Hz, 27-H); 0.98(3-H,s); 1.01(6-H, d, J=7.0Hz, isobutyrate CH_3, C-26 and C-27 methyls); 1.36 (3-H, s); 3.65(broad peak, H-29, the methylene protons on C-29 sometimes appear together with the C-11β protons at about 4.1ppm); 4.13(m, 11β-H); 4.69(m, 3α-H, 15α-H); and 5.83ppm (s).

Mass Spectrum:
LREIMS: 456(M^+ - 88, 100), 440(31), 425(9), 422(11), 407(6), 301(6), 299(3), 283(37), 265(14), 245(9), 227(11), and 161*m/e* (54).

References

T. C. McMorris; Sex Hormones of the Aquatic Fungus *Achlya*; Lipids, Vol. 13, pp. 716-722(1978).

M. W. Preus and T. C. McMorris; The Configuration at C-24 in Oogoniol (24R-3β,11α, 15β,29-Tetrahydroxystigmast-5-en-7-one) and Identification of 24(28)-Dehydro-oogoniols as Hormones in *Achlya*; J. Amer. Chem. Soc., Vol. 101, pp. 3066-3071(1979).

Common/Systematic Name

24(28)-Dehydro-oogoniol-3

Molecular Formula/Molecular Weight

$C_{31}H_{48}O_6$; MW = 516.34509

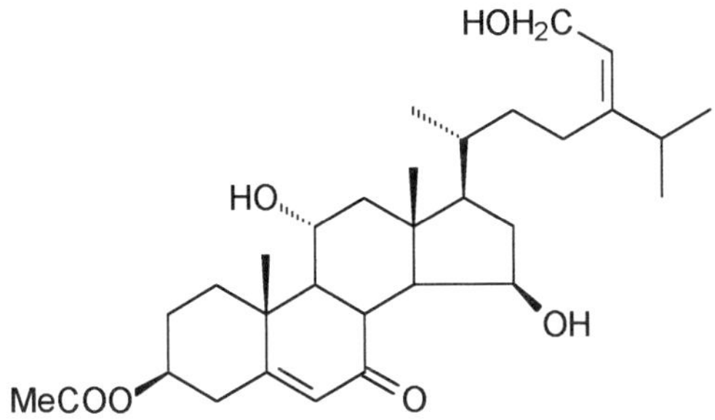

Fungal Source

Achlya heterosexualis.

Isolation/Purification

Crude active crystalline fractions from chromatography of methylene chloride extracts of the culture liquid were further purified by multiple development preparative thin-layer chromatography with ethyl acetate-petroleum ether (1:1, v/v) followed by crystallization from methanol. The dehydro derivatives occurred as minor metabolites together with parent metabolites, and final purification could only be achieved with HPLC (μ-Bondapak C_{18} column eluted with a mobile phase of 70% methanol in water and a flow rate of 2.6 ml/min at 2000psi).

Biological Activity

Sexual reproduction in *Achlya* is mediated by the steroid hormones antheridiol and the oogoniols. Sexual reproduction is initiated by the vegetative female mycelium which secretes the first hormone, antheridiol (hormone A). In response to antheridiol, male hyphae react by putting out numerous thin lateral branches which secrete another hormone, the oogoniol (hormone B). These in turn induce the development of spherical oogonial initials on female hyphae. Oogonial initials are believed to secrete relatively large amounts of antheridiol (hormone A), which directs the growth of the antheridial hyphae to the developing oogonium, leading to conjugation of the sex organs.

Spectral Data

^1H NMR:

($CDCl_3$) The dehydro derivatives of the oogoniols can be distinguished in the NMR from their parent oogoniols by the presence of a strong chemical shift at 1.01ppm (9H, d, J=7.0Hz, isobutyrate CH_3, C-26, and C-27 methyls) and the methylene protons on

<u>References</u>

T. C. McMorris; Sex Hormones of the Aquatic Fungus *Achlya*; Lipids, Vol. 13, pp. 716-722(1978).

M. W. Preus and T. C. McMorris; The Configuration at C-24 in Oogoniol (24R-3β,11α, 15β,29-Tetrahydroxystigmast-5-en-7-one) and Identification of 24(28)-Dehydro-oogoniols as Hormones in *Achlya*; J. Amer. Chem. Soc., Vol. 101, pp. 3066-3071(1979).

9

Ganoderic Acids

Ganoderic acid A
Ganoderic acid B
Ganoderic acid C1
Ganoderic acid C2
Ganoderic acid D
Ganoderic acid E
Ganoderic acid F
Ganoderic acid G
Ganoderic acid H
Ganoderic acid I
Ganoderic acid J
Ganoderic acid K
Ganoderic acid L
Compound B8
Compound B9
Ganoderic acid Ma
Ganoderic acid Mb
Ganoderic acid Mc
Ganoderic acid Md
Ganoderic acid Me
Ganoderic acid Mf
Ganoderic acid Mg
Ganoderic acid Mh
Ganoderic acid Mi
Ganoderic acid Mj
Ganoderic acid Mk
Ganoderic acid M
Ganoderic acid N
Ganoderic acid O
Ganoderic acid R
Ganoderic acid S
Ganoderic acid T
Ganoderic acid U
Ganoderic acid V
Ganoderic acid W
Ganoderic acid X
Ganoderic acid Y
Ganoderic acid Z
Ganodermic acid R
Ganodermic acid S

Common/Systematic Name

Ganoderic acid A

7β,15α-Dihydroxy-3,11,23-trioxo-5α-lanost-8-en-26-oic acid

Molecular Formula/Molecular Weight

$C_{30}H_{44}O_7$; MW = 516.30870

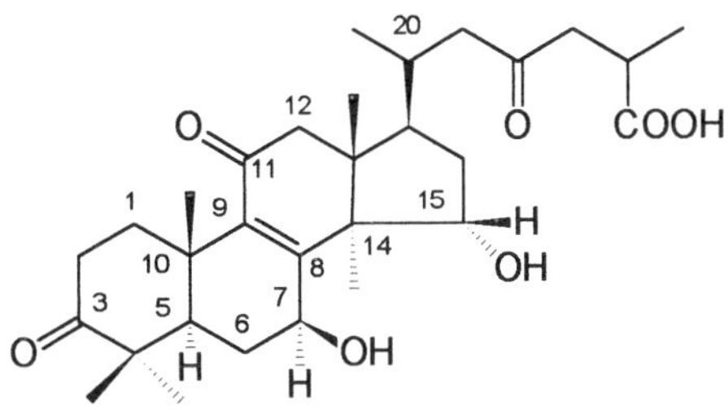

General Characteristics

An amorphous powder; $[\alpha]_D^{27}$ + 153.8° (c = 0.156, in $CHCl_3$).

Fungal Source

Isolated from the dried fruiting bodies (mushrooms) of *Ganoderma lucidum* (Polyporaceae).

Isolation/Purification

The dried chipped epidermis of *Ganoderma lucidum* was extracted twice with $CHCl_3$. The combined extracts were evaporated at 40-45°C under reduced pressure to about 1/10 of the original volume, and the resulting solution was extracted with sat. aq. $NaHCO_3$ solution. The combined extracts were acidified with 6N HCl (pH 3-4) at 0°C. The precipitate was extracted with $CHCl_3$, dried (Na_2SO_4) and the solvent removed under reduced pressure, leaving crude material. A part of the crude material was chromatographed on silica gel. Elution with $CHCl_3$:MeOH, 98:2 (v/v), gave a bitter substance as an amorphous powder (ganoderic acid A). Further elution with $CHCl_3$:MeOH, 95:5 (v/v) gave a slightly bitter component as an amorphous powder (ganoderic acid B).

Biological Activity

The mushrooms have long been used as a home remedy in China and Japan.

Spectral Data

IR:

($CHCl_3$) 3300, 2600-2400, 1700, 1650, 1250, 1100, and 1000cm^{-1}.

^{1}H NMR:

(CDCl$_3$) H-1β, 2.85; H-1α, 1.42(ddd, *J*=14, 8, and 8Hz); H-5α, 1.67(dd, *J*=12.5, 1.5Hz); H-6α, 1.70(ddd, *J*=12.5, 12.5, 9.7Hz); H-6β, 2.06(ddd, *J*=12.5, 7.6, 1.5Hz); H-7α, 4.62(dd, *J*=9.7, 7.6Hz); H-15β, 4.97(dd, *J*=9.2, 7.8Hz); H-16, 1.81; 18CH$_3$, 0.98(s); 19CH$_3$, 1.26(s); 21CH$_3$, 0.90(d, *J*=6.2Hz); H-22, 2.41(dd, *J*=17.6, 3.2Hz); H-22, 2.30(dd, *J*=16.7, 8.6Hz); H-24, 2.85(dd, *J*=18.0, 8.6Hz); H-24, 2.50(dd, *J*=18.0, 4.9Hz); H-25, 2.97(qadd, *J*=7.3, 8.6, 4.9Hz); 26CH$_3$, 1.23(d, *J*=7.3Hz); 30CH$_3$, 1.10(s); 31CH$_3$, 1.29(s); COOH(DMSO), 12.05(br, s); and OH(DMSO), 5.32 and 4.50ppm.

^{13}C NMR:

(CDCl$_3$) C-1, 35.7; C-2, 34.4; C-3, 208.4; C-4, 47.0; C-5, 49.2; C-6, 29.2; C-7, 69.1; C-8, 159.6; C-9, 140.6; C-10, 46.8; C-11, 200.0; C-12, 51.9; C-13, 38.2; C-14, 54.2; C-15, 72.6; C-16, 36.2; C-17, 48.3; C-18, 17.4; C-19, 19.6; C-20, 32.8; C-21, 19.4; C-22, 49.8; C-23, 217.3; C-24, 46.7; C-25, 34.8; C-26, 27.4; C-27, 180.1; C-30, 17.0; C-31, 20.8; and C-32, 19.8.

Mass Spectrum:

516.3013(M$^+$, 50%; calcd for $C_{30}H_{44}O_7$, 516.3087), 498(21), 139(52), and 115*m/e* (57).

Reference

T. Kubota, Y. Asaka, I. Miura, and H. Mori; Structures of Ganoderic Acid A and B, Two New Lanostane Type Bitter Triterpenes from *Ganoderma lucidum* (FR.) Karst.; Helvetica Chimica Acta, Vol. 65, pp. 611-619(1982).

Common/Systematic Name

Ganoderic acid B

Molecular Formula/Molecular Weight

$C_{30}H_{44}O_7$; MW = 516.30870

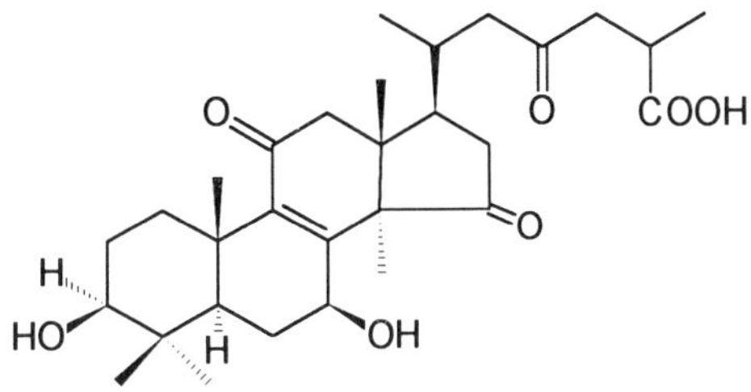

General Characteristics

Ganoderic acid B was crystallized from EtOAc-benzene as pale yellow needles; mp., 212-213°C; $[\alpha]_D^{24}$ +138° (*c*=0.2, in MeOH). Acetate derivative; pale yellow prisms from ethyl acetate; mp., 214-215°C.

Fungal Source

Fruiting bodies of *Ganoderma lucidum*.

Isolation/Purification

Dried chipped fruiting bodies of *G. lucidum* were extracted with ethanol; the extract was partitioned between chloroform and water. The chloroform layer was concentrated and separated into its acidic and neutral fractions. The acidic fraction was acidified (pH 3-4) with 6N HCl, and the resulting precipitate was applied to silica gel column chromatography using a solvent system of chloroform-methanol. The eluate was rechromatographed on the silica gel column to give two strongly bitter compounds (lucidenic acid A and ganoderic acid C) and three slightly bitter compounds (lucidenic acid B, lucidenic acid C, and ganoderic acid B).

Biological Activity

Fruiting bodies of *G. lucidum* produce some physiological effects when used as a home remedy.

Spectral Data

UV:

λ_{max}^{EtOH} 254nm (ϵ=10,740).

IR:

(KBr) 3420, 3200-2400(br), 2950, 2875, 1720, 1640, 1450, 1360, 1240, 1160, and $1025 cm^{-1}$.

CD:

λ_{max}^{MeOH} 293(- 3.8), 279(0), 254(+ 17.5), 228(θ), and 212nm (-8.6).

^1H NMR:

($CDCl_3$) 4.80(1H, dd, *J*=8.9 and 8.9Hz); 3.19(1H, dd, *J*=10.8 and 5.4Hz); 2.78(1H, d, *J*=16.8Hz); 2.69(1H, d, *J*=16.8Hz); 1.34(3H, s); 1.22(3H, d, *J*=5.5Hz); 1.21(3H s); 1.03(3H, s); 1.00(3H, s); 0.99(3H, d, *J*=5.8Hz); and 0.85ppm (3H, s).

^{13}C NMR:

($CDCl_3$) 217.3(s); 207.5(s); 197.7(s); 180.2(s); 156.7(s); 142.6(s); 78.2(d); 66.8(d); 28.1(q); 24.3(q); 19.5(q); 18.3(q); 17.3(q); 16.9(q); and 15.4ppm (q).

Mass Spectrum:

516.3110(M^+, $C_{30}H_{44}O_7$, 18%), 498(4), 488(19), 376(11), 358(32), 331(27), 149(15), 139(19), 105(27), 69(36), and 43*m/e* (100).

References

T. Kubota, Y. Asaka, I. Miura, and H. Mori; Structures of Ganoderic Acid A and B, Two New Lanostane Type Bitter Triterpenes from *Ganoderma lucidum* (FR.) Karst.; Helvetica Chimica Acta, Vol. 65, pp. 611-619(1982).

T. Nishitoba, H. Sato, T. Kasai, H. Kawagishi, and S. Sakamura; New Bitter C27 and C30 Terpenoids from the Fungus *Ganoderma lucidum* (Reishi); Agric. Biol. Chem., Vol. 48, pp. 2905-2907(1984).

T. Nishitoba, H. Sato, T. Kasai, H. Kawagishi, and S. Sakamura; New Bitter C27 and C30 Terpenoids from the Fungus *Ganoderma lucidum* (Reishi); Agric. Biol. Chem., Vol. 49, pp. 1793-1798(1985).

Common/Systematic Name

Ganoderic acid C1

7β-Hydroxy-3,11,15,23-tetraoxo-5α-lanost-8-en-26-oic acid

Molecular Formula/Molecular Weight

$C_{30}H_{42}O_7$; MW = 514.29305

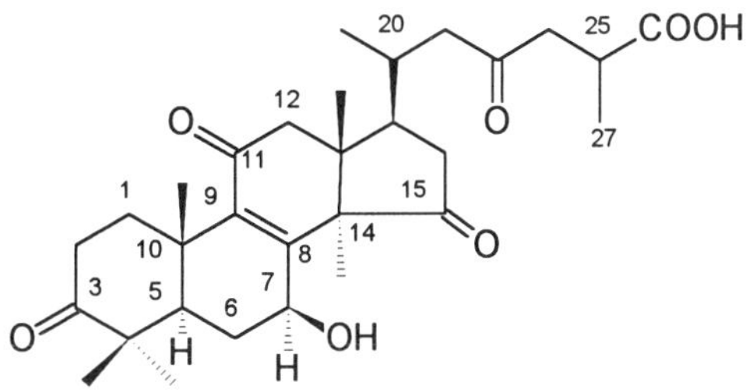

General Characteristics

Methyl ester: colorless prisms from ethanol; mp., 171-173°C; $[\alpha]_D$ + 168° (*c*=1.0, in $CHCl_3$).

Fungal Source

Fruiting bodies of *Ganoderma lucidum*.

Isolation/Purification

The dried powdered surface of the gills of the fruiting bodies of *Ganoderma lucidum* was extracted with ethyl ether. The extract was concentrated and partitioned between ethyl ether and water. The acidic fraction was obtained, methylated with diazomethane, and chromatographed on a silica gel column eluted with different ethyl acetate-benzene mixtures to give 13 fractions. Most fractions were further purified by silica gel column chromatography eluted with various mixtures of ethyl acetate-benzene or various mixtures of acetone-chloroform. Fractions 3 and 4 were combined and recrystallized from ethanol to give methyl ganoderate C1.

Biological Activity

Fruiting bodies of *G. lucidum* produce some physiological effects when used as a home remedy.

Spectral Data (Methyl ester)

^{1}H NMR:

($CDCl_3$) 1.47(1H, dt, *J*=13.5, 8.5Hz, H-1α); 2.95(1H, ddd, *J*=13.5, 7.5, 5.50Hz, H-1β); 2.46(1H, ddd, *J*=16.0, 8.0, 5.5Hz, H-2α); 2.54(1H, ddd, *J*=16.0, 8.5, 7.5Hz, H-2β); 1.56(1H, dd, *J*=13.0, 1.5Hz, H-5); 2.11(1H, ddd, *J*=13.0, 8.0, 1.5Hz, H-6α); 1.67(1H, td, *J*=13.0, 9.2Hz, H-6β); 4.87 (1H, ddd, *J*=9.2, 8.0, 4.5Hz, H-7α); 2.77(1H, br d, J=17.5Hz, H-12α); 2.72(1H, d, J=17.5Hz, H-12β); ca. 2.07(1H, H-16); 2.68(1H, H-16); ca. 2.13(1H, m, H-17); 2.16(1H, m, H-20); 2.37(2H, d, J=5.0Hz, H2-22); 2.44(1H, dd, J=17.5, 5.0Hz, H-24); 2.85(1H, dd, J=17.5, 8.8Hz, H-24); 2.96(1H, dqd, J=9.0, 7.0, 5.0Hz, H-25); 1.03(3H, s, H3-18); 1.26(3H, s, H3-19); 1.00(3H, d, J=6.0Hz, H3-21); 1.19(3H, d, J=7.0Hz, H3-27); 1.13(3H, s, H3-30); 1.11(3H, s, H3-31); 1.34(3H, s, H3-32); 4.055(1H, d, J=4.5Hz, OH); and 3.69ppm (3H, s, $COOCH_3$).

^{13}C NMR:

($CDCl_3$) 35.5, t, C-1; 34.1, t, C-2; 217.3, s, C-3; 46.7, s, C-4; 48.7, d, C-5; 27.6, t, C-6; 66.1, d, C-7; 157.7, s, C-8; 141.0, s, C-9; 38.1, s, C-10; 197.4, s, C-11; 50.0, t, C-12; 44.8, s, C-13; 59.2, s, C-14; 216.3, s, C-15; 40.8, t, C-16; 45.6, d, C-17; 17.5, q, C-18; 18.0, q, C-19; 31.8, d, C-20; 19.5, q, C-21; 48.9, t, C-22; 207.6, s, C-23; 46.6, t, C-24; 34.4, d, C-25; 175.9, s, C-26; 16.9, q, C-27; 26.8, q, C-30; 20.6, q, C-31; 24.5, q, C-32; and 51.7ppm q, CH_3O-.

Mass Spectrum:

HREIMS: 532.3395(M^+, calcd for $C_{31}H_{48}O_7$ 532.3400), 514, 496, 392, 388, 230, 171, 144, 139, and 129*m/e*.

Reference

T. Kikuchi, S. Kanomi (*née* Matsuda), S. Kadota, Y. Murai, K. Tsubono, and Z. Ogita; Constituents of the Fungus *Ganoderma lucidum* (Fr.) Karst. I. Structures of Ganoderic Acids C2, E, I, and K, Lucidenic Acid F and Related Compounds; Chem. Pharm. Bull., Vol. 34, pp. 3695-3712(1986).

Common/Systematic Name

Ganoderic acid C2

3β,7β,15α-Trihydroxy-11,23-dioxo-5α-lanost-8-en-26-oic acid

Molecular Formula/Molecular Weight

$C_{30}H_{46}O_7$; MW = 518.32435

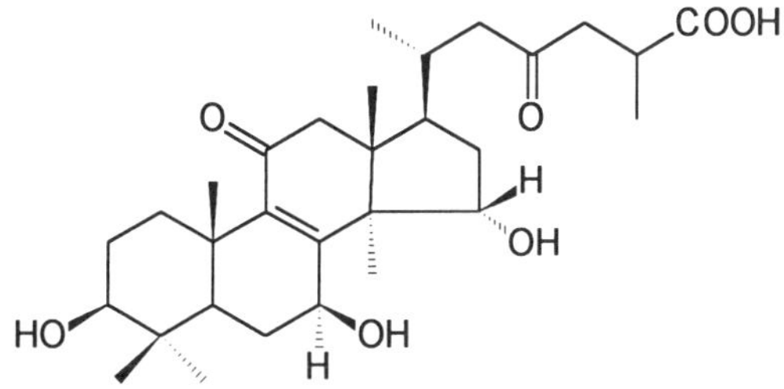

General Characteristics

Methyl ester: colorless prisms from ethyl ether; mp., 199-202°C; $[\alpha]_D$ + 98° (*c*=1.0, in $CHCl_3$).

Fungal Source

Fruiting bodies of *Ganoderma lucidum*.

Isolation/Purification

The dried powdered surface of the gills of the fruiting bodies of *Ganoderma lucidum* was extracted with ethyl ether. The extract was concentrated and partitioned between ethyl ether and water. The acidic fraction was obtained, methylated with diazomethane, and chromatographed on a silica gel column eluted with different ethyl acetate-benzene mixtures to give 13 fractions. Most fractions were further purified by silica gel column chromatography eluted with various mixtures of ethyl acetate-benzene or various mixtures of acetone-chloroform. Fraction 13 was subjected to preparative TLC (acetone-chloroform, 1:9); the more mobile band contained methyl ganoderate C2, methyl ganoderate K and compound B9. Methyl ganoderate K was crystallized as yellow prisms from ethyl ether.

Biological Activity

Fruiting bodies of *G. lucidum* produce some physiological effects when used as a home remedy.

Spectral Data (Methyl ester)

UV:

λ_{max}^{MeOH} 253nm (log ϵ = 4.01).

IR:
(KBr) 3450, 1720, 1710, 1700, and 1650cm^{-1}.

^{1}H NMR:
($CDCl_3$) ca. 0.93(1H, H-1α); 2.74(1H, dt, *J*=13.5, 4.00Hz, H-1β); ca. 1.60(1H, H-2α); ca. 1.66(1H, H-2β); 3.20(1H, dd, *J*=10.8, 5.8Hz, H-3); 0.92(1H, dd, *J*=13.0, 1.0Hz, H-5); 2.14(1H, ddd, *J*=12.5, 7.5, 1.0Hz, H-6α); 1.59(1H, td, *J*=12.5, 10.5Hz, H-6β); 4.54(1H, dd, *J*=10.3, 7.5Hz, H-7α); 2.76(1H, br d, *J*=15.5Hz, H-12α); 2.46(1H, d, *J*=15.5Hz, H-12β); 4.74(1H, br t, *J*=7.5Hz, H-15); ca. 1.78(2H, H2-16); ca. 1.80(1H, H-17); 2.02(1H, m, H-20); 2.24(1H, dd, *J*=16.5, 9.5Hz, H-22); 2.40(1H, dd, *J*=16.5, 3.0Hz, H-22); 2.46(1H, dd, *J*=16.5, 3.0Hz, H-24); 2.83(1H, dd, *J*=17.8, 9.0Hz, H-24); 2.94(1H, dqd, *J*=9.0, 7.0, 5.0Hz, H-25); 0.96(3H, s, H3-18); 1.25(3H, s, H3-19); 0.88(3H, d, *J*=6.5Hz, H3-21); 1.18(3H, d, *J*=7.5Hz, H3-27); 1.02(3H, s, H3-30); 0.84(3H, s, H3-31); 1.24(3H, s, H3-32); and 3.68ppm (3H, s, $COOCH_3$).

Mass Spectrum:
HREIMS: 532.3395(M^+, calcd for $C_{31}H_{48}O_7$ 532.3400), 514, 496, 392, 388, 230, 171, 144, 139, and 129*m/e*.

Reference

T. Kikuchi, S. Kanomi (*nee* Matsuda), S. Kadota, Y. Murai, K. Tsubono, and Z. Ogita; Constituents of the Fungus *Ganoderma lucidum* (Fr.) Karst. I. Structures of Ganoderic Acids C2, E, I, and K, Lucidenic Acid F and Related Compounds; Chem. Pharm. Bull., Vol. 34, pp. 3695-3712(1986).

Common/Systematic Name

Ganoderic acid D

7β,12β-Dihydroxy-3,11,15,23-tetraoxo-5α-lanost-8-en-26-oic acid

Molecular Formula/Molecular Weight

$C_{30}H_{42}O_8$; MW = 530.28797

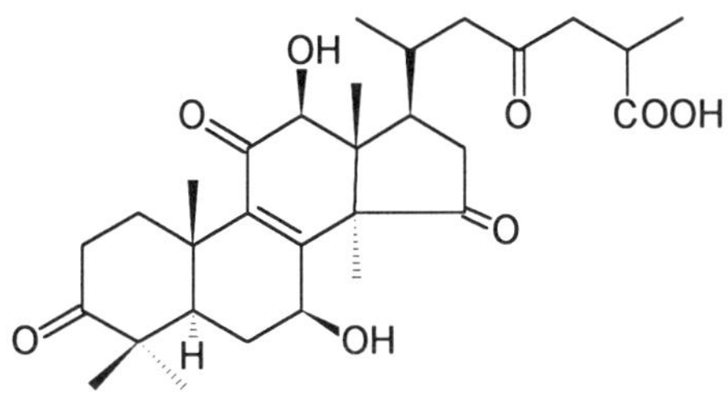

General Characteristics

Ganoderic acid D was obtained as colorless needles from EtOAc; mp., 201-203°C;$[\alpha]^{D}_{22}$ +185 (*c*=0.1, in EtOH).

Fungal Source

Fruiting bodies of *Ganoderma lucidum*.

Isolation/Purification

Dried chipped fruiting bodies of *G. lucidum* were extracted with chloroform and chromatographed on a silica gel column using various concentrations of methanol in chloroform. The bitter principles were further purified with a second silica gel column eluted in a like manner.

Biological Activity

Fruiting bodies of *G. lucidum* produce some physiological effects when used as a home remedy, also bitter tasting. The intensity of bitterness was evaluated to be lucidenic acid D, ganoderic acid C, lucidone A, lucidenic acid A, ganoderic acid B, lucidenic acid B, lucidenic acid C, and lucidenic acid E. Lucidenic acid A, ganoderic acid C, lucidenic acid D, and lucidone A had intense bitterness,and their threshold values were determined to 2.0 x 10^{-6}M, 5 x 10^{-8}M, 5 x 10^{-10}M, and 1 x 10^{-6}M, respectively, when the organoleptic test was carried out by an ascending series of concentrations in a 10% ethanol solution.

Spectral Data

UV:

λ_{max}^{EtOH} 252nm (ϵ = 8800).

IR:

3410, 3400-2500(br), 1730, 1720, and 1660cm^{-1}.

^{1}H NMR:

(C_5D_5N) 5.16(1H, dd, *J*=8.4 and 8.4Hz); 4.69(1H, s); 1.48(3H, s); 1.45(3H, s); 1.40(3H, d. *J*=6.7Hz); 1.36(3H, d, *J*=7.3Hz); 1.21(3H, s); 1.16(3H, s); and 1.10ppm (3H, s).

The ^{1}H NMR data clearly reflect a pyridine-induced solvent shift and the signals at 5.16 and 4.69 showed an upfield shift in $CDCl_3$ to 4.83 and 4.34, respectively. These ^{1}H and ^{13}C NMR spectra were very similar to those of ganoderic acid C, but the ^{13}C NMR signal at 79.2 and the ^{1}H NMR signal at 4.69(1H, s) showed the presence of a hydroxy group at C(12) by analogy with the case of lucidenic acids A and B. When the ^{1}H NMR spectrum was measured in $CDCl_3$, the signal of H-C(12) was observed at 4.38 for lucidenic acid B, which has a β-hydroxy group at C(12), and 3.83 for lucidenic acid E, which has an α-hydroxy group at C(12). On the basis of this finding, the configuration of the hydroxyl group at C(12) for ganoderic acid D was assigned as β, because the signal of H-C(12) was observed at 4.34 in $CDCl_3$. Thus, the structure of ganoderic acid D was established to be 7β,12β-dihydroxy-3,11,15,23-tetraoxo-5α-lanost-8-en-26-oic acid.

^{13}C NMR:

(C_5D_5N) (number of bonded H): 216.6(0), 215.1(0), 208.9(0), 200.9(0) 178.1(0), 159.7(0) 140.8(0), 79.2(1), and 65.7ppm (1).

Mass Spectrum:

The molecular formula, $C_{30}H_{42}O_8$, was based on HRMS (M^+, 530.2875; calcd for $C_{30}H_{42}O_8$, 530.2881).

<u>Reference</u>

T. Nishitoba, H. Sato, and S. Sakamura; New Terpenoids from *Ganoderma lucidum* and Their Bitterness; Agric. Biol. Chem., Vol. 49, pp. 1547-1549(1985).

Common/Systematic Name

Ganoderic acid E

3,7,11,15,23-Pentaoxo-5α-lanost-8-en-26-oic acid

Molecular Formula/Molecular Weight

$C_{30}H_{40}O_7$; MW = 512.27740

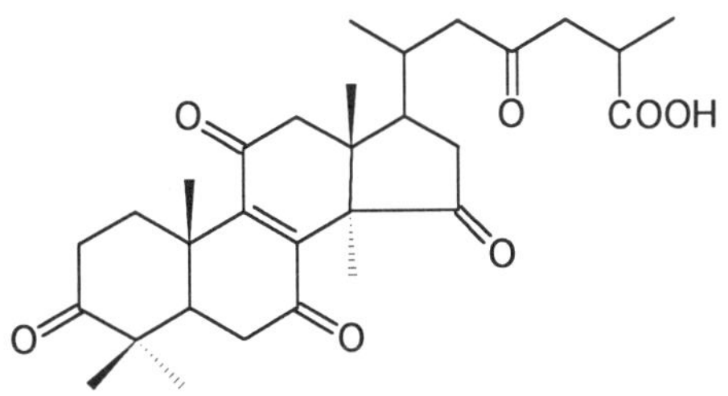

General Characteristics

Methyl ester: Yellow needles; mp., 206-208°C; $[\alpha]_D$ + 167° (in $CHCl_3$).

Fungal Source

Isolated from *Ganoderma lucidum*.

Spectral Data (Methyl ester)

^1H NMR:

($CDCl_3$) 0.89(3H, s, 18-CH_3); 1.29(3H, s, 19-CH_3); 0.99(3H, d, J=6.0Hz, 21-CH_3); 1.18(3H, d, J=7.0Hz, 27-CH_3); 1.12(3H, s, 30-CH_3); 1.15(3H, s, 31-CH_3); 1.66(3H, s, 32-CH_3); and 3.70ppm (3H, s, $COOCH_3$).

^{13}C NMR:

($CDCl_3$) 37.3, C-1, t; 34.7, C-2, t; 217.2, C-3, s; 43.9, C-4, s; 50.9, C-5, d; 33.8, C-6, t; 199.3, C-7, s; 149.7, C-8, s; 146.8, C-9, s; 39.4, C-10, s; 199.3, C-11, s; 48.9, C-12*, t; 47.0, C-13, s ; 57.2, C-14, s; 206.8, C-15, s; 39.8, C-16, t; 44.5, C-17, d; 16.1, C-18, q; 18.6, C-19, q; 32.0, C-20, d; 19.8, C-21, q; 49.1, C-22*, t; 207.6, C-23, s; 46.7, C-24, t; 34.6, C-25, d; 176.1, C-26, s; 17.1, C-27,q; 26.6, C-30, q; 20.9, C-31**, q; 20.3, C-32**, q; and 59.9ppm, CH_3O-, q.

*, ** Assignments may be reversed.

Reference

T. Kikuchi, S. Matsuda, S. Kadota, Y. Murai, and Z. Ogita; Ganoderic Acid D, E, F, and H and Lucidenic Acid D, E, and F, New Triterpenoids from *Ganoderma lucidum*; Chem. Pharm. Bull., Vol. 33, pp. 2624-2627(1985).

Common/Systematic Name

Ganoderic acid F

3,7,11,15,23-Pentaoxo-12-acetoxy-5α-lanost-8-en-26-oic acid

Molecular Formula/Molecular Weight

$C_{32}H_{42}O_9$; MW = 570.28288

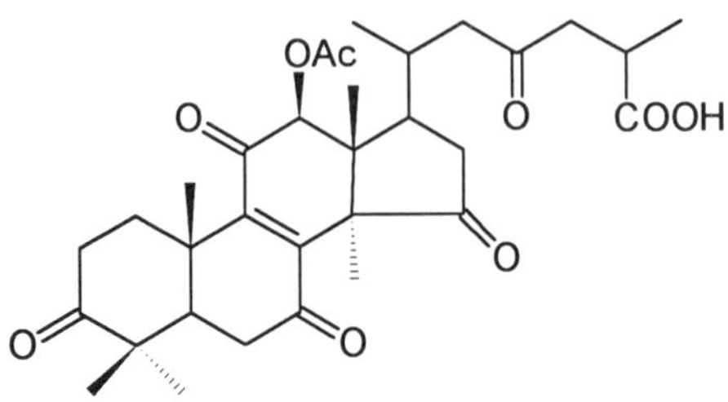

General Characteristics

Methyl ester, $C_{33}H_{44}O_9$; amorphous; $[\alpha]_D$ + 111° (in $CHCl_3$).

Fungal Source

Isolated from *Ganoderma lucidum*.

Isolation/Purification

The acidic fraction from the ether extract of the gills of dried fruit bodies of *Ganoderma lucidum* was methylated with diazomethane and the crude product was separated repeatedly by a combination of silica gel chromatography and preparative thin-layer chromatography to give pure triterpenoids.

Spectral Data (Methyl ester)

UV:

λ_{max} 252nm (log ϵ = 3.91).

IR:

1745, 1735, 1710, and 1700cm^{-1}.

^{1}H NMR:

($CDCl_3$) 0.85(3H, s, 18-CH_3); 1.33(3H, s, 19-CH_3); 0.99(3H, d, *J*=6.0Hz, 21-CH_3); 1.18(3H, d, *J*=7.0Hz, 27-CH_3); 1.12(3H, s, 30-CH_3); 1.14(3H, s, 31-CH_3); 1.80(3H, s, 32-CH_3); 3.68(3H, s, $COOCH_3$); 2.26(OAc); and 5.68ppm (1H, s, 12-H).

^{13}C NMR:

($CDCl_3$) 37.5, C-1, t; 34.1, C-2, t; 214.8, C-3, s; 46.9, C-4, s; 51.0, C-5, d; 33.7, C-6, t; 198.7, C-7, s; 149.9, C-8, s; 146.1, C-9, s; 39.3, C-10, s; 194.1, C-11, s; 79.0, C-12, d; 47.7, C-13, s ; 58.6, C-14, s; 205.4, C-15, s; 37.8, C-16, t; 44.5, C-17, d; 12.1, C-18, q; 18.7, C-19, q; 29.4, C-20, d; 21.6, C-21, q; 48.5, C-22, t; 207.4, C-23, s; 46.7, C-24, t; 34.7, C-25, d; 176.0, C-26, s; 17.1, C-27,q; 27.6, C-30, q; 20.8, C-31*, q; 20.4, C-32*, q; 51.9, CH_3O-, q; 20.9, CO<u>C</u>H_3*, s; and 170.2ppm (<u>C</u>OCH_3, s).

* Assignments may be reversed.

<u>Reference</u>

T. Kikuchi, S. Matsuda, S. Kadota, Y. Murai, and Z. Ogita; Ganoderic Acid D, E, F, and H and Lucidenic Acid D, E, and F, New Triterpenoids from *Ganoderma lucidum*; Chem. Pharm. Bull., Vol. 33, pp. 2624-2627(1985).

Common/Systematic Name
Ganoderic acid G

Molecular Formula/Molecular Weight
$C_{30}H_{44}O_8$; MW = 532.30362

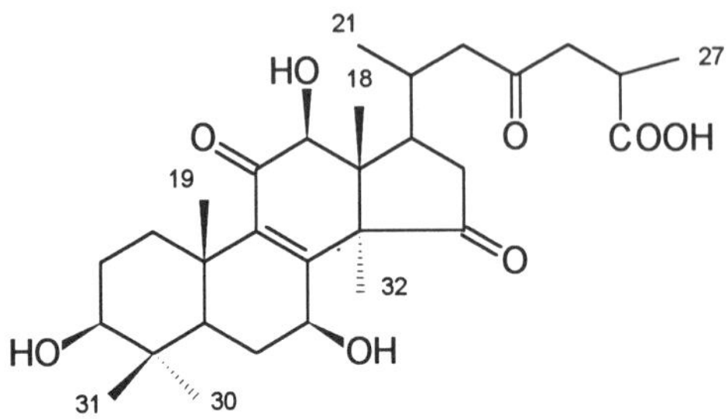

General Characteristics
Isolated and identified as methyl ester. Colorless prisms; mp., 134-135°C; $[\alpha]_D$ + 64° (in $CHCl_3$).

Fungal Source
Ganoderma lucidum.

Spectral Data (Methyl ester)

UV:
λ_{max} 252.5nm (log ϵ = 4.01)

IR:
(KBr) 3450, 1730, 1720, and 1660cm^{-1}.

^{1}H NMR:
($CDCl_3$) 0.88(3H, s, 18-CH_3); 1.32(3H, s, 19-CH_3); 1.19(3H, d, *J*=7.0Hz, 27-CH_3); 1.04(3H, s, 30-CH_3); 0.80(3H, s, 31-CH_3); 1.43(3H, s, 32-CH_3); 3.68(3H, s, $COOCH_3$); 3.22(1H, dd, *J*=10.0, 6.0Hz, 3-H); 4.80(1H, td, *J*=8.5, 4.5Hz, 7-H); 4.38(1H, d, *J*=2.5Hz, 12-H); 4.43(1H, d, *J*=4.5Hz, 7-OH); and 4.03ppm (1H, d, *J*=2.5Hz, 12-OH).

^{13}C NMR:
($CDCl_3$) C-1, 34.6, t; C-2, 27.6, t; C-3, 78.3, d; C-4, 38.6, s; C-5, 49.2, d; C-6, 26.9, t; C-7, 66.2, d; C-8, 157.4, s; C-9, 141.9, s; C-10, 38.4, s; C-11, 199.3, s; C-12, 77.9, d; C-13, 51.9, s; C-14, 60.3, s; C-15, 216.8, s; C-16, 38.4, t; C-17, 45.8, d; C-18, 12.0, q; C-19, 18.8, q; C-20, 28.7, d; C-21, 21.4, q; C-22, 48.4, t; C-23, 208.1, s; C-24, 46.4, t; C-25, 34.6, d; C-26, 176.1, s; C-27, 17.1, q; C-30, 28.1, q; C-31, 15.4, q; C-32, 23.1, q; and OCH_3, 51.9ppm.

Mass Spectrum:
LREIMS: 546(M^+), 375, 306, 139, and 129*m/e*.

Reference

T. Kikuchi, S. Matsuda, Y. Murai, and Z. Ogita; Ganoderic Acid G and I and Ganolucidic Acid A and B, New Triterpenoids from *Ganoderma lucidum*; Chem. Pharm. Bull., Vol. 33, pp. 2628-2631(1985).

Common/Systematic Name

Ganoderic acid H

3β-Hydroxy-7,11,15,23-tetraoxo-12β-acetoxy-5α-lanost-8-en-26-oic acid

Molecular Formula/Molecular Weight

$C_{32}H_{44}O_9$; MW = 572.29853

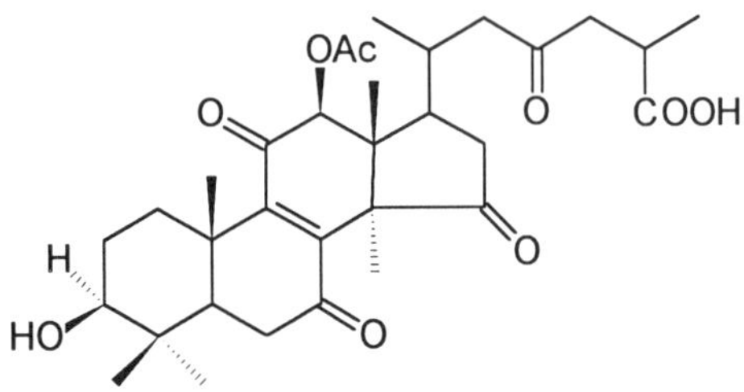

General Characteristics

Methyl ester, $C_{33}H_{46}O_9$; yellow needles; mp., 155-156°C; $[\alpha]_D$ + 55° (in $CHCl_3$).

Fungal Source

Ganoderma lucidum.

Isolation/Purification

The acidic fraction from the ether extract of the gills of dried fruit bodies of *Ganoderma lucidum* was methylated with diazomethane, and the crude product was separated repeatedly by a combination of silica gel chromatography and preparative thin-layer chromatography to give pure triterpenoids.

Spectral Data (Methyl ester)

UV:

λ_{max} 256nm (log ϵ=3.87).

^{1}H NMR:

($CDCl_3$) 0.88(3H, s, 18-CH_3); 1.33(3H, s, 19-CH_3); 0.98(3H, d, *J*=6.0Hz, 21-CH_3); 1.18(3H, d, *J*=7.0Hz, 27-CH_3); 1.03(3H, s, 30-CH_3); 0.82(3H, s, 31-CH_3); 1.73(3H, s, 32-CH_3); 3.68(3H, s, $COOCH_3$); 2.26(OAc); 3.27(1H, dd, *J*=10.0, 6.0Hz, 3-H); and 5.63ppm (1H, s, 12-H).

^{13}C NMR:

($CDCl_3$) 36.6, C-1, t; 27.3, C-2, t; 77.3, C-3, d; 40.4, C-4, s; 51.4, C-5, d; 33.2, C-6, t; 198.7, C-7, s; 151.6, C-8, s; 145.7, C-9, s; 39.1, C-10, s; 193.9, C-11, s; 79.1, C-12, d; 47.9, C-13, s ; 58.4, C-14, s; 205.5, C-15, s; 37.8, C-16, t; 44.7, C-17, d; 12.1, C-18, q; 17.9, C-19, q; 29.3, C-20, d; 21.6, C-21, q; 48.4, C-22, t; 207.4, C-23, s; 46.6, C-24, t; 34.6, C-25, d; 176.0, C-26, s; 17.0, C-27,q; 27.9, C-30, q; 15.5, C-31, q; 21.2, C-32*, q; 51.8, CH_3O-, q; 20.8, CO$\underline{C}H_3$*, s; and 170.1ppm ($\underline{C}OCH_3$, s).

* Assignments may be reversed.

Reference

T. Kikuchi, S. Matsuda, S. Kadota, Y. Murai, and Z. Ogita; Ganoderic Acid D, E, F, and H and Lucidenic Acid D, E, and F, New Triterpenoids from *Ganoderma lucidum*; Chem. Pharm. Bull., Vol. 33, pp. 2624-2627(1985).

Common/Systematic Name
Ganoderic acid I

Molecular Formula/Molecular Weight
$C_{30}H_{44}O_8$; MW = 532.30362

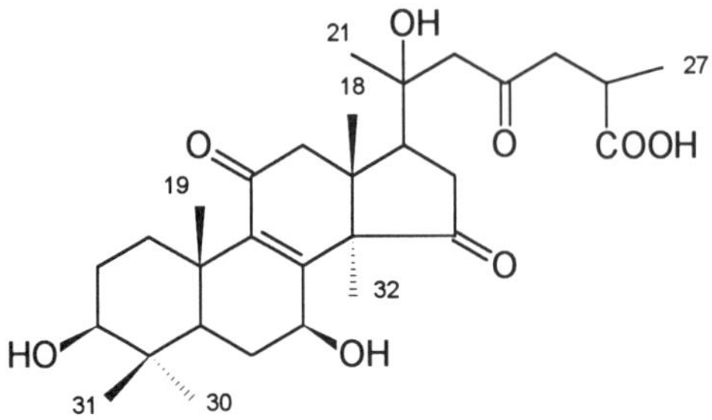

General Characteristics
Isolated and identified as methyl ester. Colorless prisms; mp., 279-281°C; $[\alpha]_D$ + 132° (in $CHCl_3$).

Fungal Source
Ganoderma lucidum.

Spectral Data (Methyl Ester)

UV:
λ_{max} 254.5nm (log ϵ = 3.86).

IR:
(KBr) 3450, 1730, 1710, and 1660cm^{-1}.

^{1}H NMR:
($CDCl_3$) 1.14(3H, s, 18-CH_3); 1.21(3H, s, 19-CH_3); 1.40(3H, d, *J*=6.0Hz, 21-CH_3); 1.19(3H, d, *J*=7.0Hz, 27-CH_3); 1.03(3H, s, 30-CH_3); 0.86(3H, s, 31-CH_3); 1.35(3H, s, 32-CH_3); 3.71(3H, s, $COOCH_3$); 3.22(1H, dd, *J*=10.0, 6.0Hz, 3-H); 4.80(1H, td, *J*=8.5, 4.5Hz, 7-H); and 4.12ppm (1H, d, *J*=4.5Hz, 7-OH).

^{13}C NMR:
($CDCl_3$) C-1, 34.9, t; C-2, 27.8, t; C-3, 78.4, d; C-4, 38.9, s; C-5, 49.2, d; C-6, 26.7, t; C-7, 66.9, d; C-8, 156.6, s; C-9, 142.3, s; C-10, 38.7, s; C-11, 197.8, s; C-12, 50.7, t; C-13, 45.7, s; C-14, 59.7, s; C-15, 217.7, s; C-16, 36.1, t; C-17, 49.3, d; C-18, 19.0, q; C-19, 18.4, q; C-20, 73.0, s; C-21, 26.7, q; C-22, 52.7, t; C-23, 210.4, s; C-24, 47.7, t; C-25, 34.5, d; C-26, 175.9, s; C-27, 17.0, q; C-30, 28.2, q; C-31, 15.5, q; C-32, 24.8, q; and OCH_3, 52.0ppm.

Reference

T. Kikuchi, S. Matsuda, Y. Murai, and Z. Ogita; Ganoderic Acid G and I and Ganolucidic Acid A and B, New Triterpenoids from *Ganoderma lucidum*; Chem. Pharm. Bull., Vol. 33, pp. 2628-2631(1985).

Common/Systematic Name

Ganoderic acid J

Molecular Formula/Molecular Weight

$C_{30}H_{42}O_7$; MW = 514.29305

Methyl ester: $C_{31}H_{44}O_7$; MW = 528.30870

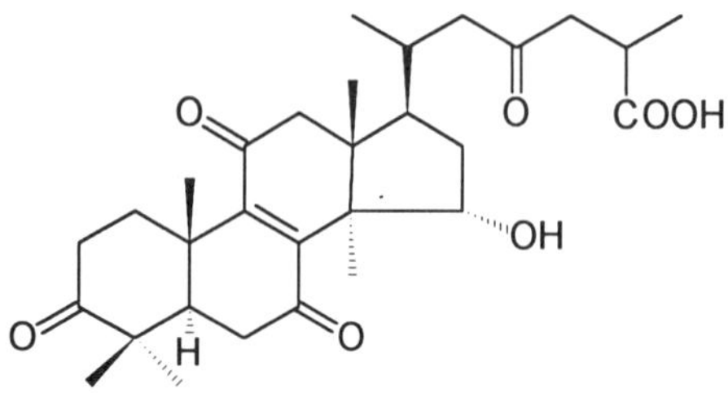

General Characteristics

Methyl ester: Obtained as a pale yellow syrup; $[\alpha]_D^{23}$ +174° (*c*=0.1, in MeOH).

Fungal Source

Isolated from the dried fruiting bodies (mushrooms) of *Ganoderma lucidum*.

Isolation/Purification

The dried chipped epidermis of *Ganoderma lucidum* was extracted with $CHCl_3$. The combined extracts were evaporated at 40-45° under reduced pressure to about 1/10 of the original volume, and the resulting solution was extracted with saturated aqueous $NaHCO_3$ solution. The combined extracts were acidified with 6N HCl (pH 3-4) at 0°C. The precipitate was extracted with $CHCl_3$, dried (Na_2SO_4) and the solvent removed under reduced pressure, leaving crude material. The acidic material was chromatographed on silica gel column chromatography [Fr. 1-11, $CHCl_3$-MeOH , 98:2 (v/v); Fr. 12, $CHCl_3$-MeOH, 9:1 (v/v), Fr. 13, MeOH]. Fractions 7-9 were combined and rechromatographed on silica gel to give a mixture of ganoderic acid J and A, which was methylated with diazomethane and separated by further silica gel column chromatography and preparative TLC. Methyl ganoderate J was obtained as a pale yellow syrup. Fraction 12 was chromatographed on a Lobal column (RP_8, Merck) and the second fraction was methylated with diazomethane. The methylated products were subjected to silica gel column chromatography and methyl ganolucidate C was obtained as colorless needles.

Biological Activity

A bitter principle; the mushrooms of *Ganoderma lucidum* (Fr.) Karst. (Polyporaceae), have long been used as a home remedy in China and Japan.

Spectral Data

UV:

λ_{max}^{MeOH} 270nm (7,500) (methyl ester).

IR:

3480, 1735, 1710, and 1670cm^{-1} (methyl ester).

^{1}H NMR:

(methyl ester) ($CDCl_3$) 0.87(3H, d, *J*= 6.4Hz); 0.91(3H, s); 1.12(3H, s); 1.14(3H, s); 1.18 =(3H, d, *J*=6.4Hz); 1.19(3H, s); 1.27(3H, s); 3.68(3H, s, $COOCH_3$); 4.29(1H, ddd, *J*=7.8, 7.8 and 1.5Hz); 4.45(1H, d, *J*=1.5Hz). The ^{1}H NMR data were very similar to those of ganoderic acid A, but the signal due to the carbinol methine proton was only observed at 4.29ppm (15β-H). This indicated the presence of a carbonyl group at C(7) of methyl ganoderate J instead of the 7-hydroxyl group for ganoderic acid A.

(Free acid)($CDCl_3$) 0.87(1H, d, *J*=6.2 Hz); 0.90(3H, s); 1.12(3H, s); 1.14(3H, s); 1.18(3H, s); 1.22(3H, d, *J*=7.0Hz); 1.27(3H, s); 1.29ppm (1H, dd, *J*=7.2 and 7.2Hz).

Mass Spectrum:

Methyl ganoderate J showed a molecular ion peak at 528.3084*m/e* (calcd for $C_{31}H_{44}O_7$, 528.3088); ganoderic acid J (M^+, 514.2968, calcd. 514.2932).

Reference

T. Nishitoba, H. Sato, and S. Sakamura; New Terpenoids, Ganoderic Acid J and Ganolucidic Acid C from the Fungus *Ganoderma lucidum*; Agric. Biol. Chem., Vol. 49, pp. 3637-3638(1985).

Common/Systematic Name

Ganoderic acid K

3β,15α-Dihydroxy-7,11,23-trioxo-5α-lanost-8-en-26-oic acid

Molecular Formula/Molecular Weight

$C_{30}H_{44}O_7$; MW = 516.30870

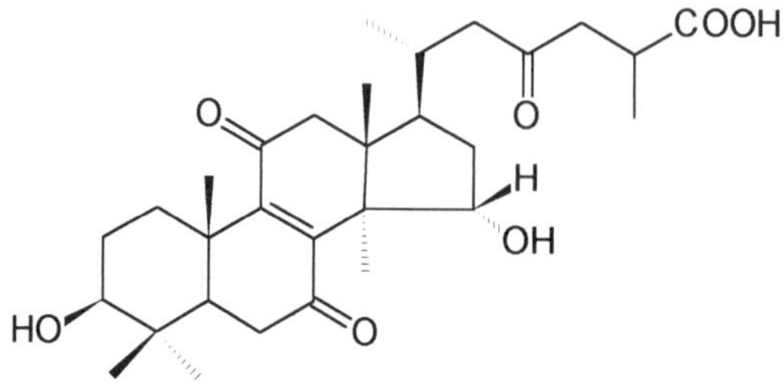

General Characteristics

Methyl ester: pale yellow prisms from ethyl ether; mp., 166-167°C; $[\alpha]_D$ + 165° (*c*=0.3, in $CHCl_3$).

Fungal Source

Fruiting bodies of *Ganoderma lucidum*.

Isolation/Purification

The dried powdered surface of the gills of the fruiting bodies of *Ganoderma lucidum* was extracted with ethyl ether. The extract was concentrated and partitioned between ethyl ether and water. The acidic fraction was obtained, methylated with diazomethane, and chromatographed on a silica gel column eluted with different ethyl acetate-benzene mixtures to give 13 fractions. Most fractions were further purified by silica gel column chromatography eluted with various mixtures of ethyl acetate-benzene or various mixtures of acetone-chloroform. Fraction 13 was subjected to preparative TLC (acetone-chloroform, 1:9, v/v) the more mobile band contained methyl ganoderate C2, methyl ganoderate K, and compound B9. Methyl ganoderate K was crystallized as yellow prisms from ethyl ether.

Biological Activity

Fruiting bodies of *G. lucidum* produce some physiological effects when used as a home remedy.

Spectral Data (Methyl ester)

UV:

λ_{max}^{MeOH} 255nm (log ϵ = 3.43).

IR:
(KBr) 3450, 1730, 1715, and 1665cm^{-1}.

^{1}H NMR:
($CDCl_3$) 1.28(1H, td, *J*=13.0, 4.50Hz, H-1α); 2.84(1H, dt, *J*=13.5, 3.50Hz, H-1β); 1.77(1H, dq, *J*=13.0, 4.5Hz, H-2α); 1.69(1H, tdd, *J*=13.0, 11.5, 3.5Hz, H-2β); 3.28(1H, br dd, *J*=11.0, 5.0Hz, H-3); 1.54(1H, dd, *J*=12.5, 4.5Hz, H-5); 2.54(1H, dd, *J*=16.0, 4.5Hz, H-6α); 2.59(1H, dd, *J*=16.0, 13.0Hz, H-6β); 2.83(1H, d, *J*=17.0Hz, H-12α); 2.54(1H, d, *J*=17.0Hz, H-12β); 4.335(1H, td, *J*=7.5, 1.8Hz, H-15); 1.84(2H, d-like, H_2-16); ca. 1.85(1H, H-17); 2.01(1H, m, H-20); 2.245(1H, dd, *J*=16.5, 9.5Hz, H-22); 2.41(1H, dd, *J*=16.6, 3.0Hz, H-22); 2.455(1H, dd, *J*=17.5, 5.0Hz, H-24); 2.823(1H, dd, *J*=17.5, 8.5Hz, H-24); 2.945(1H, dqd, *J*=8.5, 7.5Hz, H-25); 0.89(3H, s, H_3-18); 1.29(3H, s, H_3-19); 0.87(3H, d, *J*=6.5Hz, H_3-21); 1.18(3H, d, *J*=7.0Hz, H_3-27); 1.03(3H, s, H_3-30); 0.89(3H, s, H_3-31); 1.14(3H, s, H_3-32); 4.485(OH); and 3.67ppm (3H, s, $COOCH_3$).

^{13}C NMR:
($CDCl_3$) 34.3, t, C-1; 27.7, t, C-2; 77.5, d, C-3; 40.2, s, C-4; 49.8, d, C-5; 36.5, t, C-6; 205.3, s, C-7; 154.6, s, C-8; 149.8, s, C-9; 38.9, s, C-10; 201.3, s, C-11; 52.3, t, C-12; 48.0, s, C-13; 52.8, s, C-14; 72.1, d, C-15; 36.3, t, C-16; 48.2, d, C-17; 17.4, q, C-18; 17.6, q, C-19; 32.4, d, C-20; 19.5, q, C-21; 49.5, t, C-22; 208.2, s, C-23; 46.8, t, C-24; 34.7, d, C-25; 176.1, s, C-26; 17.1, q, C-27; 27.8, q, C-30; 15.4, q, C-31; 20.3, q, C-32; and 51.9ppm, q, CH_3O-.

Mass Spectrum:
HREIMS: 530.3252(M^+, calcd for $C_{31}H_{46}O_7$ 530.3244), 499, 368, 341, 303, 209, 171, 144, 139, and 129*m/e*.

Reference

T. Kikuchi, S. Kanomi (*nee* Matsuda), S. Kadota, Y. Murai, K. Tsubono, and Z. Ogita; Constituents of the Fngus *Ganoderma lucidum* (Fr.) Karst. I. Structures of Ganoderic Acids C2, E, I, and K, Lucidenic Acid F and Related Compounds; Chem. Pharm. Bull., Vol. 34, pp. 3695-3712(1986).

Common/Systematic Name

Ganoderic acid L

3β,7β,15α,20-Tetrahydroxy-11,23-dioxo-5α-lanost-8-en-26-oic acid
(Isolated and identified as the methyl ester.)

Molecular Formula/Molecular Weight

$C_{30}H_{46}O_8$; MW = 534.31927

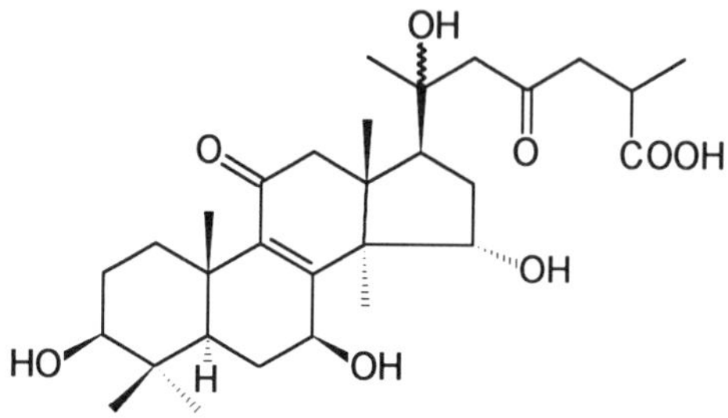

General Characteristics

Methyl ganoderate L: colorless prisms; mp., 228-230°C; $[\alpha]_D^{23}$ + 66° (*c*=0.2, in MeOH).

Fungal Source

Fruiting bodies of *Ganoderma lucidum*.

Isolation/Purification

Dried chipped fruiting bodies of *G. lucidum* were extracted with ethanol; the extract was partitioned between chloroform and water. The chloroform layer was concentrated and separated into its acidic and neutral fractions. The acidic fraction was separated into thirteen fractions; fraction 12 was subjected to Lobar column (RP_8, Merck) chromatography and the second fraction was treated with diazomethane. The resulting product was rechromatographed on silica gel and the Lobar (RP_8) column to give methyl ganoderate L, lucidone C, and methyl lucidenate G. The fourth fraction in the chromatography of Fr. 12 was purified on a silica gel column, preparative TLC and HPLC to give ganolucidic acid D.

Biological Activity

Fruiting bodies of *G. lucidum* produce some physiological effects when used as a home remedy.

Spectral Data (Monomethyl derivative)

UV:

λ_{max}^{MeOH} 256nm (ϵ=6750).

IR:
(KBr) 3430, 1720, and 1655cm^{-1}.

^{1}H NMR:
(C_5D_5N) 5.46(1H, dd, *J*=9.2 and 7.3Hz); 5.00(1H, overlapped); 3.52(1H, dd, *J*=10.6 and 5.1Hz); 1.64(3H, s); 1.59(3H, s); 1.57(3H, s), 1.54(3H, s); 1.29(3H, s); 1.15(3H, d, *J*=6.6Hz); and 1.11ppm (3H, s).

^{13}C NMR:
(C_5D_5N) (number of bonded H): 209.4(0); 200.2(0); 176.2(0); 160.3(0); 141.7(0); 77.6(1); 73.8(0); 72.4(1); and 69.5ppm (1).

Mass Spectrum:
The FD-MS showed 548(M^+, 5.3%), 404(100), and 144*m/e* (33.4); EIMS did not give a molecular ion peak.

Reference

T. Nishitoba, H. Sato, and S. Sakamura; New Terpenoids, Ganolucidic Acid D, Ganoderic Acid L, Lucidone C, and Lucidenic Acid G, from the Fungus *Ganoderma lucidum;* Agric. Biol. Chem., Vol. 50, pp. 809-811(1986).

Common/Systematic Name

Compound B8

7α,15α-Dihydroxy-3,11,23-trioxo-5α-lanost-8-en-26-oic acid

Molecular Formula/Molecular Weight

$C_{30}H_{44}O_7$; MW = 516.30870

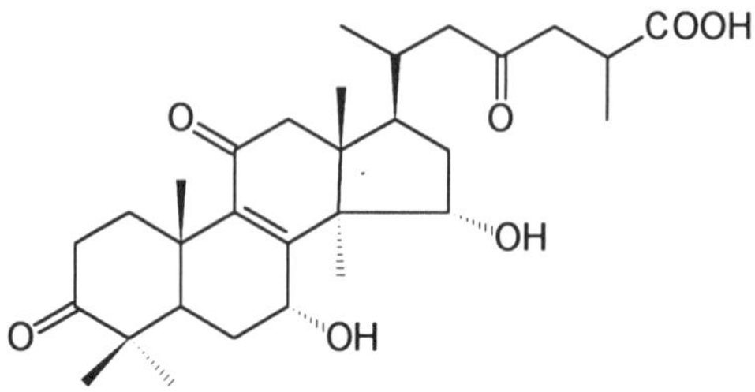

General Characteristics

Methyl ester: colorless prisms from ether; mp., 158-163°C; $[\alpha]_D$ + 128° (*c*=0.5, in $CHCl_3$).

Fungal Source

Fruiting bodies of *Ganoderma lucidum*.

Isolation/Purification

The dried powdered surface of the gills of the fruiting bodies of *Ganoderma lucidum* was extracted with ethyl ether. The extract was concentrated and partitioned between ethyl ether and water. The acidic fraction was obtained, methylated with diazomethane, and chromatographed on a silica gel column eluted with different ethyl acetate-benzene mixtures to give 13 fractions. Most fractions were further purified by silica gel column chromatography eluted with various mixtures of ethyl acetate-benzene or various mixtures of acetone-chloroform. Fractions 10-12 were combined, repeatedly chromatographed by preparative TLC (acetone-chloroform, 15:85, v/v), and recrystallized from ethyl ether to give compound B8 (7α,15α-dihydroxy-3,11,23-trioxo-5α-lanost-8-en-26-oic acid).

Biological Activity

Fruiting bodies of *G. lucidum* produce some physiological effects when used as a home remedy.

Spectral Data (Methyl ester)

UV:

λ_{max}^{EtOH} 254.5nm (log ϵ = 3.49).

IR:
(thin film) 3400, 1730, 1710, and 1650cm^{-1}.

^{1}H NMR:
($CDCl_3$) 1.71(1H, dt, *J*=15.0, 8.0Hz, H-1α); 2.97(1H, ddd, *J*=14.0, 8.5, 5.50Hz, H-1β); 2.45(1H, ddd, *J*=15.5, 9.0, 7.0Hz, H-2α); 2.61(1H, ddd, *J*=15.5, 9.5, 6.0Hz, H-2β); 2.09(1H, dd, *J*=12.5, 3.0Hz, H-5); ca. 1.72(1H, m, H-6α); ca. 1.78(1H, m, H-6β); 4.58(1H, dd, *J*=5.0, 1.8Hz, H-7β); 2.75(1H, br d, *J*=18.0Hz, H-12α); 2.42(1H, d, *J*=17.7Hz, H-12β); 4.605(1H, dd, *J*=9.0, 6.0Hz, H-15); ca. 1.82(1H, H-16); 1.95(1H, H-16); ca. 1.95(1H, H-17); 2.00(1H, m, H-20); 2.27(1H, dd, *J*=16.3, 8.5Hz, H-22); 2.38(1H, dd, *J*=16.0, 3.0Hz, H-22); 2.46(1H, dd, *J*=17.3, 5.0Hz, H-24); 2.84(1H, dd, *J*=17.5, 8.5Hz, H-24); 2.95(1H, m, H-25); 0.89(3H, s, H_3-18); 1.03(3H, s, H_3-19); 0.86(3H, d, *J*=5.8Hz, H_3-21); 1.18(3H, d, *J*=7.0Hz, H_3-27); 1.16(3H, s, H_3-30); 1.08(3H, s, H_3-31); 1.30(3H, s, H_3-32); and 3.68ppm (3H, s, $COOCH_3$).

^{13}C NMR:
($CDCl_3$) 34.8, t, C-1; 34.1, t, C-2; 217.8, s, C-3; 47.2, s, C-4; 45.2, d, C-5; 27.9, t, C-6; 66.7, d, C-7; 159.3, s, C-8; 140.0, s, C-9; 38.0, s, C-10; 199.1, s, C-11; 51.8, t, C-12; 46.4, s, C-13; 53.4, s, C-14; 72.4, d, C-15; 37.8, t, C-16; 49.0, d, C-17; 17.5, q, C-18; 17.5, q, C-19; 32.5, d, C-20; 19.3, q, C-21; 49.6, t, C-22; 208.3, s, C-23; 46.9, t, C-24; 34.7, d, C-25; 176.2, s, C-26; 17.1, q, C-27; 27.6, q, C-30; 20.5, q, C-31; 21.1, q, C-32; and 51.9ppm, q, CH_3O-.

Mass Spectrum:
HREIMS: 530.3293(M^+, calcd for $C_{31}H_{46}O_7$ 530.3244), 392, 364, 230, 171, 144, 139, and 129*m/e*.

Reference

T. Kikuchi, S. Kanomi (*nee* Matsuda), S. Kadota, Y. Murai, K. Tsubono, and Z. Ogita; Constituents of the Fungus *Ganoderma lucidum* (Fr.) Karst. I. Structures of Ganoderic Acids C2, E, I, and K, Lucidenic Acid F and Related Compounds; Chem. Pharm. Bull., Vol. 34, pp. 3695-3712(1986).

Common/Systematic Name

Compound B9

3β,7α,15α-Trihydroxy-11,23-dioxo-5α-lanost-8-en-26-oic acid

Molecular Formula/Molecular Weight

$C_{30}H_{46}O_7$; MW = 518.32435

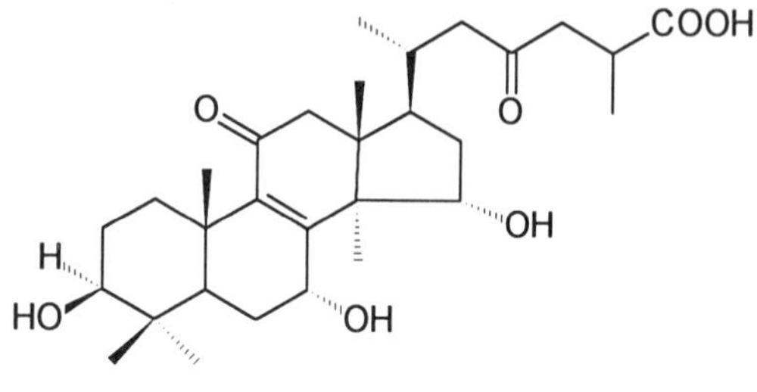

General Characteristics

Amorphous powder.

Fungal Source

Fruiting bodies of *Ganoderma lucidum*.

Isolation/Purification

The dried powdered surface of the gills of the fruiting bodies of *Ganoderma lucidum* was extracted with ethyl ether. The extract was concentrated and partitioned between ethyl ether and water. The acidic fraction was obtained, methylated with diazomethane, and chromatographed on a silica gel column eluted with different ethyl acetate-benzene mixtures to give 13 fractions. Most fractions were further purified by silica gel column chromatography eluted with various mixtures of ethyl acetate-benzene or various mixtures of acetone-chloroform. Fraction 13 was repeatedly chromatographed by preparative TLC (acetone-chloroform, 1:9, v/v) to give compound B9 (3β,7α,15α-trihydroxy-11,23-dioxo-5α-lanost-8-en-26-oic acid).

Biological Activity

Fruiting bodies of *G. lucidum* produce some physiological effects when used as a home remedy.

Spectral Data (Methyl ester)

UV:

λ_{max}^{EtOH} 255nm (log ϵ = 3.43)

IR:

(thin film) 3450, 1730, 1715, and 1650cm^{-1}.

^{1}H NMR:

(CDCl$_3$) 1.17(1H, td, *J*=13.5, 4.5Hz, H-1α); 3.00(1H, dt, *J*=14.0, 3.5Hz, H-1β); ca. 1.65(1H, m, H-2α); 1.72(1H, br d-like, H-2β); 3.31(1H, dd, *J*=11.5, 5.0Hz, H-3); 1.28(1H, dd, *J*=13.0, 2.0Hz, H-5); 1.83(1H, br d, *J*=14.0Hz, H-6α); 1.73(1H, td, *J*=13.5, 5.0Hz, H-6β); 4.56(1H, br d, *J*=5.0Hz, H-7β); 2.75(1H, br d, *J*=17.5Hz, H-12α); 2.39(1H, d, *J*=17.7Hz, H-12β); 4.57(1H, dd, *J*=9.0, 5.5Hz, H-15); ca. 1.78(1H, H-16); 1.94(1H, H-16); ca. 1.92(1H, H-17); 2.00(1H, m, H-20); 2.25(1H, dd, *J*=16.2, 9.0Hz, H-22); 2.38(1H, dd, *J*=16.2, 3.0Hz, H-22); 2.46(1H, dd, *J*=17.5, 5.0Hz, H-24); 2.83(1H, dd, *J*=17.5, 8.5Hz, H-24); 2.94(1H, dqd, *J*=9.0, 7.5, 5.5Hz, H-25); 0.87(3H, s, H_3-18); 1.04(3H, s, H_3-19); 0.85(3H, d, *J*=6.0Hz, H_3-21); 1.18(3H, d, *J*=7.0Hz, H_3-27); 1.06(3H, s, H_3-30); 0.84(3H, s, H_3-31); 1.27(3H, s, H_3-32); and 3.68ppm (3H, s, $COOCH_3$).

^{13}C NMR:

(CDCl$_3$) 34.2, C-1; 27.3, C-2; 78.5, C-3; 39.1, C-4; 47.7, C-5; 28.0, C-6; 68.0, C-7; 158.8, C-8; 141.6, C-9; 38.6, C-10; 199.4, C-11; 52.2, C-12; 46.1, C-13; 53.5, C-14; 72.3, C-15; 37.8, C-16; 49.0, C-17; 17.3, C-18; 17.3, C-19; 32.5, C-20; 19.3, C-21; 49.6, C-22; 208.3, C-23; 46.9, C-24; 34.7, C-25; 176.4, C-26; 17.1, C-27; 28.2, C-30; 15.8, C-31; 21.1, C-32; and 51.9ppm CH_3O-.

Mass Spectrum:

HREIMS: 532.3413(M^+, calcd for $C_{31}H_{48}O_7$ 532.3400), 514, 496, 364, 230, 171, 144, 139, and 129*m/e*.

Reference

T. Kikuchi, S. Kanomi (*nee* Matsuda), S. Kadota, Y. Murai, K. Tsubono, and Z. Ogita; Constituents of the Fungus *Ganoderma lucidum* (Fr.) Karst. I. Structures of Ganoderic Acids C2, E, I, and K, Lucidenic Acid F and Related Compounds; Chem. Pharm. Bull., Vol. 34, pp. 3695-3712(1986).

Common/Systematic Name

Ganoderic acid Ma

3α,7α-Diacetoxy-15α-hydroxy-5α-lanosta-8,24*E*-dien-26-oic acid

Molecular Formula/Molecular Weight

$C_{34}H_{52}O_7$; MW = 572.37130

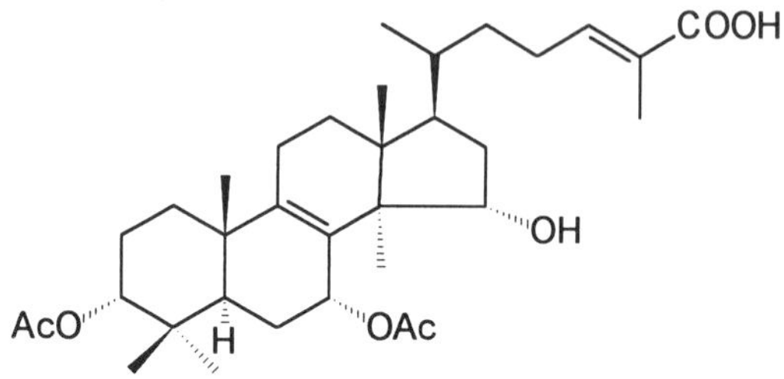

General Characteristics

Colorless syrup; $[\alpha]_D^{24}$ -16° (*c*=0.3, in MeOH).

Fungal Source

Isolated from mycelium of *Ganoderma lucidum* ("Saegusa").

Isolation/Purification

The mycelial mat grown on a solid medium was harvested and then extracted with EtOH. The extract was partitioned between $CHCl_3$ and water. The $CHCl_3$ layer was concentrated and separated on a silica gel column into seven fractions (Fr. 1-7). Fraction 5 was successively rechromatographed on a silica gel column, preparative TLC, and/or reversed-phase LC to afford ganoderic acid Ma and eight other related compounds.

Biological Activity

Possibly cytotoxic to hepatoma cells or antihepatotoxic.

Spectral Data

UV:

λ_{max}^{MeOH} 210nm (19,300).

IR:

(thin film) 3200-2500, 1715, and 1640cm^{-1} (sh).

^{1}H NMR:

($CDCl_3$) The spectrum showed the presence of five tertiary methyl groups 0.67(3H, s), 0.82(3H, s), 0.91(3H, s), 1.00(3H, s), 1.09(3H, s); one secondary methyl group 0.91 (3H, d, *J*=6.2); one olefinic methyl group 1.84(3H, d, *J*=1.1), 6.87(1H, tq, *J*=7.5, 1.1);

two secondary acetoxyl groups 2.04(3H, s), 2.10(3H, s), 4.69(1H, t, *J*=2.6; 3β-H), 5.43(1H, br. s; 7β-H); and one secondary hydroxyl group 4.26ppm (1H, dd, *J*=9.5, 5.5, 15β-H).

^{13}C NMR:
(CDCl$_3$) C-1, 30.1; C-2, 23.3; C-3, 77.3; C-4, 36.2; C-5, 40.5; C-6, 26.1; C-7, 70.0; C-8, 131.1; C-9, 145.1; C-10, 38.3; C-11, 21.4; C-12, 31.2; C-13, 45.8; C-14, 52.1; C-15, 72.4; C-16, 39.7; C-17, 48.8; C-18, 16.3; C-19, 17.7; C-20, 36.7; C-21, 18.4; C-22, 34.7; C-23, 25.8; C-24, 145.2; C-25, 126.8; C-26, 172.4; C-27, 12.0; C-28, 27.2; C-29, 22.7; C-30, 17.9; O$\underline{C}$OCH$_3$, 171.9, 170.8; OCO$\underline{C}$H$_3$, 21.9ppm.

Mass Spectrum:
FD-MS: 572*m/e* (M^+); HREIMS: 554.3629*m/e* (M^+ -H_2O, $C_{34}H_{50}O_6$, calcd 554.3609).

Reference

T. Nishitoba, H. Sato, S. Shirasu, and S. Sakamura; Novel Triterpenoids from the Mycelial Mat at the Previous Stage of Fruiting of *Ganoderma lucidum*; Agric. Biol. Chem., Vol. 51, pp. 619-622(1987).

Common/Systematic Name

Ganoderic acid Mb

3α,15α,22-Triacetoxy-7α-hydroxy-5α-lanosta-8,24*E*-dien-26-oic acid

Molecular Formula/Molecular Weight

$C_{36}H_{52}O_9$; MW = 630.37678

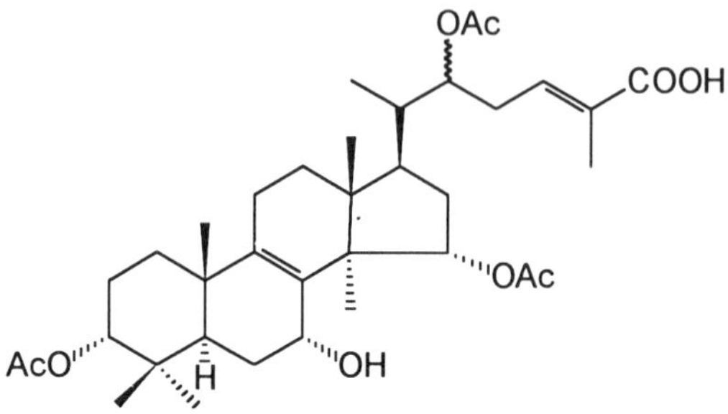

General Characteristics

Colorless syrup; $[\alpha]_D^{27}$ -4° (*c*=0.2, in MeOH).

Fungal Source

Isolated from mycelium of *Ganoderma lucidum* ("Saegusa").

Isolation/Purification

The mycelial mat grown on a solid medium was harvested and then extracted with EtOH. The extract was partitioned between $CHCl_3$ and water. The $CHCl_3$ layer was concentrated and separated on a silica gel column into seven fractions (Fr. 1-7). Fraction 5 was successively rechromatographed on a silica gel column, preparative TLC and/or reversed-phase LC to afford ganoderic acid Mb and eight other related compounds.

Biological Activity

Possibly cytotoxic to hepatoma cells or antihepatotoxic.

Spectral Data

UV:

λ_{max}^{MeOH} 210nm (22,300).

IR:

(thin film) 3450, 3200-2500, 1720, and 1640cm^{-1} (sh).

^{1}H NMR:

($CDCl_3$) 4.11(1H, br, s, 7β-H); 4.69(1H, t, *J*=2, 3β-H); 5.03(1H, t, *J*=7.0, 22-H); 5.15ppm (1H, dd, *J*=9.5, 5.5, 15β-H).

^{13}C NMR:

($CDCl_3$) C-1, 30.3; C-2, 23.3; C-3, 77.5; C-4, 36.3; C-5, 40.0; C-6, 27.4; C-7, 66.5; C-8, 133.8; C-9, 141.8; C-10, 38.4; C-11, 20.7; C-12, 31.2; C-13, 45.3; C-14, 51.3; C-15, 76.0; C-16, 36.3; C-17, 45.9; C-18, 16.3; C-19, 17.5; C-20, 39.9; C-21, 12.7; C-22, 74.4; C-23, 31.9; C-24, 138.9; C-25, 129.5; C-26, 172.0; C-27, 12.3; C-28, 27.4; C-29, 22.0; C-30, 20.2; O$\underline{C}$OCH_3, 170.9, 170.6, 170.5; and OCO$\underline{C}H_3$, 21.4, 21.2, 21.0ppm.

The ^1H and ^{13}C NMR data resembled those of ganoderic acid W and showed the presence of an additional acetoxyl group at C-22, analogous to ganoderic acids R, S, and T.

Mass Spectrum:

FD-MS: 630*m/e* (M^+); HREIMS: 612.3636*m/e* (M^+ -H_2O, $C_{36}H_{52}O_8$, calcd 612.3664).

Reference

T. Nishitoba, H. Sato, S. Shirasu, and S. Sakamura; Novel Triterpenoids from the Mycelial Mat at the Previous Stage of Fruiting of *Ganoderma lucidum*; Agric. Biol. Chem., Vol. 51, pp. 619-622(1987).

Common/Systematic Name

Ganoderic acid Mc

3α,7α,22-Triacetoxy-15α-hydroxy-5α-lanosta-8,24*E*-dien-26-oic acid

Molecular Formula/Molecular Weight

$C_{36}H_{54}O_9$; MW = 630.37678

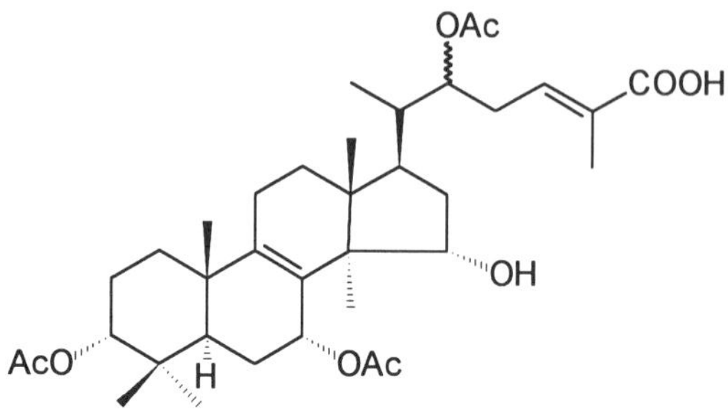

General Characteristics

Colorless syrup; $[\alpha]_D^{27}$ -23° (*c*=0.2, in MeOH).

Fungal Source

Isolated from mycelium of *Ganoderma lucidum* ("Saegusa").

Isolation/Purification

The mycelial mat grown on a solid medium was harvested and then extracted with EtOH. The extract was partitioned between $CHCl_3$ and water. The $CHCl_3$ layer was concentrated and separated on a silica gel column into seven fractions (Fr. 1-7). Fraction 5 was successively rechromatographed on a silica gel column, preparative TLC and/or reversed-phase LC to afford ganoderic acid Mc and eight other related compounds.

Biological Activity

Possibly cytotoxic to hepatoma cells or antihepatotoxic.

Spectral Data

UV:

λ_{max}^{MeOH} 210nm (17,000).

IR:

(thin film) 3450, 3200-2500, 1720, and 1640cm^{-1} (sh).

^{1}H NMR:

($CDCl_3$) The ^{1}H and ^{13}C NMR data (see below) were very similar to those of ganodermic acid Mb, but indicated that an acetyl group was attached to the 7α-hydroxyl group in ganoderic acid Mc instead of to the 15α-hydroxyl group in ganoderic acid Mb.

^{13}C NMR:

($CDCl_3$) C-1, 30.1; C-2, 23.3; C-3, 77.3; C-4, 36.2; C-5, 40.5; C-6, 26.0; C-7, 69.9; C-8, 131.1; C-9, 145.0; C-10, 38.3; C-11, 21.2; C-12, 31.2; C-13, 45.6; C-14, 52.1; C-15, 72.2; C-16, 39.3; C-17, 45.5; C-18, 16.2; C-19, 17.7; C-20, 39.5; C-21, 12.9; C-22, 74.6; C-23, 31.7; C-24, 139.1; C-25, 129.3; C-26, 171.7; C-27, 12.3; C-28, 27.2; C-29, 21.9; C-30, 18.0; O<u>C</u>OCH$_3$, 171.9, 170.8, 170.5; and OCO<u>C</u>H$_3$, 21.7, 21.4, 21.0ppm.

Mass Spectrum:

FD-MS: 630*m/e* (M^+); HREIMS: 612.3609*m/e* (M^+ -H_2O), $C_{36}H_{52}O_8$, calcd 612.3464.

<u>Reference</u>

T. Nishitoba, H. Sato, S. Shirasu, and S. Sakamura; Novel Triterpenoids from the Mycelial Mat at the Previous Stage of Fruiting of *Ganoderma lucidum*; Agric. Biol. Chem., Vol. 51, pp. 619-622(1987).

Common/Systematic Name

Ganoderic acid Md

3α,22-Diacetoxy-7α-methoxy-5α-lanosta-8,24*E*-dien-26-oic acid

Molecular Formula/Molecular Weight

$C_{35}H_{54}O_7$; MW = 586.38695

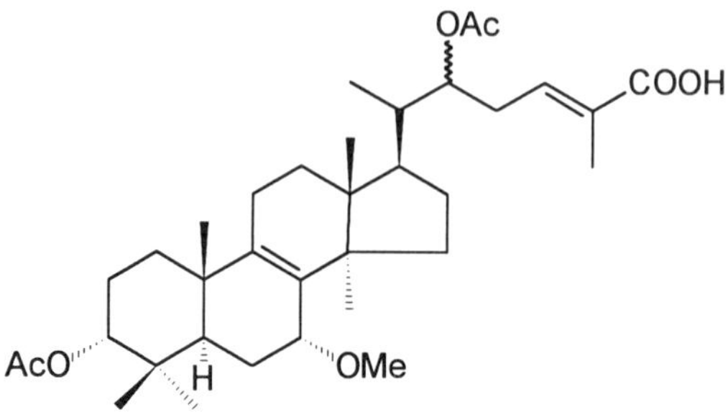

General Characteristics

Colorless needles; mp., 180-182°C; $[\alpha]_D^{24}$ -20° (*c*=0.33, in MeOH).

Fungal Source

Isolated from mycelium of *Ganoderma lucidum* ("Saegusa").

Isolation/Purification

The mycelial mat grown on a solid medium was harvested and then extracted with EtOH. The extract was partitioned between $CHCl_3$ and water. The $CHCl_3$ layer was concentrated and separated on a silica gel column into seven fractions (Fr. 1-7). Fraction 5 was successively rechromatographed on a silica gel column, preparative TLC and/or reversed-phase LC to afford ganoderic acid Md and eight other related compounds.

Biological Activity

Possibly cytotoxic to hepatoma cells or antihepatotoxic.

Spectral Data

UV:

λ_{max}^{MeOH} 210nm (17,600).

IR:

(thin film) 3400-2500, 1720, 1680, and 1640cm^{-1}.

^{1}H NMR:

($CDCl_3$) 3.29(3H, s); 3.61(1H, br, d, *J*=7.9, 7β-H); 4.69(1H, br, s, 3β-H); 5.11ppm (1H, t, *J*=7.0, 22-H). The signals at 4.69 and 5.11ppm indicated the presence of 3α-

and 22-acetoxyl groups, and the signals at 3.29 and 3.61ppm indicated the presence of a 7α-methoxy group, which were confirmed by conversion into ganoderic acid R on treatment with H_2SO_4.

^{13}C NMR:

($CDCl_3$) C-1, 30.2; C-2, 23.4; C-3, 77.6; C-4, 36.4; C-5, 40.4; C-6, 22.4; C-7, 76.5; C-8, 134.3; C-9, 141.3; C-10, 38.1; C-11, 21.0; C-12, 31.1; C-13, 45.0; C-14, 50.0; C-15, 30.2; C-16, 27.9; C-17, 47.2; C-18, 15.9; C-19, 17.5; C-20, 39.8; C-21, 12.9; C-22, 74.8; C-23, 31.9; C-24, 139.6; C-25, 129.1; C-26, 172.1; C-27, 12.3; C-28, 27.3; C-29, 22.2; C-30, 25.6; O<u>C</u>OCH$_3$, 171.1; and OCO<u>C</u>H$_3$, 21.6, 21.0ppm.

Mass Spectrum:

HREIMS: 586.3996*m/e* (M^+), $C_{35}H_{54}O_7$, calcd 586.3871.

Reference

T. Nishitoba, H. Sato, S. Shirasu, and S. Sakamura; Novel Triterpenoids from the Mycelial Mat at the Previous Stage of Fruiting of *Ganoderma lucidum*; Agric. Biol. Chem., Vol. 51, pp. 619-622(1987).

Common/Systematic Name

Ganoderic acid Me

3α,15α-Diacetoxy-5α-lanosta-7,9(11),24*E*-trien-26-oic acid

Molecular Formula/Molecular Weight

$C_{34}H_{50}O_6$; MW = 554.36074

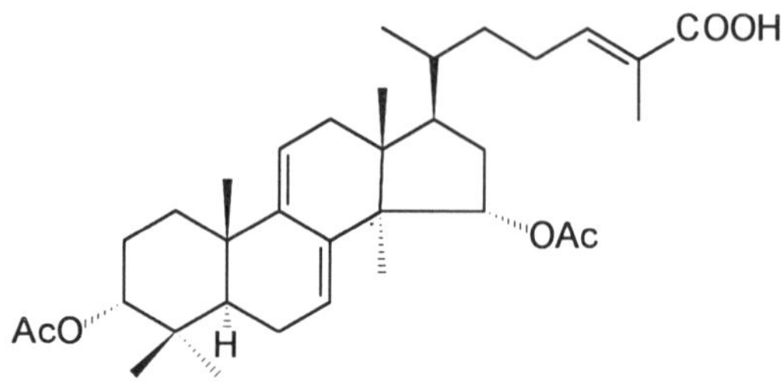

General Characteristics

Colorless syrup; $[\alpha]_D^{24}$ + 53° (*c*= 0.26, in MeOH).

Fungal Source

Isolated from mycelium of *Ganoderma lucidum* ("Saegusa").

Isolation/Purification

The mycelial mat grown on a solid medium was harvested and then extracted with EtOH. The extract was partitioned between $CHCl_3$ and water. The $CHCl_3$ layer was concentrated and separated on a silica gel column into seven fractions (Fr. 1-7). Fraction 5 was successively rechromatographed on a silica gel column, preparative TLC and/or reversed-phase LC to afford ganoderic acid Me and eight other related compounds.

Biological Activity

Possibly cytotoxic to hepatoma cells or antihepatotoxic.

Spectral Data

UV:

λ_{max}^{MeOH} 220(ε=17,400), 235(16,400), 244(16,200), and 252nm (10,900).

IR:

(thin film) 3200-2500, 1720, 1680 (sh) and 1640cm^{-1} (sh).

^{1}H NMR:

($CDCl_3$) 4.68(1H, br, s, 3β-H); 5.08(1H, dd, *J*=9.3, 4.9, 15β-H); 5.34(1H, d, *J*=5.9, 11-H); and 5.48ppm (1H, br, s, 7-H). The ^{1}H NMR data were identical with that of 3-*O*-acetyl methyl ganoderate X, except for the loss of the methyl ester signal.

^{13}C NMR:
$(CDCl_3)$ C-1, 30.9; C-2, 23.2; C-3, 78.1; C-4, 36.6; C-5, 44.0; C-6, 22.9; C-7, 121.2; C-8, 140.3; C-9, 145.9; C-10, 37.4; C-11, 115.7; C-12, 38.0; C-13, 44.2; C-14, 51.5; C-15, 77.5; C-16, 37.0; C-17, 48.9; C-18, 16.0; C-19, 22.7*; C-20, 36.0; C-21, 18.5; C-22, 34.7; C-23, 26.0; C-24, 144.9; C-25, 126.8; C-26, 172.0; C-27, 12.1; C-28, 27.8; C-29, 22.5*; C-30, 18.2; O<u>C</u>OCH$_3$, 171.2, 170.8; and OCO<u>C</u>H$_3$, 21.4, 21.3ppm.

* Assignments may be reversed.

Mass Spectrum:
HREIMS: 554.3624*m/e* (M^+); $C_{34}H_{50}O_6$, calcd 554.3609.

Reference

T. Nishitoba, H. Sato, S. Shirasu, and S. Sakamura; Novel Triterpenoids from the Mycelial Mat at the Previous Stage of Fruiting of *Ganoderma lucidum*; Agric. Biol. Chem., Vol. 51, pp. 619-622(1987).

Common/Systematic Name

Ganoderic acid Mf

3α-Acetoxy-15α-hydroxy-5α-lanosta-7,9(11),24*E*-trien-26-oic acid

Molecular Formula/Molecular Weight

$C_{32}H_{48}O_5$; MW = 512.35017

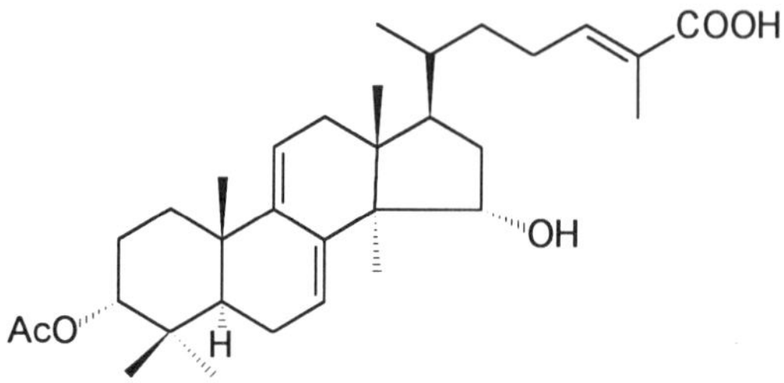

General Characteristics

Colorless syrup; $[\alpha]_D^{24}$ + 42° (*c*=0.20, in $CHCl_3$).

Fungal Source

Isolated from mycelium of *Ganoderma lucidum* ("Saegusa").

Isolation/Purification

The mycelial mat grown on a solid medium was harvested and then extracted with EtOH. The extract was partitioned between $CHCl_3$ and water. The $CHCl_3$ layer was concentrated and separated on a silica gel column into seven fractions (Fr. 1-7). Fraction 5 was successively rechromatographed on a silica gel column, preparative TLC and/or reversed-phase LC to afford ganoderic acid Mf and eight other related compounds.

Biological Activity

Possibly cytotoxic to hepatoma cells or antihepatotoxic.

Spectral Data

UV:

λ_{max}^{MeOH} 219(ε=27,900), 236(26,200), 244(26,200), and 253nm (16,900).

IR:

(thin film) 3400, 3200-2500, 1720, 1680, and 1640cm^{-1}.

^{1}H NMR:

($CDCl_3$) 4.29(1H, dd, *J*=9.5, 5.1, 15β-H); 4.68(1H, t, *J*=2.3, 3β-H); 5.32(1H, d, *J*=6.2, 11-H); 5.86(1H, br, d, *J*=3.3, 7-H). The ^{1}H and ^{13}C NMR data closely resembled those of ganoderic acid Me, but showed that ganoderic acid Mf lost the acetyl group bonded to the 15α-hydroxyl group in ganoderic acid Me.

^{13}C NMR:
($CDCl_3$) C-1, 30.7; C-2, 23.2; C-3, 78.1; C-4, 36.6; C-5, 44.1; C-6, 22.8; C-7, 121.3; C-8, 140.8; C-9, 146.2; C-10, 37.4; C-11, 115.7; C-12, 38.5; C-13, 44.5; C-14, 52.1; C-15, 74.7; C-16, 40.1; C-17, 48.9; C-18, 16.0; C-19, 22.5*; C-20, 36.0; C-21, 18.3; C-22, 34.8; C-23, 25.9; C-24, 145.2; C-25, 126.8; C-26, 172.1; C-27, 12.1; C-28, 27.8; C-29, 22.7*; C-30, 17.2; O<u>C</u>OCH$_3$, 170.8; and OCO<u>C</u>H$_3$, 21.3ppm.

* Assignments may be reversed.

Mass Spectrum:
HREIMS: 512.3531*m/e* (M^+); $C_{32}H_{48}O_5$, calcd. 512.3503.

Reference

T. Nishitoba, H. Sato, S. Shirasu, and S. Sakamura; Novel Triterpenoids from the Mycelial Mat at the Previous Stage of Fruiting of *Ganoderma lucidum*; Agric. Biol. Chem., Vol. 51, pp. 619-622(1987).

Common/Systematic Name

Ganoderic acid Mg

3α,22-Diacetoxy-15α-hydroxy-7α-methoxy-5α-lanosta-8,24*E*-dien-26-oic acid

Molecular Formula/Molecular Weight

$C_{35}H_{54}O_8$; MW = 602.38187

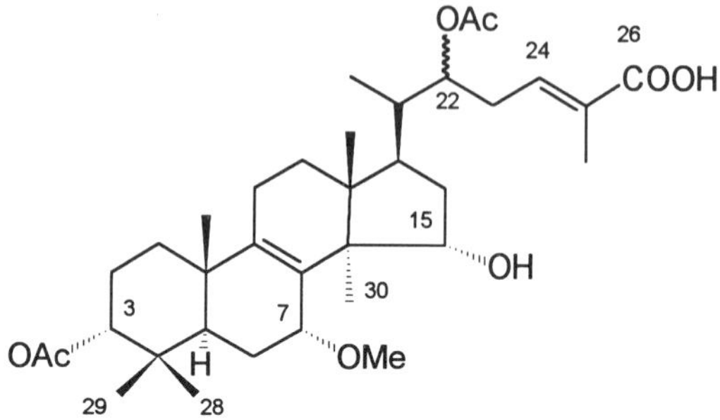

General Characteristics

An amorphous powder; mp., 126-129°C; $[\alpha]_D^{23}$ - 23° (*c*=0.2, in MeOH).

Fungal Source

Mycelial mat of *Ganoderma lucidum*.

Isolation/Purification

The mycelial mat of *G. lucidum* was extracted with EtOH, reduced in volume, partitioned between water and chloroform, and the chloroform removed under vacuum. The residue was subjected to silica gel column chromatography eluted with a gradient from benzene to ethyl acetate to yield seven fractions. Fraction six was chromatographed twice more using chloroform-methanol, 19:1 (v/v) and benzene-ethyl acetate, 1:1 to yield six more fractions (6a-6f). Fraction 6d was applied to preparative TLC developed with benzene-ethyl acetate, 1:1 (v/v) and chloroform-methanol, 19:1 (v/v) to yield purified ganoderic acid Mg.

Biological Activity

Fruiting bodies of *G. lucidum* produce some physiological effects when used as a home remedy.

Spectral Data

UV:

λ_{max}^{EtOH} 210(ε = 18,100).

IR:

(KBr) 3430, 3300-2500(br), 1720,1680(sh), and 1640cm^{-1}.

^{1}H NMR:

(CDCl$_3$) 6.81(1H, tq, *J*=7.3, 1.1Hz, H-24); 5.01(1H, td, *J*=7.0, 1.4Hz, H-22); 4.69(1H, t, *J*=2.9Hz, H-3β); 4.34(1H, t, *J*=7.1Hz, 15β-H); 4.02(1H, br d, *J*=3.3Hz, H-7β); 3.33(3H, s); 2.57(1H, ddd, *J*=14.5, 7.3, 7.0Hz, H-23); 2.08(3H, s); 2.05(3H, s); 1.87(3H, d, *J*=1.1Hz, H$_3$-27); 1.04(3H, s, H$_3$-30); 0.98(3H, s, H$_3$-28); 0.96(3H, d, *J*=6.2Hz, H$_3$-21); 0.95(3H, s, H$_3$-29); 0.92(3H, s, H$_3$-19); and 0.65ppm (3H, s, H$_3$-18).

^{13}C NMR:

(CDCl$_3$) 30.3, C-1; 23.2, C-2; 77.3, C-3; 36.5, C-4; 40.1, C-5; 21.3, C-6; 76.3, C-7; 132.8, C-8; 143.6, C-9; 38.4, C-10; 20.9, C-11; 31.4, C-12; 46.2, C-13; 52.8, C-14; 72.0, C-15; 37.3, C-16; 46.4, C-17; 16.6, C-18; 17.2, C-19; 39.3, C-20; 13.0, C-21; 74.9, C-22; 31.9, C-23; 139.1, C-24; 129.3, C-25; 172.0, C-26; 12.3, C-27; 27.3, C-28; 22.0, C-29; 18.5, C-30; 170.9, 170.4, CH$_3$O<u>C</u>-; 21.5, 21.0, <u>C</u>H$_3$OC; and 54.4ppm CH$_3$O-7.

Mass Spectrum:

HREIMS: 602.3835(M$^+$); $C_{35}H_{54}O_8$, calcd 602.3820; 4.5%, 570(-CH$_3$OH, 5.3), 495, 405, 175, and 43*m/e* (base peak).

<u>Reference</u>

T. Nishitoba, H. Sato, and S. Sakamura; Novel Mycelial Components, Ganoderic Acid Mg, Mh, Mi, Mj, and Mk from the Fungus *Ganoderma lucidum*; Agric. Biol. Chem., Vol. 51, pp. 1149-1153(1987).

Common/Systematic Name

Ganoderic acid Mh

3α,22-Diacetoxy-7α,15α-dihydroxy-5α-lanosta-8,24*E*-dien-26-oic acid

Molecular Formula/Molecular Weight

$C_{34}H_{52}O_8$; MW = 588.36622

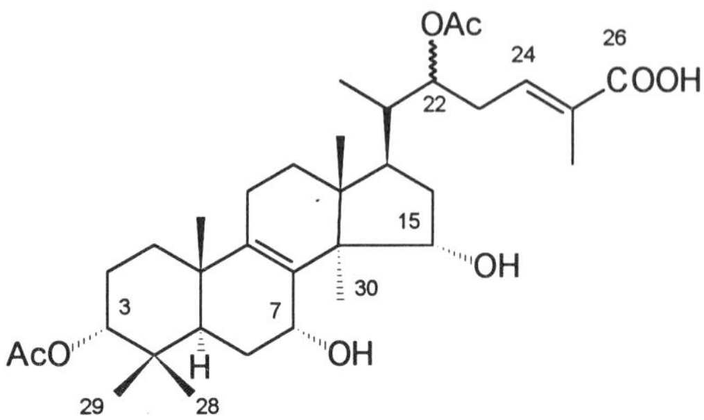

General Characteristics

A colorless syrup; $[\alpha]_D^{23}$ + 2° (*c*=0.2, in MeOH).

Fungal Source

Mycelial mat of *Ganoderma lucidum*.

Isolation/Purification

The mycelial mat of *G. lucidum* was extracted with EtOH, reduced in volume, partitioned between water and chloroform, and the chloroform removed under vacuum. The residue was subjected to silica gel column chromatography eluted with a gradient from benzene to ethyl acetate to yield seven fractions. Fraction six was chromatographed twice more using chloroform-methanol (19:1, v/v) and benzene-ethyl acetate (1:1, v/v) to yield six more fractions (6a-6f). Fractions 6e and 6f were combined and chromatographed on a silica gel column (chloroform-methanol, 9:1, v/v), reverse-phase LC (RP_{18}, methanol-water, 17:3, v/v), and preparative TLC (chloroform-methanol, 24:1, v/v) to yield purified ganoderic acid Mh.

Biological Activity

Fruiting bodies of *G. lucidum* produce some physiological effects when used as a home remedy.

Spectral Data

UV:

λ_{max}^{EtOH} 210nm (ϵ=17,900).

IR:

(KBr) 3400, 3200-2500(br), 1720,1680(sh), and 1640cm^{-1}.

^{1}H NMR:

($CDCl_3$) 6.79(1H, tq, *J*=7.3, 1.0, 24-H); 5.03(1H, t, *J*=7.0Hz, H-22); 4.69(1H, br s, H-3β); 4.42(1H, t, *J*=7.0Hz, 15β-H); 4.39(1H, br s, H-7β); 2.56(1H, ddd, *J*=14.4, 7.3, 7.0Hz, 23-H); 2.37(1H, ddd, *J*=14.4, 7.3, 7.0Hz, 23-H); 2.08(3H, s); 2.06(3H, s); 1.86(3H, s, H_3-27); 1.08(3H, s, H_3-30); 0.96(3H_3, s, H_3-28); 0.96(3H, d, *J*=6.2Hz, H_3-21); 0.93(3H, s, H_3-29); 0.92(3H, s, H_3-19); and 0.66ppm (3H, s, H_3-18).

^{13}C NMR:

($CDCl_3$) 30.3, C-1; 23.3, C-2; 77.5, C-3; 36.4, C-4; 40.5, C-5; 28.1, C-6; 66.8, C-7; 134.5, C-8; 141.7, C-9; 38.1, C-10; 20.6, C-11; 31.7, C-12; 45.7, C-13; 52.4, C-14; 72.4, C-15; 38.1, C-16; 46.4, C-17; 16.4, C-18; 17.3, C-19; 39.5, C-20; 12.9, C-21; 74.7, C-22; 31.9, C-23; 139.1, C-24; 129.3, C-25; 171.4, C-26; 12.4, C-27; 27.4, C-28; 21.9, C-29; 19.2, C-30; 170.9, 170.6, $CH_3O\underline{C}$-; and 21.5, 21.0ppm $\underline{C}H_3OC$.

Mass Spectrum:

FD-MS: 588(M^+, 30%) and 570*m/e* (M^+ - H_2O, 100%); HREIMS: 570.3524(M^+ - H_2O, $C_{30}H_{50}O_7$, calcd. 570.3558, 19%), 495, 435, and 43*m/e* (base peak).

Reference

T. Nishitoba, H. Sato, and S. Sakamura; Novel Mycelial Components, Ganoderic Acid Mg, Mh, Mi, Mj, and Mk from the Fungus *Ganoderma lucidum*; Agric. Biol. Chem., Vol. 51, pp. 1149-1153(1987).

Common/Systematic Name

Ganoderic acid Mi

3α-Acetoxy-7α-methoxy-5α-lanosta-8,24*E*-dien-26-oic acid

Molecular Formula/Molecular Weight

$C_{33}H_{52}O_6$; MW = 544.37639

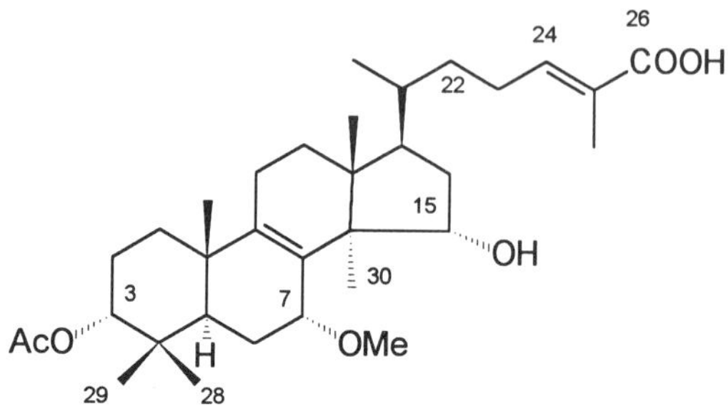

General Characteristics

A colorless syrup; $[\alpha]_D^{23}$ - 11° (*c*=0.2, in MeOH).

Fungal Source

Mycelial mat of *Ganoderma lucidum*.

Isolation/Purification

The mycelial mat of *G. lucidum* was extracted with EtOH, reduced in volume, partitioned between water and chloroform, and the chloroform removed under vacuum. The residue was subjected to silica gel column chromatography eluted with a gradient from benzene to ethyl acetate to yield seven fractions. Fraction six was chromatographed twice more using chloroform-methanol (19:1, v/v) and benzene-ethyl acetate (1:1, v/v) to yield six more fractions (6a-6f). Fraction 6b was applied to preparative TLC (benzene-ethyl acetate, 1:1, v/v; chloroform-methanol, 19:1, v/v) to yield purified ganoderic acid Mi.

Biological Activity

Fruiting bodies of *G. lucidum* produce some physiological effects when used as a home remedy.

Spectral Data

UV:

λ_{max}^{EtOH} 210nm (ϵ = 19,900).

IR:

(KBr) 3430, 3200-2500(br), 1720, 1680(sh), and 1640cm^{-1}.

^{1}H NMR:

(CDCl$_3$) 6.88(1H, tq, J=7.5, 1.1Hz, H-24); 4.27(1H, t, J=2.6Hz, H-3β); 4.34(1H, t, J=7.7Hz, 15β-H); 4.03(1H, br d, J=3.3Hz, H-7β); 3.33(3H, s); 2.08(3H, s); 1.84(3H, d, J=1.1Hz, H$_3$-27); 1.05(3H, s, H$_3$-30); 0.98(3H$_3$, s, H$_3$-28); 0.95(3H, s, H$_3$-29); 0.92(3H, s, H$_3$-19); 0.91(3H, d, J=6.6Hz, H$_3$-21); and 0.65ppm (3H, s, H$_3$-18).

Mass Spectrum:

HREIMS: 544.3739(M$^+$, $C_{33}H_{52}O_6$, calcd. 544.3766, 4.2%), 512(M$^+$ - MeOH, 6.0), 437(46), 347(53), 175(49), 95(62), and 43*m/e* (base peak).

Reference

T. Nishitoba, H. Sato, and S. Sakamura; Novel Mycelial Components, Ganoderic Acid Mg, Mh, Mi, Mj, and Mk from the Fungus *Ganoderma lucidum*, Agric. Biol. Chem., Vol. 51, pp. 1149-1153(1987).

Common/Systematic Name

Ganoderic acid Mj

22-Acetoxy-3α-hydroxy-7α-methoxy-5α-lanosta-8,24*E*-dien-26-oic acid

Molecular Formula/Molecular Weight

$C_{33}H_{52}O_6$; MW = 544.37639

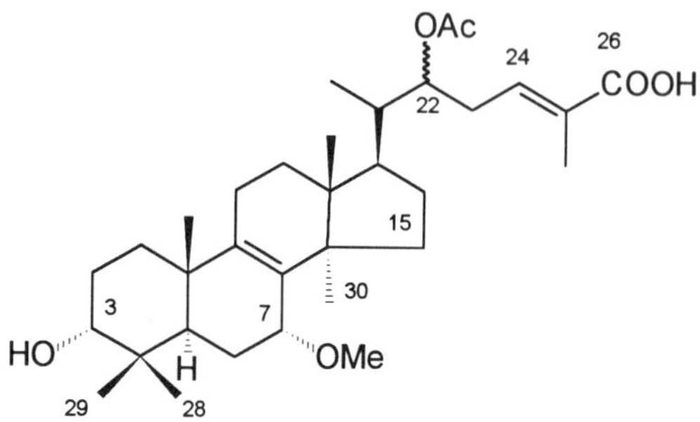

General Characteristics

A colorless syrup; $[\alpha]_D^{23}$ - 8° (*c*=0.05, in MeOH).

Fungal Source

Mycelial mat of *Ganoderma lucidum*.

Isolation/Purification

The mycelial mat of *G. lucidum* was extracted with EtOH, reduced in volume, partitioned between water and chloroform, and the chloroform removed under vacuum. The residue was subjected to silica gel column chromatography eluted with a gradient from benzene to ethyl acetate to yield seven fractions. Fraction six was chromatographed twice more using chloroform-methanol (19:1, v/v) and benzene-ethyl acetate (1:1, v/v) to yield six more fractions (6a-6f). Fraction 6b was applied to preparative TLC (benzene-ethyl acetate, 1:1, v/v; chloroform-methanol, 19:1, v/v) to yield purified ganoderic acid Mi and Mj.

Biological Activity

Fruiting bodies of *G. lucidum* produce some physiological effects when used as a home remedy.

Spectral Data

UV:

λ_{max}^{EtOH} 210nm (ϵ = 22,000).

IR:

(KBr) 3440, 3200-2500(br), 1720, 1680, and 1640cm^{-1}.

^{1}H NMR:

(CDCl$_3$) 6.81(1H, tq, J=7.3, 1.1Hz, H-24); 5.10(1H, td, J=6.9, 1.1Hz, H-22); 3.43(1H, t, J=2.9Hz, H-3β); 3.59(1H, br d, J=3.5Hz, H-7β); 3.27(3H, s); 2.57(1H, ddd, J=14.7, 7.3, 6.9, H-23); 2.37(1H, ddd, J=14.7, 7.3, 6.9Hz, H-23); 2.05(3H, s); 1.87(3H, d, J=1.1Hz, H$_3$-27); 1.02(3H$_3$, s, H$_3$-28); 1.00(3H, s, H$_3$-29); 0.98(3H, d, J=6.6Hz, H$_3$-21); 0.96(3H, s, H$_3$-30); 0.90(3H, s, H$_3$-19); and 0.61ppm (3H, s, H$_3$-18).

Mass Spectrum:

HREIMS: 544.3745(M$^+$, $C_{33}H_{52}O_6$, calcd. 544.3766, 3.4%), 529(7.9), 404(24), 262(11), 205(14), 173(18), 95(48), 55(38), and 43*m/e* (base peak).

Reference

T. Nishitoba, H. Sato, and S. Sakamura; Novel Mycelial Components, Ganoderic Acid Mg, Mh, Mi, Mj, and Mk from the Fungus *Ganoderma lucidum*, Agric. Biol. Chem., Vol. 51, pp. 1149-1153(1987).

Common/Systematic Name

Ganoderic acid Mk

3α,22-Diacetoxy-15α-hydroxy-5α-lanosta-7,9(11),24*E*-dien-26-oic acid

Molecular Formula/Molecular Weight

$C_{34}H_{50}O_7$; MW = 570.35565

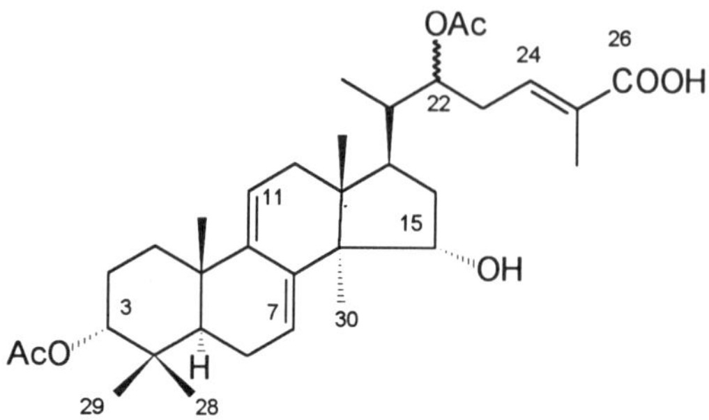

General Characteristics

A colorless syrup; $[\alpha]_D^{23}$ + 23° (*c*=0.2, in MeOH).

Fungal Source

Mycelial mat of *Ganoderma lucidum*.

Isolation/Purification

The mycelial mat of *G. lucidum* was extracted with EtOH, reduced in volume, partitioned between water and chloroform, and the chloroform removed under vacuum. The residue was subjected to silica gel column chromatography eluted with a gradient from benzene to ethyl acetate to yield seven fractions. Fraction six was chromatographed twice more using chloroform-methanol (19:1, v/v) and benzene-ethyl acetate (1:1, v/v) to yield six more fractions (6a-6f). Evaporation of fraction 6c afforded ganoderic acid Mk.

Biological Activity

Fruiting bodies of *G. lucidum* produce some physiological effects when used as a home remedy.

Spectral Data

UV:

λ_{max}^{EtOH} 218nm (ε = 11,700).

IR:

(KBr) 3450, 3300-2500(br), 1720, 1680(sh), and 1640cm^{-1}.

^{1}H NMR:

(CDCl$_3$) 6.86(1H, td, *J*=7.3, 1.1Hz, 24-H); 5.88(1H, br d, *J*=5.1Hz, H-7); 5.31(1H, d, *J*=6.2Hz, H-11); 5.04(1H, t, *J*=6.8Hz, H-22); 4.86(1H, br s, H-3β); 4.27(1H, dd, *J*=9.3, 4.9Hz, 15β-H); 2.57(1H, ddd, *J*=14.7, 7.3, 6.8Hz, 23-H); 2.07(3H, s); 2.05(3H, s); 1.87(3H, s, H$_3$-27); 1.00(3H, s, H$_3$-28); 0.99(3H, s, H$_3$-29); 0.97(3H, s, H$_3$-30); 0.97(3H, d, *J*=6.6Hz, H$_3$-21); 0.88(3H, s, H$_3$-19); and 0.62ppm (3H, s, H$_3$-18).

^{13}C NMR:

(CDCl$_3$) 30.6, C-1; 23.1, C-2; 78.1, C-3; 36.5, C-4; 44.0, C-5; 22.8, C-6; 121.6, C-7; 140.5, C-8; 146.2, C-9; 37.3, C-10; 115.3, C-11; 38.4, C-12; 44.3, C-13; 52.1, C-14; 74.5, C-15; 39.6, C-16; 45.5, C-17; 15.8, C-18; 22.7*, C-19; 39.3, C-20; 12.8, C-21; 74.7, C-22; 31.7, C-23; 139.1, C-24; 129.4, C-25; 172.2, C-26; 12.3, C-27; 27.8, C-28; 22.5*, C-29; 17.3, C-30; 170.8, 170.7, $CH_3O\underline{C}$-; and 21.3, 21.0ppm $\underline{C}H_3OC$.

* Assignments may be reversed.

Mass Spectrum:

HREIMS: 570.3549(M$^+$, $C_{34}H_{50}O_7$, calcd 570.3557, 0.9%), 510(2.8), 466(3.2), 428(8.3), 353(9.5), 95(20), 60(60), and 43*m/e* (base peak).

Reference

T. Nishitoba, H. Sato, and S. Sakamura; Novel Mycelial Components, Ganoderic Acid Mg, Mh, Mi, Mj, and Mk from the Fungus *Ganoderma lucidum*, Agric. Biol. Chem., Vol. 51, pp. 1149-1153(1987).

Common/Systematic Name

Ganoderic acid M

7β,12α-Dihydroxy-3,11,15,23-tetraoxo-5α-lanost-8-en-26-oic acid

Molecular Formula/Molecular Weight

$C_{30}H_{42}O_8$; MW = 530.28797

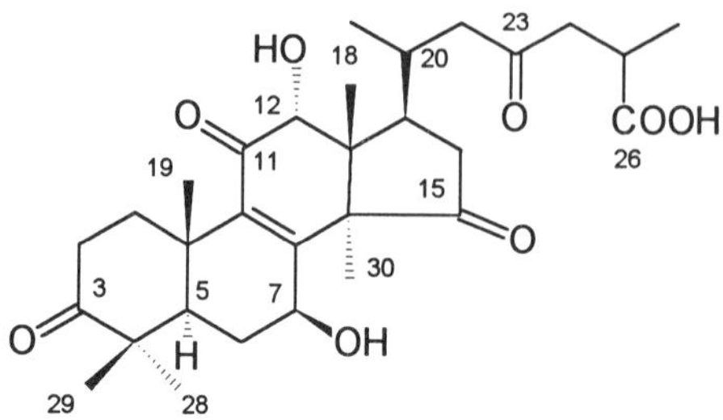

General Characteristics

Methyl ester: needles from methanol; mp., 206-210°C.

Fungal Source

Fruiting bodies of *Ganoderma lucidum*.

Isolation/Purification

Dried chipped fruiting bodies of *Ganoderma lucidum* were extracted with ethanol. The extract was concentrated and partitioned between chloroform and water. The acidic fraction was obtained and chromatographed on a silica gel column eluted with chloroform-methanol (49:1, v/v; fractions 1-11), chloroform-methanol (9:1, v/v; fraction 12), and methanol (fraction 13). Fraction 5 was rechromatographed 3 times on a silica gel column eluted with chloroform-methanol (19:1, v/v) to give several fractions. Some of the fractions were methylated and further fractionated using preparative TLC (benzene-ethyl acetate, 1:1 or 2:1, v/v), silica gel column chromatography (chloroform-methanol, 99:1, v/v), or reversed-phase LC (methanol-water, 7:3, v/v, Lobar RP_{18}). These separations gave purified methyl ganoderate M, N, O, H, E, methyl lucidenate K, L, and E_2.

Biological Activity

Fruiting bodies of *G. lucidum* produce some physiological effects when used as a home remedy.

Spectral Data (Methyl ester)

UV:

λ_{max}^{MeOH} 258nm (ε=2,940).

CD:

λ_{max}^{MeOH} Δ_{292} -1.0, Δ_{572} +7.3, Δ_{230} 0, and Δ_{218} -2.8.

IR:

(KBr) 3450, 2900, 1710, and 1660cm^{-1}.

^{1}H NMR:

($CDCl_3$) 1.82(1H, ddd, *J*=13.6, 8.7, 8.3Hz, H-1α); 2.89(1H, ddd, *J*=13.6, 7.8, 5.9Hz, H-1β); 2.50(1H, ddd, *J*=15.1, 8.3, 7.8Hz, H-2α); 2.54(1H, ddd, *J*=15.1, 8.7, 5.9Hz, H-2β); 1.74(1H, dd, *J*=13.7, 1.0Hz, H-5α); 2.11(1H, ddd, *J*=13.2, 7.3, 1.0Hz, H-6α); 1.65(1H, ddd, *J*=13.7, 13.2, 9.8Hz, H-6β); 4.85(1H, dd, *J*=9.8, 7.3Hz, H-7α); 3.78(1H, s, H-12β); 2.73(1H, dd, *J*=22.5, 8.3Hz, H-16α); 2.11(1H, dd, *J*=22.5, 12.7Hz, H-16β); 2.73(1H, m, H-17); 1.03(3H, s, H_3-18); 1.15(3H, s, H_3-19); 2.18(1H, m, H-20); 1.09(3H, d, *J*=6.8Hz, H_3-21); 2.41(1H, dd, *J*=16.1, 8.3Hz, H-22); 2.37(1H, dd, *J*=6.1, 4.4Hz, H-22); 2.87(1H, dd, *J*=17.6, 8.3Hz, H-24); 2.47(1H, dd, *J*=17.6, 5.4Hz, H-24); 2.96(1H, m, H-25); 1.19(3H, d, *J*=7.3Hz, H_3-27); 1.10(3H, s, H_3-28); 1.14(3H, s, H_3-29); 1.32(3H, s, H_3-30); and 3.69ppm (3H, s, $COOCH_3$).

Mass Spectrum:

HREIMS: 544.3058(M^+, $C_{31}H_{44}O_8$, calcd 544.3037, 3%), 526(16), 355(6), 304(100), 139(23), 129(31), 69(25), and 59*m/e* (37).

Reference

T. Nishitoba, H. Sato, and S. Sakamura; Triterpenoids from the Fungus *Ganoderma lucidum*; Phytochemistry, Vol. 26, pp. 1777-1784(1987).

Common/Systematic Name

Ganoderic acid N

7β,20-Dihydroxy-3,11,15,23-tetraoxo-5α-lanost-8-en-26-oic acid

Molecular Formula/Molecular Weight

$C_{30}H_{42}O_8$; MW = 530.28798

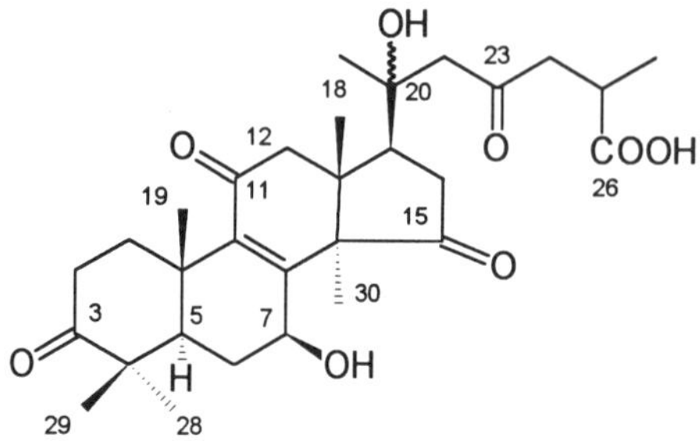

General Characteristics

Methyl ester: prisms from ethyl acetate-cyclohexane; mp., 164-167°C; $[\alpha]_D^{24}$ + 153°.

Fungal Source

Fruiting bodies of *Ganoderma lucidum*.

Isolation/Purification

Dried chipped fruiting bodies of *Ganoderma lucidum* were extracted with ethanol. The extract was concentrated and partitioned between chloroform and water. The acidic fraction was obtained and chromatographed on a silica gel column eluted with chloroform-methanol (49:1, v/v; fractions 1-11), chloroform-methanol (9:1, v/v; fraction 12), and methanol (fraction 13). Fraction 5 was rechromatographed 3 times on a silica gel column eluted with chloroform-methanol, 19:1 to give several fractions. Some of the fractions were methylated and further fractionated using preparative TLC (benzene-ethyl acetate, 1:1 or 2:1, v/v), silica gel column chromatography (chloroform-methanol, 99:1, v/v), or reversed-phase LC (methanol-water, 7:3, v/v, Lobar RP_{18}).These separations gave purified methyl ganoderate M, N, O, H, E, methyl lucidenate K, L and E_2.

Biological Activity

Fruiting bodies of *G. lucidum* produce some physiological effects when used as a home remedy.

Spectral Data (Methyl ester)

UV:

λ_{max}^{MeOH} 254nm (ε=7,180).

IR:
(KBr) 3450, 2950, 1720, 1700, and 1650cm^{-1}.

CD:
λ_{max}^{MeOH} Δ_{292} -2.5, Δ_{580} +0, Δ_{255} +18.1, Δ_{229} 0, and Δ_{215} -7.3.

^{1}H NMR:
(CDCl$_3$) 1.47(1H, ddd, *J*=13.6, 8.5, 8.5Hz, H-1α); 2.95(1H, m, H-1β); 2.50(1H, m, H-2α); 2.50(1H, m, H-2β); 1.55(1H, dd, *J*=12.6, 1.1Hz, H-5α); 2.12(1H, ddd, *J*=12.6, 7.7, 1.1Hz, H-6α); 1.68(1H, ddd, *J*=12.6, 12.6, 9.0Hz, H-6β); 4.84(1H, dd, *J*=9.0, 7.7Hz, H-7α); 2.77(1H, d, *J*=17.2Hz, H-12α); 2.85(1H, d, *J*=17.2Hz, H-12β); 2.86(1H, dd, *J*=19.4, 10.3Hz, H-16α); 2.45(1H, m, H-16β); 2.25(1H, dd, *J*=10.0, 9.0Hz, H-17); 1.17(3H, s, H$_3$-18); 1.26(3H, s, H$_3$-19); 1.41(3H, s, H$_3$-21); 2.66(1H, d, *J*=16.5Hz, H-22); 2.49(1H, d, *J*=16.5Hz, H-22); 2.95(1H, m, H-24); 2.45(1H, m, H-24); 2.95(1H, m, H-25); 1.20(3H, d, *J*=7.0Hz, H$_3$-27); 1.11(3H, s, H$_3$-28); 1.12(3H, s, H$_3$-29); 1.34(3H, s, H$_3$-30); and 3.69ppm (3H, s, COOCH$_3$).

^{13}C NMR:
(CDCl$_3$) 35.7, C-1; 34.3, C-2; 217.8, C-3; 45.3, C-4; 49.4, C-5; 27.7, C-6; 66.3, C-7; 157.6, C-8; 141.0, C-9; 38.3, C-10; 197.6, C-11; 50.5, C-12; 46.8, C-13; 59.78, C-14; 216.8, C-15; 36.3, C-16; 49.0, C-17; 19.3, C-18; 18.1, C-19; 72.9, C-20; 26.7, C-21; 52.7, C-22; 210.4, C-23; 47.7, C-24; 34.5, C-25; 175.9, C-26; 17.0, C-27; 27.0, C-28; 20.8, C-29; 25.1, C-30; and 52.0ppm, CH$_3$OOC-.

Mass Spectrum:
HREIMS: 544.3057(M$^+$, $C_{31}H_{44}O_8$, calcd 544.3037, 0.06%), 400(2), 372(5), 129(7), 112(18), 87(21), 59(210), and 43*m/e* (100).

Reference

T. Nishitoba, H. Sato, and S. Sakamura; Triterpenoids from the Fungus *Ganoderma lucidum*; Phytochemistry, Vol. 26, pp. 1777-1784(1987).

Common/Systematic Name

Ganoderic acid O

20-Hydroxy-3,7,11,15,23-pentaoxo-5α-lanost-8-en-26-oic acid

Molecular Formula/Molecular Weight

$C_{30}H_{40}O_8$; MW = 528.27232

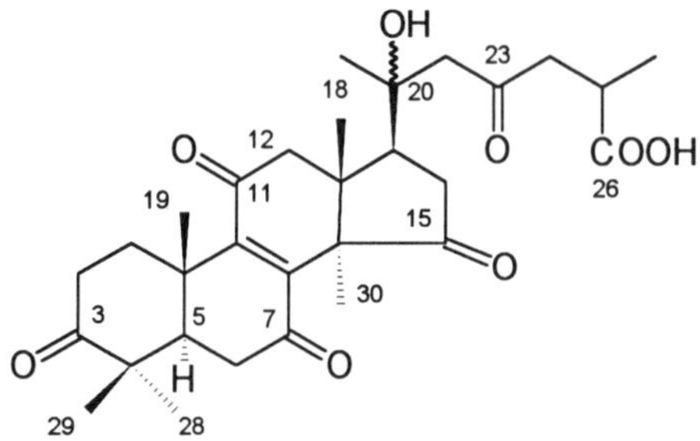

General Characteristics

Methyl ester: pale yellow needles from ethyl ether-hexane; mp., 168-171°C.

Fungal Source

Fruiting bodies of *Ganoderma lucidum*.

Isolation/Purification

Dried chipped fruiting bodies of *Ganoderma lucidum* were extracted with ethanol. The extract was concentrated and partitioned between chloroform and water. The acidic fraction was obtained and chromatographed on a silica gel column eluted with chloroform-methanol (49:1, v/v; fractions 1-11), chloroform-methanol (9:1, v/v; fraction 12), and methanol (fraction 13). Fraction 5 was rechromatographed 3 times on a silica gel column eluted with chloroform-methanol (19:1, v/v) to give several fractions. Some of the fractions were methylated and further fractionated using preparative TLC (benzene-ethyl acetate, 1:1 or 2:1, v/v), silica gel column chromatography (chloroform-methanol, 99:1, v/v), or reversed-phase LC (methanol-water, 7:3, v/v, Lobar RP_{18}). These separations gave purified methyl ganoderate M, N, O, H, E, methyl lucidenate K, L, and E_2.

Biological Activity

Fruiting bodies of *G. lucidum* produce some physiological effects when used as a home remedy.

Spectral Data (Methyl ester)

UV:

λ_{max}^{MeOH} 252nm (ϵ=6,610).

IR:
(KBr) 3475, 2950, 1740, and 1700cm^{-1}.

CD:
λ_{max}^{MeOH} Δ_{330} 0, Δ_{305} -5.0, Δ_{290} 0, Δ_{277} +5.1, Δ_{259} 0, Δ_{259} 0, Δ_{252} -1.5, Δ_{243} 0, and Δ_{212} 0.

^{1}H NMR:
($CDCl_3$) 1.79(1H, ddd, *J*=14.1, 10.4, 5.9Hz, H-1α); 3.00(1H, m, H-1β); 2.46(1H, ddd, *J*=15.6, 8.8, 5.9Hz, H-2α); 2.66(1H, ddd, *J*=15.6, 10.4, 5.9Hz, H-2β); 2.48(1H, dd, *J*=14.7, 2.4Hz, H-5α); 2.53(1H, dd, *J*=13.4, 2.4Hz, H-6α); 2.76(1H, dd, *J*=14.7, 13.2Hz, H-6β); 3.25(1H, d, *J*=16.6Hz, H-12α); 3.12(1H, d, *J*=16.6Hz, H-12β); 3.00(1H, m, H-16α); 3.00(1H, m, H-16β); 2.92(1H, dd, *J*=8.3, 8.3Hz, H-17); 1.38(3H, s, H_3-18); 1.32(3H, s, H_3-19); 1.67(3H, s, H_3-21); 2.85(1H, d, *J*=14.7Hz, H-22); 2.85(1H, d, *J*=14.7Hz, H-22); 3.00(1H, m, H-24); 2.70(1H, dd, *J*=17.0, 4.0Hz, H-24); 3.00(1H, m, H-25); 1.16(3H, d, *J*=7.3Hz, H_3-27); 1.06(3H, s, H_3-28); 1.11(3H, s, H_3-29); 1.92(3H, s, H_3-30); and 3.62ppm (3H, s, $COOCH_3$).

Mass Spectrum:
FD-MS: 542(M^+, 100%), 398(43), 355(20), 187(90), and 144*m/e* (31); LREIMS: 398($C_{24}H_{30}O_5{}^+$, 6%), 355(4), 300(42), 215(23), 87(15), 59(15), and 43*m/e* (100).

Reference
T. Nishitoba, H. Sato, and S. Sakamura; Triterpenoids from the Fungus *Ganoderma lucidum*; Phytochemistry, Vol. 26, pp. 1777-1784(1987).

Common/Systematic Name

Ganoderic acid R

(22*S*,24*E*)-3α,22-Diacetoxy-5α-lanosta-7,9(11),24-trien-26-oic acid

Molecular Formula/Molecular Weight

$C_{34}H_{50}O_6$; MW = 554.36074

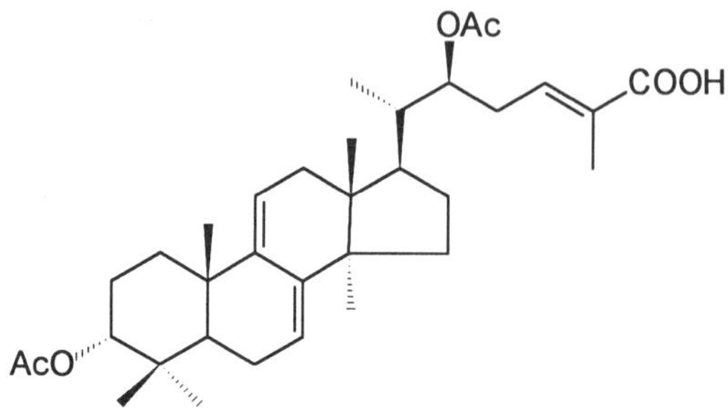

General Characteristics

Colorless needles; mp., 201-202°C; $[\alpha]_D$ +8.7° (*c*=0.092, in $CHCl_3$).

Fungal Source

Isolated from mycelium of *Ganoderma lucidum* ("Saegusa"), called Reishi in Japan.

Isolation/Purification

The mycelial mat grown on a solid medium was harvested and then extracted with EtOH. The extract was partitioned between $CHCl_3$ and water. The $CHCl_3$ layer was concentrated and separated by repeated chromatography on silica gel column and/or reversed-phase LC to afford purified ganoderic acids, R, S, and T.

Biological Activity

Strongly antihepatotoxic in galactose-induced cytotoxicity using primary cultured rat hepatocytes.

Spectral Data

UV:

λ_{max}^{MeOH} 224nm (log ε = 4.0), 234(4.0), 242(4.0), and 250nm (3.8).

IR:

(thin film) 3400(OH), 2870(CH), 1720(COO), and 1675cm^{-1} (COO).

^{1}H NMR:

($CDCl_3$) The ^{1}H NMR spectrum was very similar to that of ganoderic acid T except for the loss of the methine proton signal at 5.03ppm (15α-H) and loss of one acetoxyl methyl proton signal.

^{13}C NMR:

($CDCl_3$) C-1, 30.5; C-2, 23.0; C-3, 78.0; C-4, 36.4; C-5, 44.0; C-6, 27.5; C-7, 120.2; C-8, 142.2; C-9, 145.8; C-10, 37.1; C-11, 115.5; C-12, 37.6; C-13, 43.6; C-14, 50.3; C-15, 31.3; C-16, 22.8; C-17, 47.3; and C-18, 15.4ppm.

The ^{13}C NMR spectrum was also closely similar to that of ganoderic acid T except for the appearance of a methylene carbon signal (31.3ppm) in ganoderic acid R instead of the methine carbon signal (77.4ppm) in T.

Mass Data:

Anal. calcd for $C_{34}H_{50}O_6$: C, 73.61; H, 9.09%; found C, 73.77; H, 9.13%.

Reference

M. Hirotani, C. Ino, T. Furuya, and M. Shirob; Ganoderic Acids T, S, and R, New Triterpenoids in the Cultured Mycelia of *Ganoderma lucidum;* Chem. Pharm. Bull., Vol. 34, pp. 2282-2285(1986).

Common/Systematic Name

Ganoderic acid S

(22*S*,24*E*)-22-Acetoxy-3α-hydroxy-5α-lanosta-7,9(11),24-trien-26-oic acid

Molecular Formula/Molecular Weight

$C_{34}H_{48}O_5$; MW = 512.35017

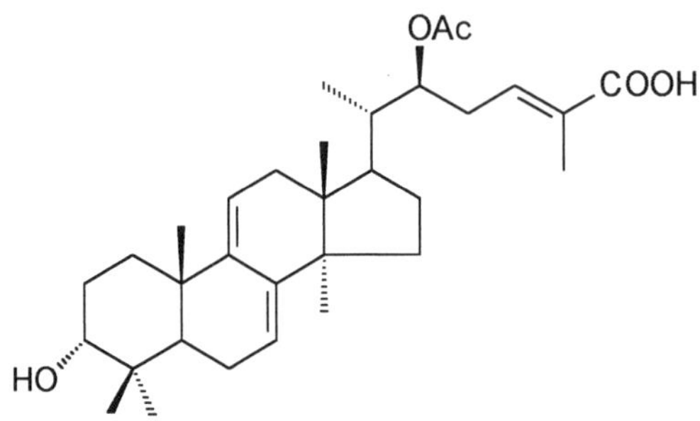

General Characteristics

Ganoderic acid S crystallized as needles; mp., 194-196°C; $[\alpha]_D$ + 19.8° (*c*= 0.085, in $CHCl_3$).

Fungal Source

Isolated from mycelium of *Ganoderma lucidum* ("Saegusa"), called Reishi in Japan.

Isolation/Purification

The mycelial mat grown on a solid medium was harvested and then extracted with ethanol. The extract was partitioned between chloroform and water. The chloroform layer was concentrated and separated by repeated chromatography on silica gel columns and/or reversed-phase LC to afford purified ganoderic acids R, S, and T.

Biological Activity

Strongly antihepatotoxic in galactose-induced cytotoxicity using primary cultured rat hepatocytes.

Spectral Data

UV:

λ_{max}^{EtOH} 231(log ∊ = 4.2), 240(4.2), and 248nm (4.0).

IR:

(Thin film) 3410(OH), 1720(COO), and 1675cm^{-1} (COO).

^{1}H NMR:

(CDCl$_3$) The proton NMR spectrum of ganoderic acid S showed a methine proton signal at 5.11ppm (ddd, *J*=6.8, 6.0, 1.7Hz, H-22) and a proton signal at 3.45ppm (dd, *J*=1.4, 1.4ppm, H-3β) which was absent in the proton spectra of T and R.

^{13}C NMR:

(CDCl$_3$) 29.8, C-1; 25.5, C-2; 76.0, C-3; 37.1*, C-4; 43.1, C-5; 27.5, C-6; 120.3, C-7; 142.2, C-8; 145.9, C-9; 37.3*, C-10; 115.5, C-11; 37.6, C-12; 43.6, C-13; 50.3, C-14; 31.2, C-15; 22.9, C-16; 47.3, C-17; 15.4, C-18; 22.7, C-19; 39.3, C-20; 12.6, C-21; 74.6, C-22; 31.7, C-23; 139.5, C-24; 129.0, C-25; 172.2, C-26; 12.2, C-27; 28.1, C-30; 22.5*, C-31; 25.7, C-32; 21.0, acetyl methyl; and 170.6ppm, C=O.

* Assignments may be reversed.

Mass Data:

Anal. Calcd. For $C_{32}H_{48}O_5$: C, 74.96; H, 9.44%; found C, 74.93; H, 9.55%.

Reference

M. Hirotani, C. Ino, T. Furuya, and M. Shirob; Ganoderic Acids T, S, and R, New Tripenoids in the Cultured Mycelia of *Ganoderma lucidum*; Chem. Pharm. Bull., Vol. 34, pp. 2282-2285(1986).

Common/Systematic Name

Ganoderic acid T

(22*S*,24*E*)3α,15α,22-Triacetoxy-5α-lanosta-7,9(11),24-trien-26-oic acid

Molecular Formula/Molecular Weight

$C_{36}H_{52}O_8$; MW = 612.36622

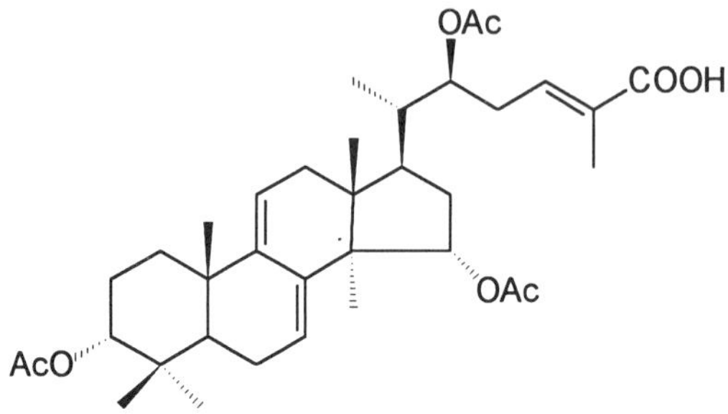

General Characteristics

Colorless needles; mp., 200-202°C; $[\alpha]_D$ +23° (*c*=0.13, in $CHCl_3$).

Fungal Source

Isolated from mycelium of *Ganoderma lucidum* ("Saegusa"), called Reishi in Japan.

Isolation/Purification

The mycelial mat grown on a solid medium was harvested and then extracted with EtOH. The extract was partitioned between $CHCl_3$ and water. The $CHCl_3$ layer was concentrated and separated by repeated chromatography on silica gel column and/or reversed-phase LC to afford purified ganoderic acids, R, S, and T.

Biological Activity

Possibly cytotoxic to hepatoma cells or antihepatotoxic.

Spectral Data

UV:

λ_{max}^{MeOH} 216nm (log ε = 4.0), 225(4.00), 234(4.0), and 242nm (4.0).

IR:

(thin film) 3420, 2940, 1720, and 1240cm^{-1}.

^{1}H NMR:

($CDCl_3$) The spectrum showed the presence of five tertiary methyl (0.66, 0.88, 0.98, 0.99, and 1.03ppm), one allyl methyl (1.86ppm) and one secondary methyl (0.97ppm, d, *J*=6.7Hz) signals. Also observed were signals due to three secondary acetoxyl groups 2.05(3H,s), 2.07(3H,s), 2.08(3H,s), 4.68(1H, dd, *J*=3.0, 3.0 H-3), 5.08(H, dd, *J*=10.0, 5.1Hz, H-15), and 5.03ppm (H, ddd, *J*=7.1, 7.0, 1.4Hz, H-22) and three olefinic proton signals 5.32ppm (1H, d, *J*=6.4Hz, H-11), 5.48(1H, bs, H-7), 6.78(1H, dd, *J*=7.5, 7.5Hz, H-24). These spectral data suggested that ganoderic acid T was a lanostane-type triterpenoid having a heteroannular diene moiety and three acetoxyl groups.

^{13}C NMR:

($CDCl_3$) C-1, 30.8; C-2, 23.3; C-3, 78.2; C-4, 36.8; C-5, 44.1; C-6, 22.7; C-7, 121.6; C-8, 140.2; C-9, 146.2; C-10, 37.5; C-11, 115.5; C-12, 38.1; C-13, 44.1; C-14, 51.6; C-15, 77.4; C-16, 36.7; C-17, 45.6; and C-18, 15.9ppm.

Reference

M. Hirotani, C. Ino, T. Furuya, and M. Shirob; Ganoderic Acids T, S, and R, New Triterpenoids in the Cultured Mycelia of *Ganoderma lucidum;* Chem. Pharm. Bull., Vol. 34, pp. 2282-2285(1986).

Common/Systematic Name

Ganoderic acid U

Molecular Formula/Molecular Weight

$C_{30}H_{48}O_4$; MW = 472.35526

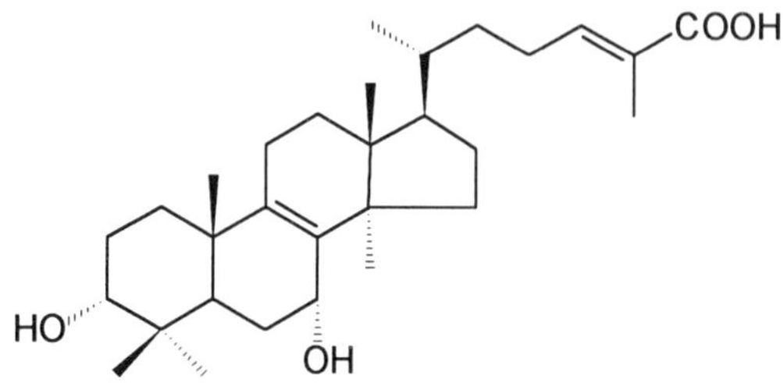

General Characteristics

Crystals; mp., 166-169°C; $[\alpha]_D^{22}$ + 35° (*c*=3.5, in EtOH).

Fungal Source

Isolated from mycelium of *Ganoderma lucidum*.

Biological Activity.

In vitro cytotoxicity on hepatoma cells.

Spectral Data

IR:

($CHCl_3$ or CCl_4) 3610, 3550-3300, 1690, and 1645cm^{-1}.

^{1}H NMR:

($CDCl_3$) 0.62(3H, s); 0.89(3H, s); 0.92(3H, d, *J*=6.0Hz); 0.97(3H, s); 1.01(3H, s); 1.04(3H, s); 1.84(3H, s); 3.48(1H, t, *J*=2.5Hz); 4.16(1H, m); and 6.91ppm (1H, t, *J*=7.0Hz)

^{13}C NMR:

($CDCl_3$) Methyl ester derivative: 29.7, C-1; 28.8, C-2; 75.8, C-3; 37.2, C-4; 39.0, C-5; 25.6, C-6; 67.3, C-7; 135.9, C-8; 141.4, C-9; 38.3, C-10; 20.9, C-11; 29.9, C-12; 45.1, C-13; 49.7, C-14; 31.1, C-15; 28.2, C-16; 50.62, C-17; 16.1, C-18; 17.3, C-19; 36.5, C-20; 18.4, C-21; 34.9, C-22; 25.9, C-23; 145.4, C-24; 126.8, C-25; 172.7, C-26; 12.0, C-27; 27.9, C-30; 22.3, C-31; 26.3, C-32; and 51.7ppm -CO_2CH.

Mass Spectrum:

LREIMS: 472(M^+), 454(M^+ - H_2O), 526(M^+ - AcOH), 508(M^+ - AcOH -H_2O), 493(M^+ - AcOH- CH_3 - H_2O), 451(M^+ - AcOH - CH_3), 433(M^+ - 2AcOH - CH_3 - H_2O), 293(M^+ - 2AcOH - H_2O - $C_9H_{15}O_2$), and 239*m/e*.

TLC Data

A rose-violet colored spot at R_f 0.55 of methyl ester using silica gel Merck HF_{254} with methylene chloride-MeOH, 19:1 as developing solvent and spraying with 50% sulfuric acid and heating.

Reference

J. O. Toth, B. Luu, J. P. Beck, and G. Ourisson; Triterpenes Cytotoxiwues de *Ganoderma lucidum*. Structures des acides Ganoderiques U-Z; J. Chem. Res.(M), pp. 2722-2787(1983).

Common/Systematic Name

Ganoderic acid V

Molecular Formula/Molecular Weight

$C_{32}H_{48}O_6$; MW = 528.34509

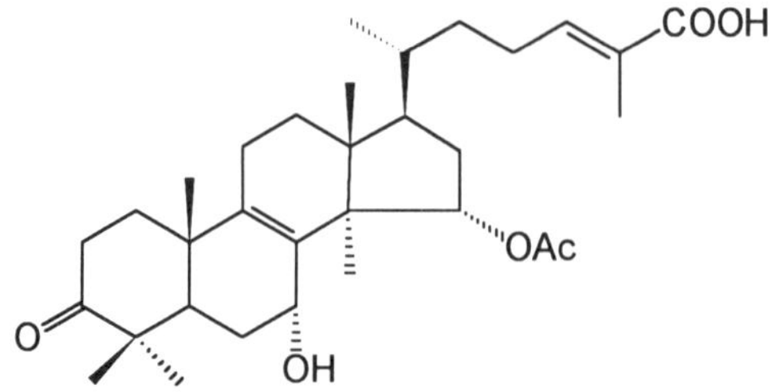

General Characteristics

Isolated as methyl ester ($C_{33}H_{50}O_6$), white amorphous powder; $[\alpha]_D^{20}$ + 83° (*c*=0.92, in $CHCl_3$).

Fungal Source

Isolated from *Ganoderma lucidum*.

Biological Activity

Cytotoxic

Spectral Data (Methyl ester)

IR:

($CHCl_3$ or CCl_4) 3600, 1713, and 1649cm^{-1}.

^{1}H NMR:

($CDCl_3$) 0.72(3H, s); 0.92(3H, d, *J*=6.0Hz); 1.05(3H, s); 1.07(3H, s); 1.13(3H, s); 1.16(3H, s); 1.82(3H, s); 3.73(3H, s); 5.10(1H, dd, *J*=9.0, 5.0Hz); and 6.72ppm (1H, t, *J*=7.0Hz)

^{13}C NMR:

($CDCl_3$) 35.2, C-1; 34.3, C-2; 217.3, C-3; 46.7, C-4; 44.8, C-5; absent, C-6; 66.2, C-7; 134.7, C-8; 140.0, C-9; 38.1, C-10; 20.8, C-11; 28.4, C-12; 45.3, C-13; 51.1, C-14; 76.6, C-15; 36.5, C-16; 49.32, C-17; 16.6, C-18; 17.3, C-19; 36.2, C-20; 18.2, C-21; 34.8, C-22; 25.7, C-23; 142.5, C-24; 127.4, C-25; 168.7, C-26; 12.4, C-27; 26.5, C-30; 21.2, C-31; 20.1, C-32; 51.7, $-CO_2CH$; 21.2, $\underline{C}H_3COO$-; and 170.6ppm $CH_3\underline{C}OO$-.

Mass Spectrum:

LREIMS: M$^+$ absent, 524(M$^+$ - H_2O), 493, 464(M$^+$ - AcOH -H_2O), 449(M$^+$ - AcOH - CH_3 - H_2O), 335, and 309*m/e* (M$^+$ - AcOH - $C_9H_{15}O_2$).

TLC Data

R_f 0.35 of methyl ester using silica gel Merck HF_{254} using ethyl ether-hexane, 1:1 as developing solvent. Appeared as yellow then chestnut-violet spot after spraying with 50% sulfuric acid and heating.

Reference

J. O. Toth, B. Luu, J. P. Beck, and G. Ourisson; Triterpenes Cytotoxiwues de *Ganoderma lucidum*. Structures des Acides Ganoderiques U-Z; J. Chem. Res. (M), pp. 2722-2787(1983).

Common/Systematic Name
Ganoderic acid W

Molecular Formula/Molecular Weight
$C_{34}H_{52}O_7$; MW = 572.37130

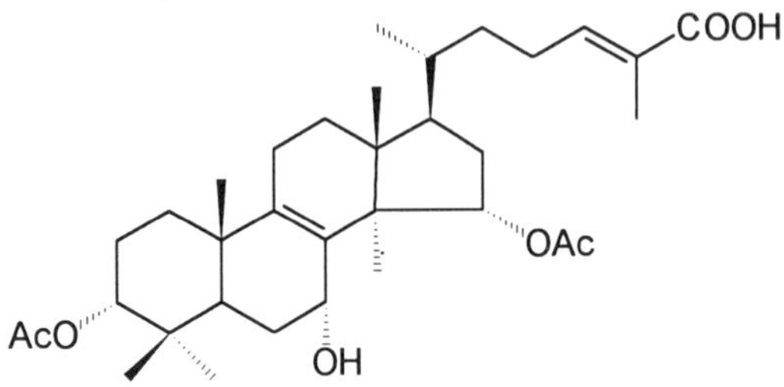

General Characteristics
Amorphous solid; mp., 114-117°C; methyl ester: $[\alpha]_D^{21}$ + 4° (*c*=0.5, in EtOH).

Fungal Source
Isolated from mycelium of *Ganoderma lucidum*.

Biological Activity.
Cytotoxic.

Spectral Data

UV:
λ_{max} 215nm (log ϵ=4.1).

IR:
($CHCl_3$ or CCl_4) 3600, 3000-2500, 1743,1690, and 1645cm^{-1}; 3600, 1742, 1728-1718, and 1650cm^{-1}

^{1}H NMR:
($CDCl_3$) 0.71(3H, s); 0.93(9H, m); 0.97(3H, s); 1.20(3H, s); 2.07(3H, s); 2.09(3H, s); 1.84(3H, s); 4.12(1H, m); 4.69(1H, t, *J*=2.5Hz); 5.09(1H, dd, *J*=9.0, 5.0Hz); and 6.86(1H, t, *J*=7.0Hz); methyl ester: 0.71(3H, s); 0.93(9H, m); 0.97(3H, s); 1.12(3H, s); 2.07(3H, s); 2.09(3H, s); 4.11(1H, m); 4.69(1H, t, *J*=2.5Hz); 5.14(1H, dd, *J*=9.0, 5.0Hz); and 6.72ppm (1H, t, *J*=7.0Hz).

^{13}C NMR:
($CDCl_3$) Methyl ester derivative: 30.2, C-1; 25.6, C-2; 77.4, C-3; 36.3, C-4; 40.0, C-5; 23.3, C-6; 66.4, C-7; 133.8, C-8; 141.7, C-9; 38.4, C-10; 20.7, C-11; 27.3, C-12; 45.4, C-13; 51.1, C-14; 76.5, C-15; 36.5, C-16; 49.2, C-17; 16.5, C-18; 17.4, C-19; 36.2, C-

20; 18.2, C-21; 34.7, C-22; 25.6, C-23; 142.6, C-24; 127.4, C-25; 168.7, C-26; 12.4, C-27; 27.4, C-30; 21.9, C-31; 20.2, C-32; 51.7 -CO_2CH_3; 21.4, $\underline{C}H_3COO$-; 171.9, $CH_3\underline{C}OO$- ; 21.2, $\underline{C}H_3COO$; and 170.5ppm $CH_3\underline{C}OO$.

Mass Spectrum:
LREIMS: Methyl ester: M^+(absent), 568(M^+ -H_2O), 526(M^+ - AcOH), 508(M^+ - AcOH -H_2O), 493(M^+ - AcOH - CH_3 - H_2O), 451(M^+ - AcOH- CH_3), 433(M^+ - 2AcOH - CH_3 - H_2O), 293(M^+ - 2AcOH - H_2O - $C_9H_{15}O_2$), and 239*m/e*.

TLC Data

R_f 0.43 of methyl ester using silica gel impregnated with silver nitrate with ethyl ether-hexane, 4:1 (v/v) as developing solvent.

Reference

J. O. Toth, B. Luu, J. P. Beck, and G. Ourisson; Triterpenes Cytotoxiwues de *Ganoderma lucidum*. Structures des acides Ganoderiques U-Z; J. Chem. Res. (M), pp. 2722-2787(1983).

Common/Systematic Name
Ganoderic acid X

Molecular Formula/Molecular Weight
$C_{32}H_{48}O_5$; MW = 512.35017

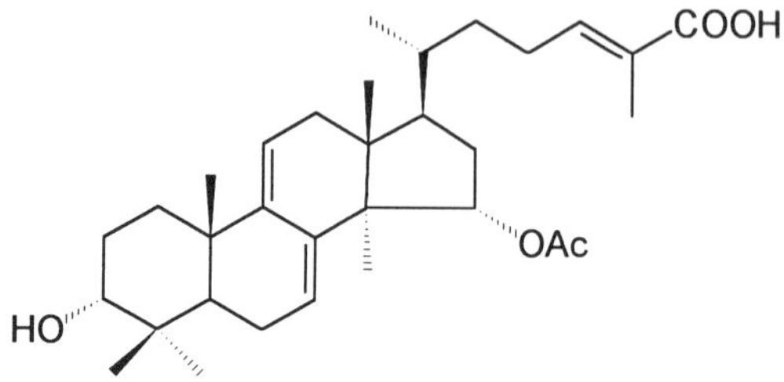

General Characteristics
Isolated as methyl ester ($C_{33}H_{50}O_5$), crystals; mp., 161-163°C; $[\alpha]_D^{23}$ + 76° (*c*=3.0, in $CHCl_3$).

Fungal Source
Isolated from *Ganoderma lucidum*.

Biological Activity
Cytotoxic.

Spectral Data (Methyl ester only)

UV:
220(log ϵ = 4.1), 233(4.0), 242(4.0), and 251nm (3.8).

IR:
($CHCl_3$ or CCl_4) 3608, 3000-2500, 1710, and 1648cm^{-1}.

^{1}H NMR:
($CDCl_3$) 0.66(3H, s); 0.92(3H, d, *J*=6.0Hz); 0.93(3H, s); 0.99(6H, s); 1.02(3H, s); 1.82(3H, s); 2.08(3H, s); 3.73(3H, s); 3.44(1H, t, *J*=2.5Hz); 5.08(1H, dd, *J*=9.0, 5.0Hz); 5.35(1H, m); 5.48(1H, m); and 6.73ppm (1H, t, *J*=7.0Hz).

^{13}C NMR:
($CDCl_3$) 30.0, C-1; 25.7, C-2; 76.0, C-3; 37.1, C-4; 43.0, C-5; 23.0, C-6; 121.3, C-7; 140.3, C-8; 146.1, C-9; 37.4, C-10; 115.6, C-11; 38.0, C-12; 44.2, C-13; 51.5, C-14; 77.5, C-15; absent, C-16; 48.9, C-17; 16.0, C-18; 22.8, C-19; 35.9, C-20; 18.5, C-21; 34.8, C-22; 25.7, C-23; 142.7, C-24; 127.4, C-25; 168.7, C-26; 12.4, C-27; 28.2, C-30; 22.7, C-31; 18.2, C-32; 51.7, -$CO_2\underline{C}H$; 21.5, $\underline{C}H_3COO$-; and 171.2ppm $CH_3\underline{C}OO$-.

Mass Spectrum:

HREIMS: 526.3661*m/e* (M^+), calcd. for ($C_{33}H_{50}O_5$) 526.3658; LREIMS: 526(M^+), 508(M^+ - H_2O), 493(M^+ - CH_3 - H_2O), 466(M^+ - AcOH), 451(M^+ - AcOH - CH_3), 448(M^+ - AcOH - H_2O), 433(M^+ - CH_3 - H_2O - AcOH), 337, 311(M^+ - AcOH - $C_9H_{15}O_2$), and 293*m/e* (M^+ - AcOH - H_2O - $C_9H_{15}O_2$).

TLC Data

R_f 0.32 of methyl ester using silica gel Merck HF_{254} using ethyl ether-hexane, 1:1 (v/v) as developing solvent. Appeared as yellow turning to maroon-violet spot after spraying with 50% sulfuric acid and heating.

Reference

J. O. Toth, B. Luu, J. P. Beck, and G. Ourisson; Triterpenes Cytotoxiwues de *Ganoderma lucidum*. Structures des acides Ganoderiques U-Z; J. Chem. Res. (M), pp. 2722-2787(1983).

Common/Systematic Name

Ganoderic acid Y

3β-Hydroxy-5α-lanosta-7,9(11),24-trien-26-oic acid

Molecular Formula/Molecular Weight

$C_{30}H_{46}O_3$; MW = 454.34470

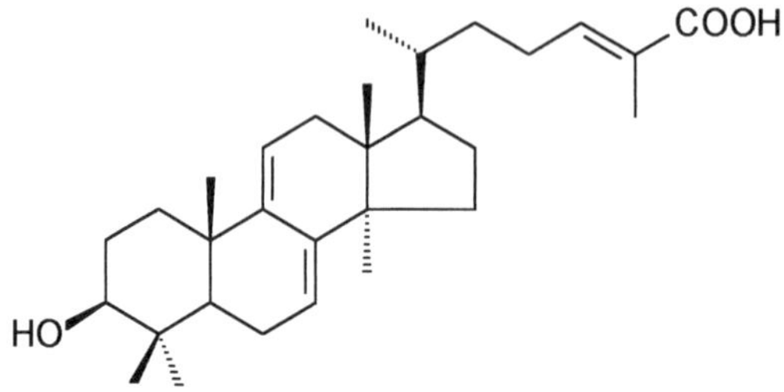

General Characteristics

Crystals from ethyl ether-ethanol, mp., 203-206°C; $[\alpha]_D^{25}$ +54° (*c*= 0.8, in ethanol); Methyl ester; mp., 139-141°C; $[\alpha]_D^{26}$ +53° (*c*= 0.6, in chloroform) .

Fungal Source

Isolated from mycelium of *Ganoderma lucidum*.

Biological Activity.

Cytotoxic.

Spectral Data

IR:

($CHCl_3$ or CCl_4) 3600, 3440, 2700-2200, 1690, and 1647cm^{-1}.

^{1}H NMR:

($CDCl_3$) 5.33 (1H, m); 5.47 (1H); and 8.88ppm (1H, t, *J*= 7.0Hz). Methyl ester; 3.73 (3H, s); 5.31 (1H, m); 5.48 (1H, m); and 6.72ppm (1H, t, *J*= 7.0Hz).

Mass Spectrum:

LREIMS: 454(M^+), 439(M^+ - CH_3), and 421m/e (M^+ - CH_3 - H_2O). Methyl ester; 468 (M^+), 453 (M^+ - CH_3), 435(M^+ - CH_3 - H_2O), and 302*m/e*.

Reference

J. O. Toth, B. Luu, J. P. Beck, and G. Ourisson; Triterpenes Cytotoxiwues de *Ganoderma lucidum*. Structures des acides Ganoderiques U-Z; J. Chem. Res. (M), pp. 2722-2787 (1983).

Common/Systematic Name
Ganoderic acid Z

Molecular Formula/Molecular Weight
$C_{30}H_{48}O_3$; MW = 456.36035

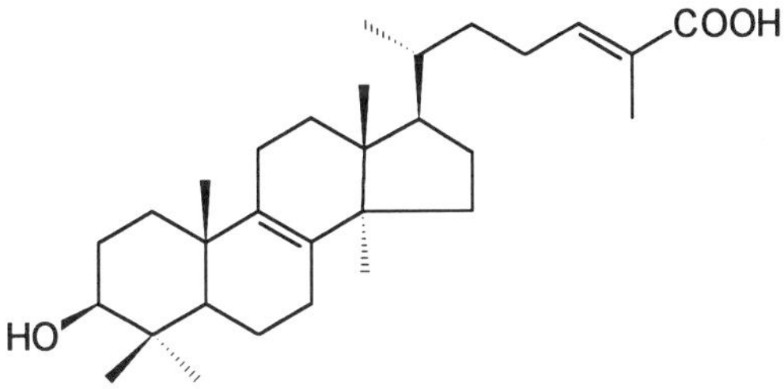

General Characteristics
Crystals; mp., 137-140°C; $[\alpha]_D^{24}$ + 59° (*c*=0.13, in $CHCl_3$); methyl ester; mp., 123-125°C; $[\alpha]_D^{22}$ + 56° (*c*=0.7, in EtOH).

Fungal Source
Isolated from mycelium of *Ganoderma lucidum*.

Biological Activity
Cytotoxic.

Spectral Data

IR:
($CHCl_3$ or CCl_4) 3600, 3470, 2700-2300, 1685, and 1645cm^{-1}.

^{1}H NMR:
($CDCl_3$) 6.90(1H, t, *J*=7.0Hz); methyl ester: 0.69(3H, s); 0.81(3H, s); 0.88(3H, s); 0.92(3H, d, *J*=6Hz); 1.00(3H, s); 1.84(3H, s); 3.24(1H, m); 3.73(3H, s); and 6.76ppm (1H, t, *J*=7.0Hz).

^{13}C NMR:
($CDCl_3$) 3β-acetylmethyl ester derivative: 35.3, C-1; 24.2, C-2; 80.9, C-3; 37.9, C-4; 50.4, C-5; 18.2, C-6; 26.4, C-7; 134.4, C-8; 134.4, C-9; 37.0, C-10; 21.0, C-11; 30.8, C-12; 44.6, C-13; 49.9, C-14; 31.0, C-15; 28.2, C-16; 50.6, C-17; 15.8, C-18; 18.6, C-19; 36.4, C-20; 19.2, C-21; 35.0, C-22; 25.7, C-23; 143.2, C-24; 127.2, C-25; 168.8, C-26; 12.3, C-27; 28.0, C-30; 16.6, C-31; 24.3, C-32; 51.6 -CO_2CH_3; 21.3, $\underline{C}H_3COO$-; and 171.0ppm $CH_3\underline{C}OO$-.

Mass Spectrum:
LREIMS: Methyl ester: 470(M^+), 455(M-CH_3), 442, 437(M-CH_3-H_2O), 427, and 409*m/e*.

TLC Data

R_f 0.28 of methyl ester using silica gel impregnated with silver nitrate with ethyl ether-hexane, 1:4 (v/v) as developing solvent.

Reference

J. O. Toth, B. Luu, J. P. Beck, and G. Ourisson; Triterpenes Cytotoxiwues de *Ganoderma lucidum*. Structures des acides Ganoderiques U-Z; J. Chem. Res. (M), pp. 2722-2787(1983).

Common/Systematic Name

Ganodermic Acid R

Lanosta-7,9(11),24-triene-3α,15α-diacetoxy-26-oic acid

Molecular Formula/Molecular Weight

$C_{34}H_{50}O_6$; MW = 554.36074

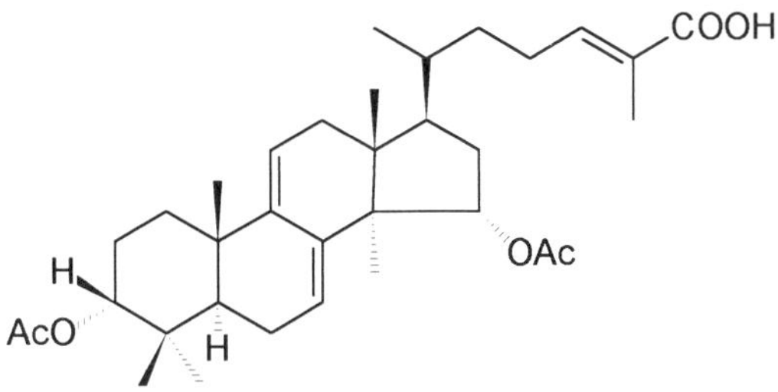

General Characteristics

Amorphous powder from *n*-hexane-$CHCl_3$; mp., 126-129°C.

Fungal Source

Ganoderma lucidum (strains TP-1 and AT- 4).

Isolation/Purification

Dried, ground mycelia were extracted with MeOH; the extract was partitioned between *n*-hexane and H_2O. The aqueous layer was re-extracted with EtOAc and chromatographed on silica gel (gradient from MeOH to $CHCl_3$) followed by reversed phase (C_{18}) HPLC (90% aqueous MeOH). Pure substance was obtained through repeated recrystallization from *n*-hexane-$CHCl_3$.

Spectral Data

UV:

λ_{max}^{MeOH} 210(log ε = 4.08), 235(4.14), 243(4.16), and 251nm (3.97).

IR:

($CHCl_3$) 3025, 2945, 1715, 1680, and 1240cm^{-1}.

CD:

(*c*=0.0001, in MeOH) $[\Theta]_{230}$ = + 17,200.

ORD:

(*c*=0.0001, in MeOH) $[\phi]_{248}$ + 8200, $[\phi]_{223}$ - 7700, $[\phi]_{211}$ -11,600, and $[\phi]_{201}$ - 20,700.

^{1}H NMR:

(CDCl$_3$) 4.64(1H, bs, H-3); 5.45(1H, m, H-7); 5.30(1H, dd, *J*=5.7Hz, H-11); 5.04(1H, dd, *J*=4.5, 10Hz, H-15); 0.63(3H, s, H-18); 0.95(3H, s, H-19); 0.88(3H, d, *J*=6.3Hz, H-21); 6.83(1H, m, H-24); 1.79(3H, s, H-27); 0.85(3H, s, H-29); 0.94(3H, s, H-30); 2.01(3H, s, AcCH$_3$) and 2.05ppm (3H, s, AcCH$_3$).

^{13}C NMR:

(CDCl$_3$) 30.51(t, C-1); 23.03(t, C-2); 77.98(d, C-3); 36.42(s, C-4); 43.82(d, C-5); 22.70(t, C-6); 121.04(d, C-7); 140.09(s, C-8); 145.40(s, C-9); 37.22(s, C-10); 115.52(d, C-11); 37.87(t, C-12); 44.04(s, C-13); 51.30(s, C-14); 77.26(d, C-15); 36.89(t, C-16); 48.76(d, C-17); 15.85(q, C-18); 22.52(q, C-19), 35.85(d, C-20); 18.08(q, C-21); 34.54(t, C-22); 25.82(t, C-23); 144.99(d, C-24); 126.72(s, C-25); 172.92(s, C-26); 11.84(q, C-27); 18.31(q, C-28); 27.65(q, C-29); 22.32(q, C-30); 21.15(q, COOCH$_3$); 21.26(q, COOCH$_3$); 170.65(s, AcCO); and 171.02ppm (AcCO).

Mass Spectrum:

554(M$^+$, $C_{34}H_{50}O_6$, 100%), 494(M$^+$ - CH$_3$COOH, 40), 479(M$^+$ -CH$_3$COOH-CH$_3$, 25), 434(M$^+$ - 2CH$_3$COOH, 33), 419(M$^+$ - 2CH$_3$COOH-CH$_3$, 90), 353(M$^+$ - CH$_3$COOH-$C_{18}H_{13}O_2$ side chain, 12), 299(D-ring cleavage-CH$_3$, 14), 293(M$^+$ - 2CH$_3$COOH-$C_{18}H_{13}O_2$ side chain, 19), and 239*m/e* (299- CH$_3$COOH, 44).

Reference

M. S. Shiao, L. J. Lin, S. F. Yen, and C. S. Chou; Two New Triterpenes of the Fungus *Ganoderma lucidum;* J. Nat. Prod., Vol. 50, No. 5, pp. 886-890(1987).

Common/Systematic Name

Ganodermic Acid S

Lanosta-7,9(11),24-triene-3β,15α-diacetoxy-26-oic acid

Molecular Formula/Molecular Weight

$C_{34}H_{50}O_6$, MW = 554.36074

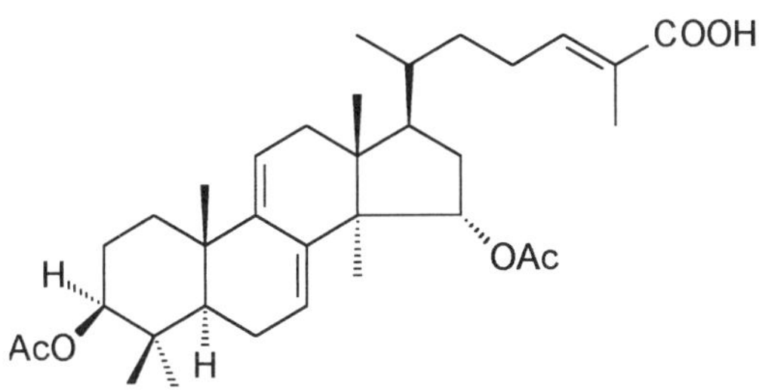

General Characteristics

Colorless needles from *n*-hexane-$CHCl_3$; mp., 123-124°C.

Fungal Source

Ganoderma lucidum (strains TP-1 and AT-4).

Isolation/Purification

Dried, ground mycelia were extracted with MeOH; the extract was partitioned between hexane and H_2O. The aqueous layer was re-extracted with EtOAc and chromatographed on silica gel (gradient from MeOH to $CHCl_3$) followed by reversed phase (C_{18}) HPLC (90% aqueous MeOH). Pure substance was obtained through repeated recrystallization from *n*-hexane-$CHCl_3$.

Spectral Data

UV:

λ_{max}^{MeOH} 211(log ε = 4.17), 235(4.21), 243(4.22), and 251nm (4.04).

IR:

($CHCl_3$) 3030, 2950, 1720, 1695, and 1245cm^{-1}.

CD:

(c=0.0001, in MeOH) [Θ]230 = +13,700.

ORD:

(c=0.0001, in MeOH) $[\Phi]_{248}$ + 7500, $[\Phi]_{225}$ - 2700, $[\Phi]_{217}$ - 10800, $[\Phi]_{209}$ - 400, and $[\Phi]_{204}$ - 4400.

^{1}H NMR:

(CDCl$_3$) 4.49(1H, dd, *J*=4.5, 11Hz, H-3); 5.44(1H, m, H-7); 5.29(1H, dd, *J*=5.7Hz, H-11); 5.04(1H, dd, *J*=4.5, 10Hz, H-15); 0.62(3H, s, H-18); 0.97(3H, s, H-19); 0.88(3H, d, *J*=6.5Hz, H-21); 6.84(1H, m, H-24); 1.80(3H, s, H-27); 0.86(3H, s, H-29); 0.92(3H, s, H-30); 2.03(3H, s, COOCH$_3$); and 2.06ppm (3H, s, COOCH$_3$).

13C NMR:

(CDCl$_3$) 35.26(t, C-1); 24.06(t, C-2); 80.60(d, C-3); 37.43(s, C-4); 48.83(d, C-5); 22.69(t, C-6); 120.93(d, C-7); 140.01(s, C-8); 145.52(s, C-9); 37.16(s, C-10); 116.00(d, C-11); 37.86(t, C-12); 43.92(s, C-13); 51.19(s, C-14); 77.15(d, C-15); 36.84(t, C-16); 48.67(d, C-17); 15.82(q, C-18); 22.69(q, C-19), 35.79(d, C-20); 18.03(q, C-21); 34.49(t, C-22); 25.78(t, C-23); 144.90(d, C-24); 126.72(s, C-25); 173.01(s, C-26); 11.78(q, C-27); 18.19(q, C-28); 27.97(q, C-29); 16.79(q, C-30); 21.06(q, COOCH$_3$); 21.18(q, COOCH$_3$); 170.83(s, AcCO); and 171.04ppm (AcCO).

Mass Spectrum:

554(M^+, $C_{34}H_{50}O_6$, 95%), 494(M^+- CH_3COOH, 35), 479(M^+ - CH_3COOH-CH_3, 25), 434(M^+- 2CH_3COOH, 23), 419(M^+ - 2CH_3COOH-CH_3, 100), 353(M^+ - CH_3COOH-$C_{18}H_{13}O_2$ side chain, 36), 299(D-ring cleavage-CH_3, 70), 293(M^+- 2CH_3COOH-$C_{18}H_{13}O_2$ side chain, 34), and 239*m/e* (299- CH_3COOH, 62).

Reference

M. S. Shiao, L. J. Lin, S. F. Yen, and C. S. Chou; Two New Triterpenes of the Fungus *Ganoderma lucidum*; J. Nat. Prod., Vol. 50, No. 5, pp. 886-890(1987).

10

Ganoderiols

- Ganoderiol A
- Ganoderiol B
- Ganoderiol C
- Ganoderiol D
- Ganoderiol E
- Ganoderiol F
- Ganoderiol G
- Ganoderiol H
- Ganoderiol I
- Ganodermanondiol
- Ganodermanontriol
- Ganodermenonol
- Ganodermadiol
- Ganodermatriol

Common/Systematic Name
Ganoderiol A
5α-Lanosta-7,9(11)-diene-3β,24,25,26-tetraol

Molecular Formula/Molecular Weight
$C_{30}H_{50}O_4$; MW = 474.37091

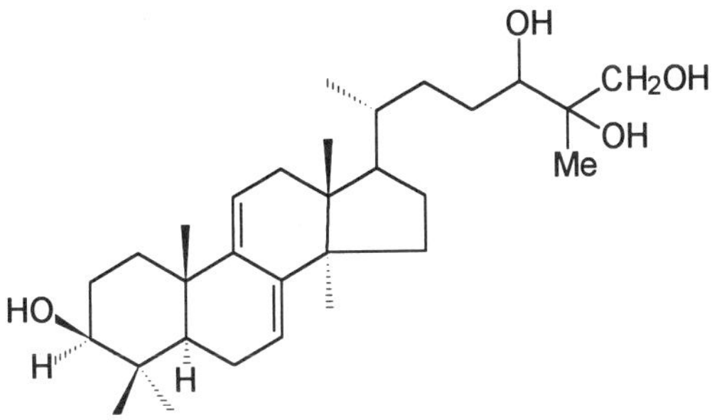

General Characteristics
Colorless needles from chloroform; mp., 232-234°C; $[\alpha]_D^{23}$ +20 (*c*=0.1, in EtOH).

Fungal Source
Ganoderma lucidum.

Isolation/Purification
Dried chipped fruiting bodies of *G. lucidum* were extracted with EtOH. The extract was concentrated and partitioned between $CHCl_3$ and water. The $CHCl_3$ layer was concentrated and extracted with sodium bicarbonate solution. The $CHCl_3$ layer was washed with water, dried with anhydrous sodium sulfate and evaporated to dryness. The neutral fraction was subjected to chromatography on a silica gel column and eluted successively with MeOH-$CHCl_3$ (2:98, 1:9, 1:1, v/v) with monitoring by TLC. Eluates were separated into three fractions. Fraction 1 was rechromatographed on a silica gel column eluted with $CHCl_3$-EtOAc (9:1, v/v) to yield five fractions. Fraction 4 from this column was subjected repeatedly to silica gel chromatography and eluted with $CHCl_3$-acetone (9:1, v/v) to yield six fractions. After evaporation of the solvent, Fraction 3 of the latter column on recrystallization from $CHCl_3$ yielded ganoderiol A as colorless needles. Fraction 2 of the latter column was rechromatographed on a silica gel column eluted with $CHCl_3$-MeOH (9:1, v/v) to yield six fractions. Further purification of Fraction 4 of this column was achieved by a Lobar column RP_{18} (MeOH-H_2O, 95 + 5, v/v), and ganodermanontriol (colorless needles) and ganodermatriol (colorless plates) were isolated. Fraction 2 of the original column was rechromatographed on a silica gel column and eluted with $CHCl_3$-MeOH (95:5, v/v) to yield seven fractions. Fractions 4 and 5 of the latter column were combined and rechromatographed on a silica gel column eluted with $CHCl_3$-EtOAc; six fractions were collected. After evaporation of the solvent, Fraction 4 of the latter column gave ganoderiol A, which was recrystallized from $CHCl_3$ to yield

colorless needles. Fraction 5 of the latter column was rechromatographed on a Lobar column RP-8 (MeOH-H_2O, 9:1, v/v). Further purification by preparative TLC with $CHCl_3$-MeOH (9:1, v/v) gave ganoderiol B as an amorphous powder.

Spectral Data

UV:
λ_{max}^{EtOH} 237(ϵ = 8058), 244(9362), and 253nm (6518).

IR:
(KBr) 3350, 2950, 2900, 2850, 1430, 1360, and 1060cm^{-1}.

^{1}H NMR:
($CDCl_3$) 5.48(1H, m, 7-H); 5.32(1H, d, *J*=5.8Hz, 11-H); 3.83, 3.47(1H each, d, *J*=10.6Hz, 26-H); 3.46(1H, t, *J*=12.1Hz, 24-H); 3.25(1H, dd, *J*=5.0 and 10.5Hz, 3-Ha); 1.11(3H, s, 27-H); 1.01(3H, s, 19-H); 0.98(3H, s, 28-H); 0.92(3H, d, *J*=6.2Hz, 21-H); 0.88(6H, s, 29 and 30-H); and 0.57ppm (3H, s, 18-H).

Mass Spectrum:
474(M^+, $C_{30}H_{50}O_4$, 100%). 459(11), 456(13), 398(22), 311(39), 271(83), 69(72), 55(91), and 43*m/e* (95).

Reference

H. Sato, T. Nishitoba, S. Shirasu, K. Oda, and S. Sakamura; Ganoderiol A and B, New Triterpenoids from the Fungus *Ganoderma lucidum* (Reishi); Agric. Biol. Chem., Vol. 50, pp. 2887-2890(1986).

Common/Systematic Name
Ganoderiol B
15α,26,27-Trihydroxy-5α-lanosta-7,9(11),24-trien-3-one

Molecular Formula/Molecular Weight
$C_{30}H_{46}O_4$; MW = 470.33961

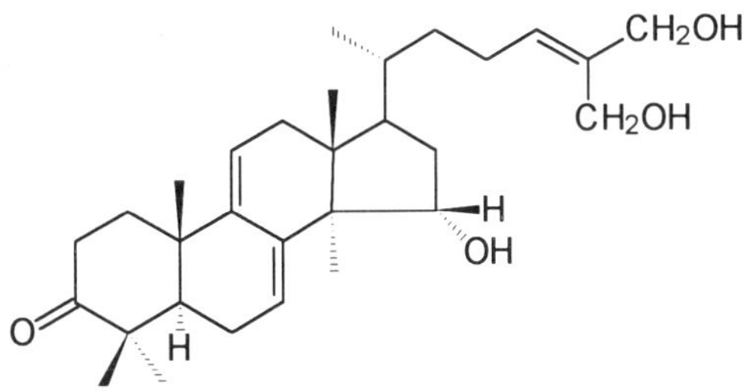

General Characteristics
Amorphous powder.

Fungal Source
Ganoderma lucidum.

Isolation/Purification
(See ganoderiol A).

Spectral Data

UV:
λ_{max}^{EtOH} 237(ϵ = 4740), 245(5400), and 253nm (2850).

^{1}H NMR:
($CDCl_3$) 5.89(1H, d, *J*=6.2Hz, 7-H); 5.54(1H, t, *J*= 7.3Hz, 24-H); 5.38(1H,d, *J*=6.6Hz, 11-H); 4.33(2H, s, 27-H); 4.29(1H, dd, *J*=5.5 and 9.9Hz, 15HB); 4.33(2H, s, 26-H); 2.77(1H, ddd, *J*=5.4, 14.4 and 14.4Hz, 2-H), 1.19(3H, s, 19-H), 1.13(3H, s, 28-H); 1.09(3H, s, 29-H); 0.93(3H, s, 30-H); 0.9(3H, d, *J*=6.6Hz, 21-H); and 0.64ppm (3H, s, 18-H).

Mass Spectrum:
HREIMS: 470.3394(M^+, $C_{30}H_{46}O_4$, 17%), 352(100), 55(59), and 43*m/e* (55).

Reference
H. Sato, T. Nishitoba, S. Shirasu, K. Oda, and S. Sakamura; Ganoderiol A and B, New Triterpenoids from the Fungus *Ganoderma lucidum* (Reishi); Agric. Biol. Chem., Vol. 50, pp. 2887-2890(1986).

Common/Systematic Name
Ganoderiol C

Molecular Formula/Molecular Weight
$C_{32}H_{54}O_5$; MW = 518.39713

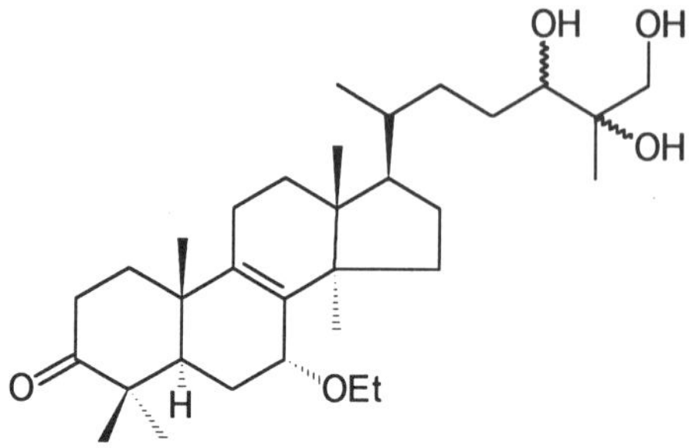

General Characteristics (Diacetyl derivative)
Colorless syrup; $[\alpha]_D^{23}$ + 54° (*c*=0.1, in EtOH).

Fungal Source
Fruiting bodies of *Ganoderma lucidum.*

Isolation/Purification
The neutral fraction obtained from the EtOH extract of *G. lucidum* (a mixture of two strains) was separated into three fractions (Fr. I, Fr. II, and Fr. III) using C_{18} Sep-Pak (MeOH; Waters Assoc.) and reversed-phase HPLC (Resolve-C_{18} column, Waters Assoc.), 2% AcOH-CH_3CN (2.7:1, v/v, 0.8ml/min., detector set at 254nm). Fraction I was further separated by silica gel and Lobar column chromatographies into several fractions (Frs. D, K, M, and P). Fractions D and K were acetylated (Ac_2O/pyridine) and subjected to silica gel column chromatography to give 24,26-di-*O*-acetylganoderiol C and 24,26-di-*O*-acetylganoderiol D, respectively. Ganoderiol F was crystallized ($CHCl_3$-MeOH) from Fr. K. Methyl ganolucidate D was obtained by preparative TLC (Kieselgel 60 F_{254} and RP_{18} F_{254} S) of Fr. P after methylation with CH_2N_2. Ganoderiols G, H, and I were obtained by repetitive column chromatography on silica gel followed by preparative TLC and recrystallizations.

Biological Activity
Fruiting bodies of *G. lucidum* produce some physiological effects when used as a home remedy; also bitter tasting.

Spectral Data (Diacetyl derivative)

IR:
(KBr, film) 3460, 2960, 2870, 1735, and 1705cm^{-1}.

^{1}H NMR:

(CDCl$_3$) 4.90(1H, dd, *J*=10.3, 2.6Hz, 24-H); 4.14(1H, d, *J*=11.4Hz, 26-H); 3.91(1H, d, *J*=11.4Hz, 26-H); 3.77(1H, br, s, 7β-H); 3.61(1H, dq, *J*=8.8, 7.0Hz, 7α-O$\underline{CH_2}$CH$_3$); 3.35(1H, dq, *J*=8.8, 7.0Hz, 7α-OCH$_2\underline{CH_3}$); 1.21(3H, dd, *J*=7.0, 7.0Hz, 7α-OCH$_2\underline{CH_3}$); 1.21(3H, s, 27-H); 1.11(3H, s, 28-H); 1.08(3H, s, 29-H); 1.06(3H, s, 19 or 30-H); 1.05(3H, s, 19 or 30-H); 0.90(3H, d, *J*=6.6Hz, 21-H); 0.62(3H, s, 18-H); 2.11(3H, s, OCOCH$_3$); and 2.07ppm (3H, s, OCOCH$_3$).

^{13}C NMR:

(CDCl$_3$) C-1, 35.2, t; C-2, 34.4, t; C-3, 217.6, s; C-4,46.8, s; C-5, 45.0, s; C-6, 24.2, t; C-7, 76.4, d; C-8, 135.5, s; C-9, 139.2, s; C-10, 37.8, s; C-11, 21.1, t; C-12, 30.3, t; C-13, 45.0, s; C-14, 50.0, s; C-15, 28.1, t; C-16, 31.1, t; C-17, 50.4, d; C-18, 16.2, q*; C-19, 17.4, q; C-20, 36.7, d; C-21, 18.7, q; C-22, 32.7, t; C-23, 26.0, t; C-24, 74.8, d; C-25, 73.3, s; C-26, 68.5, t; C-27, 20.2, q; C-28, 25.3, q; C-29, 21.4, q; C-30, 26.3, q; 7α-O$\underline{C}$H$_2$CH$_3$, 64.0, t; 7α-OCH$_2\underline{C}$H$_3$, 16.0, q*; O$\underline{C}$OCH$_3$, 171.1, s, 170.6, s; and OCO$\underline{C}$H$_3$, 21.0ppm, q, 20.8, q.

* Assignment may be reversed.

Mass Spectrum:

M$^+$, 602.4191(C$_{36}$H$_{58}$O$_7$, 2%), 587(5), 556(M$^+$ - C$_2$H$_5$OH, 8%), 496(8), 039(46), and 43*m/e* (100).

Reference

T. Nishitoba, K. Oda, H. Sato, and S. Sakamura; Novel Triterpenoids from the Fungus *Ganoderma lucidum;* Agric. Biol. Chem., Vol. 52, pp. 367-372(1988).

Common/Systematic Name
Ganoderiol D

Molecular Formula/Molecular Weight
$C_{30}H_{48}O_5$; MW = 488.35017

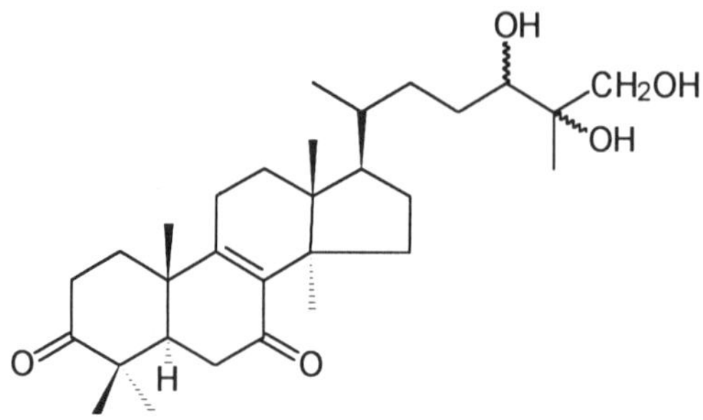

General Characteristics (Diacetyl derivative)
Colorless syrup; $[\alpha]_D^{21}$ + 8 ° (c=0.1, in MeOH).

Fungal Source
Fruiting bodies of *Ganoderma lucidum.*

Isolation/Purification
The neutral fraction obtained from the EtOH extract of *G. lucidum* (a mixture of two strains) was separated into three fractions (Fr. I, Fr. II and Fr. III) using C_{18} Sep-Pak (MeOH; Waters Assoc.) and reversed-phase HPLC (Resolve-C_{18} column, Waters Assoc.), 2% AcOH-CH_3CN (2.7:1, v/v, 0.8ml/min., detector set at 254nm). Fraction I was further separated by silica gel and Lobar column chromatographies into several fractions (Frs. D, K, M, and P). Fractions D and K were acetylated (Ac_2O/pyridine) and subjected to silica gel column chromatography to give 24,26-di-*O*-acetylganoderiol C and 24,26-di-*O*-acetylganoderiol D, respectively. Ganoderiol F was crystallized ($CHCl_3$-MeOH) from Fr. K. Methyl ganolucidate D was obtained by preparative TLC (Kieselgel 60 F_{254} and RP_{18} F_{254} S) of Fr. P after methylation with CH_2N_2. Ganoderiols G, H, and I were obtained by repetitive column chromatography on silica gel followed by preparative TLC and recrystallizations.

Biological Activity
Fruiting bodies of *G. lucidum* produce some physiological effects when used as a home remedy; they are also bitter tasting.

Spectral Data (Diacetyl derivative)

UV:
λ_{max}^{MeOH} 254nm (9730).

IR:

(KBr, film) 3440, 2950, 2860, 1730, 1705, 1655, and 1575cm^{-1}.

^{1}H NMR:

(diacetate) ($CDCl_3$) 4.90(1H, dd, *J*=9.9, 2.6Hz, 24-H); 4.13(1H, d, *J*=11.4Hz, 26-H); 3.90(1H, d, *J*=11.4Hz, 26-H); 2.70(1H, ddd, *J*=15.8, 12.1, 7.0Hz, 2β-H); 2.54(1H, dd, *J*=16.0, 14.5Hz, 6β-H); 2.46(1H, ddd, *J*=15.8, 5 9, 3.7Hz, 2α-H); 2.33(1H, dd, *J*=16.0, 3.3Hz, 6α-H); 2.14(1H, dd, *J*=14.5, 3.3Hz, 5α-H); 1.33(3H, s: 30-H); 1.20(3H, s, 27-H); 1.12(3H, s, 28-H); 1.10(3H, s, 29-H); 0.93(3H, s, 19-H); 0.92(3H, d, *J*=5.9Hz, 21-H); and 0.67ppm (3H, s, 18-H).

^{13}C NMR:

(free ganoderiol) ($CDCl_3$) C-1, 35.4, t; C-2, 34.4, t; C-3, 214.6, s; C-4, 47.3, s; C-5, 49.1, s; C-6, 37.2, t; C-7, 198.1, d; C-8, 139.6, s; C-9, 162.8, s; C-10, 39.5, s; C-11, 23.9, t; C-12, 30.2, t; C-13, 45.0, s; C-14, 47.8, s; C-15, 28.7, t; C-16, 31.9, t; C-17, 50.5, d; C-18, 16.0, q; C-19, 17.9, q; C-20, 36.6, d; C-21, 18.9, q; C-22, 33.6, t; C-23, 28.8, t; C-24, 79.1, d; C-25, 73.9, s; C-26, 67.7, t; C-27, 20.9, q; C-28, 25.4, q; C-29, 21.4, q; and C-30, 26.3ppm, q.

Mass Spectrum:

M^+, 572.3639($C_{34}H_{52}O_7$, 6%), 557(6), 339(15), 327(13), 69(65), 57(87), and 43*m/e* (100).

Reference

T. Nishitoba, K. Oda, H. Sato, and S. Sakamura; Novel Triterpenoids from the Fungus *Ganoderma lucidum;* Agric. Biol. Chem., Vol. 52, pp. 367-372(1988).

Common/Systematic Name

Ganoderiol E

Molecular Formula/Molecular Weight

$C_{30}H_{48}O_4$; MW = 472.35526

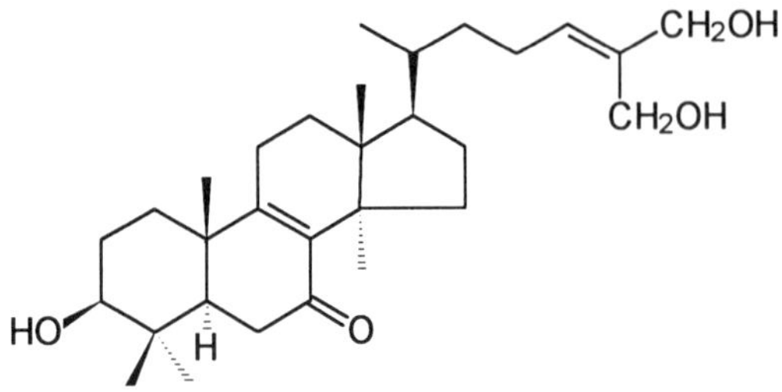

General Characteristics (Triacetyl derivative)

Isolated as 3,26,27-tri-*O*-acetylganoderiol E; colorless syrup; $[\alpha]_D^{21}$ + 18° (*c*=0.1, in MeOH).

Fungal Source

Fruiting bodies of *Ganoderma lucidum.*

Isolation/Purification

The neutral fraction obtained from the EtOH extract of *G. lucidum* (a mixture of two strains) was separated into three fractions (Fr. I, Fr. II and Fr. III) using C_{18} Sep-Pak (MeOH; Waters Assoc.) and reversed-phase HPLC (Resolve-C_{18} column, Waters Assoc.), 2% AcOH-CH_3CN (2.7:1, v/v, 0.8ml/min., detector set at 254nm). Fraction I was further separated by silica gel and Lobar column chromatographies into several fractions (Frs. D, K, M, and P). Fractions D and K were acetylated (Ac_2O/pyridine) and subjected to silica gel column chromatography to give 24,26-di-*O*-acetylganoderiol C, 24,26-di-*O*-acetylganoderiol D and 3,26,27-tri-acetyl ganoderiol E, respectively. Ganoderiol F was crystallized ($CHCl_3$-MeOH) from Fr. K. Methyl ganolucidate D was obtained by preparative TLC (Kieselgel 60 F_{254} and RP_{18} F_{254} S) of Fr. P after methylation with CH_2N_2. Ganoderiols G, H, and I were obtained by repetitive column chromatography on silica gel followed by preparative TLC and recrystallizations.

Biological Activity

Fruiting bodies of *G. lucidum* produce some physiological effects when used as a home remedy; also bitter tasting.

Spectral Data (Triacetyl derivative)

UV:

λ_{max}^{MeOH} 255nm (9160).

IR:
(KBr film) 3000, 2940, 2860, 1720, 1650, and 1570cm^{-1}.

^{1}H NMR:
($CDCl_3$) 5.77(1H, t, *J*=7.3Hz, 24-H); 4.66(2H, s, 27-H); 1.56(7H, s, 26-H); 1.51(1H, dd, *J*=11.4, 4.4Hz, 3-H); 7.46(1H, dd, *J*=16.0, 12.0Hz, 6β-H); 2.38(1H, dd, *J*=16.0, 5.0Hz, 6α-H); 1.19(3H, s, 30-H); 0.95(3H, s, 19-H); 0.93(3H, d, *J*=5.9Hz, 21-H); 0.91(3H, s, 28-H); 0.89(3H, s, 29-H); 0.65(3H, s, 18-H); and 2.06ppm (9H s, $O\underline{C}OCH_3$ x 3).

^{13}C NMR:
($CDCl_3$) C-1, 34.6, t; C-2, 23.7, t*; C-3, 79.6, s; C-4, 37.8, s; C-5, 49.0, s; C-6, 35.8, t; C-7, 198.6, s; C-8, 139.0, s; C-9, 164.6, s; C-10, 39.7, s; C-11, 23.9*, t; C-12, 30.2, t; C-13, 45.0, s; C-14, 47.8, s; C-15, 28.8, t; C-16, 32.0, t; C-17, 49.9, d; C-18, 15.8, q; C-19, 16.4, q; C-20, 36.2, d; C-21, 18.5, q*; C-22, 36.5, t; C-23, 24.7, t; C-24, 137.2, d; C-25, 128.8, s; C-26, 66.8, t; C-27, 59.9, q; C-28, 27.4, q; C-29, 21.2, q; C-30, 25.0, q; $O\underline{C}OCH_3$, 170.0, s, 170.8, s and 170.9 s; and $OCO\underline{C}H_3$, 18.7, q*, 21.0, q and 20.9ppm.

* Assignments may be reversed.

Mass Spectrum:
M^+, 598.3836($C_{36}H_{54}O_7$, 13%), 583(11), 538(15), 463(28), 369(23), 121(42), and 43*m/e* (100).

Reference

T. Nishitoba, K. Oda, H. Sato, and S. Sakamura; Novel Triterpenoids from the Fungus *Ganoderma lucidum;* Agric. Biol. Chem., Vol. 52, pp. 367-372(1988).

Common/Systematic Name

Ganoderiol F

Molecular Formula/Molecular Weight

$C_{30}H_{46}O_3$; MW = 454.34470

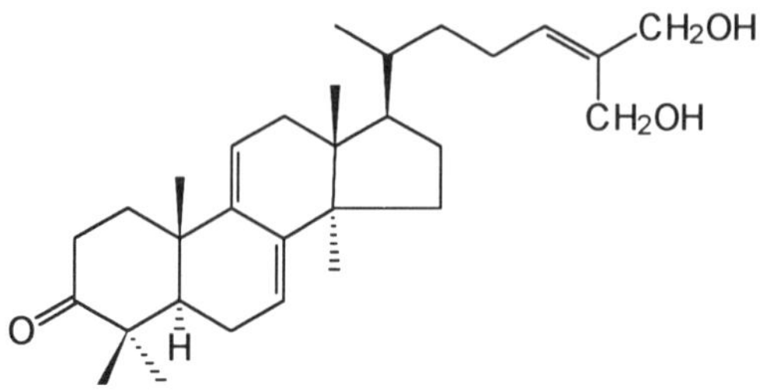

General Characteristics

Pale yellow needles from chloroform-methanol; mp., 116-120°C; $[\alpha]_D^{21}$ + 42° (*c*=0.1, in MeOH).

Fungal Source

Fruiting bodies of *Ganoderma lucidum.*

Isolation/Purification

The neutral fraction obtained from the EtOH extract of *G. lucidum* (a mixture of two strains) was separated into three fractions (Fr. I, Fr. II and Fr. III) using C_{18} Sep-Pak (MeOH; Waters Assoc.) and reversed-phase HPLC (Resolve-C_{18} column, Waters Assoc.), 2% AcOH-CH_3CN (2.7:1, v/v, 0.8ml/min., detector set at 254nm). Fraction I was further separated by silica gel and Lobar column chromatographies into several fractions (Frs. D, K, M, and P). Fractions D and K were acetylated (Ac_2O/pyridine) and subjected to silica gel column chromatography to give 24,26-di-*O*-acetylganoderiol C and 24,26-di-*O*-acetylganoderiol D, respectively. Ganoderiol F was crystallized ($CHCl_3$-MeOH) from Fr. K. Methyl ganolucidate D was obtained by preparative TLC (Kieselgel 60 F_{254} and RP_{18} F_{254} S) of Fr. P after methylation with CH_2N_2. Ganoderiols G, H, and I were obtained by repetitive column chromatography on silica gel followed by preparative TLC and recrystallizations.

Biological Activity

Fruiting bodies of *G. lucidum* produce some physiological effects when used as a home remedy; also bitter tasting.

Spectral Data

UV:

λ_{max}^{MeOH} 237(ϵ=16,600), 245(19,200), and 253nm (12,500).

IR:
(KBr film) 3350, 2900, 2850, and 1690cm^{-1}.

^{1}H NMR:
(CDCl$_3$) 5.56(1H, t, *J*=7.3Hz, 24-H); 5.51(1H, d, *J*=6.9Hz, 7-H); 5.39(1H, d, *J*=5.9Hz, 11-H); 4.33(2H, s, 27-H); 4.22(2H, s, 26-H); 2.78(1H, ddd, *J*=14.7, 14.7, 5.5Hz, 2β-H); 1.20(3H, s, 19-H); 1.13(3H, s, 28-H); 1.09(3H, s, 29-H), 0.92(3H, d, *J*=6.6Hz, 21-H); 0.88(3H, s, 30-H); and 0.59ppm (3H, s, 18-H).

^{13}C NMR:
(CDCl$_3$) C-1, 36.1, t; C-2, 34.9, t; C-3, 216.8, s; C-4, 47.5, s; C-5, 50.8, s; C-6, 23.7, t; C-7, 120.0, d; C-8, 142.9, s; C-9, 144.6, s; C-10, 37.3, s; C-11, 117.3, d; C-12, 37.9, t; C-13, 43.8, s; C-14, 50.4, s; C-15, 27.9, t; C-16, 31.5, t; C-17, 50.9, d; C-18, 15.8, q*; C-19, 22.5, q; C-20, 36.1, d; C-21, 18.4, q; C-22, 36.7, t; C-23, 24.4, t; C-24, 131.7, d; C-25, 136.8, s; C-26, 67.7, t; C-27, 60.2, t; C-28, 25.5*, q; C-29, 21.3, q; and C-30, 25.4*ppm, q.

* Assignments may be reversed.

Mass Spectrum:
M^{+}, 454.3430($C_{30}H_{46}O_3$, 4%), 436(6), 309(100), 69(24), 55(40), and 41*m/e* (31).

Reference

T. Nishitoba, K. Oda, H. Sato, and S. Sakamura; Novel Triterpenoids from the Fungus *Ganoderma lucidum*; Agric. Biol. Chem., Vol. 52, pp. 367-372(1988).

Common/Systematic Name

Ganoderiol G

Molecular Formula/Molecular Weight

$C_{31}H_{52}O_5$; MW = 504.38140

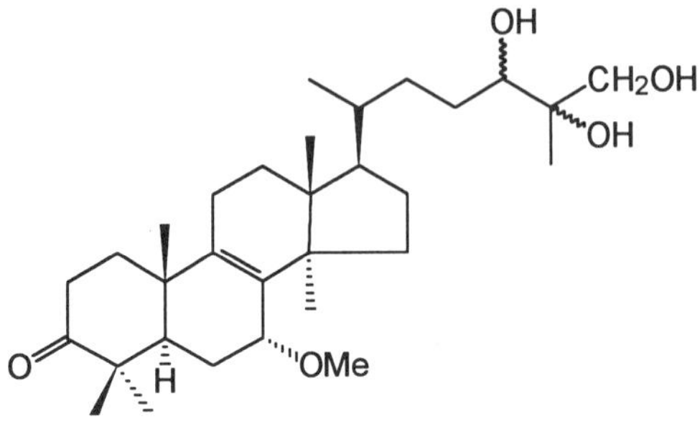

General Characteristics

A colorless syrup; $[\alpha]_D^{24}$ + 34° (*c*=0.1, in MeOH).

Fungal Source

Fruiting bodies of *Ganoderma lucidum.*

Isolation/Purification

The neutral fraction obtained from the EtOH extract of *G. lucidum* (a mixture of two strains) was separated into three fractions (Fr. I, Fr. II and Fr. III) using C_{18} Sep-Pak (MeOH; Waters Assoc.) and reversed-phase HPLC (Resolve-C_{18} column, Waters Assoc.), 2% AcOH-CH_3CN (2.7:1, v/v, 0.8ml/min., detector set at 254nm). Fraction I was further separated by silica gel and Lobar column chromatographies into several fractions (Frs. D, K, M, and P). Fractions D and K were acetylated (Ac_2O/pyridine) and subjected to silica gel column chromatography to give 24,26-di-*O*-acetylganoderiol C and 24,26-di-*O*-acetylganoderiol D, respectively. Ganoderiol F was crystallized ($CHCl_3$-MeOH) from Fr. K. Methyl ganolucidate D was obtained by preparative TLC (Kieselgel 60 F_{254} and RP_{18} F_{254} S) of Fr. P after methylation with CH_2N_2. Ganoderiols G, H and I were obtained by repetitive column chromatography on silica gel followed by preparative TLC and recrystallizations.

Biological Activity

Fruiting bodies of *G. lucidum* produce some physiological effects when used as a home remedy; they are also bitter tasting.

Spectral Data

IR:

(KBr film) 3380, 2920, 2870, and 1700cm^{-1}.

^{1}H NMR:

(CDCl$_3$) 3.84(1H, d, *J*=11.4Hz, 26-H); 3.67(1H, br, s, 7β-H); 3.19(1H, d, *J*=11.0Hz, 26-H); 3.47(1H, t, *J*=8.0Hz, 24-H); 3.32(3H, s, 7α-OCH$_3$); 1.26(3H, s, 27-H); 1.12(3H, s, 28-H); 1.09(3H, s, 29-H); 1.05(3H, s, 30-H); 1.05(3H, s, 19-H); 0.90(3H, d, *J*=8.8Hz, 21-H); and 0.64ppm (3H, s, 18-H).

Mass Spectrum:

M$^+$, 504.3860($C_{31}H_{52}O_5$, 22%), 489(43), 472(17), 457(19), 366(64), 95(55), 69(62), 55(78), and 43*m/e* (100).

<u>Reference</u>

T. Nishitoba, K. Oda, H. Sato, and S. Sakamura; Novel Triterpenoids from the Fungus *Ganoderma lucidum;* Agric. Biol. Chem., Vol. 52, pp. 367-372(1988).

Common/Systematic Name

Ganoderiol H

Molecular Formula/Molecular Weight

$C_{30}H_{50}O_5$; MW = 490.36583

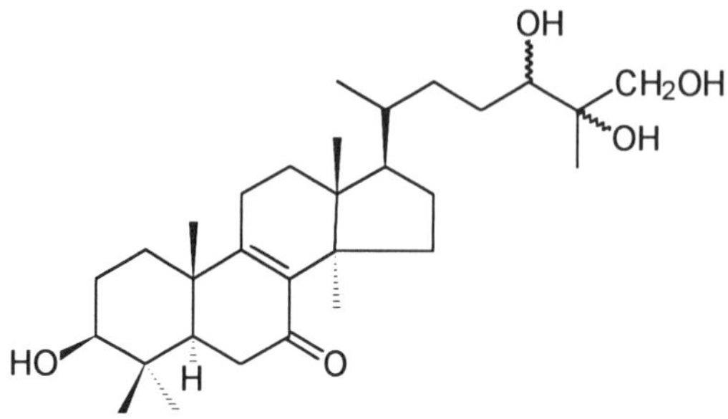

General Characteristics

Colorless needles; mp., 200-201.5°C; $[\alpha]_D^{24}$ + 22° (*c*=0.1, in MeOH).

Fungal Source

Fruiting bodies of *Ganoderma lucidum.*

Isolation/Purification

The neutral fraction obtained from the EtOH extract of *G. lucidum* (a mixture of two strains) was separated into three fractions (Fr. I, Fr. II and Fr. III) using C_{18} Sep-Pak (MeOH; Waters Assoc.) and reversed-phase HPLC (Resolve-C_{18} column, Waters Assoc.), 2% AcOH-CH_3CN (2.7:1, v/v, 0.8ml/min., detector set at 254nm). Fraction I was further separated by silica gel and Lobar column chromatographies into several fractions (Frs. D, K, M, and P). Fractions D and K were acetylated (Ac_2O/pyridine) and subjected to silica gel column chromatography to give 24,26-di-*O*-acetylganoderiol C and 24,26-di-*O*-acetylganoderiol D, respectively. Ganoderiol F was crystallized ($CHCl_3$-MeOH) from Fr. K. Methyl ganolucidate D was obtained by preparative TLC (Kieselgel 60 F_{254} and RP_{18} F_{254} S) of Fr. P after methylation with CH_2N_2. Ganoderiols G, H, and I were obtained by repetitive column chromatography on silica gel followed by preparative TLC and recrystallizations.

Biological Activity

Fruiting bodies of *G. lucidum* produce some physiological effects when used as a home remedy; they are also bitter tasting.

Spectral Data

UV:

λ_{max}^{MeOH} 255nm (9170).

IR:
(KBr film) 3380, 2950, 2860, and 1640cm^{-1}.

^{1}H NMR:
($CDCl_3$) 3.82(1H, d, *J*=11.4Hz, 26-H); 3.46(1H, d, *J*=11.4Hz, 26-H); 3.43(1H, dd, *J*=10.0, 1.5Hz, 24-H); 3.27(1H, dd, *J*=11.2, 4.9Hz, 3α-H); 2.46(1H, dd, *J*=16.5, 11.7Hz, 6α-H); 2.38(1H, dd, *J*=16.5, 6.2Hz, 6α-H); 1.17(3H, s, 30-H); 1.11(3H, s, 27-H); 1.00(3H, s, 28-H); 0.93(3H, d, *J*=7.7Hz, 21-H); 0.92(3H, s, 19-H); 0.88(3H, s, 29-H); and 0.66ppm (3H, s, 18-H).

^{13}C NMR:
($CDCl_3$) C-1, 35.2, t; C-2, 28.9, t*; C-3, 77.1, s; C-4, 40.1, s; C-5, 49.6, s; C-6, 37.2, t; C-7, 198.7, s; C-8, 138.9, s; C-9, 164.9, s; C-10, 39.5, s; C-11, 23.8, t; C-12, 30.5, t; C-13, 45.2, s; C-14, 48.2, s; C-15, 28.3, t; C-16, 32.7, t; C-17, 50.6, d; C-18, 16.1, q; C-19, 18.3, q; C-20, 37.3, d; C-21, 19.3, q; C-22, 34.5, t; C-23, 29.1, t*; C-24, 77.1, d; C-25, 74.8 s; C-26, 69.3, t; C-27, 20.1, q; C-28, 27.9, q; C-29, 16.0, q; C-30, 25.2ppm, q.

* Assignments may be reversed.

Mass Spectrum:
M^+, 490.3625($C_{30}H_{50}O_5$, 30%), 475(47), 415(46), 381(35), 329(35), 121(74), 69(59), 55(68) and 43*m/e* (100).

Reference

T. Nishitoba, K. Oda, H. Sato, and S. Sakamura; Novel Triterpenoids from the Fungus *Ganoderma lucidum;* Agric. Biol. Chem., Vol. 52, pp. 367-372(1988).

Common/Systematic Name

Ganoderiol I

Molecular Formula/Molecular Weight

$C_{31}H_{50}O_5$; MW = 502.36583

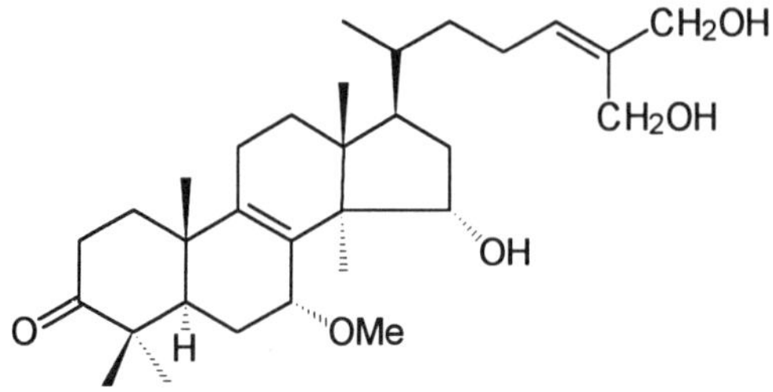

General Characteristics

Colorless syrup; $[\alpha]_D^{22}$ + 53° (*c*=0.1, in MeOH).

Fungal Source

Fruiting bodies of *Ganoderma lucidum.*

Isolation/Purification

The neutral fraction obtained from the EtOH extract of *G. lucidum* (a mixture of two strains) was separated into three fractions (Fr. I, Fr. II and Fr. III) using C_{18} Sep-Pak (MeOH; Waters Assoc.) and reversed-phase HPLC (Resolve-C_{18} column, Waters Assoc.), 2% AcOH-CH_3CN (2.7:1, 0.8ml/min., detector set at 254nm). Fraction I was further separated by silica gel and Lobar column chromatographies into several fractions (Frs. D, K, M, and P). Fractions D and K were acetylated (Ac_2O/pyridine) and subjected to silica gel column chromatography to give 24,26-di-*O*-acetylganoderiol C and 24,26-di-*O*-acetylganoderiol D, respectively. Ganoderiol F was crystallized ($CHCl_3$-MeOH) from Fr. K. Methyl ganolucidate D was obtained by preparative TLC (Kieselgel 60 F_{254} and RP_{18} F_{254} S) of Fr. P after methylation with CH_2N_2. Ganoderiols G, H, and I were obtained by repetitive column chromatography on silica gel followed by preparative TLC and recrystallizations.

Biological Activity

Fruiting bodies of *G. lucidum* produce some physiological effects when used as a home remedy; also bitter tasting.

Spectral Data

IR:

(KBr film) 3380, 2920, 2870, and 1700cm^{-1}.

^{1}H NMR:

($CDCl_3$) 5.55(1H, t, J=7.3Hz, 24-H); 4.35(1H, t, J=8.8Hz, 15β-H); 4.33(2H, s, 27-H); 4.22(2H, s, 26-H); 1.06(1H, br s, 7β-H); 3.33(3H, s, 7α-OCH_3); 2.58(1H, ddd, J=15.9, 10.6, 7.3Hz, 2β-H); 2.46(1H, ddd, J=15.9, 7.3, 3.8Hz, 2α-H); 1.13(3H, s, 28-H); 1.10(3H, s, 29- or 30-H); 1.08(3H, s, 29- or 30-H); 1.02(3H, s, 19-H); 0.89(3H, d, J=6.6Hz, 21-H); and 0.66ppm (3H, s, 18-H).

Mass Spectrum:
M^+, 502.3675($C_{31}H_{50}O_5$, 3%), 470(7), 452(2), 349(59), 325(45), 175(66), 69(60), 55(99), and 43*m/e* (100).

Reference

T. Nishitoba, K. Oda, H. Sato, and S. Sakamura; Novel Triterpenoids from the Fungus *Ganoderma lucidum*; Agric. Biol. Chem., Vol. 52, pp. 367-372(1988).

Common/Svstematic Name

Ganodermanondiol

24(*S*),25-Dihydroxy-5α-lanosta-7,9(11)-dien-3-one

Molecular Formula/Molecular Weight

$C_{30}H_{48}O_3$; MW = 456.36035

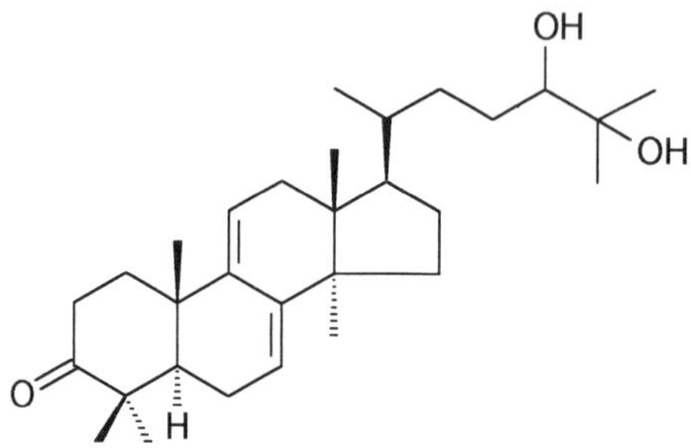

General Characteristics

Colorless needles from MeOH; mp., 182-183°C; $[\alpha]_D^{23}$ + 45.8° (*c*=0.5, in $CHCl_3$). Showed a positive Liebermann-Burchard reaction.

Fungal Source

Ganoderma lucidum.

Isolation/Purification

The fresh fruiting bodies were extracted with MeOH and the extract was partitioned between petroleum ether and MeOH-H_2O (90:10, v/v). The 90% MeOH extract was chromatographed on silica gel (0-100% gradient MeOH in $CHCl_3$). The $CHCl_3$ eluate was rechromatographed on silica gel using a gradient of EtOAc in hexane; the substance was eluted at 20% EtOAc).

Spectral Data

UV:

λ_{max}^{MeOH} 235(log ε = 4.00), 243(4.06), and 251nm (3.89).

IR:

(KBr) 3390, 2960, 2930, 1710, 1460, 1370, 1165, 1110, 1080, and 810cm^{-1}.

^{1}H NMR:

($CDCl_3$) 2.77(1H, m, H-2); 5.40(1H, dd, *J*=2.0, 6.0Hz, H-11); 5.52(1H, dd, *J*=2.0, 6.0Hz, H-7); 0.58(3H, s, H-18); 1.07(3H, s, H-29); 0.91(3H, d, *J*=6.4Hz, H-21); 3.30(1H, dd, *J*=3.0, 8.0Hz, H-24); 1.15(3H, s, H-26); 1.18(3H, s, H-19); 0.86(3H, s, H-28); 1.11(3H, s, H-30); and 1.21ppm (3H, s, H-27).

^{13}C NMR:
(CDCl$_3$) 36.63(t, C-1); 34.87(t, C-2); 216.92(s, C-3); 47.50(s, C-4); 50.70(d, C-5); 23.67(t, C-6); 119.91(d, C-7); 142.81(s, C-8); 144.46(s, C-9); 37.17(s, C-10); 117.23(d, C-11); 37.79(t, C-12); 43.75(s, C-13); 50.31(s, C-14); 31.45(t, C-15); 27.85(t, C-16); 50.94(d, C-17); 15.71(q, C-18); 22.06(q, C-19); 36.53(d, C-20); 18.62(q, C-21); 33.47(t, C-22); 28.71(t, C-23); 79.58(d, C-24); 73.26(s, C-25); 23.18(q, C-26); 26.57(q, C-27); 25.45(q, C-28); 25.32(q, C-29); and 22.46ppm (q, C-30).

Mass Spectrum:
HREIMS: 456.3610*m/e* (M^+); calcd for $C_{30}H_{48}O_{33}$, 456.3603. LREIMS: 456(M^+, 7%), 438(M^+-H_2O, 6), 424(M^+-OH-Me, 3), 309(7), 269(9), 69(11), 55(14), and 32*m/e* (100).

References

M. Arisawa, A. Fujita, T. Hayashi, M. Shimizu, and N. Morita; Revision of ^{1}H- and ^{13}C-NMR Assignments of Lanostanoids from *Ganoderma lucidum* by 2D-NMR Studies; J. Nat. Prod., Vol. 51, pp. 54-59(1988).

A. Fujita, M. Arisawa, M. Saga, T. Hayashi, and N. Morita; Two New Lanostanoids from *Ganoderma lucidum*; J. Nat. Prod., Vol. 49, pp. 1122-1125(1986).

Common/Systemic Name

Ganodermanontriol

24(*S*),25,26-Trihydroxy-5α-lanosta-7,9(11)-dien-3-one

Molecular Formula/Molecular Weight

$C_{30}H_{48}O_4$; MW = 472.35526

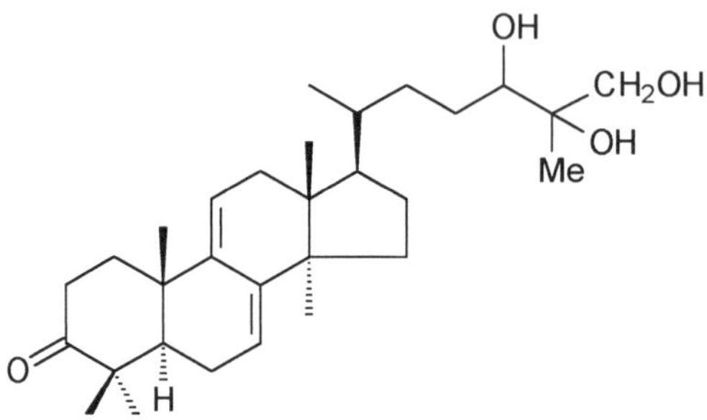

General Characteristics

Colorless needles from MeOH; mp., 161-162°C; $[\alpha]_D^{23}$ + 35.70° (*c*=1.0, in $CHCl_3$). Showed a positive Liebermann-Burchard reaction.

Fungal Source

Ganoderma lucidum.

Isolation/Purification

The fresh fruiting bodies were extracted with MeOH and the extract was partitioned between petroleum ether and MeOH-H_2O (90:10). The 90% MeOH extract was chromatographed on silica gel (0-100% gradient MeOH in $CHCl_3$). The 1% MeOH-$CHCl_3$ eluate was repeatedly chromatographed on silica gel using $CHCl_3$-EtOAc-Me_2CO (14:1:1, v/v/v) and hexane-EtOAc-Me_2CO (6:1:1, v/v/v) to give ganodermanontriol.

Spectral Data

UV:

λ_{max}^{MeOH} 236 (log ε = 4.06), 243(4.11), and 251nm (3.95).

IR:

(KBr) 3400, 2960, 2925, 2875, 1700, 1460, 1445, 1370, 1110, 1040, 1000, and 810cm^{-1}.

^{1}H NMR:

($CDCl_3$) 2.80(11, m, H-2); 5.39(11, dd, *J*=1.5, 5.0Hz, 1H-11); 5,51(11, dd, *J*=2.0, 6.1Hz, H-7); 0.58(31, s, H-18); 1.09(31, s, H-29); 0.92(31, d, *J*=6.1Hz, H-21); 3.48(11, m, H-24); 3.48, 3.84(each, 1H, d, *J*=11.3Hz, 2H-26); 1.21(31, s, H-27);

^{13}C NMR:
(CDC1$_3$) 36.63(t, C-1); 34.87(t, C-2); 216.92(s, C-3); 47.50(s, C-4); 50.70(d, C-5); 23.67(t, C-6); 119.91 (d, C-7); 142.81 (s, C-8); 144.46(s, C-9); 37.17(s, C-10); 117.23(d, C-11); 37.79(t, C-12); 43.75(s, C-13); 50.31(s, C-14); 31.45(t, C-15); 27.85(t, C-16); 50.94(d, C-17); 15.71(q, C-18); 22.06(q, C-19); 36.53(d, C-20); 18.62(q, C-21); 33.47(t, C-22); 28.71(t, C-23); 79.58(d, C-24); 73.26(s, C-25); 23.18(q, C-26); 26.57(q, C-27); 25.45(q, C-28); 25.32(q, C-29); and 22.46ppm (q, C-30).

Mass Spectrum:
HREIMS: 456.3610*m/e* (M^+); calcd for $C_{30}H_{48}O_{33}$, 456.3603. LREIMS: 456(M^+, 7%), 438(M^+-H_2O, 6), 424(M^+-OH-Me, 3), 309(7), 269(9), 69(11), 55(14), and 32*m/e* (100).

References
M. Arisawa, A. Fujita, T. Hayashi, M. Shimizu, and N. Morita; Revision of ^{1}H- and ^{13}C-NMR Assignments of Lanostanoids from *Ganoderma lucidum* by 2D-NMR Studies; J. Nat. Prod., Vol. 51, pp. 54-59(1988).

A. Fujita, M. Arisawa, M. Saga, T. Hayashi, and N. Morita; Two New Lanostanoids from *Ganoderma lucidum*; J. Nat. Prod., Vol. 49, pp. 1122-1125(1986).

Common/Systematic Name

Ganodermenonol

26-Hydroxy-5α-lanosta-7,9(11),24-trien-3-one

Molecular Formula/Molecular Weight

$C_{30}H_{46}O_2$; MW = 438.34978

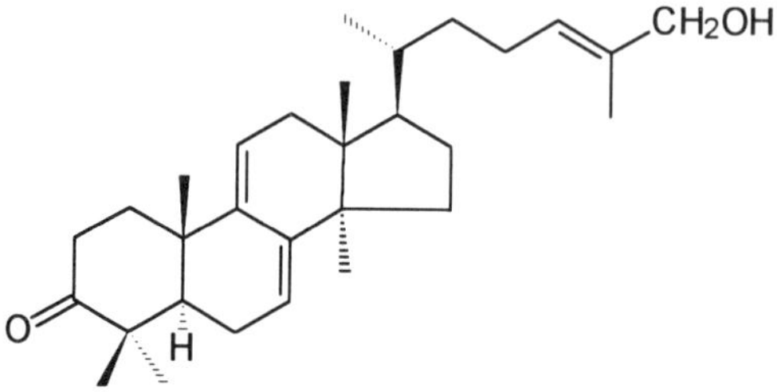

General Characteristics

Colorless needles from methanol; mp., 109-111°C; positive Liebermann-Burchard reaction; $[\alpha]_D^{23}$ +38.96° (c=1.0, in $CHCl_3$).

Fungal Source

The dried fruiting bodies of *Ganoderma lucidum*.

Isolation/Purification

The fruiting bodies of *G. lucidum* were cut into small pieces and extracted three times with MeOH at room temperature for 3 days. The MeOH extract was partitioned between H_2O and $CHCl_3$ and the $CHCl_3$ fraction was partitioned between petroleum ether and MeOH-H_2O (90:10). The 90% MeOH extract was chromatographed on a silica gel column by stepwise elution with $CHCl_3$, 1% MeOH/$CHCl_3$, 5% MeOH/$CHCl_3$, 10% MeOH/$CHCl_3$ and MeOH. The $CHCl_3$ elution was rechromatographed on a silica gel column by stepwise elution with an EtOAc/hexane solvent system to give ganodermenonol from 5% EtOAc/hexane and ganodermadiol from 10% EtOAc/hexane. The 1% eluate was repeatedly separated by silica gel column chromatography ($CHCl_3$-EtOAc-Me_2CO, 14:1:1, v/v/v, and hexane-EtOAc-Me_2CO, 6:1:1, v/v/v) to afford a mixture of colorless crystals. The mixture was treated overnight with Ac_2O and pyridine at room temperature and was separated by TLC to give ganodermatriol acetate.

Biological Activity

The dried fruiting bodies of G. *lucidum* has been prescribed in Chinese medicine as a tonic and sedative drug and have been used in the management of hepatopathy, hypertension, arthritis, neuasthenia, bronchitis, etc.

Spectral Data

UV:

λ_{max}^{MeOH} 236(log ε = 4.17), 243(4.23), and 251nm(4.06).

IR:
(KBr) 3425, 2930, 2880, 1700, 1450, 1375, 1110, 1000, and 815cm^{-1}.

^{1}H NMR:
($CDCl_3$) 0.59(s, 18-CH_3); 1.20(s, 19-CH_3); 0.92(d, *J*=6.2Hz, 21-CH_3); 0.88(s, 28-CH_3); 1.09(s, 29-CH_3); 1.13(s, 30-CH_3); 4.00(s, 26-2H); 1.67(s, 27-CH_3); 2.78, 2.35(2-CH_2); 5.51(dd, *J*=6.3Hz, 7-H); 5.41(m, 11-H); and 5.41ppm (m, 24-H).

^{13}C NMR:
($CDCl_3$) C-1, 36.59 (t); C-2, 34.82 (t); C-3, 16.95 (s); C-4, 47.43 (s); C-5, 50.66 (d); C-6, 23.62 (t); C-7, 119.83 (d); C-8, 142.81 (s); C-9, 144.47 (s); C-10, 37.17 (s); C-11, 117.24 (d); C-12, 37.76 (t); C-13, 43.68 (s); C-14, 50.27 (s); C-15, 31.44 (t); C-16, 27.86(t); C-17, 50.86 (d); C-18, 15.65 (4); C-19, 22.40 (q); C-20, 36.01 (d); C-21, 18.36 (q); C-22, 35.87 (t); C-23, 24.50 (t); C-24, 126.86 (d); C-25, 134.34 (s); C-26, 69.01 (t); C-27, 13.62 (q); C-28, 25.38 (q); C-29, 25.28 (q); and C-30, 22.43ppm (q).

Mass Spectrum:
EIMS: 438(M^+, 80%), 423(M^+ - Me, 14), 420(M^+ - H_2O, 10), 405(M^+ - Me - H_2O, 12), 309(M^+-side chain-2H, 100), 269(M^+ - side chain - 42, 50), 244(16), 199(16), 185(20), 171(30), 157(40), 145(34), 133(365), 119(36), 109(38), 95(44), 81(54), 69(64), and 55*m/e* (100); anal calcd for $C_{30}H_{46}O_2$ 438.3498; found 438.3507.

References

M. Arisawa, A. Fujita, T. Hayashi, M. Shimizu, N. Morita, T. Kikuchi, S. Kakota, and Y. Tezuka; Revision of ^{1}H and ^{13}C-NMR Assignments of Lanostanoids from *Ganoderma lucidum* by 2D-NMR Studies; J. Natural Products, Vol. 51, pp. 54-59(1988).

M. Arisawa, A. Fujita, M.Saga, H. Fukumura, T. Hayashi, M. Shimizu, and N. Morita; Three New Lanostanoids from *Ganoderma lucidum*; J. Natural Products, Vol. 49, pp. 621-625(1986).

Common/Systematic Name
Ganodermadiol
5α-Lanosta-7,9(11),24-triene-3β,26-diol

Molecular Formula/Molecular Weight
$C_{30}H_{48}O_2$; MW = 440.36543

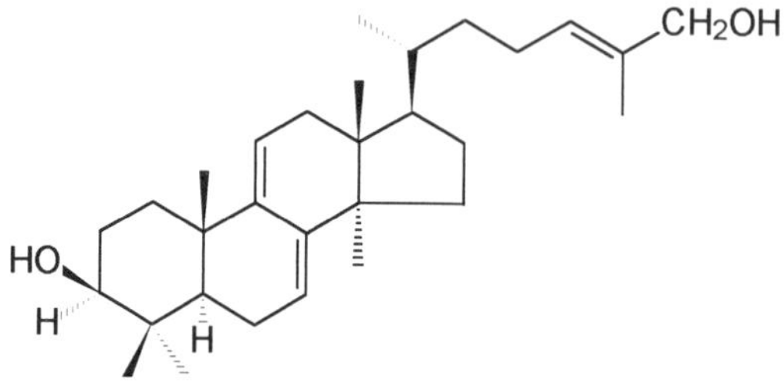

General Characteristics
Colorless needles from methanol; mp., 168-170°C; positive Liebermann-Burchard reaction; $[\alpha]_D^{23}$ + 53.0° (*c*=1.0, in $CHCl_3$).

Fungal Source
The dried fruiting bodies of *Ganoderma lucidum*.

Isolation/Purification
The fruiting bodies of *G. lucidum* were cut into small pieces and extracted three times with MeOH at room temperature for 3 days. The MeOH extract was partitioned between H_2O and $CHCl_3$ and the $CHCl_3$ fraction was partitioned between petroleum ether and MeOH-H_2O (90:10). The 90% MeOH extract was chromatographed on a silica gel column by stepwise elution with $CHCl_3$, 1% MeOH-$CHCl_3$, 5% MeOH-$CHCl_3$, 10% MeOH-$CHCl_3$, and MeOH. The $CHCl_3$ elution was rechromatographed on a silica gel column by stepwise elution with an EtOAc-hexane solvent system to give ganodermenonol from 5% EtOAc-hexane and ganodermadiol from 10% EtOAc-hexane. The 1% eluate was repeatedly separated by silica gel column chromatography ($CHCl_3$-EtOAc-Me_2CO, 14:1:1, v/v/v, and hexane-EtOAc-Me_2CO, 6:1:1, v/v/v) to afford a mixture of colorless crystals. The mixture was treated overnight with Ac_2O and pyridine at room temperature and was separated by TLC to give ganodermatriol acetate.

Biological Activity
The dried fruiting bodies of *G. lucidum* have been prescribed in Chinese medicine as a tonic and sedative drug and have been used in the management of hepatopathy, hypertension, arthritis, neuasthenia, bronchitis, etc.

Spectral Data

UV:
λ_{max}^{MeOH} 236(log ε = 4.08), 243(4.12), and 251nm (3.95).

IR:
(KBr) 3340, 2930, 1440, 1430, 1370, 1070, 1040, and 1010cm^{-1}.

^{1}H NMR:
($CDCl_3$) 0.57(s, 18-CH_3); 0.98(s, 19-CH_3); 0.92(d, *J*=6.3Hz, 21-CH_3); 0.88(s, 28-CH_3); 1.00(s, 29-CH_3); 0.88(s, 30-CH_3); 4.00(s, 26-2H); 1.67(s, 27-CH_3); 3.25(d, *J*=5.4, 10.1Hz); 5.47(m, 7-H); 5.34(m, 11-H); and 5.42ppm (m, 24-H).

^{13}C NMR:
($CDCl_3$) C-1, 35.76 (t); C-2, 27.82 (t); C-3, 78.98 (d); C-4, 38.73 (s); C-5, 49.15 (d); C-6, 23.13 (t); C-7, 120.25 (d); C-8, 142.69 (s); C-9, 145.93 (s); C-10, 37.36 (s); C-11, 116.26 (d); C-12, 37.85 (t); C-13, 43.84 (s); C-14, 50.35 (s); C-15, 31.53 (t); C-16, 27.95(t); C-17, 50.94 (d); C-18, 15.71 (4); C-19, 22.79 (q); C-20, 36.12 (d); C-21, 18.44 (q); C-22, 35.98 (t); C-23, 24.97 (t); C-24, 126.99 (d); C-25, 134.33 (s); C-26, 69.11 (t); C-27, 13.65 (q); C-28, 26.60 (q); C-29, 28.17 (q); and C-30, 15.80ppm (q).

Mass Spectrum:
EIMS: 440(M^+, 100%), 425(M^+-Me, 12), 422(M^+-H_2O, 8), 407(M^+-Me-H_2O, 12), 311(M^+-side chain, 50), 271(M^+-side chain-40, 44), 253(30), 171(28), 157(32), 145(34), 119(38), 107(40), 95(48), 81 (54), 69(70), 55(98), 109(38), 95(44), 81(54), 69(64), and 55*m/e* (100); anal calcd for $C_{30}H_{48}O_2$ 440.3654; found 440.3612.

References

M. Arisawa, A. Fujita, T. Hayashi, M. Shimizu, N. Morita, T. Kikuchi, S. Kakota, and Y. Tezuka; Revision of ^{1}H and ^{13}C-NMR Assignments of Lanostanoids from *Ganoderma lucidum* by 2D-NMR Studies; J. Natural Products, Vol. 51, pp. 54-59(1988).

M. Arisawa, A. Fujita, M.Saga, H. Fukumura, T. Hayashi, M. Shimizu, and N. Morita; Three New Lanostanoids from *Ganoderma lucidum*; Journal of Natural Products, Vol. 49, pp. 621-625(1986).

Common/Systematic Name
Ganodermatriol

Molecular Formula/Molecular Weight
$C_{30}H_{48}O_3$; MW = 456.36035

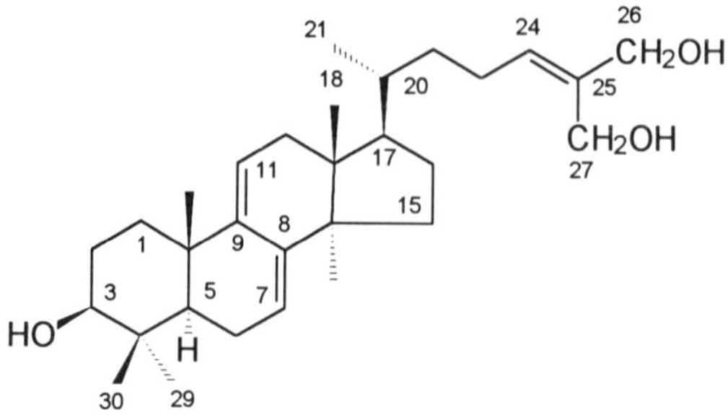

Fungal Source
Fresh fruiting body of the fungus *Ganoderma lucidum*.

Spectral Data (Acetate derivative)

^{1}H NMR:
($CDCl_3$) 18-H_3, 0.56(s); 19-H_2, 1.01(s), 0.92(d, *J*=7.4Hz); 28-H_3, 0.87(s); 29-H_3, 0.89(s); 30-H_3, 0.96(s); 26-H_2, 4.57(s, 2H); 27-H_2, 4.66(s, 2H); 3α-H, 4.50(s); 7-H, 5.46(m); 11-H, 5.33(m); 24-H, 5.78(dd, *J*=7.3, 7.4Hz); and 3 X OAc, 2.06(3H), 2.07ppm (6H).

^{13}C NMR:
($CDCl_3$) C-1, 35.40 (t); C-2, 24.33 (t); C-3, 80.82 (d); C-4, 37.22 (s); C-5, 49.27 (d); C-6, 22.90 (t); C-7, 120.01 (d); C-8, 142.63 (s); C-9, 145.64 (s); C-10, 37.82 (s); C-11, 116.50 (d); C-12, 37.95 (t); C-13, 43.81 (s); C-14, 50.32 (s); C-15, 31.50 (t); C-16, 27.91 (t); C-17, 50.83 (d); C-18, 15.69 (q); C-19, 22.87 (q); C-20, 36.07 (d); C-21, 18.41 (q); C-22, 35.78 (t); C-23, 24.81 (t); C-24, 137.83 (d); C-25, 128.64 (s); C-26, 66.83 (t); C-27, 59.84 (t); C-28, 25.56 (q); C-29, 28.10 (q); and C-30, 16.97ppm (q).

Reference
M. Arisawa, A. Fujita, T. Hayashi, M. Shimizu, N. Morita, T. Kikuchi, S. Kadota, and Y. Tezuka; Revision of ^{1}H- and ^{13}C-NMR Assignments of Lanostanoids from *Ganoderma lucidum* by 2D-NMR Studies; J. Nat. Products, Vol. 51, pp. 54-59(1988).

11

Ganolucidic Acids

- Ganolucidic acid A
- Ganolucidic acid B
- Ganolucidic acid C
- Ganolucidic acid D
- Ganolucidic acid E

Common/Systematic Name
Ganolucidic acid A

Molecular Formula/Molecular Weight
$C_{30}H_{44}O_6$; MW = 500.31379

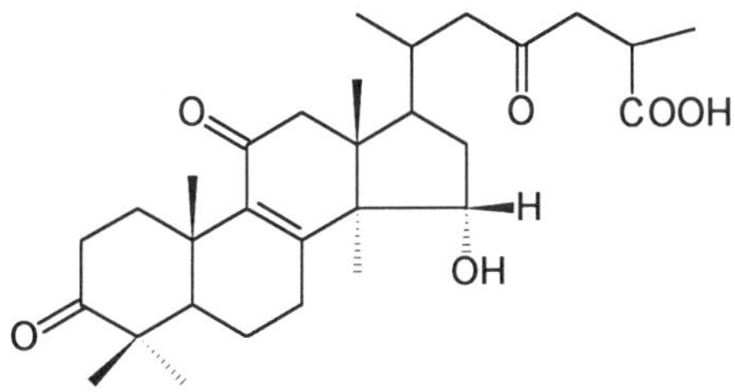

General Characteristics
Isolated and identified as methyl ester. Colorless needles; mp., 192-194°C; $[\alpha]_D$ + 188° (in $CHCl_3$).

Fungal Source
Ganoderma lucidum.

Spectral Data (Methyl ester)

UV:
λ_{max} 256.5nm (log ϵ = 4.00).

IR:
(KBr) 3450, 1730, 1710, and 1645cm^{-1}.

^{1}H NMR:
($CDCl_3$) 0.91(3H, s, 18-CH_3); 1.12(3H, s, 19-CH_3); 1.18(3H, d, *J*=7.0Hz, 27-CH_3); 1.08(3H, s, 30-CH_3); 1.12(3H, s, 31-CH_3); 1.18(3H, s, 32-CH_3); 3.70(3H, s, $COOCH_3$); and 4.41ppm (1H, dd, *J*=9.0, 5.5Hz, 15-H).

^{13}C NMR:
($CDCl_3$) C-1, 35.1, t; C-2, 34.2, t; C-3, 217.7, s; C-4, 47.0*, s; C-5, 51.7, d; C-6, 18.7, t; C-7, 29.6, t; C-8, 163.2, s; C-9, 138.6, s; C-10, 37.1, s; C-11, 198.1, s; C-12, 51.7, t; C-13, 46.8*, s; C-14, 53.6, s; C-15, 72.9, d; C-16, 38.6, t; C-17, 48.7, d; C-18, 17.2, q; C-19, 18.8**, q; C-20, 32.6, d; C-21, 19.4, q; C-22, 49.6, t; C-23, 208.3, s; C-24, 46.8, t; C-25, 34.7, d; C-26, 176.2, s; C-27, 17.1, q; C-30, 27.8, q; C-31, 20.6, q; C-32, 19.0**, q; and OCH_3, 51.9ppm.

*, ** Assignments may be reversed.

Mass Spectrum:
LREIMS: 514(M^+), 417, 371, 287, 171, 139, and 129*m/e*.

Reference

T. Kikuchi, S. Matsuda, Y. Murai, and Z. Ogita; Ganoderic Acid G and I and Ganolucidic Acid A and B, New Triterpenoids from *Ganoderma lucidum*; Chem. Pharm. Bull., Vol. 33, pp. 2628-2631(1985).

Common/Systematic Name
Ganolucidic acid B

Molecular Formula/Molecular Weight
$C_{30}H_{46}O_6$; MW = 502.32944

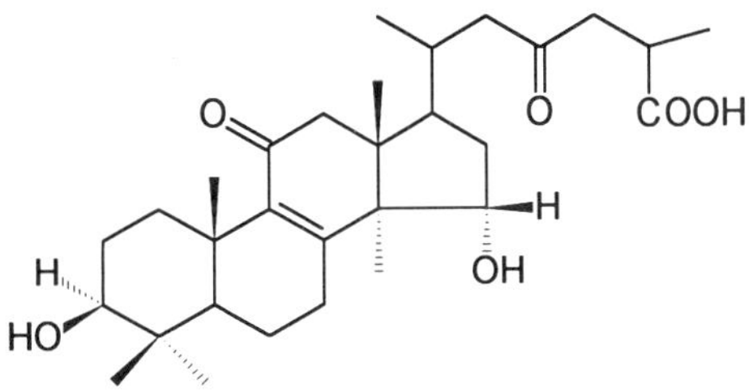

General Characteristics
Isolated and identified as methyl ester. Colorless needles; mp., 167-169°C; $[\alpha]_D$ + 114° (in $CHCl_3$).

Fungal Source
Ganoderma lucidum.

Spectral Data (Methyl ester)

UV:
λ_{max} 256.5nm (log ϵ = 4.00).

IR:
(KBr) 3450, 1730, 1710, and 1645cm^{-1}.

^{1}H NMR:
($CDCl_3$) 0.89(3H, s, 18-CH_3); 1.11(3H, s, 19-CH_3); 1.18(3H, d, *J*=7.0Hz, 27-CH_3); 1.01(3H, s, 30-CH_3); 0.82(3H, s, 31-CH_3); 1.16(3H, s, 32-CH_3); 3.68(3H, s, $OOCH_3$); 3.25(1H, dd, *J*=10.0, 6.0Hz, 3-H); and 4.39ppm (1H, dd, *J*=9.0, 5.5Hz, 15-H).

^{13}C NMR:
($CDCl_3$) C-1, 34.4, t; C-2, 28.0, t; C-3, 78.7, d; C-4, 39.0, s; C-5, 51.8, d; C-6, 17.4, t; C-7, 30.4, t; C-8, 162.9, s; C-9, 140.0, s; C-10, 37.8, s; C-11, 198.3, s; C-12, 52.1, t; C-13, 47.2, s; C-14, 53.5, s; C-15, 73.0, d; C-16, 38.7, t; C-17, 48.7, d; C-18, 17.1, q; C-19, 18.8*, q; C-20, 32.5, d; C-21, 19.4, q; C-22, 49.7, t; C-23, 208.3, s; C-24, 46.8, t; C-25, 34.6, d; C-26, 176.1, s; C-27, 17.1, q; C-30, 28.3, q; C-31, 15.7, q; C-32, 19.0*, q; and OCH_3, 51.9ppm.

* Assignments may be reversed.

Mass Spectrum:
LREIMS: 516(M^+), 373, 289, 171, 139, and 129*m/e*.

Reference

T. Kikuchi, S. Matsuda, Y. Murai, and Z. Ogita; Ganoderic Acid G and I and Ganolucidic Acid A and B, New Triterpenoids from *Ganoderma lucidum*; Chem. Pharm. Bull., Vol. 33, pp. 2628-2631(1985).

Common/Systematic Name

Ganolucidic acid C

3β,15α,29-Trihydroxy-11,23-dioxo-5α-lanost-8-en-26-oic acid

Molecular Formula/Molecular Weight

$C_{30}H_{46}O_7$; MW = 518.32435

Methyl ester: $C_{31}H_{48}O_7$; MW = 532.34000

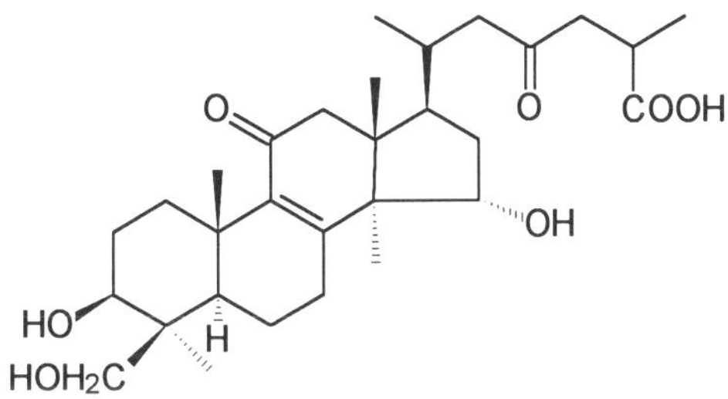

General Characteristics

Methyl ester: Crystallized from methanol as colorless needles; mp., 230-231°C; $[\alpha]_D^{25}$ + 124° (*c*=0.1, in EtOH).

Fungal Source

Isolated from the dried fruiting bodies (mushrooms) of *Ganoderma lucidum* (Fr.) Karst. (Polyporaceae).

Isolation/Purification

The dried chipped epidermis of *Ganoderma lucidum* was extracted twice with $CHCl_3$. The combined extracts were evaporated at 40-45° under reduced pressure to about 1/10 of the original volume and the resulting solution was extracted with sat. aq. $NaHCO_3$ solution. The combined extracts were acidified with 6N HCl (pH 3-4) at 0°. The precipitate was extracted with $CHCl_3$, dried (Na_2SO_4) and the solvent removed under reduced pressure, leaving crude material. The acidic material was chromatographed on silica gel column chromatography (Fr. 1-11, $CHCl_3$-MeOH = 98:2, v/v; Fr. 12, $CHCl_3$-MeOH = 9:1, v/v, Fr. 13, MeOH). Fractions 7-9 were combined and rechromatographed on silica gel to give a mixture of ganoderic acid J and A, which was methylated with diazomethane and separated by further silica gel column chromatography and preparative TLC. Methyl ganoderate J was obtained as a pale yellow syrup. Fraction 12 was chromatographed on a Lobal column (RP_8, Merck) and the second fraction was methylated with diazomethane. The methylated products were subjected to silica gel column chromatography and methyl ganolucidate C was obtained as colorless needles.

Biological Activity

A bitter principle; the mushrooms of *Ganoderma lucidum* (Fr.) Karst. (Polyporaceae) have long been used as a home remedy in China and Japan.

Spectral Data (Methyl ester)

UV:

λ_{max}^{EtOH} 258nm (7,930).

IR:

(KBr) 3400, 1720 and 1620cm^{-1}.

^{1}H NMR:

(C_5D_5N) 0.92(3H, d, *J*=5.9Hz); 0.96(3H, s); 1.19(3H, d, *J*=7.3 Hz); 1.36(3H, s); 1.48(3H, s); 1.56(3H, s), 3.70(1H, dd, *J*=12.0 and 4.6Hz); 3.74(1H, *J*=11.2 Hz); 4.59(1H, d, *J*=11.2Hz); and 1.63ppm (1H, dd, *J*=9.3 and 5.9Hz).

(Free acid) (C_5D_5N) 0.96(3H, s); 0.97(3H, d, *J*=5.9 Hz); 1.36(3H, d, *J*=6.9 Hz); 1.37(3H, s); 1.48(3H, s); 1.56(3H, s); 3.71(1H, dd, *J*=11.1 and 4.4Hz); 3.75(1H, d, *J*=10.8Hz); 4.60(1H, d, *J*=10.8Hz); and 4.64ppm (1H, dd, *J*=9.2 and 5.6Hz).

^{13}C NMR:

(C_5D_5N)(number of bonded H): 208.6(0); 198.2(0); 176.2(0); 164.5(0); 139.7(0); 79.8(1); 72.0(1); 64.4(2); 51.6(3); 23.8(3); 20.0(3); 19.7(3); 19.4(3); 17.2(3); and 17.1ppm (3).

Mass Spectrum:

Methyl ganolucidate C showed a molecular ion peak at 532.3408*m/e* (calcd. for $C_{31}H_{48}O_7$, 532.3401); ganolucidic acid C (M^+, 518.3268, calcd. 518.3245).

Reference

T. Nishitoba, H. Sato, and S. Sakamura; New Terpenoids, Ganoderic Acid J and Ganolucidic Acid C, from the Fungus *Ganoderma lucidum*; Agric. Biol. Chem., Vol. 49, pp. 3637-3638(1985).

Common/Systematic Name

Ganolucidic acid D

15α,23-Dihydroxy-3,11-dioxo-5α-lanosta-8,24*E*-dien-26-oic acid

Molecular Formula/Molecular Weight

$C_{30}H_{44}O_6$; MW = 500.31379

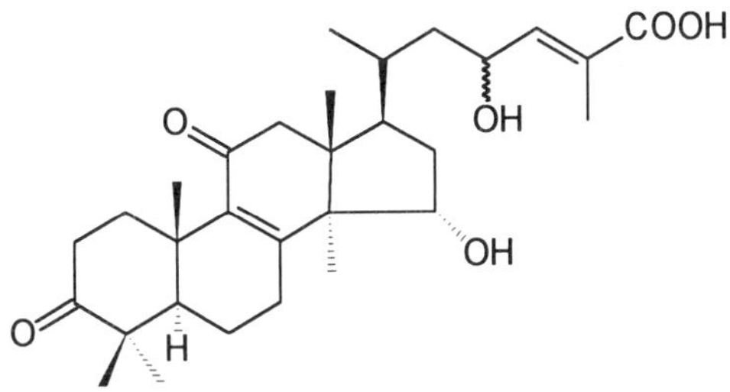

General Characteristics

Ganolucidic acid D was a crystalline solid; $[\alpha]_D^{23}$ + 192° (*c*=0.1, in EtOH).

Fungal Source

Fruiting bodies of *Ganoderma lucidum*.

Isolation/Purification

Dried chipped fruiting bodies of *G. lucidum* were extracted with ethanol; the extract was partitioned between chloroform and water. The chloroform layer was concentrated and separated into its acidic and neutral fractions. The acidic fraction was separated into thirteen fractions; Fraction 12 was subjected to Lobar column (RP_8, Merck) chromatography and the second fraction was treated with diazomethane. The resulting product was rechromatographed on silica gel and the Lobar (RP_8) column to give methyl ganoderate L, lucidone C, and methyl lucidenate G. The fourth fraction in the chromatography of Fr. 12 was purified on a silica gel column, preparative TLC and HPLC to give ganolucidic acid D.

Biological Activity

Fruiting bodies of *G. lucidum* produce some physiological effects when used as a home remedy.

Spectral Data (Monomethyl derivative)

UV:

λ_{max}^{EtOH} 257nm (ε=7800).

IR:

(KBr) 3400, 1700, and 1640cm^{-1}.

^{1}H NMR:

(C_5D_5N) 7.17(1H, dq, *J*=8.8 and 1.5Hz); 4.95(1H, overlapped); 4.62(1H, ddd, *J*=8.8, 5.9, and 5.4Hz); 2.01(3H, d, *J*=1.5Hz); 1.47(3H, s); 1.26(3H, s), 1.13(3H, s); 1.12(3H, s); 1.09(3H, d, *J*=6.8Hz); and 0.95ppm (3H, s).

^{13}C NMR:

(C_5D_5N) (number of bonded H): 217.1(0); 198.3(0); 168.6(0); 165.5(0); 146.2(1); 138.2(0); 127.3(0); 72.0(1); and 66.7ppm (1). These data indicated that methyl ganolucidate D was different from methyl ganolucidate A in the side chain moiety, having a double bond between C-24 and C-25 and a hydroxyl group at C-23. In the ^{1}H NMR spectrum of methyl ganolucidate D, the signal due to H-24 resonated at 6.59ppm in $CDCl_3$, which resembled those of tiglic acid and ganoderic acids U-Z, so the configuration of the double bond between C-24 and C-25 was assigned as E. From these observations, the structure of ganolucidic acid D was concluded to be 15α,23-dihydroxy-3,11-dioxo-5α-lanosta-8,24*E*-dien-26-oic acid.

Mass Spectrum:

HREIMS: M^+ 500.3135; $C_{30}H_{44}O_6$.

Reference

T. Nishitoba, H. Sato, and S. Sakamura; New Terpenoids, Ganolucidic Acid D, Ganoderic Acid L, Lucidone C, and Lucidenic Acid G, from the Fungus *Ganoderma lucidum;* Agric. Biol. Chem., Vol. 50, pp. 809-811(1986).

Common/Systematic Name

Ganolucidic acid E

15α-Hydroxy-3,11-dioxo-5α-lanosta-8,24*E*-dien-26-oic acid

Molecular Formula/Molecular Weight

$C_{30}H_{44}O_5$; MW = 484.31887

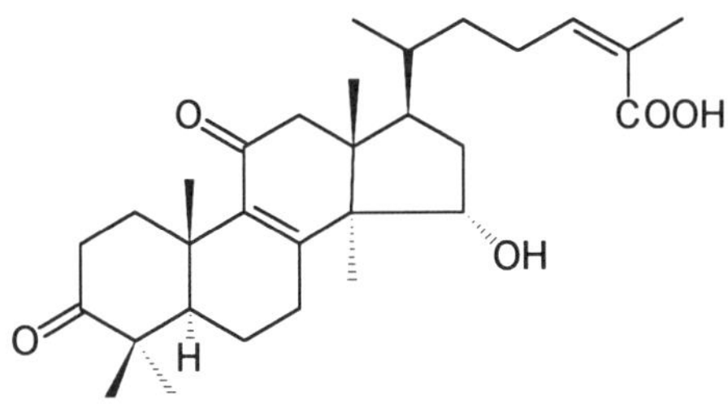

General Characteristics

Methyl ganolucidate E was a colorless syrup; $[\alpha]_D^{24}$ + 154 (*c*=0.1, in MeOH).

Fungal Source

Fruiting bodies of *Ganoderma lucidum*.

Isolation/Purification

The neutral fraction obtained from the EtOH extract of *G. lucidum* (a mixture of two strains) was separated into three fractions (Fr. I, Fr. II and Fr. III) using C_{18} Sep-Pak (MeOH; Waters Assoc.) and reversed-phase HPLC (Resolve-C_{18} column, Waters Assoc.), 2% AcOH-CH_3CN (2.7:1, 0.8ml/min., detector set at 254nm). Fraction I was further separated by silica gel and Lobar column chromatographies into several fractions (Frs. D, K, M, and P). Methyl ganolucidate E was obtained by preparative TLC (precoated Kieselgel 60 F_{254} and RP_{18} F_{254}) of Fr. P after methylation with diazomethane.

Biological Activity

Fruiting bodies of *G. lucidum* produce some physiological effects when used as a home remedy

Spectral Data (Methyl ganolucidate)

UV:

λ_{max}^{EtOH} 219(ϵ = 12,100) and 257nm (7,080).

IR:

(KBr) 3450, 2940, 2920, 2870, 1700, 1635, and 1570cm^{-1}.

^{1}H NMR:
(CDCl$_3$) 6.73(1H, tq, *J*=7.5,1.5Hz, 24-H); 4.39(1H, dd, *J*=9.2, 5.7 Hz, 15β-H); 3.01(1H, ddd, *J*=13.9, 8.2, 5.6 Hz, 1β-H); 2.72(1H, d, *J*=17.2 Hz, 12α-H); 2.45(1H, d, *J*=16.9 Hz, 12β-H); 1.83(3H, s, 27-H); 1.18(3H, s, 30-H); 1.13(3H, s, 19- or 28-H); 1.12(3H, s, 19- or 28-H); 1.08(3H, s, 29-H); 0.89(3H, s, 18-H); 0.89(3H, d, *J*=6.2, 21-H); and 3.74ppm (3H, s, COOCH$_3$).

Mass Spectrum:
HREIMS: 498.3335(M$^+$, $C_{31}H_{46}O_5$), 470(18%), 401(100), 287(69), 95(34), 69(28), 55(36), and 41*m/e* (29).

References

T. Nishitoba, K. Oda, H. Sato, and S. Sakamura; Novel Triterpenoids from the Fungus *Ganoderma lucidum*; Agric. Biol. Chem., Vol. 52, pp. 367-372(1988).

T. Nishitoba, H. Sato, and S. Sakamura; New Terpenoids, Ganolucidic Acid D, Ganoderic Acid L, Lucidone C, and Lucidenic Acid G, from the Fungus *Ganoderma lucidum*; Agric. Biol. Chem., Vol. 50, pp. 809-811(1986).

12

Lucidenic Acids

- Lucidenic acid A
- Lucidenic acid B
- Lucidenic acid C
- Lucidenic acid D1
- Lucidenic acid D2
- Lucidenic acid E1
- Lucidenic acid E2
- Lucidenic acid F
- Lucidenic acid G
- Lucidenic acid H
- Lucidenic acid I
- Lucidenic acid J
- Lucidenic acid K
- Lucidenic acid L
- Lucidenic acid M

Common/Systematic Name

Lucidenic acid A

3,11,15-Trioxo-7β-hydroxy-5α-lanost-8-en-24-oic acid

Molecular Formula/Molecular Weight

$C_{27}H_{38}O_6$; MW = 458.26684

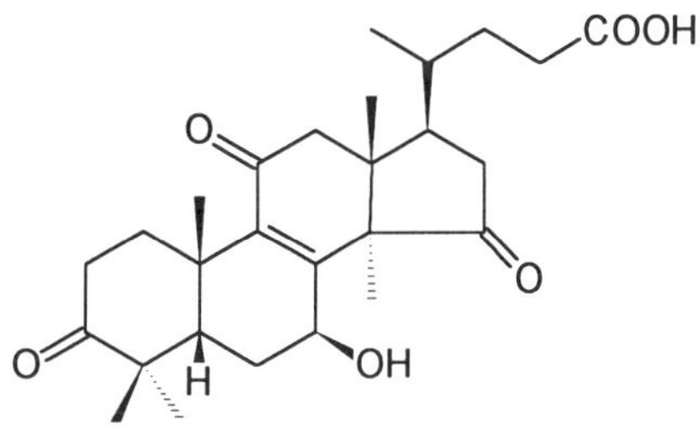

General Characteristics

Crystallized from EtOAc as colorless needles; mp., 194-195°C; $[\alpha]_D^{24}$ +173° (*c*=0.2, in $CHCl_3$).

Fungal Source

Fruiting bodies of *Ganoderma lucidum*.

Isolation/Purification

Dried chipped fruiting bodies of *G. lucidum* were extracted with ethanol; the extract was partitioned between chloroform and water. The chloroform layer was concentrated and separated into its acidic and neutral fractions. The acidic fraction was acidified (pH 3-4) with 6N HCl, and the resulting precipitate was applied to silica gel column chromatography using a solvent system of chloroform-methanol. The eluate was rechromatographed on the silica gel column to give two strongly bitter compounds (lucidenic acid A and ganoderic acid C) and three slightly bitter compounds (lucidenic acid B, lucidenic acid C, and ganoderic acid B).

Biological Activity

Fruiting bodies of *G. lucidum* produce some physiological effects when used as a home remedy.

Spectral Data

UV:

λ_{max}^{EtOH} 254nm (ε=7,680).

IR:

(KBr) 3440, 3300-2500(br), 2975, 2875, 1720, 1700, 1680, 1380, 1270, 1170, 1060, and 750cm^{-1}.

CD:

(MeOH) 295(- 1.9), 282(0), 255(+ 15.0), 227(θ), and 212nm (-7.2).

^{1}H NMR:

($CDCl_3$) 4.86(1H, dd, *J*=9.5 and 7.7Hz, 7α-H); 2.94(1H, ddd, *J*=13.6, 7.7 and 5.5Hz, 1β-H); 2.81(1H, dd, *J*=19.6 and 8.2Hz, 16α-H); 2.78(1H, d, *J*=17.2Hz, 12α-H); 2.75(1H, d, *J*=17.2 Hz, 12β-H); 2.52(1H, 2-H); 2.45(2H, 2-H and 23-H); 2.34(1H, ddd, *J*=16.4. 8.2 and 8.2Hz, 23-H); 2.17(1H, dd, *J*=19.6 and 9.9Hz, 16β-H); 2.12(1H, ddd, *J*=13.0, 7.7 and 1.2 Hz, 6α-H); 2.01(1H, ddd, *J*=9.9, 9.0 and 8.2Hz, 17-H); 1.80(1H, m, 22-H); 1.68(1H, ddd, *J*=13.0, 13.0 and 9.5Hz, 6β-H); 1.60(1H, m, 20-H); 1.58(1H, dd, *J*=13.0 and 1.2Hz, 5α-H); 1.48(1H, ddd, *J*=13.6, 8.3 and 8.3Hz, 1α-H); 1.37(1H, 22-H); 1.35(3H, s, 27-H); 1.26(3H, s, 19-H); 1.14(3H, s, 25-H); 1.11(3H, s, 26-H); 1.01(3H, s, 18-H) and 0.99ppm (3H, d, *J*=6.6Hz, 21-H).

Mass Spectrum:

458.2698(M^+, $C_{27}H_{38}O_6$, 4%), 430(2), 329(10), 320(5), 55(36), 44(100), 43(36), and 40*m/e* (71).

References

T. Nishitoba, H. Sato, T. Kasai, H. Kawagishi, and S. Sakamura; New Bitter C_{27} and C_{30} Terpenoids from the Fungus *Ganoderma lucidum* (Reishi); Agric. Biol. Chem., Vol. 48, pp. 2905-2907(1984).

T. Nishitoba, H. Sato, T. Kasai, H. Kawagishi, and S. Sakamura; New Bitter C_{27} and C_{30} Terpenoids from the Fungus *Ganoderma lucidum* (Reishi); Agric. Biol. Chem., Vol. 49, pp. 1793-1798(1985).

Common/Systematic Name

Lucidenic acid B

3,11,15-Trioxo-7β,12β-dihydroxy-5α-lanost-8-en-24-oic acid

Molecular Formula/Molecular Weight

$C_{27}H_{38}O_7$; MW = 474.26175

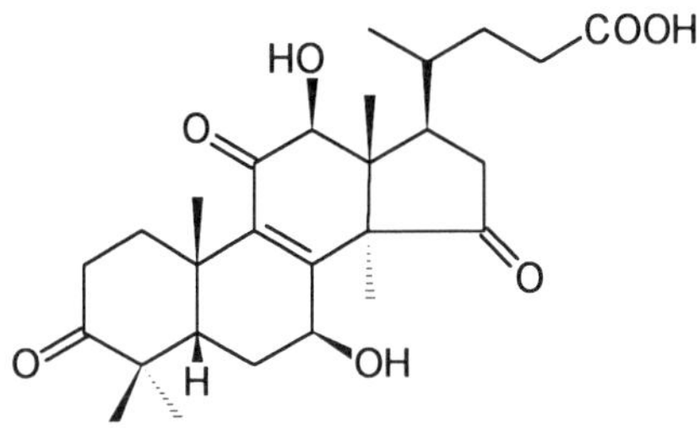

General Characteristics

Crystallized from EtOAc as colorless needles; mp., 179-181°C; $[\alpha]_D^{24}$ + 168.9° (c=0.22, in MeOH).

Fungal Source

Fruiting bodies of *Ganoderma lucidum*.

Isolation/Purification

Dried chipped fruiting bodies of *G. lucidum* were extracted with ethanol; the extract was partitioned between chloroform and water. The chloroform layer was concentrated and separated into its acidic and neutral fractions. The acidic fraction was acidified (pH 3-4) with 6N HCl, and the resulting precipitate was applied to silica gel column chromatography using a solvent system of chloroform-methanol. The eluate was rechromatographed on the silica gel column to give two strongly bitter compounds (lucidenic acid A and ganoderic acid C) and three slightly bitter compounds (lucidenic acid B, lucidenic acid C, and ganoderic acid B).

Biological Activity

Fruiting bodies of *G. lucidum* produce some physiological effects when used as a home remedy.

Spectral Data

UV:

λ_{max}^{EtOH} 252nm (ε=6,010).

CD:
(MeOH) 290(- 7.0), 277(0), 248(+ 25.6), 224(θ), and 208nm (-12.1).

IR:
(KBr) 3380, 3200-2400(br), 2950, 2925, 2860, 1730, 1700, 1660, 1380, 1270, 1160, and 1050cm^{-1}.

^{1}H NMR:
($CDCl_3$) 4.83(1H, dd, *J*=8.7 and 8.7Hz, 7α-H); 4.38(1H, s, 12α-H); 2.79(1H, ddd, *J*=13.7, 7.0 and 4.2Hz, 1β-H); 2.68(1H, dd, *J*=19.5 and 8.5Hz, 16α-H); 2.64(1H, ddd, *J*=15.6, 11.0 and 7.0Hz, 2β-H); 2.45(2H, 17-H and 23-H); 2.40(1H, 23-H); 2.40(1H, ddd, *J*=15.6, 6.6 and 4.2Hz, 2α-H); 2.32(1H, dd, *J*=19.5 and 10.2Hz, 16β-H); 2.18(1H, ddd, *J*=13.7. 8.7 and 1.8Hz, 6α-H); 1.96(1H, m, 20-H); 1.86(1H, m, 22-H); 1.79(1H, ddd, *J*=13.7, 13.7 and 8.7Hz, 6β-H); 1.48(1H, dd, *J*=13.7 and 1.8Hz, 5α-H); 1.44(3H, s, 27-H); 1.43(3H, s, 19-H); 1.35(1H, ddd, *J*=13.7, 11.0 and 6.6Hz, 1α-H); 1.30(1H, m, 22-H); 1.15(3H, d, *J*=6.7Hz, 21-H); 1.13(3H, s, 25-H); 1.13(3H, s, 26-H); and 0.83ppm (3H, s, 18-H).

^{13}C NMR:
($CDCl_3$) C-1, 35.1(t); C-2, 34.3(t); C-3, 215.7(s); C-4, 46.9(s); C-5, 49.4(d); C-6, 27.7(t); C-7, 65.8(d); C-8, 158.1(s); C-9, 140.1(s); C-10, 37.8(s); C-11, 199.6(s); C-12, 78.3(t); C-13, 451.6(s); C-14, 60.2(s); C-15, 217.1(s); C-16, 37.6(t); C-17, 46.6(d); C-18, 12.1(q); C-19, 18.3(q); C-20, 31.5(d); C-21, 20.5(q); C-22, 29.5(t); C-23, 31.8(t); C-24, 178.7(s); C-25, 26.3(q); C-26, 21.2(q); and C-27, 23.3ppm (q).

The ^{1}H and ^{13}C NMR data of lucidenic acid B were very similar to those of lucidenic acid A, but lucidenic acid B showed the new signals at 4.38ppm (singlet) in the ^{1}H NMR and at 78.3ppm (doublet) in the ^{13}C NMR, and lacked the signals of the methylene protons at C(12) in the ^{1}H NMR. This indicated the presence of a secondary hydroxy group at C(12) in lucidenic acid B.

Mass Spectrum:
474.2579(M^+, $C_{27}H_{38}O_7$, 4%), 456(3), 304(100), 55(9), 44(96), 43(10), and 40*m/e* (14).

References

T. Nishitoba, H. Sato, T.Kasai, H. Kawagishi, and S. Sakamura; New Bitter C_{27} and C_{30} Terpenoids from the Fungus *Ganoderma lucidum* (Reishi); Agric. Biol. Chem., Vol. 48, pp. 2905-2907(1984).

T. Nishitoba, H. Sato, T. Kasai, H. Kawagishi, and S. Sakamura; New Bitter C_{27} and C_{30} Terpenoids from the Fungus *Ganoderma lucidum* (Reishi); Agric. Biol. Chem., Vol. 49, pp. 1793-1798(1985).

Common/Systematic Name

Lucidenic acid C

11,15-Dioxo-3β,7β,12β-trihydroxy-5α-lanost-8-en-24-oic acid

Molecular Formula/Molecular Weight

$C_{27}H_{40}O_7$; MW = 476.27740

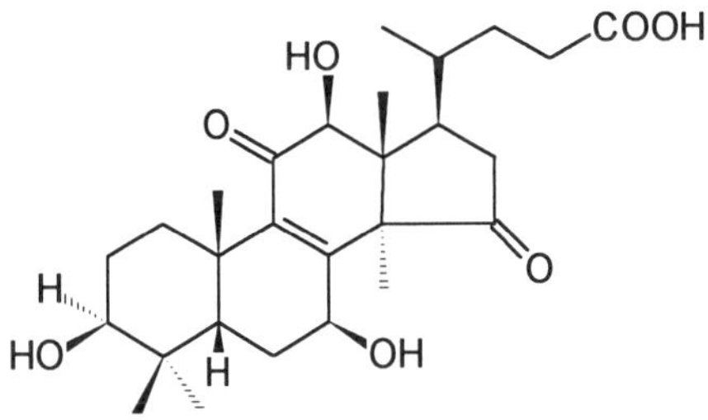

General Characteristics

Crystallized from $CHCl_3$-hexane as pale yellow needles; mp., 199-200°C; $[\alpha]_D^{24}$ +140° (*c*=0.2, in MeOH).

Fungal Source

Fruiting bodies of *Ganoderma lucidum*.

Isolation/Purification

Dried chipped fruiting bodies of *G. lucidum* were extracted with ethanol; the extract was partitioned between chloroform and water. The chloroform layer was concentrated and separated into its acidic and neutral fractions. The acidic fraction was acidified (pH 3-4) with 6N HCl, and the resulting precipitate was applied to silica gel column chromatography using a solvent system of chloroform-methanol. The eluate was rechromatographed on the silica gel column to give two strongly bitter compounds (lucidenic acid A and ganoderic acid C) and three slightly bitter compounds (lucidenic acid B, lucidenic acid C, and ganoderic acid B).

Biological Activity

Fruiting bodies of *G. lucidum* produce some physiological effects when used as a home remedy.

Spectral Data

UV:

λ_{max}^{EtOH} 254nm (ε=5,960).

CD:
(MeOH) 290(- 5.8), 276(0), 248(+ 15.4), 224(θ), and 207nm (-7.8).

IR:
(KBr) 3350, 3200-2400(br), 2950, 2850, 1720, 1675, 1260, 1090, 1040, and 1020cm^{-1}.

^{1}H NMR:
($CDCl_3$) The NMR spectrum of lucidenic acid C was very similar to that of lucidenic acid B, but it showed three signals at 3.21(double doublet *J*=6.4, 10.1Hz), 4.73(singlet) and 4.78ppm (double doublet, *J*=8.7, 8.7Hz) due to methine groups bearing the hydroxyl groups. This provided support for a secondary hydroxyl group at C-3 in lucidenic acid C, instead of a carbonyl group.

^{13}C NMR:
($CDCl_3$) C-1, 33.8(t); C-2, 23.8(t); C-3, 79.0(d); C-4, 37.7(s); C-5, 49.1(d); C-6, 25.0(t); C-7, 68.8(d); C-8, 154.4(s); C-9, 146.0(s); C-10, 38.2(s); C-11, 192.0(s); C-12, 80.2(t); C-13, 50.2(s); C-14, 58.2(s); C-15, 210.5(s); C-16, 37.9(t); C-17, 46.4(d); C-18, 13.4(q); C-19, 18.2(q); C-20, 31.8(d); C-21, 20.3(q); C-22, 29.6(t); C-23, 31.7(t); C-24, 178.3(s); C-25, 28.1(q); C-26, 16.5(q); and C-27, 25.5(q).

Mass Spectrum:
476.2753(M^+, $C_{27}H_{40}O_7$, 2%), 458(2), 307(14), 306(100), 55(17), 44(12), and 43*m/e* (16).

References

T. Nishitoba, H. Sato, T. Kasai, H. Kawagishi, and S. Sakamura; New Bitter C_{27} and C_{30} Terpenoids from the Fungus *Ganoderma lucidum* (Reishi); Agric. Biol. Chem., Vol. 48, pp. 2905-2907(1984).

T. Nishitoba, H. Sato, T. Kasai, H. Kawagishi, and S. Sakamura; New Bitter C_{27} and C_{30} Terpenoids from the Fungus *Ganoderma lucidum* (Reishi); Agric. Biol. Chem., Vol. 49, pp. 1793-1798(1985).

Common/Systematic Name

Lucidenic acid D1

3,7,11,15-Pentaoxo-5α-lanost-8-en-24-oic acid

Molecular Formula/Molecular Weight

$C_{27}H_{34}O_7$; MW = 470.23045

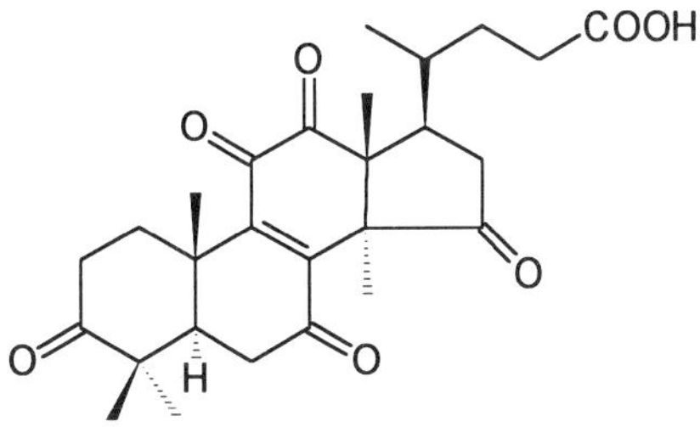

General Characteristics

Crystallized from EtOAc-cyclohexane as pale yellow needles; mp., 229-231°C; $[\alpha]_D^{22}$ + 84° (*c*=0.1, in EtOH).

Fungal Source

Fruiting bodies of *Ganoderma lucidum*.

Isolation/Purification

Dried chipped fruiting bodies of *G. lucidum* were extracted with chloroform and chromatographed on a silica gel column using various concentrations of methanol in chloroform. Lucidenic acid D1 and lucidone A were further purified with a second silica gel column eluted in a like manner.

Biological Activity

Fruiting bodies of *G. lucidum* produce some physiological effects when used as a home remedy; also bitter tasting. The intensity of bitterness was evaluated to be lucidenic acid D, ganoderic acid C, lucidone A, lucidenic acid A, ganoderic acid B, lucidenic acid B, lucidenic acid C, and lucidenic acid E. Lucidenic acid A, ganoderic acid C, lucidenic acid D, and lucidone A had intense bitterness and their threshold values were determined to 2.0 x 10^{-6}M, 5 x 10^{-8}M, 5 x 10^{-10}M, and 1 x 10^{-6}M, respectively, when the organoleptic test was carried out by an ascending series of concentrations in a 10% ethanol solution.

Spectral Data

UV:

λ_{max}^{EtOH} 224(sh, ε=4,160) and 266nm (5,170).

IR:

3400-2400(br), 2960, 2875, 1755(sh), 1730, and 1685cm^{-1}.

^{1}H NMR:

($CDCl_3$) The ^{1}H NMR spectrum showed the presence of six methyl groups; 1.56(s), 1.36(s), 1.17(s), 1.16(s), 1.14(s), and 0.91ppm (d, *J*=6.1Hz), but did not show any signal due to the methine group bearing the hydroxy group, which was observed in the case of lucidenic acids A, B, and C. These data led to the structure of lucidenic acid D being a pentaketo acid.

^{13}C NMR:

($CDCl_3$) (number of bonded H); 214.4(0), 204.2(0), 198.2(0). 197.2(0), 192.8(0), 178.4(0), 150.0(0), 149.6(0), 61.0(0), 59.0(0), 51.0(1), 47.0(0). 39.3(0). 39.1(2). 39.0(1), 37.3(2), 35.5(1), 33.6(2), 33.5(2), 30.9(2), 30.1(2), 27.5(3), 23.3(3), 20.3(3), 18.6(3), 17.8(3), and 12.5ppm (3).

Mass Spectrum:

Field-desorption mass spectrum gave a molecular ion peak at 470*m/e*.

<u>Reference</u>

T. Nishitoba, H. Sato, and S. Sakamura; New Terpenoids from *Ganoderma lucidum* and Their Bitterness; Agric. Biol. Chem., Vol. 49, pp. 1547-1549(1985).

Common/Systematic Name

Lucidenic acid D2

3,7,11,15-Tetraoxo-12-acetoxy-5α-lanost-8-en-24-oic acid

Molecular Formula/Molecular Weight

$C_{29}H_{38}O_8$; MW = 514.25667

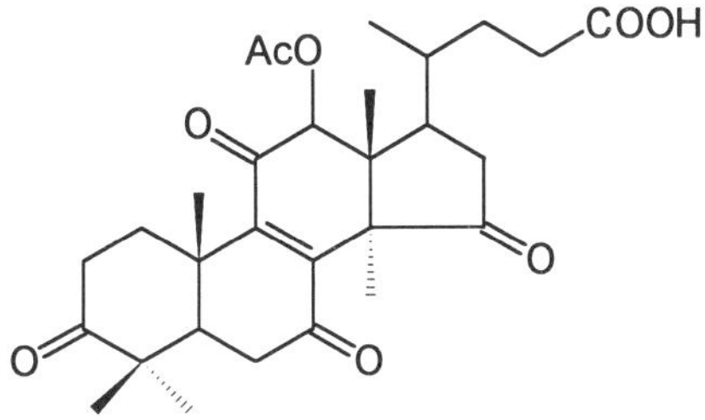

General Characteristics

Isolated and identified as methyl ester. Yellow amorphous powder; $[\alpha]_D$ + 136° (*c*=1.0, in $CHCl_3$).

Fungal Source

Ganoderma lucidum.

Isolation/Purification

The acidic fraction from the ether extract of the gills of dried fruit bodies of *Ganoderma lucidum* was methylated with diazomethane. The crude product was separated repeatedly by a combination of silica gel chromatography and preparative thin-layer chromatography to give pure triterpenoids.

Spectral Data (Methyl ester)

UV:

λ_{max} 252.5nm (log ε=3.96).

IR:

(KBr) 1745, 1735(sh), 1700, 1680, and 1220cm^{-1}.

^{1}H NMR:

($CDCl_3$) 173(1H, ddd, *J*=14.5, 10.0, 6.5Hz, H-1α); 2.77(1H, ddd, *J*=14.0, 8.5, 6.0Hz, H-1β); 2.425(1H, ddd, *J*=15.5, 8.0, 5.5Hz, H-2α); 2.61(ddd, *J*=15.5, 9.5, 6.0Hz, H-2β); 2.31(1H, dd, *J*=15.0, 2.5Hz, H-5); 2.49(1H, dd, *J*=15.0, 2.5Hz, H-6α); 2.73(1H, dd, *J*=14.8, 14.0Hz, H-6β); 5.67(1H, s, H-12α); 2.83(1H, dd, *J*=18.5, 10.0Hz, H-16α); 2.08(1H, dd, *J*=18.5, 8.0Hz, H-16β); 2.48(1H, dt, *J*=10.0, 8.0Hz, H-17); 1.64(1H, m, H-20); 1.30(1H, dtd, *J*=13.5, 8.5, 5.0Hz, H-22); 1.84(1H, dtd, *J*=13.5, 8.5, 2.5Hz, H-22); 2.28(1H, dt, *J*=16.0, 8.5Hz, H-23); 2.41(1H, ddd, *J*=16.0, 8.5, 5.5Hz, H-23);

0.86(3H, s, 18-CH_3); 1.33(3H, s, 19-CH_3); 1.00(3H, d, *J*=6.8Hz, 21-CH_3); 1.14(3H, s, 30-CH_3); 1.12 (3H, s, 31-CH_3); 1.81(3H, s, 32-CH_3); and 3.68ppm (3H, s, COO$\underline{CH_3}$).

^{13}C NMR:

($CDCl_3$) C-1, 34.0, t; C-2, 33.8, t; C-3, 214.8, s; C-4, 46.9, s; C-5, 51.0, d; C-6, 37.4, t; C-7, 198.5*, s; C-8, 149.7, s; C-9, 146.2, s; C-10, 39.3, s; C-11, 194.1*, s; C-12, 79.1, d; C-13, 47.6, s; C-14, 58.6, s; C-15, 205.8, s; C-16, 37.4, t; C-17, 45.2, d; C-18, 12.0, q; C-19, 18.6, q; C-20, 33.0, d; C-21, 20.1, q; C-22, 30.1, t; C-23, 31.6, t; C-24, 173.6, s; C-30, 27.6, q; C-31, 20.45, q; C-32, 20.8, q; O$\underline{C}H_3$, 51.6, q; OCO$\underline{C}H_3$, 20.8, q; and O$\underline{C}OCH_3$, 170.0ppm, s.

* Assignments may be reversed.

Mass Data:

LREIMS: 528(M^+, 14%), 486(M^+ - CH_2CO-, 22), 468(27), 413(5), 371(11), 353(8), 302(100), 227(30), 185(84), 153(32), 167(95), 167(95), 135(22), 115(9), and 55*m/e* (49); HREIMS: 528.2722*m/e* (calcd for $C_{30}H_{40}O_8$ 528.2722); anal. calcd for $C_{30}H_{40}O_8$ • ½H_2O C, 67.03; H, 7.68; found, C, 67.20; H, 7.47.

References

T. Kikuchi, S. Kanomi (*nee* Matsuda), Y. Murai, S. Kadota, K. Tsubono, and Z. Ogita; Constituents of the Fungus *Ganoderma lucidum* (Fr.) Karst. II. Structures of Ganoderic Acids F, G, and H, Lucidenic Acids D2 and E2, and Related Compounds; Chem. Pharm. Bull., Vol. 34, pp. 4018-4029(1985).

T. Kikuchi, S. Matsuda, S. Kadota, Y. Murai, and Z. Ogita; Ganoderic Acid D, E, F, and H and Lucidenic Acid D, E, and F, New Triterpenoids from *Ganoderma lucidum*; Chem. Pharm. Bull. Vol. 33, pp. 2624-2627(1985).

Common/Systematic Name

Lucidenic acid E1

7β,12α-Dihydroxy-3,11,15-trioxo-5α-lanost-8-en-24-oic acid

Molecular Formula/Molecular Weight

$C_{27}H_{38}O_7$; MW = 474.26175

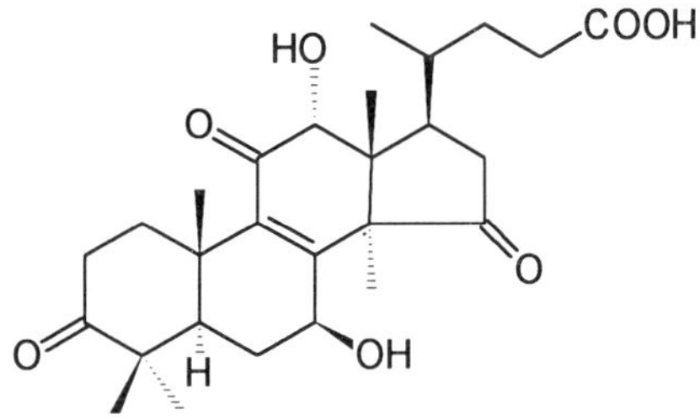

General Characteristics

Lucidenic acid E was crystallized from EtOAc as colorless plates; mp., 216-219°C; $[\alpha]_D^{20}$ + 229° (*c*=0.2, in MeOH).

Fungal Source

Fruiting bodies of *Ganoderma lucidum.*

Isolation/Purification

Dried chipped fruiting bodies of *G. lucidum* were extracted with chloroform and chromatographed on a silica gel column using various concentrations of methanol in chloroform. The bitter principles were further purified with a second silica gel column eluted in a like manner.

Biological Activity

Fruiting bodies of *G. lucidum* produce some physiological effects when used as a home remedy; also bitter tasting. The intensity of bitterness was evaluated to be lucidenic acid D, ganoderic acid C, lucidone A, lucidenic acid A, ganoderic acid B, lucidenic acid B, lucidenic acid C, and lucidenic acid E. Lucidenic acid A, ganoderic acid C, lucidenic acid D, and lucidone A had intense bitterness and their threshold values were determined to 2.0 x 10^{-6}M, 5 x 10^{-8}M, 5 x 10^{-10}M, and 1 x 10^{-6}M, respectively, when the organoleptic test was carried out by an ascending series of concentrations in a 10% ethanol solution.

Spectral Data

UV:

λ_{max}^{MeOH} 258(ϵ = 6,640).

IR:
(KBr) 3450, 3400-2400(br), 2950, 2880(sh), 1725, 1705, and 1640cm^{-1}.

^{1}H NMR:
($CDCl_3$) 4.87(1H, dd, *J*=9.5 and 7.3Hz); 3.83(1H, s); 1.34(3H, s), 1.17(3H, s); 1.14(3H, s); 1.10(3H, s); 1.08(3H, d, *J*=6.7Hz); and 0.98ppm (3H, s).

^{13}C NMR:
($CDCl_3$)(number of bonded H) 218.8(0), 216.6(0), 201.1(0), 178.3(0), 160.7(0), 139.2(0), 78.9(1), 66.2(1), The ^{13}C and ^{1}H NMR data of lucidenic acid D were very similar to those of lucidenic acid B; however, the signal of H-C(12) was observed at 3.83(1H, s), in contrast to 4.38ppm (1H, s) for lucidenic acid B.

Mass Spectrum:
$C_{27}H_{38}O_7$ by HR-MS (M^+, 474.2624; calcd for $C_{27}H_{38}O_7$, 474.2618).

<u>Reference</u>

T. Nishitoba, H. Sato, and S. Sakamura; New Terpenoids from *Ganoderma lucidum* and Their Bitterness; Agric. Biol. Chem., Vol. 49, pp. 1547-1549(1985).

Common/Systematic Name

Lucidenic acid E2

Molecular Formula/Molecular Weight

$C_{29}H_{40}O_8$; MW = 516.27232

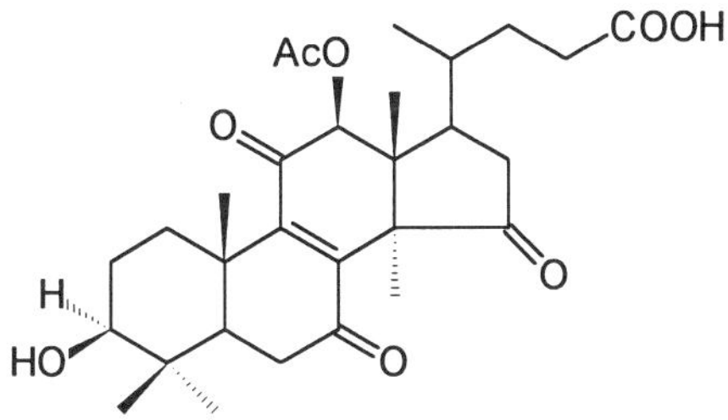

General Characteristics

Isolated and identified as methyl ester. Yellow needles from ether; mp., 141-144°C; $[\alpha]_D$ + 86° (*c*=0.5, in $CHCl_3$).

Fungal Source

Ganoderma lucidum.

Spectral Data (Methyl ester)

UV:

λ_{max} 256nm (log ϵ = 3.75).

IR:

(KBr) 3400, 1745, 1735(sh), 1720(sh), 1690, 1680(sh), and 1220cm^{-1}.

^{1}H NMR:

($CDCl_3$) 1.18(1H, td, *J*=13, 4.5Hz, H-1α); 2.73(1H, dt, *J*=13.5, 3.5Hz, H-1β); 1.72(2H, m, H-2α and β); 3.26(1H, dd, *J*=11.0, 5.0Hz, H-3); 1.56(1H, dd, *J*=14.2, 2.5Hz, H 5); 2.58(1H, dd, *J*=14.0, 2.5Hz, H 6α); 267(1H, t, *J*=14.0Hz, H-6β); 5.62(1H, s, H-12α); 2.79(1H, dd, *J*=18.2, 9.5Hz, H-16α); 2.08(1H, dd, *J*=18.2, 8.1Hz, H-16β); 2.46(1H, dt, *J*=9.0, 8.0Hz, H-17); 1.66(1H, m, H-20); 1.29(1H, dtd, *J*=14.0, 8.5, 5.0Hz, H-22); 1.84(1H, dtd, *J*=13.5, 8.0, 2.5Hz, H-22); 2.27(1H, dt, *J*=16.0, 8.0Hz, H-23); 2.40(1H, ddd, *J*=16.0, 8.8, 5.5Hz, H-23); 0.83(3H, s, 18-CH_3); 1.33(3H, s, 19-CH_3); 1.00(3H, d, *J*=6.7Hz, 21-CH_3); 1.03(3H, s, 30-CH_3); 0.89(3H, s, 31-CH_3); 1.73(3H, s, 32-CH_3); 3.67(3H, s, $COOCH_3$); and 2.22ppm (3H, CH_3COO).

^{13}C NMR:

($CDCl_3$) C-1, 33.3, t; C-2, 27.4, t; C-3, 77.5, d; C-4, 40.5, s; C-5, 51.4, d; C-6, 36.7, t; C-7, 198.8*, s; C-8, 151.6, s; C-9, 146.0, s; C-10, 39.2, s; C-11, 194.1*, s; C-12, 79.4, d; C-13, 48.0, s; C-14, 58.5, s; C-15, 206.0, s; C-16, 37.6, t; C-17, 45.5, d; C-18, 12.1, q; C-19, 18.0, q; C-20, 33.0, d; C-21, 20.2, q; C-22, 30.2, t; C-23, 31.8, t; C-24, 173.7, s; C-30, 27.9, q; C-31, 15.6, q; C-32, 21.4, q; O<u>C</u>H_3, 51.6, q; OCO<u>C</u>H_3, 20.9, q; and O<u>C</u>OCH_3, 170.1ppm, s.

* Assignments may be reversed.

Mass Data:

LREIMS: 530(M^+, 5%), 488(M^+ - CH_2CO-, 2), 470(10), 415(2), 373(5), 355(2), 304(100), 227(10), 185(23), 167(30), 153(10), 135(8), 115(3), and 55*m/e* (10); HREIMS: 530.2879*m/e* (calcd for $C_{30}H_{42}O_8$, 530.2879); anal. calcd for $C_{30}H_{42}O_8$ • ½H_2O: C, 66.78; H, 8.03%; found; C, 66.98; H, 7.94%.

<u>Reference</u>

T. Kikuchi, S. Kanomi (*nee* Matsuda), Y. Murai, S. Kadota, K. Tsubono, and Z. Ogita; Constituents of the Fungus *Ganoderma lucidum* (Fr.) Karst. II. Structures of Ganoderic Acids F, G, and H, Lucidenic Acids D2 and E2, and Related Compounds; Chem. Pharm. Bull., Vol. 34, pp. 4018-4029(1985).

Common/Systematic Name

Lucidenic acid F

3,7,11,15-Tetraoxo-5α-lanost-8-en-24-oic acid

Molecular Formula/Molecular Weight

$C_{27}H_{36}O_6$; MW = 456.25119

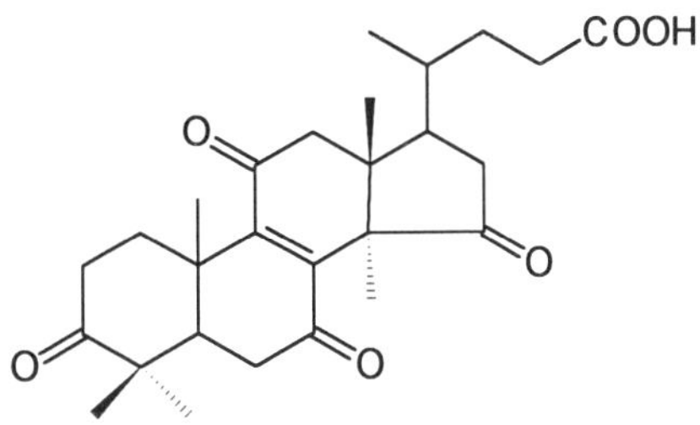

General Characteristics

Yellow needles; mp., 208-211°C; $[\alpha]_D$ + 195° (in $CHCl_3$).

Fungal Source

Isolated from *Ganoderma lucidum*.

Isolation/Purification

The acidic fraction from the ether extract of the gills of dried fruit bodies of *Ganoderma lucidum* was methylated with diazomethane and the crude product was separated repeatedly by a combination of silica gel chromatography and preparative thin-layer chromatography to give pure triterpenoids.

Spectral Data (Methyl ester)

^{1}H NMR:

($CDCl_3$) 0.85(3H, s, 18-CH_3); 1.27 (3H, s, 19-CH_3); 0.96 (3H, d, *J*=6.0Hz, 21-CH_3); 1.11 (3H, s, 30-CH_3); 1.14 (3H, s, 31-CH_3); 1.65H, s, 32-CH_3); and 3.68ppm (3H, s $COOCH_3$)

^{13}C NMR:

($CDCl_3$) 37.3, C-1, t; 34.6, C-2, t; 215.3, C-3, s; 43.9, C-4, s; 51.0, C-5, d; 33.9, C-6, t; 199.5, C-7, s; 149.7, C-8, s; 146.9, C-9, s; 39.4, C-10, s; 194.4, C-11, s; 49.0, C-12, t; 47.0, C-13, s ; 57.2, C-14, s; 207.3, C-15, s; 39.9, C-16, t; 45.2, C-17, d; 16.1, C-18, q; 18.6, C-19, q; 35.4, C-20, d; 18.3, C-21, q; 30.8, C-22, t; 31.0, C-23, t; 173.8, C-24, s; 27.7, C-30, q; 20.9, C-31* q; 20.3, C-32*, q; and 51.7ppm, CH_3O-, q.

* Assignments may be reversed.

Reference

T. Kikuchi, S. Matsuda, S. Kadota, Y. Murai, and Z. Ogita; Ganoderic Acid D, E, F, and H and Lucidenic Acid D, E, and F, New Triterpenoids from *Ganoderma lucidum*; Chem. Pharm. Bull., Vol. 33, pp. 2624-2627(1985).

Common/Systematic Name

Lucidenic acid G

3,11-Dioxo-4β,7β,15α-trihydroxy-5α-lanost-8-en-24-oic acid

Molecular Formula/Molecular Weight

$C_{27}H_{40}O_7$; MW = 476.27740

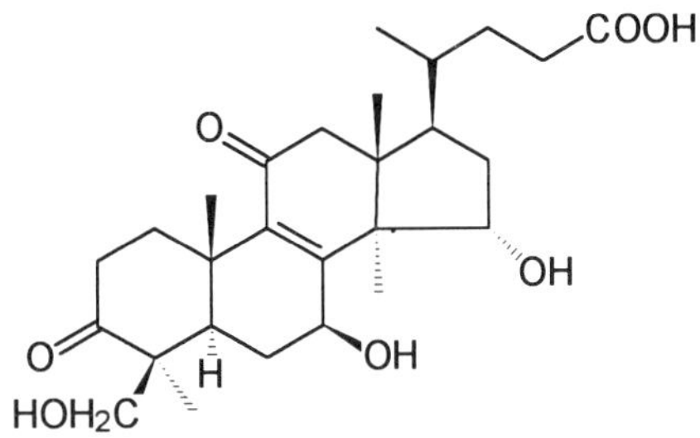

General Characteristics

Colorless syrup; $[\alpha]_D^{21}$ + 127° (*c*=0.2, in MeOH).

Fungal Source

Fruiting bodies of *Ganoderma lucidum*.

Isolation/Purification

Dried chipped fruiting bodies of *G. lucidum* were extracted with ethanol; the extract was partitioned between chloroform and water. The chloroform layer was concentrated and separated into its acidic and neutral fractions. The acidic fraction was separated into thirteen fractions; Fraction 12 was subjected to Lobar column (RP_8, Merck) chromatography and the second fraction was treated with diazomethane. The resulting product was rechromatographed on silica gel and the Lobar (RP_8) column to give methyl ganoderate L, lucidone C, and methyl lucidenate G. The fourth fraction in the chromatography of Fr. 12 was purified on a silica gel column, preparative TLC and HPLC to give ganolucidic acid D.

Biological Activity

Fruiting bodies of *G. lucidum* produce some physiological effects when used as a home remedy.

Spectral Data (Methyl ester)

UV:

λ_{max}^{MeOH} 254nm (ε=8040).

IR:
(KBr) 3400, 1730(sh), 1700, and 1660cm^{-1}.

^{1}H NMR:
(C_5D_5N) 5.24(1H, dd, *J*=9.3 and 7.1 Hz); 4.99(1H, dd, *J*=9.9 and 7.3Hz); 4.37(1H, d, *J*=11.0Hz); 3.99(1H, d, *J*=11.0Hz); 1.72(3H, s); 1.52(3H, s); 1.51(3H, s); 1.07(3H, s); and 0.81ppm (3H, d, *J*=5.9Hz).

^{13}C NMR:
(C_5D_5N) 214.4(0); 199.8(0); 174.1(0); 161.3(0); 140.6(0); 72.2(1); 69.2(1); and 65.1ppm (2).

Mass Spectrum:
HREIMS: 490.2905*m/e* (M^+).

Reference

T. Nishitoba, H. Sato, and S. Sakamura; New Terpenoids, Ganolucidic Acid D, Ganoderic Acid L, Lucidone C and Lucidenic Acid G from the Fungus *Ganoderma lucidum*; Agric. Biol. Chem., Vol. 50, pp. 809-811(1986).

Common/Systematic Name

Lucidenic acid H

11,15-Dioxo-3β,7β-dihydroxy-5α-lanost-8-en-24-oic acid

Molecular Formula/Molecular Weight

$C_{27}H_{40}O_7$; MW = 476.27740

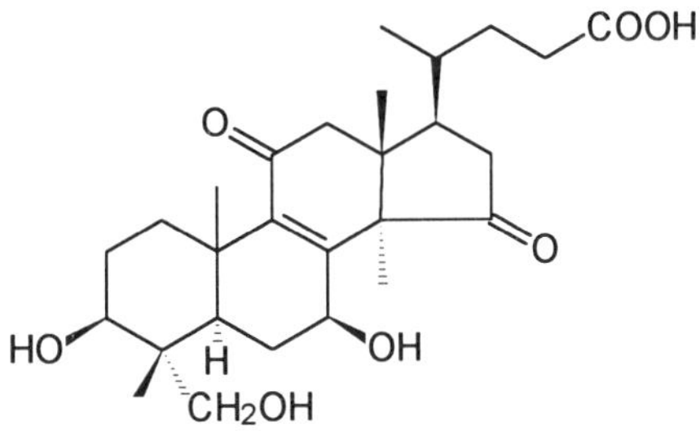

General Characteristics

Methyl ester: pale yellow prisms from ethyl acetate-cyclohexane; mp., 190-192°C; $[\alpha]_D^{23}$ + 136° (*c*=0.2, in MeOH).

Fungal Source

Fruiting bodies of *Ganoderma lucidum*.

Isolation/Purification

Dried chipped fruiting bodies of *Ganoderma lucidum* were extracted with ethanol. The extract was concentrated and partitioned between chloroform and water. The acidic fraction was obtained and chromatographed on a silica gel column eluted with chloroform-methanol (49:1, v/v; fractions 1-11), chloroform-methanol (9:1, v/v; fraction 12) and methanol (fraction 13). Fraction 5 was rechromatographed 3 times on a silica gel column eluted with chloroform-methanol (19:1, v/v) to give several fractions. Some of the fractions were methylated and further fractionated using silica gel column chromatography (chloroform-methanol, 19:1, v/v; chloroform-acetone, 2:1, v/v; benzene-ethyl acetate, 7:1, v/v) and reversed-phase LC (methanol-water, 7:3, v/v , Lobar RP_{18}) to yield methyl lucidenate H, I, and J.

Biological Activity

Fruiting bodies of *G. lucidum* produce some physiological effects when used as a home remedy.

Spectral Data (Methyl ester)

UV:

λ_{max}^{MeOH} 255nm (ϵ = 7,550).

IR:
(thin film) 3400, 2925, 1720, and 1655cm^{-1}.

CD:
λ_{max}^{MeOH} Δ_{293} - 3.3, Δ_{280} 0, Δ_{255} + 14.9, Δ_{230} 0, and Δ_{210} - 8.3.

^{1}H NMR:
($CDCl_3$) 1.29(1H, ddd, *J*=13.6, 12.5, 6.9Hz, H-1α); 3.27(1H, ddd, *J*=13.6, 3.5, 3.5Hz, H-1β); 2.05(1H, m, H-2α); 2.05(1H, m, H-2β); 4.30(1H, dd, *J*=11.7, 5.5Hz, H-3α); 1.91(1H, d, *J*=12.7Hz, H-5α); 2.58(1H, dd, *J*=12.1, 7.5Hz, H-6α); 1.95(1H, m, H-6β); 5.25(1H, dd, *J*=7.5, 7.5Hz, H-7α); 2.95(1H, dq, *J*=16.8, 1.0Hz, H-12α); 2.81(1H, d, *J*=16.8Hz, H-12β); 2.77(1H, dd, *J*=19.2, 8.1Hz, H-16α); 2.21(1H, dd, *J*=19.2, 9.9Hz, H-16β); 2.00(1H, m, H-17); 1.10(3H, s, H_3-18); 1.54(3H, s, H_3-19); 1.50(1H, m, H-20); 0.86(3H, d, *J*=6.2Hz, H_3-21); 1.80(1H, m, H-22); 1.30(1H, m, H-22); 2.43(1H, ddd, *J*=15.0, 9.5, 5.5Hz, H-23); 2.30(1H, ddd, *J*=15.0, 8.8, 7.0Hz, H-23); 4.24(1H, d, *J*=10.4Hz, H-25); 3.78(1H, d, *J*=10.4Hz, H-25); 1.11(3H, s, H_3-26); 1.30(3H, s, H_3-27); and 3.65ppm (3H, s, $COOCH_3$).

^{13}C NMR:
($CDCl_3$) 35.4, C-1; 28.3, C-2; 72.0, C-3; 43.0, C-4; 42.5, C-5; 27.8, C-6; 66.7, C-7; 158.8, C-8; 142.9, C-9; 39.2, C-10; 198.3, C-11; 51.1, C-12; 45.6, C-13; 59.2, C-14; 216.8, C-15; 41.4, C-16; 46.2, C-17; 17.8, C-18; 19.2, C-19; 35.4, C-20; 18.1, C-21; 31.0, C-22; 31.0, C-23; 174.0, C-24; 66.6, C-25; 13.2, C-26; 24.8, C-27; and 51.4ppm CH_3OOC-.

Mass Spectrum:
HREIMS: 490.2932(M^+, $C_{28}H_{42}O_7$, 38%; calcd 490.2932), 472(22), 462(30), 347(100), 334(63), 322(36), 107(61), 95(35), 81(36), 69(43), 55(66), and 43*m/e* (88).

Reference

T. Nishitoba, H. Sato, and S. Sakamura; Triterpenoids from the Fungus *Ganoderma lucidum*; Phytochemistry, Vol. 26, pp. 1777-1784(1987).

Common/Systematic Name
Lucidenic acid I
7,11,15-Trioxo-3β-hydroxy-5α-lanost-8-en-24-oic acid

Molecular Formula/Molecular Weight
$C_{27}H_{38}O_7$; MW = 474.26175

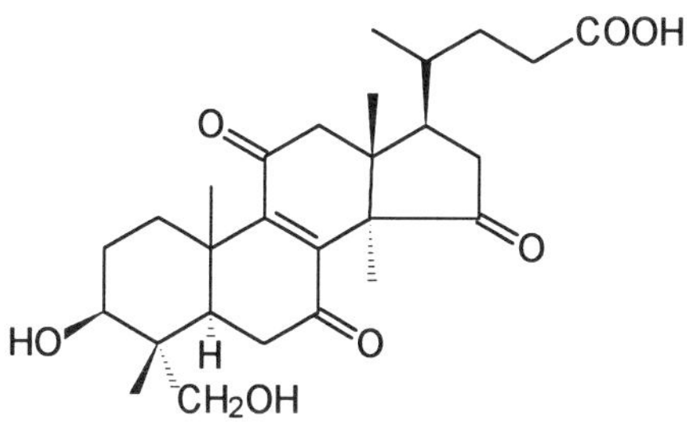

General Characteristics
Methyl ester: pale yellow syrup; $[\alpha]_D^{23}$ + 118° (*c*=0.2, in MeOH).

Fungal Source
Fruiting bodies of *Ganoderma lucidum*.

Isolation/Purification
Dried chipped fruiting bodies of *Ganoderma lucidum* were extracted with ethanol. The extract was concentrated and partitioned between chloroform and water. The acidic fraction was obtained and chromatographed on a silica gel column eluted with chloroform-methanol (49:1, v/v; fractions 1-11), chloroform-methanol (9:1, v/v; fraction 12) and methanol (fraction 13). Fraction 5 was rechromatographed 3 times on a silica gel column eluted with chloroform-methanol (19:1, v/v) to give several fractions. Some of the fractions were methylated and further fractionated using silica gel column chromatography (chloroform-methanol, 19:1, v/v; chloroform-acetone, 2:1, v/v; benzene-ethyl acetate, 7:1, v/v) and reversed-phase LC (methanol-water, 7:3, v/v, Lobar RP_{18}) to yield methyl lucidenate H, I, and J.

Biological Activity
Fruiting bodies of *G. lucidum* produce some physiological effects when used as a home remedy.

Spectral Data (Methyl ester)

UV:
λ_{max}^{MeOH} 262nm (ε = 6,830).

CD:
λ_{max}^{MeOH} Δ_{333} 0, Δ_{305} - 4.9, Δ_{292} 0, Δ_{275} + 8.3, Δ_{245} 0, Δ_{231} + 0.2, Δ_{225} 0, and Δ_{207} - 4.4.

IR:
(thin film) 3400, 2925, 1735, and 1675cm^{-1}.

^{1}H NMR:
($CDCl_3$) 1.45(1H, m, H-1α); 3.10(1H, ddd, *J*=13.6, 3.5, 3.5Hz, H-1β); 2.05(1H, m, H-2α); 2.05(1H, m, H-2β); 4.29(1H, dd, *J*=11.2, 5.3Hz, H-3α); 2.65(1H, dd, *J*=13.6, 2.6Hz, H-5α); 2.92(1H, dd, *J*=13.9, 2.6Hz, H-6α); 2.80(1H, dd, *J*=13.9, 13.6Hz, H-6β); 3.09(1H, d, *J*=15.3Hz, H-12α); 2.81(1H, d, *J*=15.3Hz, H-12β); 2.78(1H, m, H-16α); 2.10(1H, m, H-16β); 2.05(1H, m, H-17); 0.93(3H, s, H_3-18); 1.47(3H, s, H_3-19); 1.45(1H, m, H-20); 0.86(3H, d, *J*=6.6Hz, H_3-21); 1.80(1H, m, H-22); 1.40(1H, m, H-22); 2.45(1H, ddd, *J*=15.1, 9.3, 5.2Hz, H-23); 2.31(1H, ddd, *J*=15.1, 8.8, 7.0Hz, H-23); 4.16(1H, d, *J*=10.8Hz, H-25); 3.61(1H, d, *J*=10.8Hz, H-25); 1.05(3H, s, H_3-26); 1.63(3H, s, H_3-27); and 3.65ppm (3H, s, $COOCH_3$).

^{13}C NMR:
($CDCl_3$) 36.7, C-1; 27.9, C-2; 70.7, C-3; 43.6, C-4; 44.5, C-5; 34.1, C-6; 200.4, C-7; 152.2, C-8; 147.4, C-9; 40.8, C-10; 199.8, C-11; 49.9, C-12; 44.5, C-13; 57.6, C-14; 208.0, C-15; 40.5, C-16; 45.5, C-17; 16.4, C-18; 18.6, C-19; 35.5, C-20; 18.2, C-21; 31.0, C-22; 31.1, C-23; 174.0, C-24; 65.1, C-25; 13.1, C-26; 21.6, C-27; and 51.4ppm CH_3OOC-.

Mass Spectrum:
HREIMS: 488.2796(M^+, $C_{28}H_{40}O_7$, 30%; calcd. 490.2775), 470(12), 265(20), 141(33), 109(42), 95(36), 81(33), 69(40), 55(73), and 44*m/e* (100).

Reference

T. Nishitoba, H. Sato, and S. Sakamura; Triterpenoids from the Fungus *Ganoderma lucidum*; Phytochemistry, Vol. 26, pp. 1777-1784(1987).

Common/Systematic Name

Lucidenic acid J

3,11,15-Trioxo-3β,12β-dihydroxy-5α-lanost-8-en-24-oic acid

Molecular Formula/Molecular Weight

$C_{27}H_{38}O_8$; MW = 490.25667

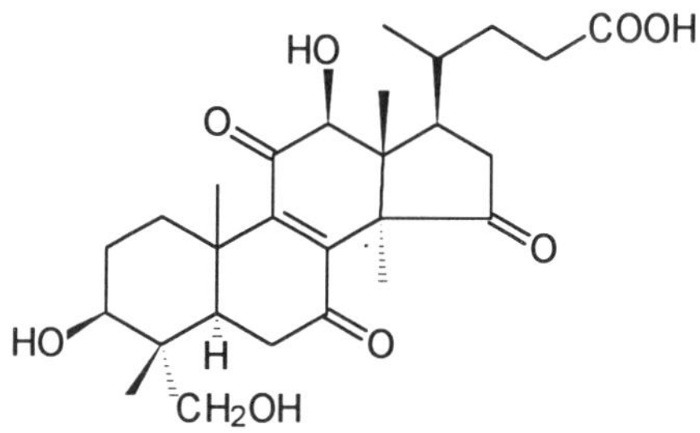

General Characteristics

Methyl ester: pale yellow syrup; $[\alpha]_D^{23}$ + 78° (*c*=0.1, in MeOH).

Fungal Source

Fruiting bodies of *Ganoderma lucidum*.

Isolation/Purification

Dried chipped fruiting bodies of *Ganoderma lucidum* were extracted with ethanol. The extract was concentrated and partitioned between chloroform and water. The acidic fraction was obtained and chromatographed on a silica gel column eluted with chloroform-methanol (49:1, v/v; fractions 1-11), chloroform-methanol (9:1, v/v; fraction 12) and methanol (fraction 13). Fraction 5 was rechromatographed 3 times on a silica gel column eluted with chloroform-methanol (19:1, v/v) to give several fractions. Some of the fractions were methylated and further fractionated using silica gel column chromatography (chloroform-methanol, 19:1, v/v; chloroform-acetone, 2:1, v/v; benzene-ethyl acetate, 7:1, v/v) and reverse-phase LC (methanol-water, 7:3, v/v, Lobar RP_{18}) to yield methyl lucidenate H, I, and J.

Biological Activity

Fruiting bodies of *G. lucidum* produce some physiological effects when used as a home remedy.

Spectral Data (Methyl ester)

UV:

λ_{max}^{MeOH} 255nm (ϵ = 6,400).

CD:

λ_{max}^{MeOH} Δ_{346} 0, Δ_{308} - 6.0, Δ_{292} 0, Δ_{279} + 6.0, Δ_{262} 0, Δ_{255} - 1.2, Δ_{248} 0, Δ_{215} 0, and Δ_{228} + 5.7.

IR:

(thin film) 3430, 2930, 1740, and 1680cm^{-1}.

^{1}H NMR:

($CDCl_3$) 1.35(1H, m, H-1α); 2.97(1H, ddd, *J*=13.6, 3.5, 3.5Hz, H-1β); 2.05(1H, m, H-2α); 2.05(1H, m, H-2β); 4.27(1H, dd, *J*=11.5, 5.2Hz, H-3α); 2.65(1H, dd, *J*=13.6, 2.6Hz, H-5α); 2.93(1H, dd, *J*=13.6, 2.6Hz, H-6α); 2.83(1H, dd, *J*=13.6, 13.6Hz, H-6β); 4.92(1H, s, H-12α); 2.87(1H, dd, *J*=17.6, 9.2Hz, H-16α); 2.26(1H, dd, *J*=17.6, 8.1Hz, H-16β); 2.70(1H, ddd, *J*=8.0, 8.0, 8.0Hz, H-17); 0.98(3H, s, H_3-18); 1.52(3H, s, H_3-19); 1.95(1H, m, H-20); 1.28(3H, d, *J*=6.6Hz, H_3-21); 1.95(1H, m, H-22); 1.41(1H, m, H-22); 2.48(1H, ddd, *J*=16.2, 9.2, 5.8Hz, H-23); 2.37(1H, ddd, *J*=16.2, 8.7, 6.9Hz, H-23); 4.16(1H, d, *J*=10.6Hz, H-25); 3.61(1H, d, *J*=10.6Hz, H-25); 1.04(3H, s, H_3-26); 1.78(3H, s, H_3-27); and 3.65ppm (3H, s, $COOCH_3$).

Mass Spectrum:

HREIMS: 504.2746(M^+, $C_{28}H_{40}O_8$, 51%; calcd. 504.2724), 458(25), 371(65), 353(33), 155(39), 115(42), 95(34), 55(73), and 44*m/e* (100).

Reference

T. Nishitoba, H. Sato, and S. Sakamura, Triterpenoids from the Fungus *Ganoderma lucidum;* Phytochemistry, Vol. 26, pp. 1777-1784(1987).

Common/Systematic Name

Lucidenic acid K

3,7,11,15-Tetraoxo-12α-hydroxy-5α-lanost-8-en-24-oic acid

Molecular Formula/Molecular Weight

$C_{27}H_{36}O_7$; MW = 472.24610

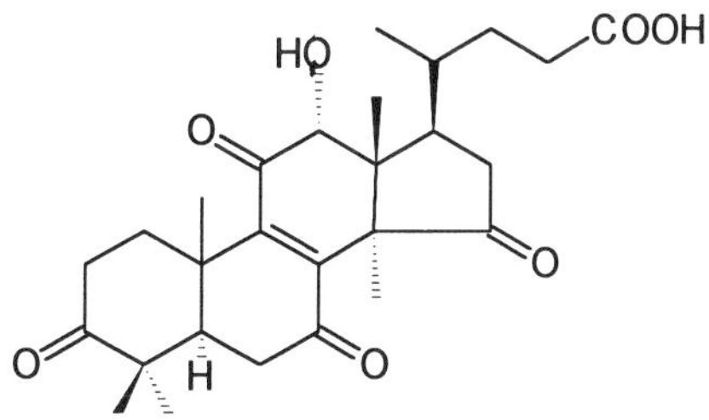

General Characteristics

Methyl ester: pale yellow syrup.

Fungal Source

Fruiting bodies of *Ganoderma lucidum*.

Isolation/Purification

Dried chipped fruiting bodies of *Ganoderma lucidum* were extracted with ethanol. The extract was concentrated and partitioned between chloroform and water. The acidic fraction was obtained and chromatographed on a silica gel column eluted with chloroform-methanol (49:1, v/v; fractions 1-11), chloroform-methanol (9:1, v/v; fraction 12) and methanol (fraction 13). Fraction 5 was rechromatographed 3 times on a silica gel column eluted with chloroform-methanol (19:1, v/v) to give several fractions. Some of the fractions were methylated and further fractionated using preparative TLC (benzene-ethyl acetate, 1:1 or 2:1, v/v), silica gel column chromatography (chloroform-methanol, 99:1, v/v), or reversed-phase LC (methanol-water, 7:3, v/v, Lobar RP_{18}).These separations gave purified methyl ganoderate M, N, O, H, E, methyl lucidenate K, L, and E_2.

Biological Activity

Fruiting bodies of *G. lucidum* produce some physiological effects when used as a home remedy.

Spectral Data (Methyl ester)

UV:

λ_{max}^{MeOH} 258nm (ϵ = 5,440).

CD:
λ_{max}^{MeOH} Δ_{337} 0, Δ_{308} - 4.1, Δ_{291} 0, Δ_{272} + 4.4, Δ_{233} + 3.4, and Δ_{218} 0.

IR:
(thin film) 3450, 2920, 1735, 1700, and 1680cm^{-1}.

^{1}H NMR:
($CDCl_3$) 2.08(1H, m, H-1α); 2.78(1H, ddd, *J*=14.2, 8.3, 5.8Hz, H-1β); 2.55(1H, ddd, *J*=15.6, 8.3, 7.8Hz, H-2α); 2.61(1H, ddd, *J*=15.6, 8.8, 5.8Hz, H-2β); 2.39(1H, dd, *J*=14.6, 3.4Hz, H-5α); 2.54(1H, dd, *J*=15.1, 3.4Hz, H-6α); 2.68(1H, dd, *J*=15.1, 14.6Hz, H-6β); 3.94(1H, s, H-12β); 2.84(1H, dd, *J*=18.6, 9.3Hz, H-16α); 2.07(1H, dd, *J*=18.6, 9.3Hz, H-16β); 2.07(1H, ddd, *J*=9.3, 9.3, 9.3Hz, H-17); 0.84(3H, s, H_3-18); 1.23(3H, s, H_3-19); 1.55(1H, m, H-20); 1.06(3H, d, *J*=6.8Hz, H_3-21); 1.82(1H, m, H-22); 1.43(1H, m, H-22); 2.43(1H, ddd, *J*=16.1, 9.3, 5.4Hz, H-23); 2.30(1H, ddd, *J*=16.1, 8.8, 7.3Hz, H-23); 1.12(3H, s, H_3-25); 1.15(3H, s, H_3-26); 1.58(3H, s, H_3-27); and 3.68ppm (3H, s, $COOCH_3$).

Mass Spectrum:
HREIMS: 486.2601(M^+, $C_{28}H_{38}O_7$, 63%, calcd 486.2618), 458(19), 371(39), 353(22), 302(100), 287(35), 55(55), and 43*m/e* (57).

Reference

T. Nishitoba, H. Sato, and S. Sakamura, Triterpenoids from the Fungus *Ganoderma lucidum;* Phytochemistry, Vol. 26, pp. 1777-1784(1987).

Common/Systematic Name
Lucidenic acid L
7,11,15-Trioxo-3β,12β-dihydroxy-5α-lanost-8-en-24-oic acid

Molecular Formula/Molecular Weight
$C_{27}H_{38}O_7$; MW = 474.26175

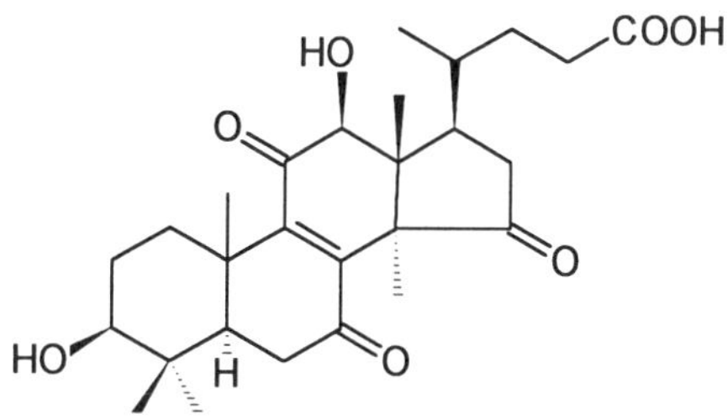

General Characteristics
Methyl ester: a pale yellow syrup.

Fungal Source
Fruiting bodies of *Ganoderma lucidum*.

Isolation/Purification
Dried chipped fruiting bodies of *Ganoderma lucidum* were extracted with ethanol. The extract was concentrated and partitioned between chloroform and water. The acidic fraction was obtained and chromatographed on a silica gel column eluted with chloroform-methanol (49:1, v/v; fractions 1-11), chloroform-methanol (9:1, v/v; fraction 12) and methanol (fraction 13). Fraction 5 was rechromatographed 3 times on a silica gel column eluted with chloroform-methanol (19:1, v/v) to give several fractions. Some of the fractions were methylated and further fractionated using preparative TLC (benzene-ethyl acetate, 1:1 or 2:1, v/v), silica gel column chromatography (chloroform-methanol, 99:1, v/v), or reversed-phase LC (methanol-water, 7:3, v/v, Lobar RP_{18}). These separations gave purified methyl ganoderate M, N, O, H, E, methyl lucidenate K, L, and E_2.

Biological Activity
Fruiting bodies of *G. lucidum* produce some physiological effects when used as a home remedy.

Spectral Data (Methyl ester)

UV:
λ_{max}^{MeOH} 255nm (ε = 6,930).

CD:

λ^{MeOH}_{max} Δ_{345} 0, Δ_{307} - 7.1, Δ_{291} 0, Δ_{278} + 6.2, Δ_{264} 0, Δ_{256} - 2.1, Δ_{246} 0, Δ_{227} + 8.3, and Δ_{213} 0.

IR:

(thin film) 3450, 2920, 1740, and 1680cm^{-1}.

^{1}H NMR:

($CDCl_3$) 1.14(1H, ddd, *J*=13.6, 13.6, 5.8Hz, H-1α); 2.76(1H, ddd, *J*=13.6, 3.7, 3.5Hz, H-1β); 1.75(1H, m, H-2α); 1.75(1H, m, H-2β); 3.27(1H, dd, *J*=9.9, 6.2Hz, H-3α); 1.55(1H, dd, *J*=14.3, 2.6Hz, H-5α); 2.58(1H, dd, *J*=14.3, 2.6Hz, H-6α); 2.70(1H, dd, *J*=14.3, 14.3Hz, H-6β); 4.51(1H, s, H-12α); 2.74(1H, dd, *J*=18.0, 9.9Hz, H-16α); 2.06(1H, dd, *J*=18.0, 8.1Hz, H-16β); 2.55(1H, ddd, *J*=9.9, 8.1, 8.3Hz, H-17); 0.64(3H, s, H_3-18); 1.37(3H, s, H_3-19); 1.80(1H, m, H-20); 1.11(3H, d, *J*=7.0Hz, H_3-21); 1.80(1H, m, H-22); 1.30(1H, m, H-22); 2.41(1H, ddd, *J*=15.8, 9.2, 5.5Hz, H-23); 2.27(1H, ddd, *J*=15.8, 8.8, 7.0Hz, H-23); 1.04(3H, s, H_3-25); 0.90(3H, s, H_3-26); 1.69(3H, s, H_3-27); and 3.66ppm (3H, s, $COOCH_3$).

Mass Spectrum:

HREIMS: 488.2766(M^+, $C_{28}H_{40}O_7$, 10%; calcd 488.2775), 304(100), 191(21), 129(35), 95(21), 69(26), 55(43), and 43*m/e* (78).

Reference

T. Nishitoba, H. Sato, and S. Sakamura; Triterpenoids from the Fungus *Ganoderma lucidum*; Phytochemistry, Vol. 26, pp. 1777-1784(1987).

Common/Systematic Name

Lucidenic acid M

11-Oxo-3β,7α,15α-trihydroxy-5α-lanost-8-en-24-oic acid

Molecular Formula/Molecular Weight

$C_{27}H_{42}O_6$; MW = 462.29814

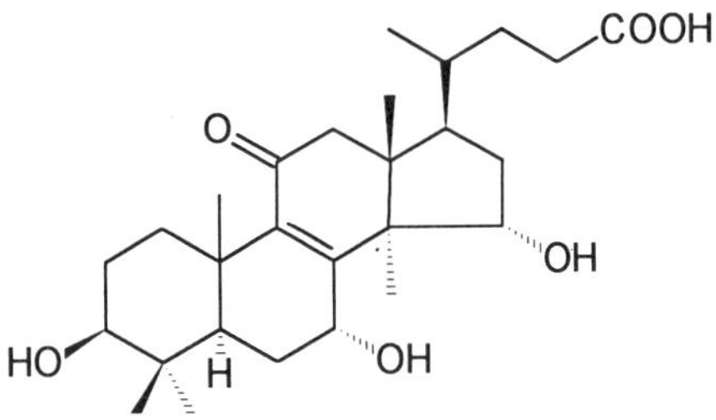

General Characteristics

Methyl ester: a syrup.

Fungal Source

Fruiting bodies of *Ganoderma lucidum*.

Isolation/Purification

Dried chipped fruiting bodies of *Ganoderma lucidum* were extracted with ethanol. The extract was concentrated and partitioned between chloroform and water. The acidic fraction was obtained and chromatographed on a silica gel column eluted with chloroform-methanol (49:1, v/v; fractions 1-11), chloroform-methanol (9:1, v/v; fraction 12) and methanol (fraction 13). Fraction 12 was subjected to reversed-phase LC (methanol-water, 13:7, v/v, Lobar RP_{18}) and separated into four fractions (12a-12d). Methyl ganoderate M was isolated from the methylated fraction 12d after a combination of silica gel column chromatography (chloroform-methanol, 9:1), preparative TLC (chloroform-acetone, 2:1, v/v) and reversed-phase HPLC (methanol-water, 3:2, v/v, μBondapak C_{18}).

Biological Activity

Fruiting bodies of *G. lucidum* produce some physiological effects when used as a home remedy.

Spectral Data (Methyl ester)

UV:

λ_{max}^{MeOH} 255nm (ϵ = 7,900).

^{1}H NMR:

(CDCl$_3$) 3.00(1H, ddd, J=13.2, 3.5, 3.5Hz, H-1β); 3.32(1H, dd, J=11.1, 4.7Hz, H-3α); 4.55(1H, m, H-6β); 2.76(1H, d, J=17.6Hz, H-12α); 2.40(1H, d, J=17.6Hz, H-12β); 4.58(1H, dd, J=19.9, 5.1Hz, H-15β); 0.84(3H, s, H$_3$-18); 1.06(3H, s, H$_3$-19); 0.85(3H, d, J=5.8Hz, H$_3$-21); 1.05(3H, s, H$_3$-25); 0.84(3H, s, H$_3$-26); 1.28(3H, s, H$_3$-27); and 3.67ppm (3H, s, COOCH$_3$).

Mass Spectrum:

HREIMS: 476.3105(M$^+$, $C_{28}H_{44}O_6$, 100%; calcd 476.3139), 458(37), 336(56), 236(53), 69(58), 55(92), and 43*m/e* (96).

Reference

T. Nishitoba, H. Sato, and S. Sakamura; Triterpenoids from the Fungus *Ganoderma lucidum;* Phytochemistry, Vol. 26, pp. 1777-1784(1987).

13

Lucidones

Lucidone A
Lucidone B
Lucidone C

Common/Systematic Name
Lucidone A

Molecular Formula/Molecular Weight
$C_{24}H_{34}O_5$; MW = 402.24062

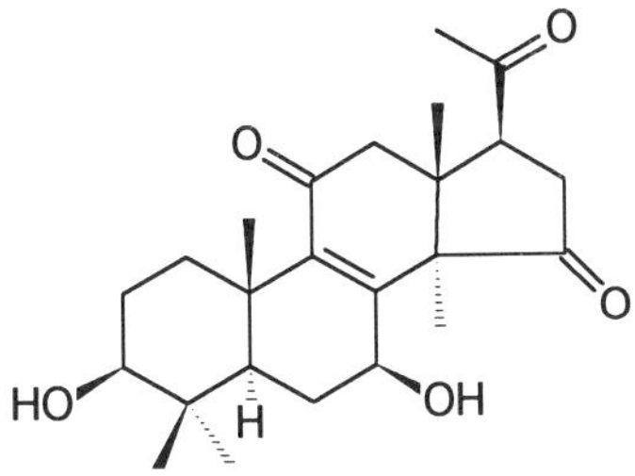

General Characteristics
Lucidone A was isolated as a pale yellow amorphous powder; mp., 280.3°C; $[\alpha]_D^{24}$ +210 (c=0.2, in $CHCl_3$).

Fungal Source
Fruiting bodies of *Ganoderma lucidum*.

Isolation/Purification
Dried chipped fruiting bodies of *G. lucidum* were extracted with chloroform and chromatographed on a silica gel column using various concentrations of methanol in chloroform. The bitter principles were further purified with a second silica gel column eluted in a like manner.

Biological Activity
Fruiting bodies of *G. lucidum* produce some physiological effects when used as a home remedy, also bitter tasting.

Spectral Data

UV:
λ_{max}^{EtOH} 256nm (ε=11,320).

IR:
(KBr) 3450, 2960, 2925, 1720(sh), 1715, and 1660cm^{-1}.

^{1}H NMR:
($CDCl_3$) 4.82(1H, dd, J=9.1 and 8.5 Hz); 3.21(1H, dd, J=11.0 and 5.2Hz); 3.02(1H, dq, J=16.2 and 0.7 Hz); 2.77(1H, d, J=16.2); 2.22(3H, s); 1.41(3H, s); 1.22(3H, s); 1.04(3H, s); 0.88(3H, s); and 0.85ppm (3H, s).

^{13}C NMR:

($CDCl_3$) 215.8 (s); 204.8 (s); 196.3 (s); 156.2 (s); 142.6 (s); 78.1 (d); 66.6 (d); 31.2 (q); 28.1 (q), 24.5 (q); 19.0 (q); 18.3 (q); and 15.3ppm (q).

These data closely resemble those of ganoderic acids and lucidenic acids, although lucidone A differs from them in having twenty-four carbons. The signals at 2.22(3H, s) in the ^{1}H NMR spectrum and 31.2 (q) in the ^{13}C NMR spectrum indicated the presence of an acetyl group at C(17) instead of the side chains of ganoderic acids and lucidenic acids.

Mass Spectrum:

The molecular formula was assigned as $C_{24}H_{34}O_5$ based on HRMS (M^+, 402.2428; calcd for $C_{24}H_{34}O_5$, 402.2407).

Reference

T. Nishitoba, H. Sato, and S. Sakamura; New Terpenoids from *Ganoderma lucidum* and Their Bitterness; Agric. Biol. Chem., Vol. 49, pp. 1547-1549(1985).

Common/Systematic Name

Lucidone B

7β-Hydroxy-4,4,14α-trimethyl-3,11,15,20-tetraoxo-5α-pregn-8-ene

Molecular Formula/Molecular Weight

$C_{24}H_{32}O_5$; MW = 400.22497

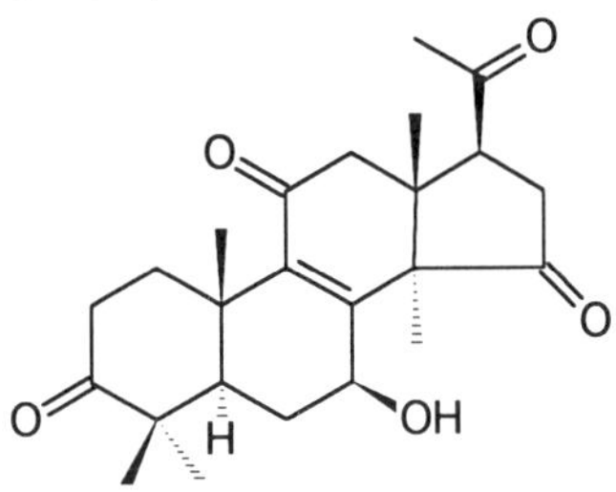

General Characteristics

Lucidone B was crystallized from EtOAc as colorless prisms; mp., 270-271°C; $[\alpha]_D^{22}$ +278° (c=0.2, in $CHCl_3$).

Fungal Source

Fruiting bodies of *Ganoderma lucidum*.

Isolation/Purification

Dried chipped fruiting bodies of *G. lucidum* were extracted with chloroform and chromatographed on a silica gel column using various concentrations of methanol in chloroform. The bitter principles were further purified with a second silica gel column eluted in a like manner.

Biological Activity

Fruiting bodies of *G. lucidum* produce some physiological effects when used as a home remedy, also bitter tasting. The intensity of bitterness was evaluated to be lucidenic acid D, ganoderic acid C, lucidone A, lucidenic acid A, ganoderic acid B, lucidenic acid B, lucidenic acid C, and lucidenic acid E. Lucidenic acid A, ganoderic acid C, lucidenic acid D, and lucidone A had intense bitterness and their threshold values were determined to be 2.0×10^{-6}M, 5×10^{-8}M, 5×10^{-10}M, and 1×10^{-6}M, respectively, when the organoleptic test was carried out by an ascending series of concentrations in a 10% ethanol solution.

Spectral Data

UV:

λ_{max}^{MeOH} 255nm (ε=7,900).

IR:

(KBr) 3500, 2960, 1730, 1705, and 1660cm^{-1}.

^{1}H NMR and ^{13}C NMR:

The ^{1}H and ^{13}C NMR spectra of lucidone B were very similar to those of lucidone A, but the ^{1}H-NMR data for lucidone B did not show any signal around 3.21(H-3) of lucidone A. The ^{13}C-NMR data also did not show any signal around 78.1(C-3) of lucidone A], but gave a new signal at 216.2. These facts indicated the presence of a carbonyl group at C-3 instead of the hydroxy group. Thus, lucidone B was confirmed to be 7β-hydroxy-4,4,14α-trimethyl-3,11,15,20-tetraoxo-5α-pregn-8-ene.

Mass Spectrum:

It had a molecular formula $C_{24}H_{32}O_5$ from HRMS data (M^+, 400.2267; calcd for $C_{24}H_{32}O_5$, 400.2251).

Reference

T. Nishitoba, H. Sato, and S. Sakamura; New Terpenoids from *Ganoderma lucidum* and Their Bitterness; Agric. Biol. Chem., Vol. 49, pp. 1547-1549(1985).

Common/Systematic Name

Lucidone C

3β,7β,15α-Trihydroxy-4,4,14α-trimethyl-11,20-dioxo-5α-preg-8-ene

Molecular Formula/Molecular Weight

$C_{24}H_{36}O_5$; MW = 404.25627

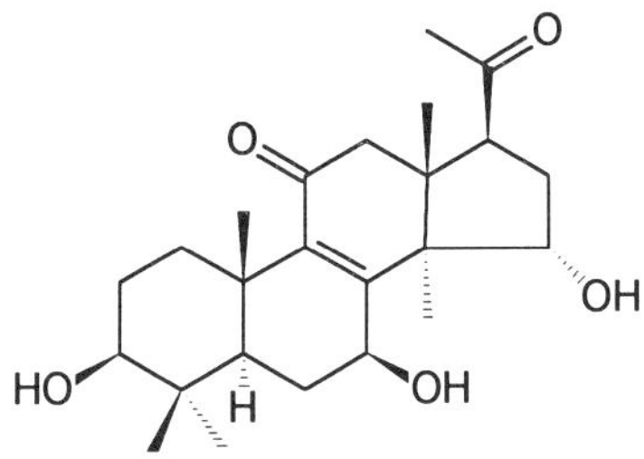

General Characteristics

Colorless syrup; $[\alpha]_D^{24}$ + 145° (*c*=0.2, in MeOH).

Fungal Source

Fruiting bodies of *Ganoderma lucidum*.

Isolation/Purification

Dried chipped fruiting bodies of *G. lucidum* were extracted with ethanol; the extract was partitioned between chloroform and water. The chloroform layer was concentrated and separated into its acidic and neutral fractions. The acidic fraction was separated into thirteen fractions. Fraction 12 was subjected to Lobar column (RP_8, Merck) chromatography and the second fraction was treated with diazomethane. The resulting product was rechromatographed on silica gel and the Lobar (RP_8) column to give methyl ganoderate L, lucidone C, and methyl lucidenate G. The fourth fraction in the chromatography of Fr. 12 was purified on a silica gel column, preparative TLC and HPLC to give ganolucidic acid D.

Biological Activity

Fruiting bodies of *G. lucidum* produce some physiological effects when used as a home remedy.

Spectral Data

UV:

λ_{max}^{MeOH} 255nm (ϵ=7680).

IR:

(KBr) 3500(sh), 3430, 1700, and 1660cm^{-1}.

^{1}H NMR:
(C_5D_5N) 5.34(1H, dd, *J*=9.5 and 7.3 Hz); 4.98(1H, dd, *J*=9.5 and 7.3Hz); 3.51(1H, dd, *J*=10.8 and 5.3Hz); 2.11(3H, s); 1.59(3H, s); 1.51(3H, s), 1.29(3H, s); 1.11(3H, s); and 1.07ppm (3H, s).

^{13}C NMR:
(C_5D_5N) 207.6(0); 198.9(0); 159.9(0); 141.9(0); 77.5(1); 72.4(1); and 69.4ppm (1).

Mass Spectrum:
HREIMS: 404.2527*m/e* (M^+).

Reference

T. Nishitoba, H. Sato, and S. Sakamura; New Terpenoids, Ganolucidic Acid D, Ganoderic Acid L, Lucidone C, and Lucidenic Acid G from the Fungus *Ganoderma lucidum*; Agric. Biol. Chem., Vol. 50, pp. 809-811(1986).

14

Azasterols

Antibiotic A25822A
Antibiotic A25822B
Antibiotic A25822D
Antibiotic A25822L
Antibiotic A25822M
Antibiotic A25822N
Antibiotic A25822H

Common/Systematic Name

Antibiotic A25822A

4,4-Dimethyl-3β-hydroxy-24-methylene-14α-aza-*D*-homo-5α-cholesta-8(9),14(14α)-diene

Molecular Formula/Molecular Weight

$C_{30}H_{49}NO$; MW = 439.38142

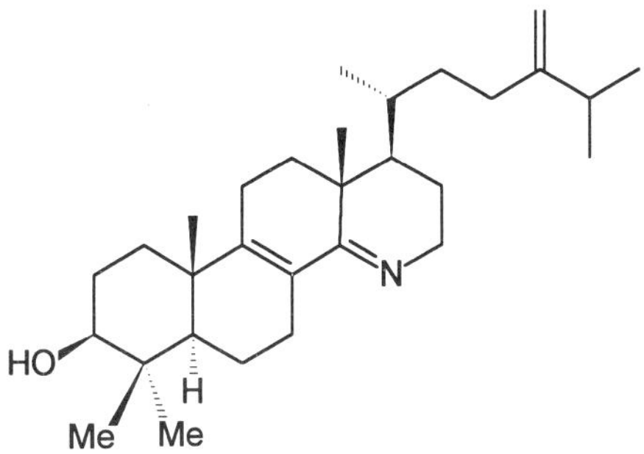

General Characteristics

Crystals; mp., 147°C; $[\alpha]_D^{25}$ - 72° (*c*=1.15, in MeOH). $pK_{a'}$ = 8.0 [66% DMF]. Dihydro-diacetate derivative: crystals; mp., 130-132°C.

Fungal Source

Geotrichum flavo-brunneum (NRRL 3862).

Isolation/Purification

The mycelium of *Geotrichum flavo-brunneum* was extracted with methanol, concentrated, adjusted to pH 10.0, and extracted with ethyl acetate. The organic phase was concentrated, partitioned between 2.5N aqueous sodium hydroxide and ethyl ether, washed with water, and extracted with 0.25N sulfuric acid. The aqueous acidic phase was adjusted to pH 10.0 with sodium hydroxide and extracted with *n*-hexane (extract A) followed by ethyl acetate (extract B). Extract A contained antibiotic A25822A, B, M, and N. Extract B contained antibiotics A25822D, H, and L. Extract A was dissolved in acetone (50ml) and *n*-hexane (150ml) and the clear solution was adjusted to one liter with *n*-hexane. Antibiotic A25822A crystallized in small white needles. The filtrate was combined with extract B and reduced in volume to give extract C. Extract C was dissolved in a minimal volume of ethyl acetate-*n*-hexane-water (80:16:4, v/v/v) and chromatographed on a column of basic alumina eluted with the same solvent combination. The later fractions containing antibiotic A25822D were dissolved in chloroform and chromatographed on a silica gel column eluted with a linear gradient from chloroform to chloroform-40% methanol. The antibiotic A25822D containing fractions were combined and concentrated to yield the antibiotic as a dry foam of amorphous material that could not be crystallized.

Biological Activity

Broad spectrum antifungal activity.

Spectral Data

UV:
λ_{max}^{EtOH} 239nm (ε=12,600).

IR:
($CHCl_3$) 3584(OH) and 1621cm^{-1}.

Mass Spectrum:
HREIMS: 439.3776*m/e* (M^+).

References

J. W. Chamberlin, M. O. Chaney, S. Chen, P. V. Demarco, N. D. Jones, and J. L. Occolowitz; Structure of Antibiotic A25822 B, A Novel Nitrogen-containing C_{28}-Sterol With Antifungal Properties; J. Antibiotics, Vol. 27, pp. 992-993(1974).

W. B. Turner and D. C. Aldridge; Fungal Metabolites II; Academic Press, New York, p. 229(1983).

Common/Systematic Name

Antibiotic A25822B

3β-Hydroxy-24-methylene-14α-aza-*D*-homo-5α-cholesta-8(9),14(14α)-diene

Molecular Formula/Molecular Weight

$C_{28}H_{45}NO$; MW = 411.35012

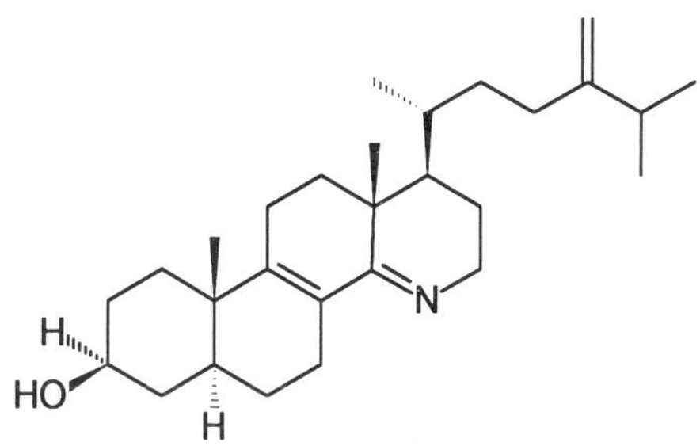

General Characteristics

Crystals; mp., 115-118° C; $[\alpha]_D^{25}$ - 20° (*c*=0.775, in MeOH). $pK_{a'}$ = 8.4 [66% DMF]. Dihydro-diacetate derivative: crystals; mp., 130-132°C.

Fungal Source

Geotrichum flavo-brunneum (NRRL 3862)

Isolation/Purification

The mycelium of *Geotrichum flavo-brunneum* was extracted with methanol, concentrated, adjusted to pH 10.0, and extracted with ethyl acetate. The organic phase was concentrated, partitioned between 2.5N aqueous sodium hydroxide and ethyl ether, washed with water, and extracted with 0.25N sulfuric acid. The aqueous acidic phase was adjusted to pH 10.0 with sodium hydroxide and extracted with *n*-hexane (extract A) followed by ethyl acetate (extract B). Extract A contained antibiotic A25822A, B, M, and N. Extract B contained antibiotics A25822D, H, and L. Extract A was dissolved in acetone (50ml) and *n*-hexane (150ml) and the clear solution was adjusted to one liter with *n*-hexane. Antibiotic A25822A crystallized in small white needles. The filtrate was combined with extract B and reduced in volume to give extract C. Extract C was dissolved in a minimal volume of ethyl acetate-*n*-hexane-water (80:16:4, v/v/v) and chromatographed on a column of basic alumina eluted with the same solvent combination. The later fractions containing antibiotic A25822D were dissolved in chloroform and chromatographed on a silica gel column eluted with a linear gradient from chloroform to chloroform-40% methanol. The antibiotic A25822D containing fractions were combined and concentrated to yield the antibiotic as a dry foam of amorphous material that could not be crystallized.

Biological Activity

Broad spectrum antifungal activity, primarily against *Candida* and *Trichophyton*.

Spectral Data

UV:
λ_{max}^{EtOH} 238nm (ϵ=12,300); shift to 277nm (ϵ=12,400) in acidic solution.

IR:
($CHCl_3$) 3571 and 1618cm^{-1}.

Mass Spectrum:
HREIMS: 411.3497*m/e* (M^+).

References

L. D. Boeck, M. M. Hoehn, J. E. Westhead, R. K. Wolter, and D. N. Thomas, New Azasteroidal Antifungal Antibiotics from *Geotrichum flavo-brunneum*. I. Discovery and Fermentation Studies; J. Antibiotics, Vol. 28, pp. 95-101(1975).

J. W. Chamberlin, M. O. Chaney, S. Chen, P. V. Demarco, N. D. Jones, and J. L. Occolowitz; Structure of Antibiotic A25822 B, A Novel Nitrogen-containing C_{28}-Sterol with Antifungal Properties; J. Antibiotics; Vol. 27, pp. 992-993(1974).

W. B. Turner and D. C. Aldridge; Fungal Metabolites II; Academic Press, New York, p. 229 (1983).

Common/Systematic Name

Antibiotic A25822D

3β,15-Dihydroxy-24-methylene-14α-aza-*D*-homo-5α-cholesta-8(9),14(14α)-diene

Molecular Formula/Molecular Weight

$C_{28}H_{45}NO_2$; MW = 427.34503

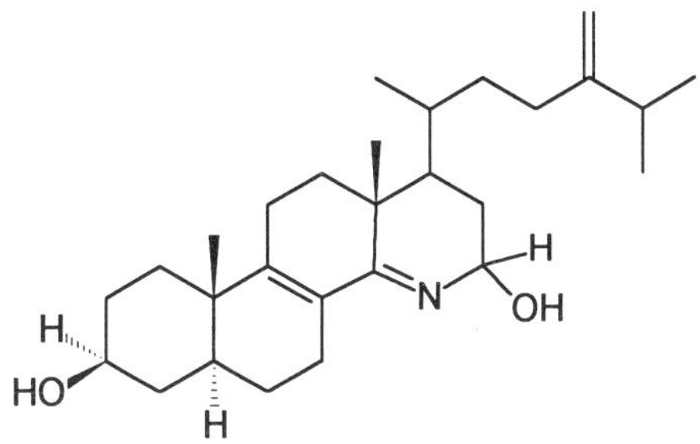

General Characteristics

An amorphous material that could not be crystallized; $[\alpha]_D^{25}$ + 39° (*c*=0.722, in MeOH). pK_a 7.7 (66% DMF).

Fungal Source

Geotrichum flavo-brunneum.

Isolation/Purification

The mycelium of *Geotrichum flavo-brunneum* was extracted with methanol, concentrated, adjusted to pH 10.0, and extracted with ethyl acetate. The organic phase was concentrated, partitioned between 2.5N aqueous sodium hydroxide and ethyl ether, washed with water, and extracted with 0.25N sulfuric acid. The aqueous acidic phase was adjusted to pH 10.0 with sodium hydroxide and extracted with *n*-hexane (extract A) followed by ethyl acetate (extract B). Extract A contained antibiotic A25822A, B, M, and N. Extract B contained antibiotics A25822D, H, and L. Extract A was dissolved in acetone (50ml) and *n*-hexane (150ml) and the clear solution was adjusted to one liter with *n*-hexane. Antibiotic A25822A crystallized in small white needles. The filtrate was combined with extract B and reduced in volume to give extract C. Extract C was dissolved in a minimal volume of ethyl acetate-*n*-hexane-water (80:16:4, v/v/v) and chromatographed on a column of basic alumina eluted with the same solvent combination. The later fractions containing antibiotic A25822D were dissolved in chloroform and chromatographed on a silica gel column eluted with a linear gradient from chloroform to chloroform-40% methanol. The antibiotic A25822D containing fractions were combined and concentrated to yield the antibiotic as a dry foam of amorphous material that could not be crystallized.

Biological Activity

Antibiotic activity against a broad spectrum of pathogenic fungi, including *Candida* spp. and *Trichophyton* spp.

Spectral Data

UV:

λ^{EtOH}_{max} 235nm (neutral and basic pH, ϵ = 10,400); 270nm (acidic pH, ϵ = 8,600).

IR:

($CHCl_3$) 3600(OH), 1640(-C=C-, -C=N-), and 890cm^{-1} (exocyclic methylene).

Mass Data:

LREIMS: 427(M^+). Found: C, 78.40; H, 10.28; N; 3.17, O, 8.06%; calcd for $C_{28}H_{45}NO_2$: C, 78.40; H, 10.54; N, 3.28; O, 7.49%.

TLC Data

Silica gel plates developed with diethyl ether-ethanol (3:1, v/v); R_f = 0.27.

GC Data

3% OV-17 coated onto Chromosorb W; column temperature 275°C; flash heater 285°C; flow rates, hydrogen 47ml/min, helium 75ml/min, and air 350ml/min; relative retention time of 8.8 (relative to cholesterol).

References

K. H. Michel, R. L. Hamill, S. H. Larsen, and R. H. Williams; New Azasteroidal Antifungal Antibiotics from *Geotrichum flavo-brunneum*; J. of Antibiotics, Vol. 28, pp. 102-111(1975).

W. B. Turner and D. C. Aldridge; Fungal Metabolites II; Academic Press, New York, pp. 329 and 332(1983).

Common/Systematic Name

Antibiotic A25822L

3β-Hydroxy-24-methylene-14α-aza-*D*-homo-5α-cholesta-8(9),14(14α)-dien-15-one

Molecular Formula/Molecular Weight

$C_{28}H_{43}NO_2$; MW = 425.42938

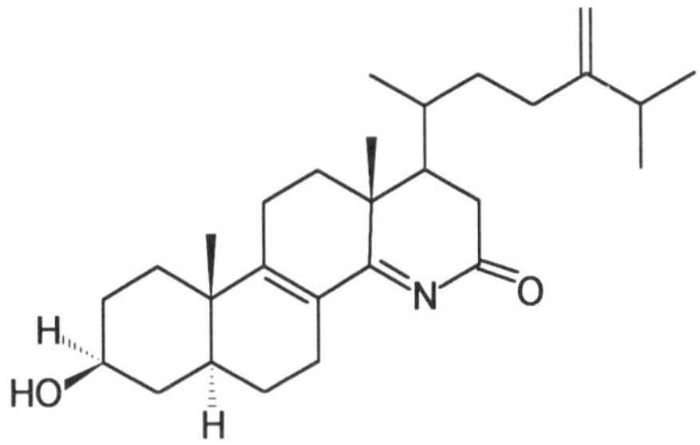

General Characteristics

An amorphous powder that could not be crystallized; $[\alpha]_D^{25}$ + 75° (*c*=0.072, in MeOH). pK_a 4.75 (66% DMF).

Fungal Source

Geotrichum flavo-brunneum.

Isolation/Purification

The mycelium of *Geotrichum flavo-brunneum* was extracted with methanol, concentrated, adjusted to pH 10.0, and extracted with ethyl acetate. The organic phase was concentrated, partitioned between 2.5N aqueous sodium hydroxide and ethyl ether, washed with water, and extracted with 0.25N sulfuric acid. The aqueous acidic phase was adjusted to pH 10.0 with sodium hydroxide and extracted with *n*-hexane (extract A) followed by ethyl acetate (extract B). Extract A contained antibiotic A25822A, B, M, and N. Extract B contained antibiotics A25822D, H and L. Extract A was dissolved in acetone (50 ml) and *n*-hexane (150ml) and the clear solution was adjusted to one liter with *n*-hexane. Antibiotic A25822A crystallized in small white needles. The filtrate was combined with extract B and reduced in volume to give extract C. Extract C was dissolved in a minimal volume of ethyl acetate-*n*-hexane-water (80:16:4, v/v/v) and chromatographed on a column of basic alumina eluted with the same solvent combination. The later fractions containing antibiotic A25822D were dissolved in chloroform and chromatographed on a silica gel column eluted with a linear gradient from chloroform to chloroform-40% methanol. The antibiotic A25822L containing fractions were dissolved in a minimal amount of chloroform and chromatographed on a Woelm activity III column eluted with chloroform and the A25822L containing fractions concentrated to yield the antibiotic as an amorphous powder.

Biological Activity

Antibiotic activity against a broad spectrum of pathogenic fungi, including *Candida* spp. and *Trichophyton* spp.

Spectral Data

UV:

λ_{max}^{EtOH} 262nm (neutral and basic pH, ϵ = 10,200); 278nm (acidic pH, ϵ = 10,800).

IR:

($CHCl_3$) 3620(OH), 1670-1685(amide), and 890cm^{-1} (exocyclic methylene).

ORD:

$[\alpha]_{360}^{25}$ -270°, $[\alpha]_{272}^{25}$ +7550°, $[\alpha]_{235}^{25}$ -3500°, and $[\alpha]_{206}^{25}$ +16,700° (*c*=0.072, in MeOH).

Mass Data:

LREIMS: 425(M^+); found: C, 78.66; H, 9.74; N; 3.06, O, 8.37%; calcd for $C_{28}H_{43}NO_2$: C, 79.01; H, 10.18; N, 3.29; O, 7.52%.

TLC Data

Silica gel plates developed with diethyl ether-ethanol (3:1, v/v); R_f = 0.67.

GC Data

3% OV-17 coated onto Chromosorb W; column temperature 275°C; flash heater 285°C; flow rates, hydrogen 47ml/min, helium 75ml/min, and air 350ml/min; relative retention time of 8.4 (relative to cholesterol).

References

K. H. Michel, R. L. Hamill, S. H. Larsen, and R. H. Williams; New Azasteroidal Antifungal Antibiotics from *Geotrichum flavo-brunneum*; J. of Antibiotics, Vol. 28, pp. 102-111(1975).

W. B. Turner and D. C. Aldridge; Fungal Metabolites II; Academic Press, New York, pp. 329 and 332(1983).

Common/Systematic Name

Antibiotic A25822M

3β-Acetoxy-24-methylene-14α-aza-*D*-homo-5α-cholesta-8(9),14(14α)-diene

Molecular Formula/Molecular Weight

$C_{30}H_{47}NO_2$; MW = 453.36068

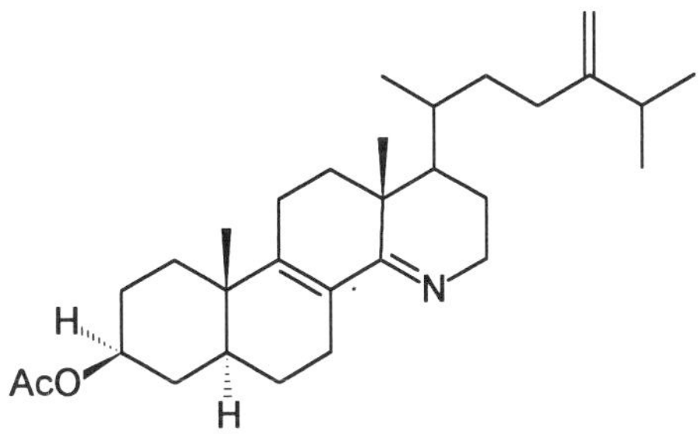

General Characteristics

A yellowish amorphous powder; $[\alpha]_D^{25}$ - 15° (*c*=0.021, in MeOH). pK_a 7.9 (66% DMF).

Fungal Source

Geotrichum flavo-brunneum.

Isolation/Purification

The mycelium of *Geotrichum flavo-brunneum* was extracted with methanol, concentrated, adjusted to pH 10.0, and extracted with ethyl acetate. The organic phase was concentrated, partitioned between 2.5N aqueous sodium hydroxide and ethyl ether, washed with water, and extracted with 0.25N sulfuric acid. The aqueous acidic phase was adjusted to pH 10.0 with sodium hydroxide and extracted with *n*-hexane (extract A) followed by ethyl acetate (extract B). Extract A contained antibiotic A25822A, B, M, and N. Extract B contained antibiotics A25822D, H, and L. Extract A was dissolved in acetone (50ml) and *n*-hexane (150ml) and the clear solution was adjusted to one liter with *n*-hexane. Antibiotic A25822A crystallized in small white needles. The filtrate was combined with extract B and reduced in volume to give extract C. Extract C was dissolved in a minimal volume of ethyl acetate-*n*-hexane-water (80:16:4, v/v/v) and chromatographed on a column of basic alumina eluted with the same solvent combination. The antibiotic A25822M and A25822N containing fraction was dissolved in acetone and placed at -20°C that formed a white crystalline precipitate of A25822N. The filtrate was reduced to an oily residue which was dissolved in a minimal amount of chloroform and chromatographed on a column containing Woelm basic alumina deactivated with 6% water and eluted with chloroform. The antibiotic A25822M containing fraction was dissolved in benzene and lyophilized to give a yellowish amorphous powder that could not be crystallized.

Biological Activity

Antibiotic activity against a broad spectrum of pathogenic fungi, including *Candida* spp. and *Trichophyton* spp.

Spectral Data

UV:

λ^{EtOH}_{max} 239nm (neutral and basic pH, ϵ = 13,900); 278nm (acidic pH, ϵ = 14,200).

IR:

($CHCl_3$) 1730(carbonyl), 1625(-C=C-, -C=N-), and 890cm^{-1} (exocyclic methylene).

Mass Data:

LREIMS: 453(M^+); found: C, 79.57; H, 10.26; N; 2.92, O, 7.21%; calcd for $C_{30}H_{47}NO_2$: C, 79.42; H, 10.44; N, 3.09; O, 7.05%.

TLC Data

Silica gel plates developed with diethyl ether-ethanol (3:1, v/v); R_f = 0.42.

GLC Data

3% OV-17 coated onto Chromosorb W; column temperature 275°C; flash heater 285°C; flow rates, hydrogen 47ml/min, helium 75ml/min, and air 350ml/min; relative retention time of 9.2 (relative to cholesterol).

References

K. H. Michel, R. L. Hamill, S. H. Larsen, and R. H. Williams; New Azasteroidal Antifungal Antibiotics from *Geotrichum flavo-brunneum*; J. of Antibiotics, Vol. 28, pp. 102-111(1975).

W. B. Turner and D. C. Aldridge; Fungal Metabolites II; Academic Press, New York, pp. 329 and 332(1983).

Common/Systematic Name

Antibiotic A25822N

3-Keto-24-methylene-14α-aza-*D*-homo-5α-cholesta-8(9),14(14α)-diene

Molecular Formula/Molecular Weight

$C_{28}H_{43}NO$; MW = 409.33447

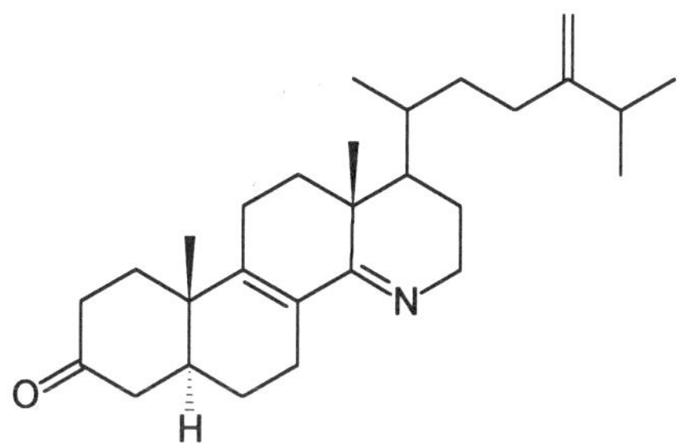

General Characteristics

A white crystalline precipitate; mp., 165°C; $[\alpha]_D^{25}$ -14° (*c*=0.05, in MeOH). pK_a 7.45 (66% DMF).

Fungal Source

Geotrichum flavo-brunneum.

Isolation/Purification

The mycelium of *Geotrichum flavo-brunneum* was extracted with methanol, concentrated, adjusted to pH 10.0, and extracted with ethyl acetate. The organic phase was concentrated, partitioned between 2.5N aqueous sodium hydroxide and ethyl ether, washed with water, and extracted with 0.25N sulfuric acid. The aqueous acidic phase was adjusted to pH 10.0 with sodium hydroxide and extracted with *n*-hexane (extract A) followed by ethyl acetate (extract B). Extract A contained antibiotic A25822A, B, M, and N. Extract B contained antibiotics A25822D, H, and L. Extract A was dissolved in acetone (50ml) and *n*-hexane (150ml) and the clear solution was adjusted to one liter with *n*-hexane. Antibiotic A25822A crystallized in small white needles. The filtrate was combined with extract B and reduced in volume to give extract C. Extract C was dissolved in a minimal volume of ethyl acetate-*n*-hexane-water (80:16:4, v/v/v) and chromatographed on a column of basic alumina eluted with the same solvent combination. The antibiotic A25822M and A25822N containing fraction was dissolved in acetone and placed at -20°C that formed a white crystalline precipitate of A25822N. The filtrate was reduced to an oily residue which was dissolved in a minimal amount of chloroform and chromatographed on a column containing Woelm basic alumina deactivated with 6% water and eluted with chloroform. The antibiotic A25822M containing fraction was dissolved in benzene and lyophilized to give a yellowish amorphous powder that could not be crystallized.

Biological Activity

Antibiotic activity against a broad spectrum of pathogenic fungi, including *Candida* spp. and *Trichophyton* spp.

Spectral Data

UV:

λ^{EtOH}_{max} 239nm (neutral and basic pH, ϵ = 13,900); 278nm (acidic pH, ϵ = 14,100).

IR:

($CHCl_3$) 1715(carbonyl), 1620(-C=C-, -C=N-), and 890cm^{-1} (exocyclic methylene).

Mass Data:

LREIMS: 409(M^+); found: C, 81.82; H, 10.61; N; 3.53, O, 3.91%; calcd for $C_{28}H_{43}NO$: C, 82.16; H, 10.52; N, 3.43; O, 3.89%.

TLC Data

Silica gel plates developed with diethyl ether-ethanol (3:1, v/v); R_f = 0.40.

GLC Data

3% OV-17 coated onto Chromosorb W; column temperature 275°C; flash heater 285°C; flow rates, hydrogen 47ml/min, helium 75ml/min, and air 350ml/min; relative retention time of 7.1 (relative to cholesterol).

References

K. H. Michel, R. L. Hamill, S. H. Larsen, and R. H. Williams; New Azasteroidal Antifungal Antibiotics from *Geotrichum flavo-brunneum*; J. of Antibiotics, Vol. 28, pp. 102-111(1975).

W. B. Turner and D. C. Aldridge; Fungal Metabolites II; Academic Press, New York, pp. 329 and 332 (1983).

Common/Systematic Name

Antibiotic A25822H (isomeric with Antibiotic A25822L)

Molecular Formula/Molecular Weight

$C_{28}H_{43}NO_2$; MW = 425.32938

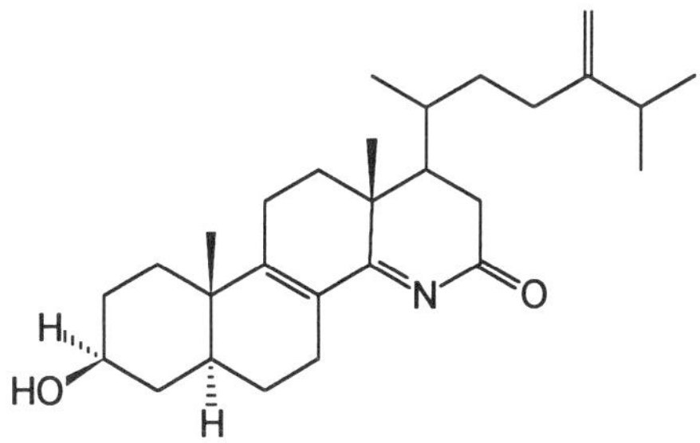

General Characteristics

An amorphous powder that could not be crystallized; $[\alpha]_D^{25}$ + 15° (*c*=0.147, in MeOH). pK_a 6.9 (66% DMF).

Fungal Source

Geotrichum flavo-brunneum.

Isolation/Purification

The mycelium of *Geotrichum flavo-brunneum* was extracted with methanol, concentrated, adjusted to pH 10.0, and extracted with ethyl acetate. The organic phase was concentrated, partitioned between 2.5N aqueous sodium hydroxide and ethyl ether, washed with water, and extracted with 0.25N sulfuric acid. The aqueous acidic phase was adjusted to pH 10.0 with sodium hydroxide and extracted with *n*-hexane (extract A) followed by ethyl acetate (extract B). Extract A contained antibiotic A25822A, B, M, and N. Extract B contained antibiotics A25822D, H, and L. Extract A was dissolved in acetone (50ml) and *n*-hexane (150ml) and the clear solution was adjusted to one liter with *n*-hexane. Antibiotic A25822A crystallized in small white needles. The filtrate was combined with extract B and reduced in volume to give extract C. Extract C was dissolved in a minimal volume of ethyl acetate-*n*-hexane-water (80:16:4, v/v/v) and chromatographed on a column of basic alumina eluted with the same solvent combination. The fractions containing antibiotic A25822B and H were dissolved in chloroform and chromatographed on a Woelm basic alumina column eluted with chloroform. The antibiotic A25822H containing fractions were dissolved in a minimal amount of acetone and chromatographed on preparative silica gel F-254 plates developed with diethyl ether-ethanol (3:1) followed by silica gel column chromatography eluted with ethyl acetate to yield the antibiotic as a white amorphous powder.

Biological Activity

Antibiotic activity against a broad spectrum of pathogenic fungi, including *Candida* spp. and *Trichophyton* spp.

Spectral Data

UV:

λ_{max}^{EtOH} 235nm (neutral and basic pH, ϵ = 10,000); 270nm (acidic pH, ϵ = 8,700).

IR:

($CHCl_3$) 3550(OH), 1625(-C=C-, -C=N-), and 890cm^{-1} (exocyclic methylene).

Mass Data:

LREIMS: 443($M^+ \bullet H_2O$, 425*m/e* + H_2O); found: C, 74.50; H, 10.15; N; 3.54, O, 11.36%; calcd for $C_{28}H_{43}NO_2 + H_2O$: C, 75.84; H, 10.20; N, 3.16; O, 10.80%.

TLC Data

Silica gel plates developed with diethyl ether-ethanol (3:1, v/v); R_f = 0.54.

GLC Data

3% OV-17 coated onto Chromosorb W; column temperature 275°C; flash heater 285°C; flow rates, hydrogen 47ml/min, helium 75ml/min, and air 350ml/min; relative retention time of 8.4 (relative to cholesterol).

References

K. H. Michel, R. L. Hamill, S. H. Larsen, and R. H. Williams; New Azasteroidal Antifungal Antibiotics from *Geotrichum flavo-brunneum*; J. of Antibiotics, Vol. 28, pp. 102-111(1975).

W. B. Turner and D. C. Aldridge; Fungal Metabolites II; Academic Press, New York, pp. 329 and 332 (1983).

15

Fasciculols

Fasciculol A
Fasciculol B
Fasciculol C
Fasciculol C depsipeptide E
Fasciculol C depsipeptide F
Fasciculol B depsipeptide

Common/Systematic Name

Fasciculol A

2α,3β,24(*R*),25-Tetrahydroxylanost-8-ene

Molecular Formula/Molecular Weight

$C_{30}H_{52}O_4$; MW = 476.38656

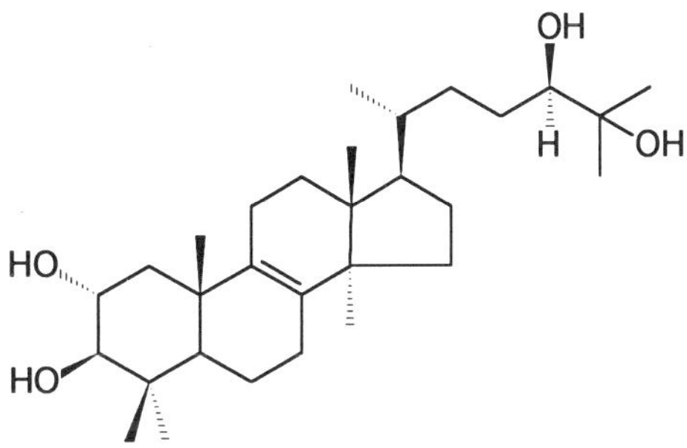

General Characteristics

Crystals; mp., 209°C; $[\alpha]_D^{21}$ +50.7° (*c*=1.0, in MeOH). Triacetate ($C_{36}H_{58}O_7$); mp., 177°C; $[\alpha]_D^{21}$ +7.7 (*c*=1.0, in MeOH).

Fungal Source

Fruit bodies of *Neamatoloma fasciculare*.

Isolation/Purification

Fruit bodies were extracted with ethyl acetate and the neutral fraction was applied to a silica gel column (chloroform-ethanol, 100:4, v/v), followed by preparative TLC with Kieselgel GF_{254} plates and chloroform-ethanol, 10:1 (v/v) as developing solvent.

Also see R. C. Heupel; Isolation and Primary Characterization of Sterols, In Analysis of Sterols and Other Biologically Significant Steroids; W. D. Nes and E. J. Parish (eds.), Academic Press Inc., New York, NY, 341 pp. (1989).

Biological Activity

Inhibited the growth of Chinese cabbage seedlings.

Spectral Data

UV:

λ_{max}^{MeOH} 207(ϵ = 5,700).

IR:
(KBr) 3380cm^{-1} (no carbonyl bands); triacetate 3560 and 1736cm^{-1}

^{1}H NMR:
(pyridine-d_5) 0.77; 0.90; 1.08; 1.11; 1.22; (each 3H, s); 0.93; (3H, d, *J*=6Hz); 1.46(6H, s); 3.30(1H, d, *J*=10Hz); 3.62(1H, br s); and 3.96ppm (1H, td, *J*=10, 10, 4.5Hz). Triacetate: ($CDCl_3$) eight methyl signals in the region from 0.72 to 1.20ppm; three acetyl methyl signals at 1.99, 2.01, and 2.10; acetoxy methine protons at 5.14(1H, td, *J*=10, 10, 4.5ppm); 4.72(1H, d, *J*=10Hz) and 5.0-4.7ppm (1H, m, overlapping with other signals)..

Mass Spectrum:
LREIMS: 476(M$^+$), 331, 289, 215, 187, 173, and 145*m/e*.

TLC Data

Kieselgel GF_{254} plates were developed with chloroform-ethanol, 10:1 (v/v). Fasciculol A had an R_f of 0.85 and was detected by spraying plates with 0.5% vanillin in concentrated H_2SO_4, followed by heating at 120°C.

References

M. Ikeda, Y. Sato, M. Izawa, T. Sassa, and Y. Miura; Isolation and Structure of Fasciculol A, a New Plant Growth Inhibitor from *Neamatoloma fasciculare*; Agric. Biol. Chem., Vol. 41, pp. 1539-1541(1977).

W. D. Nes and E. J. Parish (eds.); Analysis of Sterols and Other Biologically Significant Steroids; Academic Press Inc. New York, NY, 341 pp. (1989).

W. B. Turner and D. C. Aldridge; Fungal Metabolites II; Academic Press, Inc., New York, NY, p. 319 (1983).

J. D. Weete; Structure and Function of Sterols in Fungi, Advances in Lipid Research, Vol. 23, pp. 115-167(1989).

Common/Systematic Name
Fasciculol B
2α,3β,12α,24(*R*),25-Pentahydroxylanost-8-ene

Molecular Formula/Molecular Weight
$C_{30}H_{52}O_5$; MW = 492.38148

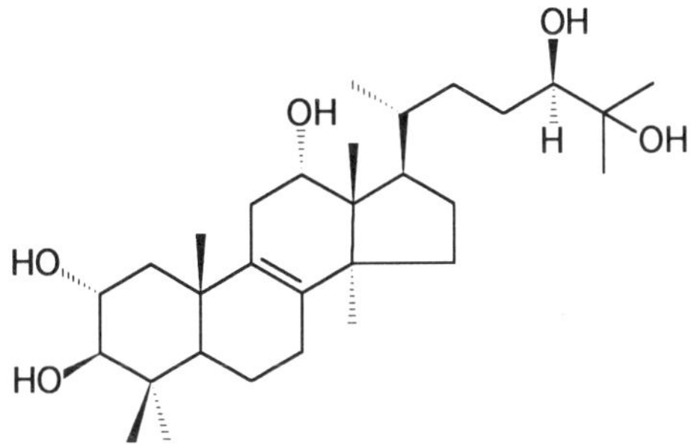

General Characteristics
Crystals; mp., 235°C; $[\alpha]_D^{21}$ +76.5° (*c*=1.0, in MeOH). Positive Liebermann-Burchard reaction and tetranitromethane test. Triacetate ($C_{36}H_{56}O_8$); mp., 191°C.

Fungal Source
Fruit bodies of *Neamatoloma fasciculare*.

Isolation/Purification
Fruit bodies were extracted with ethyl acetate and the neutral fraction was applied to a silica gel column (chloroform-ethanol, 100:7, v/v). Further purification and separation of fasciculol B from fasciculol D was achieved by silica gel column chromatography by elution with ethyl acetate-ethanol (100:4, v/v) for fasciculol B and (100:13, v/v) for fasciculol D. Final purification was by preparative TLC with Kieselgel GF_{254} plates and chloroform-ethanol, 20:3 (v/v) as developing solvent.

Also see R. C. Heupel; Isolation and Primary Characterization of Sterols, In Analysis of Sterols and Other Biologically Significant Steroids; W. D. Nes and E. J. Parish (eds.), Academic Press Inc., New York, NY, 341 pp. (1989).

Biological Activity
Inhibited the growth of Chinese cabbage seedlings.

Spectral Data

UV:
λ_{max}^{MeOH} 207(ϵ = 6,000).

IR:
(KBr) 3400cm^{-1} (no carbonyl bands); triacetate 3560, and 1736cm^{-1}

^{1}H NMR:
(pyridine-d_5) 0.78; 1.13; 1.16; 1.26; 1.43; (each 3H, s); 1.39; (3H, d, *J*=6Hz); 1.48(6H, s); 3.34(1H, d, *J*=10Hz); 3.74 (1H, br s); 4.09(1H, td, *J*=10, 10, 4.5Hz); and 4.22ppm (1H, d, *J*=6.0Hz). Triacetate: ($CDCl_3$) Three acetoxy methine protons at 4.72(1H, d, *J*=10Hz), 5.14(1H, td, *J*=10, 10, 4.5Hz), and 5.0-4.7ppm (1H, m, overlapping with other signals).

^{13}C NMR:
(pyridine-d_5) C-1, 44.7 t; C-2, 68.9 d, C-3, 83.5 d; C-4, 39.7 s*; C-5, 51.0 d; C-6, 18.7 t; C-7, 28.3 t; C-8, 134.9 s; C-9, 133.6 s; C-10, 38.3 s*; C-11, 26.7 t; C-12, 72.0 d; C-13, 49.9 s; C-14, 49.9 s; C-15, 32.6 t**; C-16, 29.1 t**; C-17, 43.4 d; C-18, 16.7 q***; C-19, 20.4 q; C-20, 36.8 d; C-21, 18.0 q; C-22, 34.7 t****; C-23, 34.2 t; C-24, 79.1 d; C-25, 72.7 s; C-26, 26.1 q*****; C-27, 25.8 q*****; C-28, 25.2 q*****; C-29, 29.1 q*****; and C-30, 17.4ppm q****.

*, **, ***, ****, ***** Assignments may be interchanged.

Mass Spectrum:
LREIMS: 492(M$^+$), 187, and 145*m/e*.

TLC Data

Kieselgel GF_{254} plates were developed with chloroform-ethanol, 20:3 (v/v). Fasciculol B had an R_f of 0.3 (dark green spot) and fasciculol D was at R_f 0.7 (dark green spot). Detection was by spraying plates with 0.5% vanillin in concentrated H_2SO_4, followed by heating at 120°C.

References

M. De Bernardi, G. Mellerio, G. Vidari, P. Vita-Finzi, G. Fronza, M. Kocòr, and J. St. Pyrek; Fungal Metabolites. IX. Triterpenes from *Neamatoloma sublateritium*; J. Natural Products, Vol. 44, pp. 351-356(1981).

M. Ikeda, Y. Sato, M. Izawa, T. Sassa, and Y. Miura; Isolation and Structure of Fasciculol A, a New Plant Growth Inhibitor from *Neamatoloma fasciculare*; Agric. Biol. Chem., Vol. 41, pp. 1539-1541(1977).

W. D. Nes and E. J. Parish (eds.); Analysis of Sterols and Other Biologically Significant Steroids; Academic Press Inc. New York, NY, 341 pp. (1989).

W. B. Turner and D. C. Aldridge; Fungal Metabolites II; Academic Press, Inc., New York, NY, p. 319 (1983).

J. D. Weete; Structure and Function of Sterols in Fungi, Advances in Lipid Research, Vol. 23, pp. 115-167(1989).

Common/Systematic Name

Fasciculol C

2α,3β,12α,21,24(*R*),25-Hexahydroxylanost-8-ene

Molecular Formula/Molecular Weight

$C_{30}H_{52}O_6$; MW = 508.37639

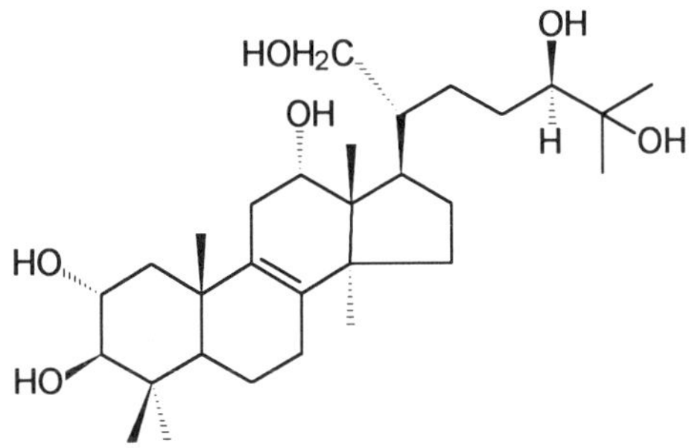

General Characteristics

Crystals; mp., 244°C; $[\alpha]_D^{21}$ + 67.5°; triacetate ($C_{36}H_{56}O_8$), mp., 96-97°C.

Fungal Source

Fruit bodies of *Neamatoloma fasciculare*.

Isolation/Purification

Fruit bodies were extracted with ethyl acetate and the neutral fraction was applied to a silica gel column (chloroform-ethanol, 100:7, v/v). Further purification and separation was achieved by silica gel column chromatography by elution with ethyl acetate-ethanol (100:4, v/v) and (100:13, v/v). Final purification was by preparative TLC with Kieselgel GF_{254} plates and chloroform-ethanol, 20:3 (v/v) as developing solvent; detected at R_f 0.2 as a dark purple spot by spraying with 0.5% vanillin in concentrated sulfuric acid and heating at 120°C.

Biological Activity

Inhibited the growth of Chinese cabbage seedlings.

Spectral Data

UV:

λ_{max}^{MeOH} 207(ϵ = 6,000).

IR:

(KBr) 3400cm^{-1} (no carbonyl bands); tetraacetate 3400 and 1730cm^{-1}.

^{1}H NMR
(pyridine-d_5) 0.72; 1.10; 1.13; 1.26; 1.37(each 3H, s); 1.48(6H, s); 3.34(1H, d, J=10Hz); and 3.6-4.4ppm (5H, m).

^{13}C NMR:
(pyridine-d_5) C-1, 44.4 t; C-2, 68.9 d; C-3, 83.5 d; C-4, 39.6 s*; C-5, 51.0 d; C-6, 18.7 t; C-7, 28.3 t**; C-8, 134.7 s; C-9, 133.9 s; C-10, 38.4 s*; C-11, 26.8 t; C-12, 73.1 d; C-13, 50.6 s***; C-14, 50.2 s***; C-15, 32.3 t**; C-16, 29.7 t**; C-17, 38.6 d; C-18, 17.1 q***; C-19, 20.3 q; C-20, 44.0 d; C-21, 61.2 q; C-22, 28.1 t; C-23, 33.1 t; C-24, 78.9 t; C-25, 72.6 d; C-26, 26.0 q*****; C-27, 25.7 q*****; C-28, 24.1 q*****; C-29, 29.0 q*****; and C-30, 17.4ppm q****.

*, **, ***, ****, ***** Assignments may be interchanged.

Mass Spectrum:
LREIMS: 508(M$^+$), 347, 305, 203, and 161*m/e*.

TLC Data

Kieselgel GF_{254} plates were developed with chloroform-ethanol, 20:3 (v/v). Fasciculol C had an R_f of 0.2 and appeared as a dark purple spot after spraying the TLC plate with 0.5% vanillin in concentrated sulfuric acid and heating at 120°C.

References

M. De Bernardi, G. Mellerio, G. Vidari, P. Vita-Finzi, G. Fronza, M. Kocòr, and J. St. Pyrek; Fungal Metabolites. IX. Triterpenes from *Neamatoloma sublateritium*; J. Natural Products, Vol. 44, pp. 351-356(1981).

M. Ikeda, G. Niwa, K. Tohyama, T. Sassa, and Y. Miura; Structures of Fasciculol C and Its Depsipeptides, New Biologically Active Substances From *Neamatoloma fasciculare*; Agric. Biol. Chem., Vol. 41, pp. 1803-1805(1977).

W. B. Turner and D.C. Aldridge; Fungal Metabolites II; Academic Press, Inc., New York, New York, p. 319 (1983).

Common/Systematic Name
Fasciculol C depsipeptide E

Molecular Formula/Molecular Weight
$C_{39}H_{65}NO_{11}$; MW = 723.45576

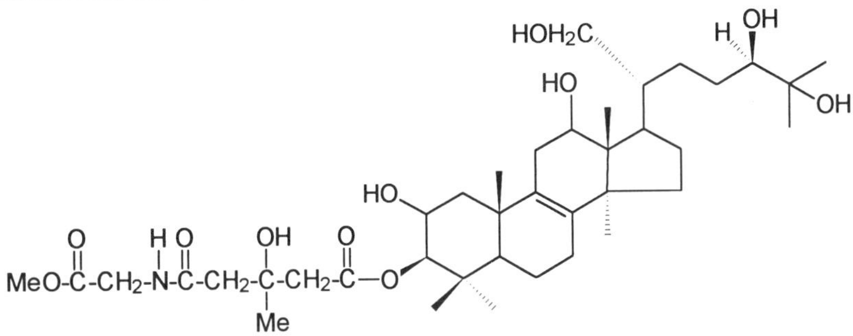

General Characteristics
Crystals; mp., 105-106°C; $[\alpha]_D^{21}$ +30.6°. Positive ninhydrin test.

Fungal Source
Fruit bodies of *Neamatoloma fasciculare*.

Isolation/Purification
Fruit bodies were extracted with ethyl acetate and the neutral fraction was applied to a silica gel column (chloroform-ethanol, 100:7, v/v). Further purification and separation was achieved by silica gel column chromatography by elution with ethyl acetate-ethanol (100:4, v/v) and (100:13, v/v). Final purification was by preparative TLC with Kieselgel GF_{254} plates and chloroform-ethanol, 20:3, v/v, as developing solvent; detected at R_f 0.2 as a dark purple spot by spraying with 0.5% vanillin in concentrated sulfuric acid and heating at 120°C.

Biological Activity
Inhibited the growth of Chinese cabbage seedlings.

Spectral Data

UV:
λ_{max}^{EtOH} 208(ϵ = 5,000).

IR:
(KBr) 3600-3300, 1730, 1655, and 1530cm^{-1}.

^{1}H NMR:
($CDCl_3$) (acetonide) 0.59(3H, s); 0.87(6H, s); 1.06(9H, s); 1.22(3H, s); 1.31(3H, s); 1.39(6H, s); 2.65(2H, br d, *J*=1-2Hz); 2.73(2H, s); 3.72(3H, s); 4.00(2H, d, *J*=5Hz); 4.57(1H, d, *J*=10Hz), and 7.51ppm (1H, t, *J*=5Hz).

TLC Data

Kieselgel GF_{254} plates were developed with chloroform-ethanol, 20:3 (v/v). Fasciculol C depsipeptides had an R_f of 0.25 (dark purple spot) and R_f 0.5 (dark purple spot). Detection was by spraying plates with 0.5% vanillin in concentrated H_2SO_4, followed by heating at 120°C.

Reference

M. Ikeda, G. Niwa, K. Tohyama, T. Sassa, and Y. Miura; Structures of Fasciculol C and Its Depsipeptides, New Biologically Active Substances From *Neamatoloma fasciculare*; Agric. Biol. Chem., Vol. 41, pp. 1803-1805(1977).

Common/Systematic Name

Fasciculol C depsipeptide F

Molecular Formula/Molecular Weight

$C_{39}H_{65}NO_{11}$; MW = 723.45576

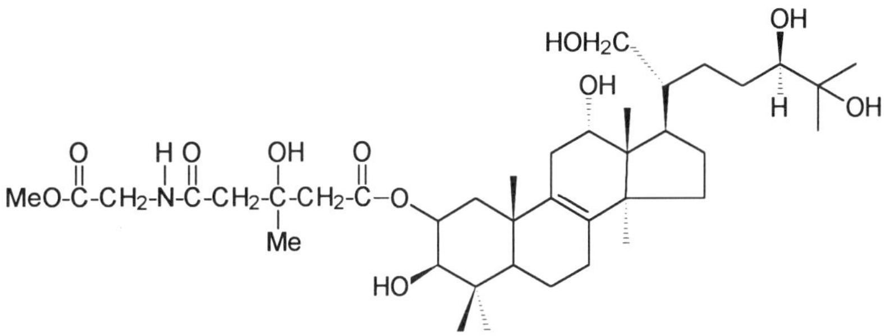

General Characteristics

Crystals; mp., 102-103°C; $[\alpha]_D^{21}$ +9.9°. Positive ninhydrin test.

Fungal Source

Fruit bodies of *Neamatoloma fasciculare*.

Isolation/Purification

Fruit bodies were extracted with ethyl acetate and the neutral fraction was applied to a silica gel column and eluted with chloroform-ethanol, 100:7 (v/v). Further purification and separation was achieved by silica gel column chromatography by elution with ethyl acetate-ethanol (100:4, v/v) and (100:13, v/v). Final purification was by preparative TLC with Kieselgel GF_{254} plates and chloroform-ethanol, 20:3 (v/v) as developing solvent; detected at R_f 0.5 as a dark purple spot by spraying with 0.5% vanillin in concentrated sulfuric acid and heating at 120°C.

Biological Activity

Inhibited the growth of Chinese cabbage seedlings.

Spectral Data

UV:

λ_{max}^{EtOH} 208(ϵ = 5,000).

IR:

(KBr) 3600-3300, 1730, 1655, and 1530cm^{-1}.

^1H NMR:

(CDCl$_3$) (acetonide) 0.59(3H, s); 0.90(3H, s); 1.09(12H, s); 1.25(3H, s); 1.32(3H, s); 1.40(6H,s); 2.65(2H, br d, *J*=12Hz); 2.73(2H,s); 3.24(1H, d, *J*=10Hz); 4.02(2H, d, *J*=5Hz); 5.00(1H, td, *J*=10, 4.5Hz); and 7.40ppm (1H, t, *J*=5Hz).

TLC Data

Kieselgel GF$_{254}$ plates were developed with chloroform-ethanol, 20:3 (v/v). Fasciculol C depsipeptide F had an R_f of 0.5 (dark purple spot). Detection was by spraying plates with 0.5% vanillin in concentrated H_2SO_4, followed by heating at 120°C.

Reference

M. Ikeda, G. Niwa, K. Tohyama, T. Sassa, and Y. Miura; Structures of Fasciculol C and Its Depsipeptides, New Biologically Active Substances from *Neamatoloma fasciculare*; Agric. Biol. Chem., Vol. 41, pp. 1803-1805(1977).

Common/Systematic Name

Fasciculol B depsipeptide

Molecular Formula/Molecular Weight

$C_{39}H_{65}NO_{10}$; MW = 707.46085

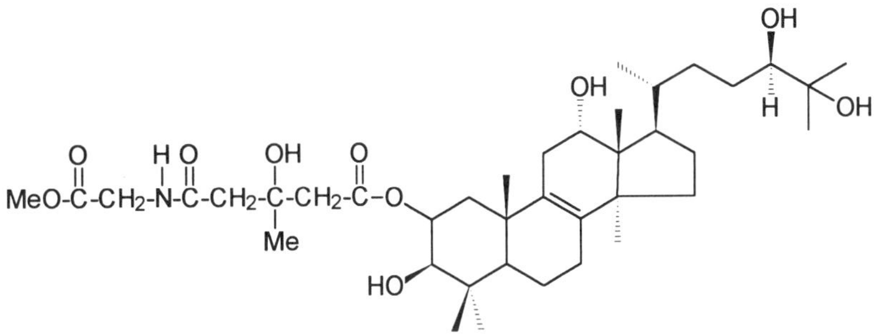

General Characteristics

Crystals; mp., 95-97°C; $[\alpha]_D^{21}$ +14.1° (*c*=1.0, in MeOH). Positive ninhydrin test.

Fungal Source

Fruit bodies of *Neamatoloma fasciculare*.

Isolation/Purification

Fruit bodies were extracted with ethyl acetate and the neutral fraction was applied to a silica gel column (chloroform-ethanol, 100:7, v/v). Further purification and separation of fasciculol B from fasciculol D was achieved by silica gel column chromatography by elution with ethyl acetate-ethanol (100:4, v/v) for fasciculol B and (100:13, v/v) for fasciculol D. Final purification was by preparative TLC with Kieselgel GF_{254} plates and chloroform-ethanol, 20:3 (v/v) as developing solvent.

Biological Activity

Inhibited the growth of Chinese cabbage seedlings and showed weak antimicrobial activity against *Staphylococcus aureus* and *Klebsiella pneumoniae*.

Spectral Data

UV:

λ_{max}^{EtOH} 208(ϵ = 3,900).

IR:

(KBr) 3600-3400, 1655, and 1530cm^{-1}.

^{1}H NMR:

($CDCl_3$) nine methyl groups in the region 0.62-1.42; a methoxy group at 3.78; a secondary methylene amino group at 4.07(2H, d, *J*=5.0Hz); 7.32(1H, t, *J*=5.0Hz, disappeared on addition of D_2O); an acyloxy methine proton at 5.14(1H, td, *J*=10, 10, 4.5Hz); and a hydroxy methine proton at 3.24ppm (1H, d, *J*=10Hz). Six protons in the region of 2.4-4.2 were unassigned.

TLC Data

Kieselgel GF_{254} plates were developed with chloroform-ethanol, 20:3, v/v. Fasciculol B had an R_f of 0.3 (dark green spot) and fasciculol D was at R_f 0.7 (dark green spot). Detection was by spraying plates with 0.5% vanillin in concentrated H_2SO_4, followed by heating at 120°C.

References

M. Ikeda, Y. Sato, M. Izawa, T. Sassa, and Y. Miura; Isolation and Structure of Fasciculol A, a New Plant Growth Inhibitor from *Neamatoloma fasciculare*; Agric. Biol. Chem., Vol. 41, pp. 1539-1541(1977).

W. D. Nes and E. J. Parish (eds.); Analysis of Sterols and Other Biologically Significant Steroids; Academic Press Inc. New York, NY, 341 pp. (1989).

W. B. Turner and D. C. Aldridge; Fungal Metabolites II; Academic Press, Inc., New York, NY, p. 319 (1983).

J. D. Weete; Structure and Function of Sterols in Fungi; Advances in Lipid Research, Vol. 23, pp. 115-167(1989).

16

Fusidanes and Protostanes

Virdominic acid A
Virdominic acid B
Virdominic acid C
Cephalosporin P1
3-Ketofusidic acid
11-Ketofusidic acid
3-Epifusidic acid
11-Epifusidic acid
9,11-Anhydrofusidic acid
9,11-Anhydro-12-hydroxyfusidic acid
9,11- Anhydro-9α,11α-epoxyfusidic acid
7,8-Dehydropseudofusidic acid
Fusilactidic acid
Fusidic acid
Helvolic acid

Common/Systematic Name

Viridominic acid A

Molecular Formula/Molecular Weight

$C_{33}H_{48}O_9$; MW = 588.32983

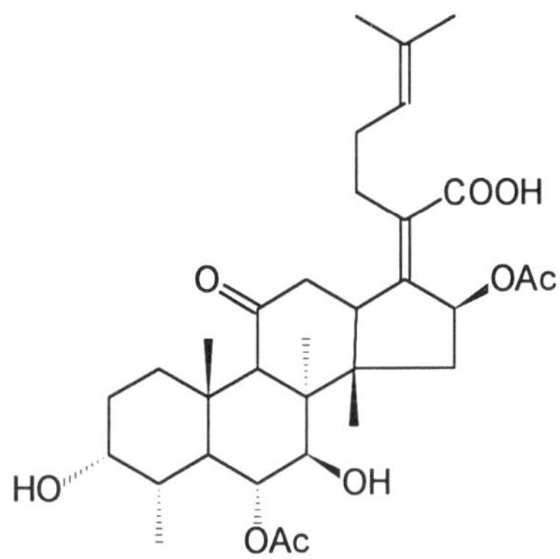

General Characteristics

Crystals; mp., 151-155°C; $[\alpha]_D^{22}$ + 49° (*c*=0.1, in MeOH); diketo derivative; mp., 124-127.5°C; dihydrool derivative; mp., 119-121°C; monoethyl ester; $[\alpha]_D^{25}$ +23° (*c*=0.1, in MeOH).

Fungal Source

Cladosporium sp. (501-7γ).

Isolation/Purification

The culture filtrate was acidified to pH 3 and extracted with ethyl acetate. The extract was washed with water, concentrated and extracted with saturated $NaHCO_3$ solution. The aqueous layer was acidified, extracted with ethyl acetate, dried over anhydrous sodium sulfate and evaporated to a viscous residue. The chloroform soluble fraction was chromatographed on a silicic acid column eluted with benzene and benzene-ethanol (25:1, v/v). Biologically active fractions from the benzene-ethanol eluate were chromatographed on a silicic acid column eluted with benzene-ethanol (100:1, 50:1, and 25:1, v/v). Final purification was by repetitive silica gel column chromatography eluted with chloroform and chloroform-ethanol (50:1, v/v).

Biological Activity

Chlorosis-inducing activity in higher plants (six day old seedlings of Chinese cabbage *Brassica campestris* L. var. *pekinensis*; seven day old seedlings of barnyard grass *Panicum crus-galli* L. var. *frumentaceum*).

Spectral Data

UV:

λ_{max}^{EtOH} 220(sh, ϵ = 6,500) and 290nm (103); monoethyl ester: λ_{max}^{EtOH} 217nm (ϵ=8,100).

IR:

(KBr) 3420(OH), 2660(H-bonded OH), and 1729cm^{-1} (carboxylic C=O); monoethyl ester: 3500, 1710(br), 1245, 1020, and 900cm^{-1}.

^{1}H NMR:

($CDCl_3$) Methyl ester, 3.70(br s, H-3); 4.48(d, *J*=11Hz, H-6); 3.50(s, H-7); 3.28(s, H-9); 5.84(d, *J*=8.0Hz, H-16); 5.05(br t, *J*=8.0Hz, H-24); 0.92(d, *J*=7.0Hz, secondary methyl); 1.03, 1.28, and 1.47(tertiary methyls); 1.60, 1.67ppm (vinyl methyls); and 1.95, 2.03ppm (each 3H, s, acetoxy groups).

Mass Spectrum:

LREIMS: methyl ester, 533*m/e* (M^+ -69).

References

H. Kaise, K. Munakata, and T. Sassa; Structures of Viridominic Acids A and B, New Chlorosis-inducing Metabolites of a Fungus; Tet. Lett., pp. 3789-3782(1972).

H. Kaise, Y. Ogawa, T. Sassa and K. Munakata, Studies on the Chlorosis-Inducing Substances Produced by a Fungus. Part I. Isolation and Biological Activities of Viridominic Acids A, B, C and Cephalosporin P_1; Agr. Biol. Chem., Vol. 36, No. 1, pp. 120-124(1972).

Common/Systematic Name
Viridominic acid B

Molecular Formula/Molecular Weight
$C_{33}H_{48}O_{10}$; MW = 604.32475

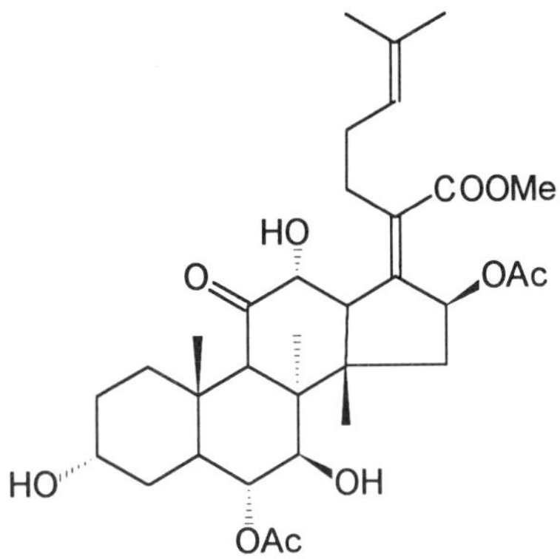

General Characteristics
Amorphous; $[\alpha]_D^{17}$ + 64° (*c*=0.37, in MeOH); diketo acetate derivative; mp., 236-240°C.

Fungal Source
Cladosporium sp. (501-7γ).

Isolation/Purification
The culture filtrate was acidified to pH 3 and extracted with ethyl acetate. The extract was washed with water, concentrated and extracted with saturated $NaHCO_3$ solution. The aqueous layer was acidified, extracted with ethyl acetate, dried over anhydrous sodium sulfate and evaporated to a viscous residue. The chloroform soluble fraction was chromatographed on a silicic acid column eluted with benzene and benzene-ethanol (25:1, v/v). Biologically active fractions from the benzene-ethanol eluate were chromatographed on a silicic acid column eluted with benzene-ethanol (100:1, 50:1, and 25:1, v/v). Final purification was by repetitive silica gel column chromatography eluted with chloroform and chloroform-ethanol (50:1, v/v).

Biological Activity
Chlorosis-inducing activity in higher plants (six day old seedlings of Chinese cabbage *Brassica campestris* L. var. *pekinensis*; seven day old seedlings of barnyard grass *Panicum crus-galli* L. var. *frumentaceum*).

Spectral Data

UV:
λ_{max}^{EtOH} 213nm (ϵ=8,000).

IR:
(KBr) 3580, 3400, 1728(br.), 1250, 1120, and 930cm^{-1}.

^{1}H NMR:
($CDCl_3$) Methyl ester, 3.73(br s, H-3); 4.45(d, *J*=10Hz, H-6); 3.50(s, H-7); 3.39(s, H-9); 4.28(d, *J*=8.0Hz, H-12); 5.73(d, *J*=8.0Hz, H-16); 5.12(br t, *J*=7.5Hz, H-24); 0.92(d, *J*=6.5Hz, secondary methyl); 0.90, 1.32, and 1.56(tertiary methyls); 1.58, 1.65ppm (vinyl methyls); 1.58 and 1.65(each 3H, s), 5.14(br t), isopentenyl group; and 1.98, 2.03ppm (each 3H, s, acetoxy groups).

Mass Spectrum:
LREIMS: methyl ester, 594*m/e* (M^+ -69).

TLC Data
Silicic acid plates developed with chloroform-ethanol (8:1, v/v), R_f 0.45.

References

H. Kaise, K. Munakata, and T. Sassa; Structures of Viridominic Acids A and B, New Chlorosis-inducing Metabolites of a Fungus; Tet. Lett., pp. 3789-3782(1972).

H. Kaise, Y. Ogawa, T. Sassa, and K. Munakata; Studies on the Chlorosis-inducing Substances Produced by a Fungus. Part I. Isolation and Biological Activities of Viridominic Acids A, B, C and Cephalosporin P_1; Agr. Biol. Chem., Vol. 36, No. 1, pp. 120-124(1972).

Common/Systematic Name

Viridominic acid C

Molecular Formula/Molecular Weight

$C_{33}H_{50}O_9$; MW = 590.34548

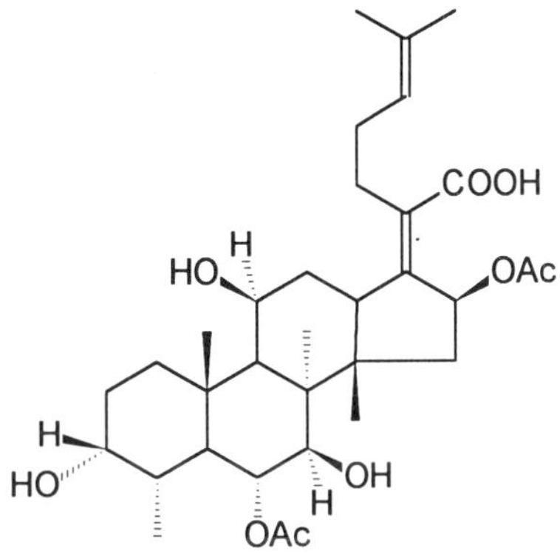

General Characteristics

Crystals; mp., 168-171°C; $[\alpha]_D^{25}$ +38° (*c*=0.1, in MeOH); diacetate derivative; mp., 196-198°C; lactone; mp., 248-250°C.

Fungal Source

Cladosporium sp. (501-7γ).

Isolation/Purification

The culture filtrate was acidified to pH 3 and extracted with ethyl acetate. The extract was washed with water, concentrated and extracted with saturated $NaHCO_3$ solution. The aqueous layer was acidified, extracted with ethyl acetate, dried over anhydrous sodium sulfate and evaporated to a viscous residue. The chloroform soluble fraction was chromatographed on a silicic acid column eluted with benzene and benzene-ethanol (25:1, v/v). Biologically active fractions from the benzene-ethanol eluate were chromatographed on a silicic acid column eluted with benzene-ethanol (100:1, 50:1, and 25:1, v/v). Final purification was by repetitive silica gel column chromatography eluted with chloroform and chloroform-ethanol (50:1, v/v).

Biological Activity

Chlorosis-inducing activity in higher plants (six day old seedlings of Chinese cabbage *Brassica campestris* L. var. *pekinensis*; seven day old seedlings of barnyard grass *Panicum crus-galli* L. var. *frumentaceum*).

Spectral Data

UV:
λ^{EtOH}_{max} 220nm (ϵ=8,100).

IR:
(KBr) 3450, 1715(br), 1265, and 915cm^{-1}.

^{1}H NMR:
($CDCl_3$/DMSO-d_6) 3.51(br s, H-3); 4.51(d, *J*=12Hz, H-6); 3.30(s, H-7); 3.70(m, H-11); 5.74 (d, *J*=7.5Hz, H-16); 5.05(br t, *J*=7.5Hz, H-24); 0.77(d, *J*=6.5Hz, secondary methyl); 0.99, 1.07 and, 1.16(tertiary methyls); 1.57, 1.64ppm (vinyl methyls); 3.56; methyl ester derivative (3H, s, methyl ester); 4.83(1H, d, *J*=5.0Hz); 4.34, D_2O exchangeable = sec. OH); (1H, d, *J*=7.0Hz, D_2O exchangeable, sec. OH); 3.14ppm (1H, d, *J*=3.5Hz, D_2O exchangeable, sec. OH).

Mass Spectrum:
LREIMS: methyl ester, 535*m/e* (M^+ -69).

TLC Data
Silicic acid plates developed with chloroform-ethanol (8:1, v/v), R_f 0.35.

References
H. Kaise, K. Munakata, and T. Sassa; Structures of Viridominic Acids A and B, New Chlorosis-inducing Metabolites of A Fungus; Tet. Lett., pp. 3789-3782(1972).

H. Kaise, K. Munakata, and T. Sassa; Structure of Viridominic Acid C, A New Steroidal Metabolite of a Fungus Having Chlorosis-inducing Activity; Tet. Lett., pp. 199-202(1972).

H. Kaise, Y. Ogawa, T. Sassa, and K. Munakata; Studies on the Chlorosis-Inducing Substances Produced by a Fungus. Part I. Isolation and Biological Activities of Viridominic Acids A, B, C and Cephalosporin P_1; Agr. Biol. Chem., Vol. 36, No. 1, pp. 120-124(1972).

W. B. Turner and D. C. Aldridge; Fungal Metabolites II; Academic Press, New York, p. 332(1983).

Common/Systematic Name

Cephalosporin P_1

Molecular Formula/Molecular Weight

$C_{33}H_{50}O_8$; MW = 574.35057

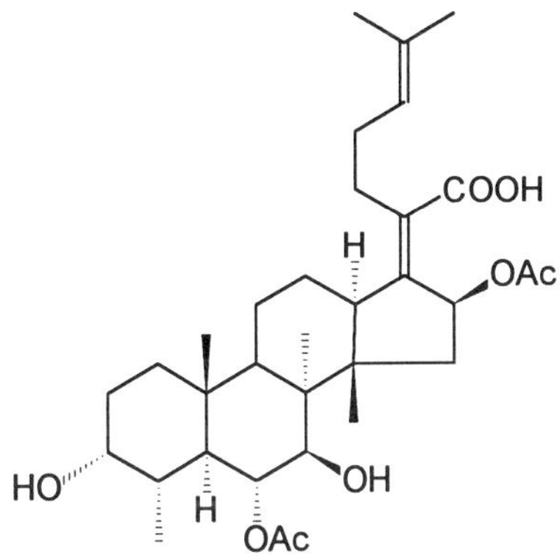

General Characteristics

Colorless powder; monomethyl ester: crystals from acetone-hexane; mp., 99-101.5°C; $[\alpha]_D$ +25° (*c*=0.5, in MeOH).

Fungal Source

Cladosporium sp. (501-7γ) and *C. acremonium*.

Isolation/Purification

The culture filtrate was acidified to pH 3 and extracted with ethyl acetate. The extract was washed with water, concentrated and extracted with saturated $NaHCO_3$ solution. The aqueous layer was acidified, extracted with ethyl acetate, dried over anhydrous sodium sulfate and evaporated to a viscous residue. The chloroform soluble portion of the residue was chromatographed on a silicic acid column eluted with benzene and benzene-ethanol (25:1, v/v). Biologically active fractions from the benzene-ethanol eluate were chromatographed on a silicic acid column eluted with benzene-ethanol (100:1, 50:1, and 25:1, v/v). Final purification was by repetitive silica gel column chromatography eluted with chloroform and chloroform-ethanol (50:1, v/v).

Biological Activity

Antibiotic and chlorosis-inducing activity in higher plants (six day old seedlings of Chinese cabbage *Brassica campestris* L. var. *pekinensis*; seven day old seedlings of barnyard grass *Panicum crus-galli* L. var. *frumentaceum*).

Spectral Data

UV:

Monomethyl ester, λ_{max}^{EtOH} 218nm (ϵ = 8,300); monodesacetylcephalosporin P_1; λ_{max}^{EtOH} 20nm (ϵ = 7,000).

IR:

Monodesacetylcephalosporin P_1; (KBr) 3400, 2920, 1700, 1440, 1370, and 1260cm^{-1}.

^{1}H NMR:

($CDCl_3$/DMSO-d_6) 0.80(3H, d, J=6.5Hz, secondary methyl); 0.99, 1.09, and 1.09ppm (3 X tertiary methyls); 1.52 and 1.60ppm (vinyl methyls); 3.52(br s, H-3); 4.50(d, J=10Hz, H-6); 3.38(s, H-7); 5.67(d, J=8.0Hz, H-16); and 5.00ppm (br t, J=7.5Hz, H-24).

TLC Data

Silicic acid plates developed with chloroform-ethanol (8:1, v/v), R_f 0.50.

References

T. S. Chou, E. J. Eisenbraun, and R. T. Rapala; The Chemistry of Cephalosporin P_1; Tet. Lett., pp. 409-414(1967).

T. G. Halsall, E. R. H. Jones, G. Lowe, and C. E. Newall; Cephalosporin P_1; Chem. Commun., pp. 685-687(1966).

H. Kaise, K. Munakata, and T. Sassa; Structure of Viridominic Acid C, A New Steroidal Metabolite of a Fungus Having Chlorosis-inducing Activity; Tet. Lett., pp. 199-202 (1972).

H. Kaise, Y. Ogawa, T. Sassa, and K. Munakata; Studies on the Chlorosis-Inducing Substances Produced by a Fungus. Part I. Isolation and Biological Activities of Viridominic Acids A, B, C and Cephalosporin P_1; Agr. Biol. Chem., Vol. 36, No. 1, pp. 120-124(1972).

P. Oxley; Cephalosporin P_1 and Helvolic Acid; Chem. Commun., pp. 729-730(1966).

W. B. Turner and D. C. Aldridge; Fungal Metabolites II; Academic Press, New York, p. 332 (1983).

Common/Systematic Name

3-Ketofusidic acid

Molecular Formula/Molecular Weight

$C_{31}H_{46}O_6$; MW = 514.32944

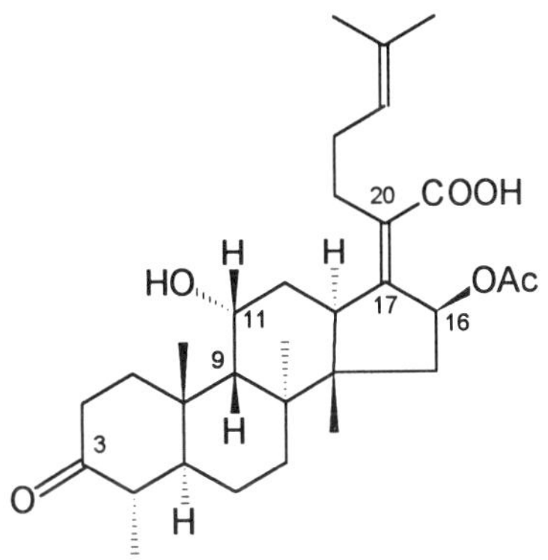

General Characteristics

Crystals from ethyl ether, dimorphic forms; mp., 177-178°C and 191-192°C; $[\alpha]_D^{20}$ + 23°; methyl ester: crystals from ether-light petroleum; mp., 125-126°C.

Fungal Source

Fusidium coccineum.

Isolation/Purification

Following the usual procedure for the isolation of fusidic acid (see Godtfredsen, *et al.*, 1979), the benzene filtrate after collection of the fusidic acid-benzene solvate was evaporated yielding an oil. This oil was dissolved in acetone, diethanolamine was added and, after seeding and standing, the precipitate of crystalline diethanolamine salts was collected and washed with cold acetone. The mixture of crystalline diethanolamine salts was suspended in ethanol and acidified to pH 3. Benzene and water were added; the benzene layer was concentrated to give a fusidic acid-benzene solvate. Evaporation of the benzene filtrate and solution of the residue in acetone and addition of dicyclohexylamine gave a mixture of crystalline dicyclohexylamine salts. The acetone filtrate was evaporated to dryness and the residue dissolved in ethanol, followed by acidification to pH 3. Addition of ether completed the precipitation of dicyclohexamine hydrochloride. The ether filtrate was washed with water, dried and evaporated to yield a solid mixture of acidic metabolites. This mixture was fractionated on a column containing silicic acid eluted with ether-light petroleum ether-acetic acid (60:40:0.5, v/v/v). An additional fractionation with a silicic acid column eluted with ether-light petroleum ether-acetic acid (50:50:0.5, v/v/v) gave a mixture which yielded 3-ketofusidic acid after crystallization from ether solution.

Biological Activity

Antibiotic activity against Gram-positive bacteria and, in particular, a strain of *Staphylococcus aureus*; no activity against the Gram-negative bacterium, *Escherichia coli*.

Spectral Data

UV:

λ_{max}^{EtOH} 204(ϵ = 10,500) and 225nm (infl ϵ = 7,500); methyl ester: λ_{max}^{EtOH} 204(ϵ = 10,200) and 225nm (ϵ = 8,900) .

IR:

(KBr) 3550(hydroxyl), 1730, 1255(acetoxy), and 1700cm^{-1} (carboxyl); methyl ester: 3540, 1720, 1700, and 1250cm^{-1}.

^{1}H NMR:

($CDCl_3$) methyl ester: 4.42(1H, m, 11α-H); 3.05(1H, m, 13α-H); 5.92(1H, d, *J*=7.5-8.5Hz, 16α-H); 5.14(1H, m, H-24); 1.63 and 1.68(26- and 27-methyls); 0.97, 1.18, and 1.33(18-, 19- and 32-tertiary methyls); 1.99(16-acetate group); and 3.68ppm (21-methoxy group).

Mass Data:

Anal. calcd for $C_{31}H_{46}O_6$: C, 72.34; H, 9.01; found: C, 72.22; H, 8.95%; methyl ester, anal. calcd for $C_{32}H_{48}O_6$: C, 72.69; H, 9.15; found: C, 72.75; H, 9.17%.

TLC Data

Silica gel (Merck HF_{254}) plates developed with (A) ether-acetic acid (100:0.5, v/v), (B) ether-dichloromethane-acetic acid (50:50:0.5, v/v/v), (C) dichloromethane-methanol-acetic acid (95:5:0.5, v/v/v), and (D) cyclohexane-chloroform-methanol-acetic acid (20:80:2.5:10, v/v/v/v). R_f = 0.42, 0.22, 0.35, and 0.60, respectively. Detected by spraying plates with concentrated H_2SO_4 and heating; initial color was crimson changing to blue-violet.

References

W. O. Godtfredsen, N. Rastrup-Andersen, S. Vangedal, and W. D. Ollis, Metabolites of *Fusidium coccineum;* Tetrahedron, Vol. 35, pp. 2419-2431(1979).

W. B. Turner and D. C. Aldridge; Fungal Metabolites II; Academic Press, New York, p. 333(1983).

Common/Systematic Name
11-Ketofusidic acid

Molecular Formula/Molecular Weight
$C_{31}H_{46}O_6$; MW = 514.32944

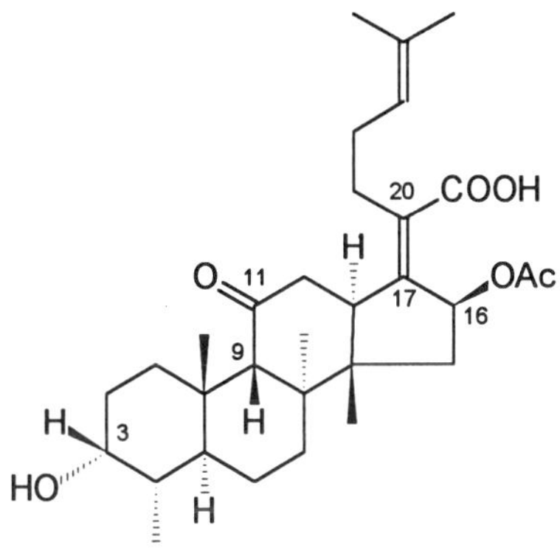

General Characteristics
Crystals from ethyl ether; mp., 196-197°C; $[\alpha]_D^{20}$ + 55°; methyl ester: crystals from ether-light petroleum; mp., 160-162°C.

Fungal Source
Fusidium coccineum.

Isolation/Purification
Following the usual procedure for the isolation of fusidic acid (see Godtfredsen, *et al.*, 1979), the benzene filtrate after collection of the fusidic acid-benzene solvate was evaporated yielding an oil. This oil was dissolved in acetone, diethanolamine was added and after seeding and standing, and the precipitate of crystalline diethanolamine salts was collected and washed with cold acetone. The mixture of crystalline diethanolamine salts was suspended in ethanol and acidified to pH 3. Benzene and water were added; the benzene layer was concentrated to give a fusidic acid-benzene solvate. Evaporation of the benzene filtrate and solution of the residue in acetone and addition of dicyclohexylamine gave a mixture of crystalline dicyclohexylamine salts. The acetone filtrate was evaporated to dryness and the residue dissolved in ethanol, followed by acidification to pH 3. Addition of ether completed the precipitation of dicyclohexamine hydrochloride. The ether filtrate was washed with water, dried and evaporated to yield a solid mixture of acidic metabolites. This mixture was fractionated on a column containing silicic acid eluted with ether-light petroleum ether-acetic acid (60:40:0.5, v/v/v). An additional fractionation with a silicic acid column eluted with ether-light petroleum ether-acetic acid (50:50:0.5, v/v/v) gave a mixture which yielded 11-ketofusidic acid after crystallization from ether solution.

Biological Activity

Antibiotic activity against Gram-positive bacteria and, in particular, a strain of *Staphylococcus aureus*; no activity against the Gram-negative bacterium, *Escherichia coli*.

Spectral Data

UV:

λ_{max}^{EtOH} 204(ϵ = 10,900) and 225nm (infl ϵ = 6,950); methyl ester: λ_{max}^{EtOH} 204(ϵ=10,300) and 225nm (ϵ=7,700).

IR:

(KBr) 3450, 1710, 1260, and 1260cm^{-1}; methyl ester: 3560, 1720, 1250, and 1720cm^{-1}.

^{1}H NMR:

($CDCl_3$) Methyl ester: 3.79(1H, m, 3β-H); 2.62(9β-H); 2.80(3H, m, 12α-, 12β- and 13α-H); 5.93(1H, d, *J*=7.5-8.5Hz, 16α-H); 5.12(1H, m, H-24); 0.92(3H, d, *J*=7.5-8.5Hz, 30-secondary methyl); 1.62 and 1.68(26- and 27-methyls); 1.03, 1.18, and 1.18(18, 19 and 32-tertiary methyls); 2.01(16-acetate group); and 3.68ppm (21-methoxy group).

Mass Data:

HREIMS: methyl ester: 528.3436*m/e* (M^+); anal. calcd for $C_{31}H_{46}O_6$: C, 72.34; H, 9.01; found: C, 72.19; H, 8.92%; methyl ester, anal. calcd for $C_{32}H_{48}O_6$: C, 72.69; H, 9.15; found: C, 72.66; H, 9.04%.

TLC Data

Silica gel (Merck HF_{254}) plates developed with (A) ether-acetic acid (100:0.5, v/v), (B) ether-dichloromethane-acetic acid (50:50:0.5, v/v/v), (C) dichloromethane-methanol-acetic acid (95:5:0.5, v/v/v), and (D) cyclohexane-chloroform-methanol-acetic acid (20:80:2.5:10, v/v/v/v). R_f = 0.68, 0.54, 0.50, and 0.73, respectively. Detected by spraying plates with concentrated H_2SO_4 and heating; initial color was orange changing to brown.

References

W. O. Godtfredsen, N. Rastrup-Andersen, S. Vangedal, and W. D. Ollis; Metabolites of *Fusidium coccineum*; Tetrahedron, Vol. 35, pp. 2419-2431(1979).

W. B. Turner and D. C. Aldridge; Fungal Metabolites II; Academic Press, New York, p. 333(1983).

Common/Systematic Name

3-Epifusidic acid

Molecular Formula/Molecular Weight

$C_{31}H_{48}O_6$; MW = 516.34509

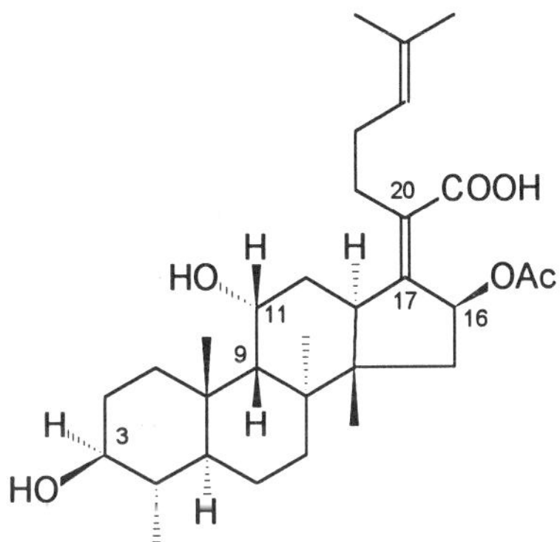

General Characteristics

Colorless crystals from chloroform; mp., 211°C; $[\alpha]_D^{20}$ + 10° (in pyridine); methyl ester: crystals from ether-light petroleum; mp., 142-143°C.

Fungal Source

Fusidium coccineum.

Isolation/Purification

Following the usual procedure for the isolation of fusidic acid (see Godtfredsen, *et al.*, 1979), the benzene filtrate after collection of the fusidic acid-benzene solvate was evaporated yielding an oil. This oil was dissolved in acetone, diethanolamine was added and, after seeding and standing, the precipitate of crystalline diethanolamine salts was collected and washed with cold acetone. The mixture of crystalline diethanolamine salts was suspended in ethanol and acidified to pH 3. Benzene and water were added, the benzene layer was concentrated to give a fusidic acid-benzene solvate. Evaporation of the benzene filtrate and solution of the residue in acetone and addition of dicyclohexylamine gave a mixture of crystalline dicyclohexylamine salts. The acetone filtrate was evaporated to dryness and the derived acids were fractionated on a column containing silicic acid eluted with cyclohexane-chloroform-methanol-acetic acid (20:80:2.5:0.5, v/v/v/v). 3-Epifusidic acid separated on standing in etheral solution and was recrystallized from chloroform.

Biological Activity

Antibiotic activity against Gram-positive bacteria and, in particular, a strain of *Staphylococcus aureus*; no activity against the Gram-negative bacterium, *Escherichia coli*.

Spectral Data

UV :

λ_{max}^{EtOH} 204(ϵ = 9,800) and 225nm (7,350); methyl ester: λ_{max}^{EtOH} 204(ϵ = 10,000) and 225nm (infl ϵ = 8,400).

IR:

(KBr) 3560, 1715, 1690, and 1260cm^{-1} ; methyl ester: 3510, 3410, 1740, 1240, and 1700cm^{-1}.

^{1}H NMR:

($CDCl_3$) methyl ester: 3.05(1H, m, 3α-H); 4.37(1H, m, 11β-H); 3.05(1H, m, 13α-H); 5.87(1H, d, *J*=7.5-8.5Hz, 16α-H); 5.10(1H, m, H-24); 1.61 and 1.67(26- and 27-methyls); 0.93, 1.02, and 1.33(18-, 19- and 32-tertiary methyls); 1.97(16-acetate group); and 3.65ppm (21-methoxy group).

Mass Data:

Anal. calcd for $C_{31}H_{48}O_6$: C, 72.06; H, 9.36%; found: C, 71.89; H, 9.48%; methyl ester: anal. calcd for $C_{32}H_{50}O_6$: C, 72.41; H, 9.50%; found: C, 72.32; H, 9.48%.

TLC Data

Silica gel (Merck HF_{254}) plates developed with (A) ether-acetic acid (100:0.5, v/v), (B) ether-dichloromethane-acetic acid (50:50:0.5, v/v/v), (C) dichloromethane-methanol-acetic acid (95:5:0.5, v/v/v), and (D) cyclohexane-chloroform-methanol-acetic acid (20:80:2.5:10, v/v/v/v). R_f = 0.45, 0.35, 0.34, and 0.58, respectively. Detected by spraying plates with concentrated H_2SO_4 and heating; initial color was grayish changing to red.

References

W. O. Godtfredsen, N. Rastrup-Andersen, S. Vangedal, and W. D. Ollis; Metabolites of *Fusidium coccineum*; Tetrahedron, Vol. 35, pp. 2419-2431(1979).

W. B. Turner and D. C. Aldridge; Fungal Metabolites II; Academic Press, New York, p. 333(1983).

Common/Systematic Name

11-Epifusidic acid

Molecular Formula/Molecular Weight

$C_{31}H_{48}O_6$; MW = 516.34509

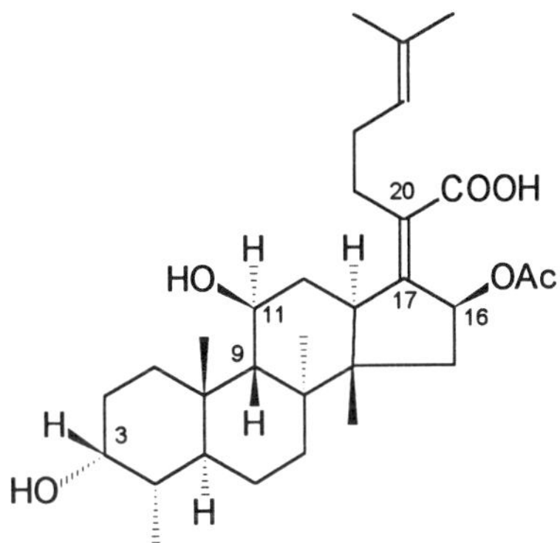

General Characteristics

Colorless crystals from acetone; mp., 202-203 °C; cyclohexamine salt; mp., 172-173 °C.

Fungal Source

Fusidium coccineum.

Isolation/Purification

Following the usual procedure for the isolation of fusidic acid (see Godtfredsen, *et al.*, 1979), the benzene filtrate after collection of the fusidic acid-benzene solvate was evaporated yielding an oil. This oil was dissolved in acetone, diethanolamine was added and, after seeding and standing, the precipitate of crystalline diethanolamine salts was collected and washed with cold acetone. The mixture of crystalline diethanolamine salts was suspended in ethanol and acidified to pH 3. Benzene and water were added; the benzene layer was concentrated to give a fusidic acid-benzene solvate. Evaporation of the benzene filtrate and solution of the residue in acetone and addition of dicyclohexylamine gave a mixture of crystalline dicyclohexylamine salts. The mixture of dicyclohexylamine salts was acidified and extracted with ether. 11-Epifusidic acid was purified from the acidic fraction by chromatography on silicic acid eluted with ether-dichloromethane-acetic acid, 25:75:0.5 (v/v/v) followed by crystallization from acetone.

Biological Activity

Antibiotic activity against Gram-positive bacteria and, in particular, a strain of *Staphylococcus aureus*; no activity against the Gram-negative bacterium, *Escherichia coli*.

Spectral Data

UV:

λ_{max}^{EtOH} 204(ϵ = 9,750) and 225nm (infl 7,050); methyl ester: λ_{max}^{EtOH} 204(ϵ = 10,000) and 225nm (infl ϵ = 8,200).

IR:

(KBr) 3600, 3400, 1710, and 1270cm^{-1}; methyl ester: 3460, 3440, 1740, 1700, and 1230cm^{-1}.

^{1}H NMR:

($CDCl_3$) Methyl ester: 3.80(1H, m, 3β-H); 3.88(1H, m, 11α-H); 5.87(1H, d, *J*=7.5-8.5Hz, 16α-H); 5.10(1H, m, H-24); 5.13(1H, m 24-H); 0.92(1H, d, X component of ABX system J_{AX} 6.0-6.5Hz and J_{BX}=0Hz, 30-secondary methyl); 1.62 and 1.68(26- and 27-methyls); 0.99, 1.12, and 1.17(18-, 19- and 32-tertiary methyls); 1.98(16-acetate group); and 3.67ppm (21-methoxy group).

Mass Data:

Anal. calcd. for $C_{31}H_{48}O_6 \cdot ½H_2O$, C, 70.85; H, 9.40%; found: C, 70.68; H, 9.32%; methyl ester: anal. calcd. for $C_{32}H_{50}O_6$: C, 72.41; H, 9.50%; found: C, 72.30; H, 9.46%.

TLC Data

Silica gel (Merck HF_{254}) plates developed with (A) ether-acetic acid (100:0.5, v/v), (B) ether-dichloromethane-acetic acid (50:50:0.5, v/v/v), (C) dichloromethane-methanol-acetic acid (95:5:0.5, v/v/v), and (D) cyclohexane-chloroform-methanol-acetic acid (20:80:2.5:10, v/v/v/v). R_f = 0.58, 0.44, 0.40, and 0.64, respectively. Detected by spraying plates with concentrated H_2SO_4 and heating; color was red.

References

W. O. Godtfredsen, N. Rastrup-Andersen, S. Vangedal, and W. D. Ollis; Metabolites of *Fusidium coccineum*; Tetrahedron, Vol. 35, pp. 2419-2431(1979).

W. B. Turner and D. C. Aldridge; Fungal Metabolites II; Academic Press, New York, p. 333(1983).

Common/Systematic Name
9,11-Anhydrofusidic acid

Molecular Formula/Molecular Weight
$C_{31}H_{46}O_5$; MW = 498.33452

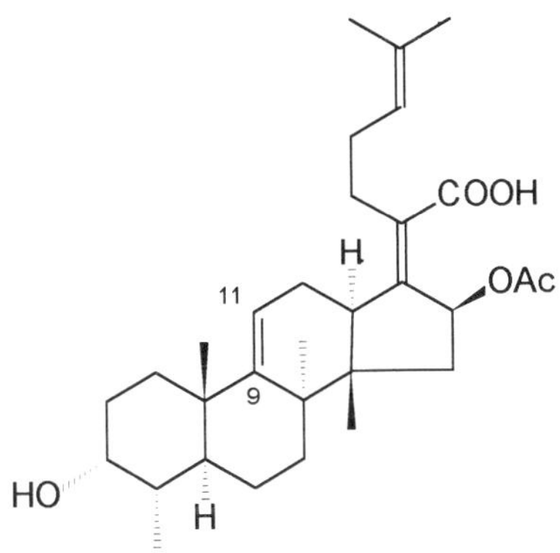

General Characteristics
Crystals from ether-light petroleum; mp., 135-137°C; methyl ester: an amorphous solid, dicyclohexylamine salt; mp., 163-164°C; $[\alpha]_D^{20}$ -34°.

Fungal Source
Fusidium coccineum

Isolation/Purification
Following the usual procedure for the isolation of fusidic acid (see Godtfredsen, *et al.*, 1979), the benzene filtrate after collection of the fusidic acid-benzene solvate was evaporated yielding an oil. This oil was dissolved in acetone, diethanolamine was added and, after seeding and standing, the precipitate of crystalline diethanolamine salts was collected and washed with cold acetone. The mixture of crystalline diethanolamine salts was suspended in ethanol and acidified to pH 3. Benzene and water were added, the benzene layer was concentrated to give a fusidic acid-benzene solvate. Evaporation of the benzene filtrate and solution of the residue in acetone and addition of dicyclohexylamine gave a mixture of crystalline dicyclohexylamine salts. The mixture of dicyclohexylamine salts was acidified and extracted with ether. 9,11-Anhydrofusidic acid was purified from the acidic fraction by chromatography on silicic acid eluted with ether-dichloromethane-acetic acid, 25:75:0.5 (v/v/v) followed by crystallization of its dicyclohexylamine salt from early fractions from the silicic acid column.

Biological Activity
Antibiotic activity against Gram-positive bacteria and, in particular, a strain of *Staphylococcus aureus*; no activity against the Gram-negative bacterium, *Escherichia coli*.

Spectral Data

UV:

λ_{max}^{EtOH} 204(ϵ = 15,400) and 225nm (infl 8,250).

IR:

(KBr) 3440, 1740, 1690, 1625, and 1250cm^{-1}.

^{1}H NMR:

($CDCl_3$) methyl ester: 3.73(1H, m, 3β-H); 5.47(1H, t, *J*=3.5Hz, 11-olefinic H); 5.95(1H, d, *J*=7.5-8.5Hz, 16α-H); 5.12(1H, m, H-24); 1.60 and 1.68(26- and 27-methyls); 0.88, 0.98, and 1.18(18-, 19- and 32-tertiary methyls); 1.97(16-acetate group); and 3.65ppm (21-methoxy group).

Mass Data:

Anal. calcd for $C_{31}H_{46}O_5$: C, 74.66; H, 9.30%; found: C, 74.53; H, 9.25%; methyl ester: anal. calcd for $C_{32}H_{48}O_6$: C, 74.96; H, 9.44%; found: C, 74.55; H, 9.32%.

TLC Data

Silica gel (Merck HF_{254}) plates developed with (A) ether-acetic acid (100:0.5, v/v), (B) ether-dichloromethane-acetic acid (50:50:0.5, v/v/v), (C) dichloromethane-methanol-acetic acid (95:5:0.5, v/v/v), and (D) cyclohexane-chloroform-methanol-acetic acid (20:80:2.5:10, v/v/v/v). R_f = 0.71, 0.53, 0.48, and 0.75, respectively. Detected by spraying plates with concentrated H_2SO_4 and heating; initial color was crimson changing to blue-violet.

References

W. O. Godtfredsen, N. Rastrup-Andersen, S. Vangedal, and W. D. Ollis; Metabolites of *Fusidium coccineum*; Tetrahedron, Vol. 35, pp. 2419-2431(1979).

W. B. Turner and D. C. Aldridge; Fungal Metabolites II; Academic Press, New York, p. 333(1983).

Common/Systematic Name

9,11-Anhydro-12-hydroxyfusidic acid

Molecular Formula/Molecular Weight

$C_{31}H_{46}O_6$; MW = 514.32944

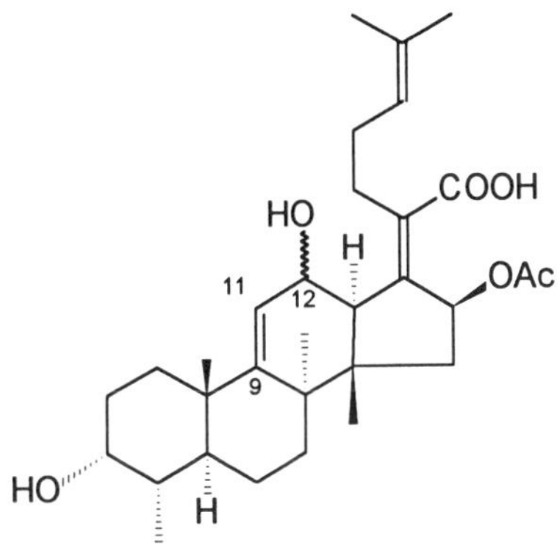

General Characteristics

An amorphous solid; $[\alpha]_D^{20}$ + 32°; methyl ester: an amorphous solid.

Fungal Source

Fusidium coccineum.

Isolation/Purification

Following the usual procedure for the isolation of fusidic acid (see Godtfredsen, *et al.*, 1979), the benzene filtrate after collection of the fusidic acid-benzene solvate was evaporated yielding an oil. This oil was dissolved in acetone, diethanolamine was added and, after seeding and standing, the precipitate of crystalline diethanolamine salts was collected and washed with cold acetone. The mixture of crystalline diethanolamine salts was suspended in ethanol and acidified to pH 3. Benzene and water were added; the benzene layer was concentrated to give a fusidic acid-benzene solvate. Evaporation of the benzene filtrate and solution of the residue in acetone and addition of dicyclohexylamine gave a mixture of crystalline dicyclohexylamine salts. The mixture of dicyclohexylamine salts was acidified and extracted with ether. 9,11-Anhydro-12-hydrofusidic acid was purified by chromatography on silicic acid eluted with dichloromethane-ethyl acetate-acetic acid, 70:30:0.5 (v/v/v) followed by preparative TLC (ether-acetic acid, 100:0.5, v/v) of the more polar column fractions.

Biological Activity

Antibiotic activity against Gram-positive bacteria and, in particular, a strain of *Staphylococcus aureus*; no activity against the Gram-negative bacterium, *Escherichia coli*.

Spectral Data

UV:

λ_{max}^{EtOH} 205(ϵ = 15,700) and 225nm (infl 6,300).

IR:

(KBr) 3450, 1720, and 1250cm^{-1}; ($CHCl_3$) 3640, 3600, 3020, 1725, 1700, and 1240cm^{-1}.

^{1}H NMR:

($CDCl_3$) methyl ester: 3.75(1H, m, 3β-H); 5.53(1H, m, olefinic 7-H); 4.37(2H, dd, 12α- and 12β-H); 2.67(1H, d, *J*=8.0Hz, 13α-H); 5.89(1H, d, *J*=7.5-8.5Hz, 16α-H); 5.18(1H, m, H-24); 1.60 and 1.69(C-26 and C-27 methyls); 0.90, 1.02, and 1.23(18-, 19-, and 32-tertiary methyls); 1.98(16-acetate group); and 3.68ppm (21-methoxy group).

Mass Data:

Methyl ester: anal. calcd for $C_{31}H_{46}O_6 \cdot H_2O$: C, 71.12; H, 9.05%; found: C, 71.35; H, 9.10%; methyl ester: HREIMS: 510.3360*m/e* (M^+ - H_2O, $C_{32}H_{46}O_5$); metastable transition detected (M^+, 528 → *m/e* 510).

TLC Data

Silica gel (Merck HF_{254}) plates developed with (A) ether-acetic acid (100:0.5, v/v), (B) ether-dichloromethane-acetic acid (50:50:0.5, v/v/v), (C) dichloromethane-methanol-acetic acid (95:5:0.5, v/v/v), and (D) cyclohexane-chloroform-methanol-acetic acid (20:80:2.5:10, v/v/v/v). R_f = 0.43, 0.27, 0.30, and 0.50, respectively. Detected by spraying plates with concentrated H_2SO_4 and heating; initial color was brown-violet changing to blue-violet.

References

W. O. Godtfredsen, N. Rastrup-Andersen, S. Vangedal; and W. D. Ollis; Metabolites of *Fusidium coccineum*; Tetrahedron, Vol. 35, pp. 2419-2431(1979).

W. B. Turner and D. C. Aldridge; Fungal Metabolites II; Academic Press, New York, p. 333(1983).

Common/Systematic Name

9,11-Anhydro-9α,11α-epoxyfusidic acid

Molecular Formula/Molecular Weight

$C_{31}H_{46}O_6$; MW = 514.32944

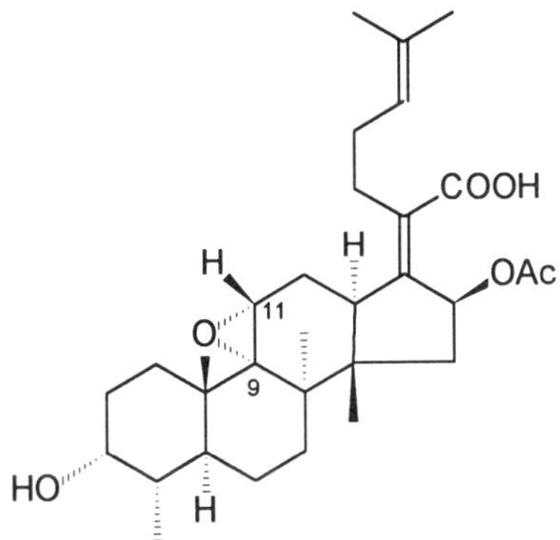

General Characteristics

An amorphous solid; $[\alpha]_D^{20}$ - 40°; methyl ester: an amorphous solid, dicyclohexylamine salt; mp., 135-138°C.

Fungal Source

Fusidium coccineum.

Isolation/Purification

Following the usual procedure for the isolation of fusidic acid (see Godtfredsen, *et al.*, 1979), the benzene filtrate after collection of the fusidic acid-benzene solvate was evaporated yielding an oil. This oil was dissolved in acetone, diethanolamine was added and, after seeding and standing, the precipitate of crystalline diethanolamine salts was collected and washed with cold acetone. The mixture of crystalline diethanolamine salts was suspended in ethanol and acidified to pH 3. Benzene and water were added; the benzene layer was concentrated to give a fusidic acid-benzene solvate. Evaporation of the benzene filtrate and solution of the residue in acetone and addition of dicyclohexylamine gave a mixture of crystalline dicyclohexylamine salts. The mixture of dicyclohexylamine salts was acidified and extracted with ether. 9,11-Anhydro-9α,11α-epoxyfusidic acid was purified from the mother liquor of the acidic fraction by chromatography on silicic acid eluted with ether-light petroleum-acetic acid, 60:40:0.5 (v/v/v). The 9,11-anhydro-9α,11α-epoxyfusidic acid containing fraction was converted to its dicyclohexylamine salt and chromatographed on silicic acid eluted with ether-light petroleum-acetic acid, 50:50:0.5 (v/v/v) to give the purified salt.

Biological Activity

Antibiotic activity against Gram-positive bacteria and, in particular, a strain of *Staphylococcus aureus*; no activity against the Gram-negative bacterium, *Escherichia coli*.

Spectral Data

UV:

λ_{max}^{EtOH} 204(ϵ = 10,900) and 225nm (infl 8,000).

IR:

(KBr) 3450, 1730, 1710, and 1240cm^{-1}.

^{1}H NMR:

($CDCl_3$) Methyl ester: 3.72(1H, m, 3β-H); 3.17(1H, m, 11β-H); 5.95(1H, d, *J*=7.5-8.5Hz, 16α-H); 5.13(1H, m, H-24); 0.97(3H, d, X component of an ABX system, *J*AX=6.0-6.5Hz, 30-secondary methyl); 1.62 and 1.70(26- and 27-methyls); 0.97, 1.07, and 1.12(18-, 19-, and 32-tertiary methyls); 1.98(16-acetate group); and 3.68ppm (21-methoxy group).

Mass Data:

Anal. calcd. for $C_{31}H_{46}O_6$: C, 71.12; H, 9.05; found: C, 71.17; H, 9.14%; methyl ester: HREIMS: 468.3250*m/e* (M^+ - HOAc, $C_{30}H_{44}O_4$); metastable transition detected (M^+, 528 → *m/e* 468).

TLC Data

Silica gel (Merck HF_{254}) plates developed with (A) ether-acetic acid (100:0.5, v/v), (B) ether-dichloromethane-acetic acid (50:50:0.5, v/v/v), (C) dichloromethane-methanol-acetic acid (95:5:0.5, v/v/v), and (D) cyclohexane-chloroform-methanol-acetic acid (20:80:2.5:10, v/v/v/v). R_f = 0.62, 0.49, 0.45, and 0.68, respectively. Detected by spraying plates with concentrated H_2SO_4 and heating; initial color was brown-violet changing to blue-violet.

References

W. O. Godtfredsen, N. Rastrup-Andersen, S. Vangedal, and W. D. Ollis; Metabolites of *Fusidium coccineum*; Tetrahedron, Vol. 35, pp. 2419-2431(1979).

W. B. Turner and D. C. Aldridge; Fungal Metabolites II; Academic Press, New York, p. 333(1983).

Common/Systematic Name

7,8-Dehydropseudofusidic acid

Molecular Formula/Molecular Weight

$C_{31}H_{46}O_6$; MW = 514.32944

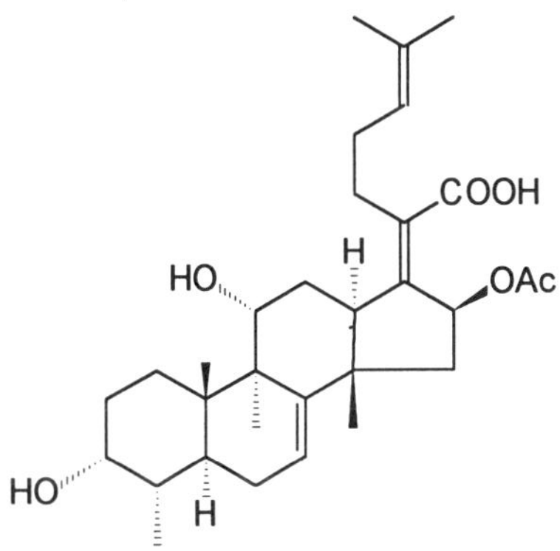

General Characteristics

An amorphous solid; $[\alpha]_D^{20}$ - 33°; methyl ester: an amorphous solid.

Fungal Source

Fusidium coccineum.

Isolation/Purification

Following the usual procedure for the isolation of fusidic acid, the benzene filtrate after the collection of the fusidic acid-benzene solvate was evaporated yielding an oil. This oil was dissolved in acetone, diethanolamine was added and, after seeding and standing, the precipitate of crystalline diethanolamine salts was collected and washed with cold acetone. The mixture of crystalline diethanolamine salts was suspended in ethanol and acidified to pH 3. Benzene and water were added; the benzene layer was concentrated to give a fusidic acid-benzene solvate. Evaporation of the benzene filtrate and solution of the residue in acetone and addition of dicyclohexylamine gave a mixture of crystalline dicyclohexylamine salts. The mixture of dicyclohexylamine salts was acidified and extracted with ether. 7,8-Dehydrofusidic acid was purified by chromatography on silicic acid eluted with dichloromethane-ethyl acetate-acetic acid, 70:30:0.5 (v/v/v), followed by preparative TLC (ether-acetic acid, 100:0.5, v/v) of the more polar column fractions.

Biological Activity

Antibiotic activity against Gram-positive bacteria and, in particular, a strain of *Staphylococcus aureus*; no activity against the Gram-negative bacterium, *Escherichia coli*.

Spectral Data

UV:

λ_{max}^{EtOH} 204(ϵ=15,300) and 225nm (infl 8,200).

IR:
(KBr) 3450, 1720, and 1255cm^{-1}.

^{1}H NMR:
($CDCl_3$) methyl ester: 3.73(1H, m, 3β-H); 5.35(1H, m, olefinic 7-H); 4.48(1H, m, 11β-H); 5.95(1H, d, J=7.5-8.5Hz, 16α-H); 5.17(1H, m, H-24); 0.95(3H, d, X component of an ABX system, J_{AX}=6.0-6.5Hz, J_{BX}=0, 30-secondary methyl); 1.62 and 1.70(26- and 27-methyls); 0.96, 1.06, and 1.25(18-, 19- and 32-tertiary methyls); 1.98(16-acetate group); and 3.68ppm (21-methoxy group).

Mass Data:
Anal. calcd. for $C_{31}H_{46}O_6 \cdot H_2O$: C, 69.89; H, 9.08%; found: C, 70.17; H, 9.00%; methyl ester: HREIMS: 468.3250*m/e* (M^+ - HOAc, $C_{30}H_{44}O_4$); metastable transition detected (M^+, 528 → *m/e* 468).

TLC Data

Silica gel (Merck HF_{254}) plates developed with (A) ether-acetic acid (100:0.5, v/v), (B) ether-dichloromethane-acetic acid (50:50:0.5, v/v/v), (C) dichloromethane-methanol-acetic acid (95:5:0.5, v/v/v), and (D) cyclohexane-chloroform-methanol-acetic acid (20:80:2.5:10, v/v/v/v). R_f = 0.52, 0.38, 0.37, and 0.60, respectively. Detected by spraying plates with concentrated H_2SO_4 and heating; initial color was brown-violet changing to blue-violet.

References

W. O. Godtfredsen, N. Rastrup-Andersen, S. Vangedal; and W. D. Ollis; Metabolites of *Fusidium coccineum*; Tetrahedron, Vol. 35, pp. 2419-2431(1979).

W. B. Turner and D. C. Aldridge; Fungal Metabolites II; Academic Press, New York, p. 333(1983).

Common/Systematic Name

Fusilactidic acid

Molecular Formula/Molecular Weight

$C_{31}H_{46}O_7$; MW = 530.32435

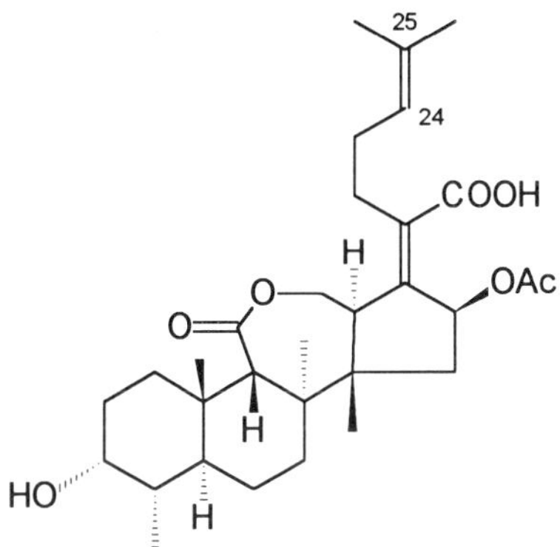

General Characteristics

Crystals from ether; mp., 192-193 °C; $[\alpha]_D^{20}$ - 14°; methyl ester: mp., 192-193 °C.

Fungal Source

Fusidium coccineum.

Isolation/Purification

Following the usual procedure for the isolation of fusidic acid, the benzene filtrate after the collection of the fusidic acid-benzene solvate was evaporated yielding an oil. This oil was dissolved in acetone, diethanolamine was added and, after seeding and standing, the precipitate of crystalline diethanolamine salts was collected and washed with cold acetone. The mixture of crystalline diethanolamine salts was suspended in ethanol and acidified to pH 3. Benzene and water were added; the benzene layer was concentrated to give a fusidic acid-benzene solvate. Evaporation of the benzene filtrate and solution of the residue in acetone and addition of dicyclohexylamine gave a mixture of crystalline dicyclohexylamine salts. The mixture of dicyclohexylamine salts was acidified and extracted with ether. Fusilactidic acid was purified from the mother liquor of the acidic fraction by chromatography on silicic acid eluted with ether-light petroleum-acetic acid, 60:40:0.5 (v/v/v). The fusilactidic acid containing fraction was crystallized from ether on standing.

Biological Activity

Antibiotic activity against Gram-positive bacteria and, in particular, a strain of *Staphylococcus aureus*; no activity against the Gram-negative bacterium, *Escherichia coli*.

Spectral Data

UV:
λ_{max}^{EtOH} 208(ϵ=9,900) and 225nm (infl 6,800); methyl ester: λ_{max}^{EtOH} 205(ϵ=10,800) and 225nm (infl 7,450).

IR:
(KBr) 3460, 3540, 1735, 1720, 1680, and 1240cm^{-1}; methyl ester: 3560, 1735, 1725, 1705, and 1250cm^{-1}.

^{1}H NMR:
($CDCl_3$) methyl ester, 3.70(1H, m, 3β-H); 2.83(1H, 9β-H); 4.82 and 4.47(2H, m, 12α- and 12β-H); 2.97(1H, m, 13α-H); 5.83(1H, d, *J*=7.5-8.5Hz, 16α-H); 5.09(1H, m, H-24); 0.98(3H, d, X component of an ABX system, J_{AX}=6.0-6.5Hz; J_{BX} = 0); 1.62 and 1.69(C-26 and C-27 methyls); 1.13, 1.13 and 1.18(18-, 19- and 32-tertiary methyls); 1.99(16-acetate group); and 3.67ppm (21-methoxy group).

Mass Data:
Anal. calcd for $C_{31}H_{46}O_7$: C, 70.16%; H, 8.74; found: C, 70.16; H, 8.73%; methyl ester: anal. calcd for $C_{31}H_{48}O_7$: C, 70.56; H, 8.88%; found: C, 70.45; H, 8.97%; HREIMS: 544.3391*m/e*; 3-*O*-acetate, 586.3503*m/e*.

TLC Data

Silica gel (Merck HF_{254}) plates developed with (A) ether-acetic acid (100:0.5, v/v), (B) ether-dichloromethane-acetic acid (50:50:0.5, v/v/v), (C) dichloromethane-methanol-acetic acid (95:5:0.5, v/v/v), and (D) cyclohexane-chloroform-methanol-acetic acid (20:80:2.5:10, v/v/v/v). R_f = 0.32, 0.35, 0.41, and 0.61, respectively. Detected by spraying plates with concentrated H_2SO_4 and heating; color was yellow.

References

W. O. Godtfredsen, N. Rastrup-Andersen, S. Vangedal and W. D. Ollis; Metabolites of *Fusidium coccineum;* Tetrahedron, Vol. 35, pp. 2419-2431(1979).

W. B. Turner and D. C. Aldridge; Fungal Metabolites II; Academic Press, New York, p. 333(1983).

Common/Systematic Name
Fusidic acid

Molecular Formula/Molecular Weight
$C_{31}H_{48}O_6$; MW = 516.34509

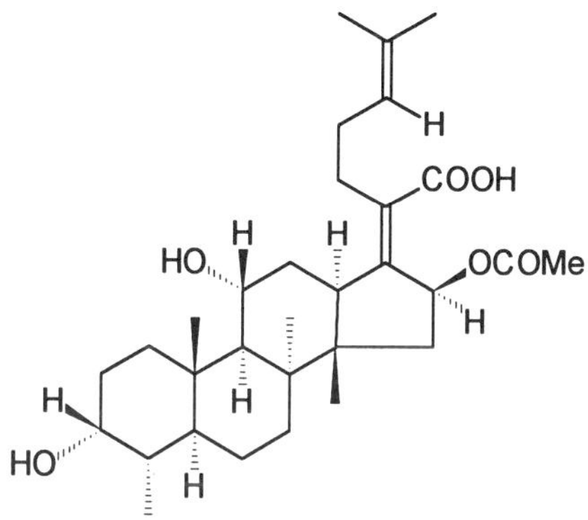

General Characteristics
Colorless needles from benzene; mp., 133.5-135°C.

Fungal Source
Isaria kogane (a mushroom parasitic mainly on larvae and imagos of beetles), *Acremonium fusidioides, Calcarisporium arbuscula, Cephalosporium acremonium,* and *Gabarnaudia tholispora.*

Isolation/Purification
The culture filtrate was extracted with ethyl acetate, evaporated to dryness and chromatographed on silica gel twice. The biologically active fractions were combined and fusidic acid was crystallized from benzene.

Biological Activity
Antibiotic activity against Gram-positive bacteria and, in particular, a strain of *Staphylococcus aureus*; no activity against the Gram-negative bacterium, *Escherichia coli*.

Spectral Data

UV:
λ_{max}^{EtOH} 220nm (log ϵ = 3.86).

IR:
(KBr) 3400(hydroxyl), 1725, 1250(acetoxy), and 1715cm^{-1} (carboxyl).

^{1}H NMR:

($CDCl_3$) 0.90, 0.96, and 1.36ppm (3 methyl singlets, C-18, C-19 and C-32); 0.91(3H, d, *J*=6Hz, C-30-methyl); 1.58, 1.66ppm (3H, d, *J*=4.0Hz, C-26 and C-27 methyls); 1.94ppm (3H, s, C-16-$OCOCH_3$); 3.73(1H, m, C-3-H); 4.30(1H, m, C-11-H); ~5.1(1H, m, C-24-H); and 5.84ppm (1H, d, *J*=9.0 Hz, C-16-H).

Mass Data:

LREIMS: 516(M^+), 456, 438, 420, 405, 351, 283, 232, 189, 95, 93, 69, 60, 45, 43, and 41*m/e*; anal. calcd for $C_{31}H_{48}O_6 \cdot \frac{1}{2} H_2O$: C, 69.94; H, 9.40%; found: C, 70.12; H, 9.34%.

References

H. Hikino, Y. Asada, S. Arihara, and T. Takemoto; Fusidic Acid, a Steroidal Antibiotic from *Isaria kogane*; Chem. Pharm. Bull., Vol. 20, pp. 1067-1069(1972).

W. B. Turner and D. C. Aldridge; Fungal Metabolites II; Academic Press, New York, p. 334(1983).

Common/Systematic Name
Helvolic acid; Fumigacin

Molecular Formula/Molecular Weight
$C_{33}H_{44}O_8$; MW = 568.30362

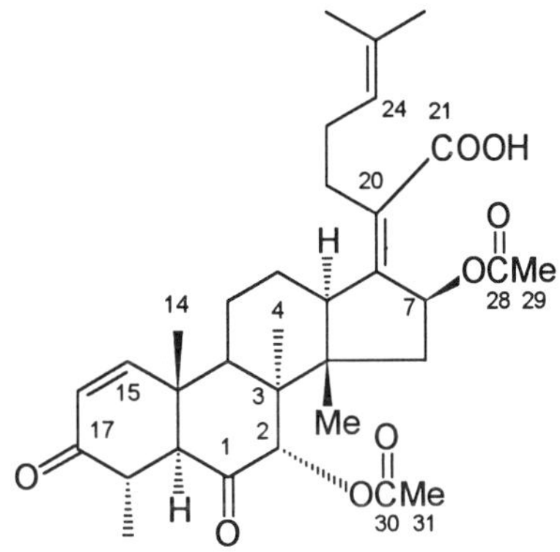

General Characteristics
Crystals from aqueous methanol; mp., 208-212°C; $[\alpha]_D^{25}$ - 121° (in $CHCl_3$).

Fungal Source
Aspergillus fumigatus, *Cephalosporium caerulens*, and *Emericellopsis terricola*.

Biological Activity
Antibiotic activity towards bacteria; Gram-positive bacteria are more sensitive than Gram-negative bacteria. Early mammalian toxicity reports for helvolic acid were in error due to contamination by gliotoxin.

Spectral Data

UV:
λ_{max}^{EtOH} 231nm (ϵ = 17,400).

^{13}C NMR:
($CDCl_3$) 13.0; 17.7; 17.9; 18.2; 20.4; 20.6; 23.9; 25.6; 25.9; 27.4; 28.2; 28.5; 30.1; 40.4; 40.7; 41.8; 46.5; 47.2; 49.4; 52.6; 73.4; 73.7; 122.6; 127.6; 130.2; 132.6; 147.4; 156.8; 168.5; 169.8; 200.9; and 208.3ppm.

Mass Spectrum:
LREIMS: The largest detectable ion in the mass spectrum was 508(due to M^+ - 60), 490, 479, 473, 465, 439(100), 430, 415, 387, 343, and 283*m/e*.

References

R. J. Cole and R. H. Cox; Handbook of Toxic Fungal Metabolites; Academic Press, New York, pp. 806-809(1981).

W. B. Turner and D. C. Aldridge; Fungal Metabolites II; Academic Press, New York, p. 324-328(1983).

17

LL-Z1271 Antibiotics and Related Metabolites

LL-Z1271α
LL-Z1271β
LL-Z1271γ
Wentilactone A
Wentilactone B

Common/Systematic Name
LL-Z1271α

Molecular Formula/Molecular Weight
$C_{17}H_{20}O_5$; MW = 304.13107

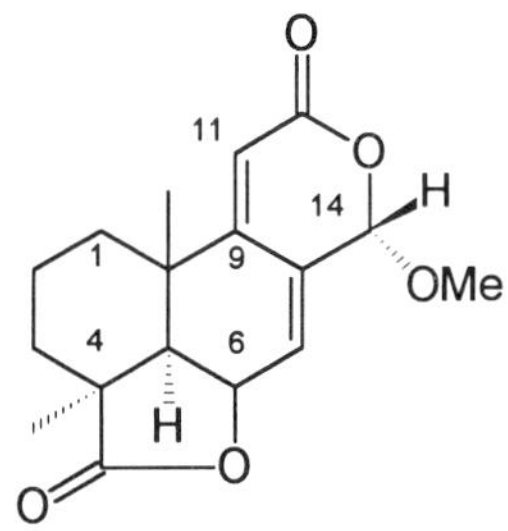

General Characteristics
Recrystallization from either acetone-hexane or methylene chloride-hexane gave crystals with mp. 214-215°C; $[\alpha]_D^{25}$ -203° (*c*=0.29, in MeOH).

Fungal Source
Acrostalagmus sp.

Isolation/Purification
The culture was adjusted to pH 3.0 with hydrochloric acid and extracted with ethyl acetate. Concentration of the extract *in vacuo* followed by drying over sodium sulfate and the addition of hexane caused essentially pure LL-Z1271α to crystallize. Final purification was by recrystallization from either acetone-hexane or methylene chloride-hexane.

Biological Activity
Antifungal activity.

Spectral Data

UV:
λ_{max}^{MeOH} 257nm (ϵ=13,500).

IR:
(KBr) 1775(γ-lactone) and 1730cm^{-1} (δ-lactone).

^{1}H NMR:
($CDCl_3$) 1.11 and 1.33(s, CH_3's at C-10 and C-4); 1.93(d, H-5, $J_{5,6}$=5.3 Hz); 3.70(s, OMe); 5.07(t, H-6); 5.76(2H's, H-11 and -14); and 6.53ppm (m, H-7); (benzene-d_6) 1.08 and 1.16(s, CH_3's); 3.60(s, OMe); 4.73(t, H-6); 5.52(t, H-14); 5.70(d, H-11); and 6.43ppm (m, H-7).

Mass Spectrum:
EIMS: 304*m/e* (M^+, $C_{17}H_{20}O_5$).

Reference

G. A. Ellestad, R. H. Evans, Jr., M. P. Kunstmann, J. E. Lancaster, and G. O. Morton; Structure and Chemistry of Antibiotic LL-Z1271α, an Antifungal Carbon-17 Terpene; J. Am. Chem. Soc., Vol. 92, pp. 5483-5489(1970).

Common/Systematic Name

LL-Z1271β

Molecular Formula/Molecular Weight

$C_{16}H_{24}O_5$; MW = 296.16237

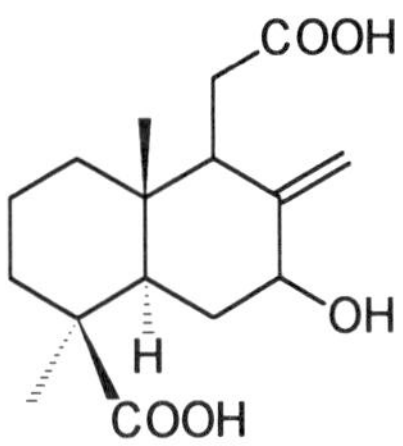

General Characteristics

Crystals; mp., 200°C (dec); $[\alpha]_D$ +22.5 (in MeOH).

Fungal Source

Acrostalagmus sp.

Spectral Data

IR:

(KBr) 3500(hydroxyl), 2750, 1710(carboxyl), 1650, and 900cm^{-1} (exocyclic methylene).

^{1}H NMR:

($CDCl_3$) The NMR spectrum showed signals for two tertiary C-methyl groups at 0.53 and 1.15ppm and characteristic broadened doublet for an exocyclic methylene group at 4.44 and 5.08ppm. Also present was a broad one-proton signal at 3.75 assigned to an axial carbinol proton.

Mass Spectrum:

EIMS: 296(M^+), 167, 139, and 121*m/e*. The latter three were the result of cleavage of the allylic bonds at C-6 and C-9. The base peak was at 123*m/e*.

Reference

G. A. Ellestad, R. H. Evans, Jr., and M. P. Kunstmann; LL-Z1271β, and Additional C_{16} Terpenoid Metabolite from an *Acrostalagmus* Species; Tetrahedron Letters, pp. 497-500(1971).

Common/Systematic Name

LL-Z1271γ

Molecular Formula/Molecular Weight

$C_{16}H_{18}O_5$; MW = 290.11542

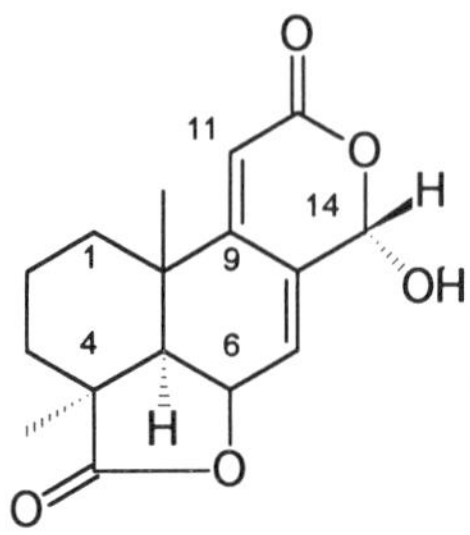

General Characteristics

Crystals from ether; mp., 241-247°C; $[\alpha]_D^{25}$ -256° (c=0.43, in MeOH).

Fungal Source

Acrostalagmus sp.

Isolation/Purification

The methanol extracts of the mycelium pads were combined and concentrated *in vacuo* to an aqueous phase, which was extracted twice with one-half volumes of chloroform. The chloroform layer was separated, dried over sodium sulfate and concentrated to an oily residue. Silica gel chromatography (slurry packed in methylene chloride) by elution with methylene chloride and ether-methylene chloride (1:4, v/v) (monitored at 255nm) gave two bands. The antibiotic was crystallized from both bands from ethyl ether solution.

Biological Activity

Antifungal activity.

Spectral Data

UV:

λ_{max}^{MeOH} 257nm (ε=15,400).

IR:

(KBr) 1780(γ-lactone) and 1680cm^{-1} (δ-lactol, occurs at 1720cm^{-1} in $CHCl_3$).

^{1}H NMR:

($CDCl_3$ with 2 drops of $DMSO_4$) 1.16 and 1.33ppm (s, CH_3's at C-10 and C-4); 1.95(d, H-5, $J_{5,6}$=5.0Hz); 5.07(t, H-6); 5.77(d, H-11, $J_{7,11}$=2.0Hz); 6.08(broadened s, H-14); and 6.46ppm (m, H-6).

Mass Data:

EIMS: 290*m/e* (M^+, $C_{16}H_{18}O_5$); anal. calcd for $C_{16}H_{18}O_5$, C, 66.19; H, 6.25%; found: C, 66.46; H, 6.31%.

Reference

G. A. Ellestad, R. H. Evans, Jr., M. P. Kunstmann, J. E. Lancaster and G. O. Morton; Structure and Chemistry of Antibiotic LL-Z1271α, an Antifungal Carbon-17 Terpene; J. Am. Chem. Soc., Vol. 92, pp. 5483-5489(1970).

Common/Systematic Name

Wentilactone A

Molecular Formula/Molecular Weight

$C_{16}H_{16}O_6$; MW = 304.09469

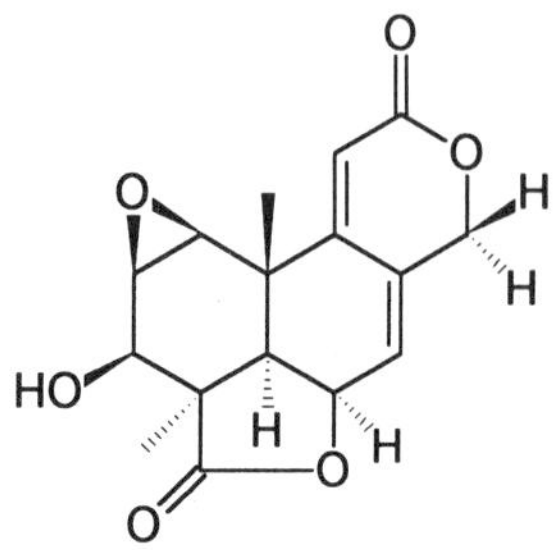

General Characteristics

Crystalline cubes from methanol-ethyl acetate (1:1); mp., 258°C. Soluble in water and acetone, sparingly soluble in methanol and insoluble in hexane, benzene, ethyl ether, chloroform, and ethyl acetate.

Fungal Source

Aspergillus wentii (NRRL 6435).

Isolation/Purification

Fungal cultures were extracted with water saturated chloroform, dried over sodium sulfate and concentrated to near dryness. The crude extract was fractionated using a series of silica gel column chromatography systems (elution with a linear gradient from benzene to ethyl acetate) followed by a final purification using Florisil column chromatography (elution with a linear gradient from benzene to ethyl acetate) to separate wentilactone A from B. Purification was monitored with a one-day-old chick bioassay.

Biological Activity

Wentilactone A had an LD_{50} of 7mg/kg in day-old chicks; it was phytotoxic showing growth inhibition of 81, 81, and 48% in wheat coleoptiles at 10^{-3}, 10^{-4} and 10^{-5} M concentrations, respectively.

Spectral Data

UV:

λ_{max}^{MeOH} 259nm (1.34 X 10^4 M^{-1} cm^{-1}) (indicates diene lactone system).

IR:
(KBr) 3500, 1780, and 1710cm^{-1}.

CD:
(MeOH) 259nm = -8.23 X 10^3 deg cm^2/dmol.

^{1}H NMR:
(DMSO-d_6) H-1, 3.63(*J*=4.4Hz); H-2, 3.38(*J*=4.4, 6.1Hz); H-3, 4.30(*J*=6.1Hz); H-5, 2.22(*J*=5.6Hz); H-6, 5.09(*J*=5.6, 4.4Hz); H-7, 6.36(*J* = 4.4, 1.5Hz); H-11, 6.22(*J*=1.5Hz); H-14, 5.04; H-18, 1.10; and H-20, 1.38ppm.

^{13}C NMR:
(DMSO-d_6) C-1, 50.20; C-2, 54.69; C-3, 66.78; C-4, 48.21; C-5, 47.72; C-6, 69.49; C-7, 121.82; C-8, 131.15; C-9, 156.45; C-10, 35.57; C-11, 111.91; C-12, 162.95; C-14, 68.81; C-18, 18.65; C-19, 175.84; and C-20, 24.85ppm.

Mass Spectrum:
HREIMS: 304.0914*m/e* (M^+, consistent with a molecular formula of $C_{16}H_{16}O_6$; calculated mass 304.0946).

TLC Data

R_f 0.18 appeared as a brown quenching spot under long-wave UV light after spraying with Ehrlich's reagent and heating (silica gel GH-R, solvent system, chloroform-acetone-methanol, 17:2:1, v/v/v).

Reference

J. W. Dorner, R. J. Cole, J. P. Springer, R. H. Cox, H. G. Cutler, and D. T. Wicklow; Isolation and Identification of Two New Biologically Active Norditerpene Dilactones from *Aspergillus wentii*; Phytochemistry, Vol. 19, pp. 1157-1161(1980).

Common/Systematic Name
Wentilactone B

Molecular Formula/Molecular Weight
$C_{16}H_{18}O_5$; MW = 290.11542

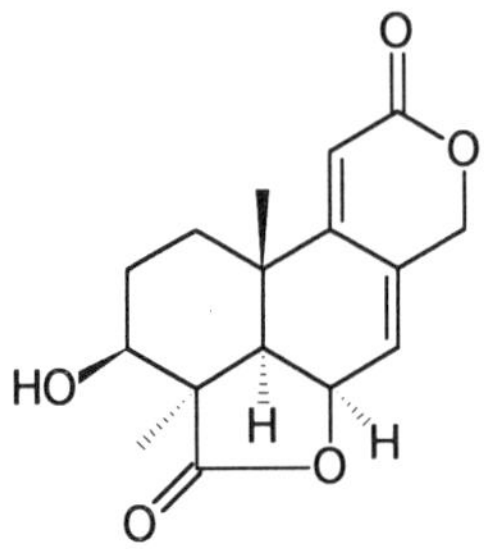

General Characteristics
Crystalline cubes from methanol-ethyl acetate (1:1); mp., 260°C. Soluble in water and acetone, sparingly soluble in methanol and insoluble in hexane, benzene, ethyl ether, chloroform, and ethyl acetate.

Fungal Source
Aspergillus wentii (NRRL 6435).

Isolation/Purification
Fungal cultures were extracted with water saturated chloroform, dried over sodium sulfate and concentrated to near dryness. The crude extract was fractionated using a series of silica gel column chromatography systems (elution with a linear gradient from benzene to ethyl acetate) followed by a final purification using Florisil column chromatography (elution with a linear gradient from benzene to ethyl acetate) to separate wentilactone A from B. Purification was monitored with a one-day-old chick bioassay.

Biological Activity
Wentilactone B was not toxic to day-old chicks dosed up to 250mg/kg; it was less phytotoxic than A showing growth inhibition of 100 and 48% in wheat coleoptiles at 10^{-3} and 10^{-4} M concentrations respectively.

Spectral Data

UV:
λ_{max}^{MeOH} 259nm (1.40 X 10^4 M^{-1} cm^{-1}) (indicates diene lactone system).

IR:
(KBr) 3500, 1780, and 1710cm^{-1}.

CD:
(MeOH) 260nm = -8.23 X 10^3 deg cm^2/dmol.

^{1}H NMR:
(DMSO-d_6) H-1, 1.3-2.3; H-2, 1.3-2.3; H-3, 4.98; H-5, 2.07(J=5.4Hz); H-6, 5.19(J=5.4, 4.4Hz); H-7, 6.35(J=4.4, 1.7Hz); H-11, 5.74(J=1.7Hz); H-14, 4.98; H-18, 1.11; and H-20, 1.30ppm.

^{13}C NMR:
(DMSO-d_6) C-1, 39.63; C-2, 36.33; C-3, 63.22; C-4, 42.24; C-5, 45.53; C-6, 71.30; C-7, 121.64; C-8, 131.38; C-9, 158.17; C-10, 35.56; C-11, 111.85; C-12, 163.16; C-14, 69.32; C-18, 23.15; C-19, 180.70; and C-20, 27.37ppm.

Mass Spectrum:
HREIMS: 290.1166*m/e* (M^+, consistent with a molecular formula of $C_{16}H_{18}O_5$; calculated mass 290.1153).

TLC Data

R_f 0.12 appeared as a yellow fluorescent spot under long-wave UV light after spraying with Ehrlich's reagent and heating (silica gel GH-R, solvent system, chloroform-acetone-methanol, 17:2:1, v/v/v).

Reference

J. W. Dorner, R. J. Cole, J. P. Springer, R. H. Cox, H. G. Cutler, and D. T. Wicklow; Isolation and Identification of Two New Biologically Active Norditerpene Dilactones from *Aspergillus wentii*; Phytochemistry, Vol. 19, pp. 1157-1161(1980).

18

LL-Z1272 Antibiotics and Related Metabolites

LL-Z1272α
LL-Z1272β
LL-Z1272δ
LL-Z1272ε
LL-Z1272ζ
LL-Z1272γ (Ascochlorin)
Ascofuranone
Ascofuranol
Chloronectrin
Deacetylchloronectrin

Common/Systematic Name

Antibiotic LL-Z1272α

Molecular Weight/Molecular Formula

$C_{23}H_{31}O_3Cl$; MW = 390.19617

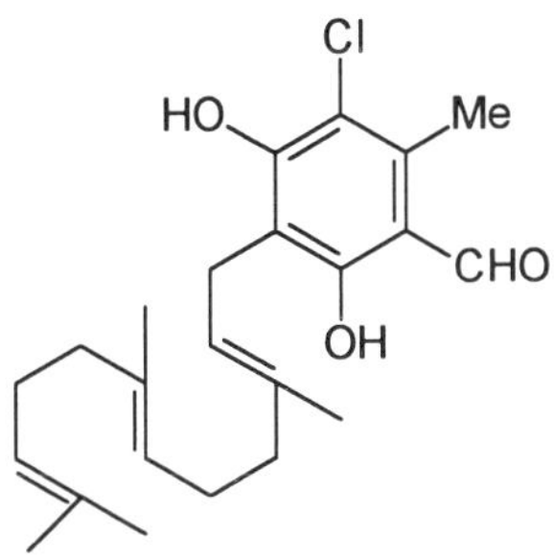

Fungal Source

Unknown *Fusarium* species designated LL-Z1272.

Isolation/Purification

Fungal culture extracted with ethyl acetate, chromatographed on a silica gel column, eluted with CH_2Cl_2, followed by a second silica gel column eluted with a gradient of CH_2Cl_2-ethyl ether, 1:1 (v/v). Final purification was by countercurrent distribution using a MeOH-heptane system.

General Characteristics

Crystallization from MeOH-isooctane and then MeOH-water; mp., 72.5-73°.

Biological Activity

This compound inhibited the growth of the protozoan, *Tetrahymena pyriformis.*

Spectral Data

UV:

λ_{max} 228, 293, and 345nm (ε = 11,150, 10,400, and 7800, respectively).

IR:

Spectrum is characterized by strong carbonyl absorption around 1630cm^{-1} in accord with a 2,4-dihydroxybenzaldehyde system.

^{1}H NMR:

759(C-2, chelated OH); 389(C-4, non-chelated OH); 605(C-1, aldehyde H); 153(C-6, Ar-CH_3); 203(C-1', AR-CH_2); 315(C-2', -CH=); and 108Hz (C-3', CH_3). The non-aromatic portions which covered the residual $C_{15}H_{25}$ portion of the molecule suggested the presence of a farnesyl moiety. The NMR spectrum exhibited an aldehydic proton signal at 600-607Hz and the chelated phenolic hydrogen signal at 757-761Hz. The remaining para phenolic hydroxyl signal was at 389-405Hz by exchange with CD_3OD. The 3-proton singlet at 146-155Hz was assigned to the aromatic Me group which is considerably deshielded by the adjacent carbonyl.

Mass Data:

Found: C, 70.73; H, 8.05; Cl, 9.65%; $C_{23}H_{31}O_3Cl$ requires: C, 70.66; H, 7.99; Cl, 9.07%.

Reference

G. A. Ellestad, R. H. Evans, Jr., and M. P. Kunstmann; Some New Terpenoid Metabolites from an Unidentified *Fusarium* Species; Tetrahedron, Vol. 25, pp. 1323-1334(1969).

Common/Systematic Name
Antibiotic LL-Z1272β

Molecular Weight/Molecular Formula
$C_{23}H_{32}O_3$; MW = 356.23515

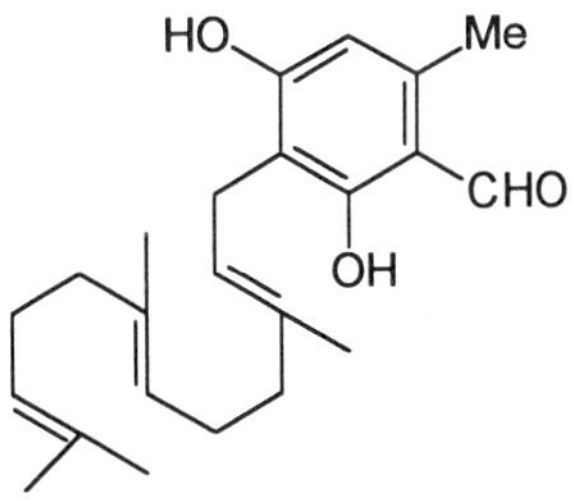

Fungal Source
Unknown *Fusarium* species designated LL-Z1272.

Isolation/Purification
Fungal culture extracted with ethyl acetate, chromatographed on a silica gel column eluted with CH_2Cl_2, followed by a second silica gel column eluted with a gradient of CH_2Cl_2-ethyl ether, 1:1 (v/v). Final purification was by countercurrent distribution using a MeOH-heptane system.

General Characteristics
A semicrystalline material was recrystallized several times from MeOH-water; mp., 97.5°C.

Biological Activity
This compound inhibited the growth of the protozoan, *Tetrahymena pyriformis*.

Spectral Data

UV:
λ_{max} 223, 233 sh, 297 and 340nm sh (ϵ=15,100, 11,500, 16,000, and 3900, respectively).

IR:
Spectrum was characterized by strong carbonyl absorption around 1630cm^{-1} in accord with a 2,4-dihydroxybenzaldehyde system.

^{1}H NMR:
759(C-2, chelated OH); 396(C-4, non-chelated OH); 600(C-1, aldehyde H); 148(C-6, AR-CH); 372(C-5, AR-H); 203(C-1', AR-CH_2); 316(C-2', -CH=); and 108ppm (C-3',

CH_3). The non-aromatic portions which cover the residual $C_{15}H_{25}$ portion of the molecule suggested the presence of a farnesyl moiety. The NMR spectrum exhibited an aldehydic proton signal at 600-607Hz and the chelated phenolic hydrogen signal at 757-761Hz. The remaining para phenolic hydroxyl signal was observed at 389-405Hz by exchange with CD_3OD. The 3-proton singlet at 146-155Hz was assigned to the aromatic Me group which was considerably deshielded by the adjacent carbonyl.

Mass Data:
Found: C, 77.08; H, 8.96%. $C_{23}H_{32}O_3$ requires: C, 77.49; H, 9.05%.

Reference

G. A. Ellestad, R. H. Evans, Jr., and M. P. Kunstmann; Some New Terpenoid Metabolites from an Unidentified *Fusarium* Species; Tetrahedron, Vol. 25, pp. 1323-1334(1969).

Common/Systematic Name
Antibiotic LL-Z1272δ

Molecular Weight/Molecular Formula
$C_{23}H_{31}O_4Cl$; MW = 406.19109

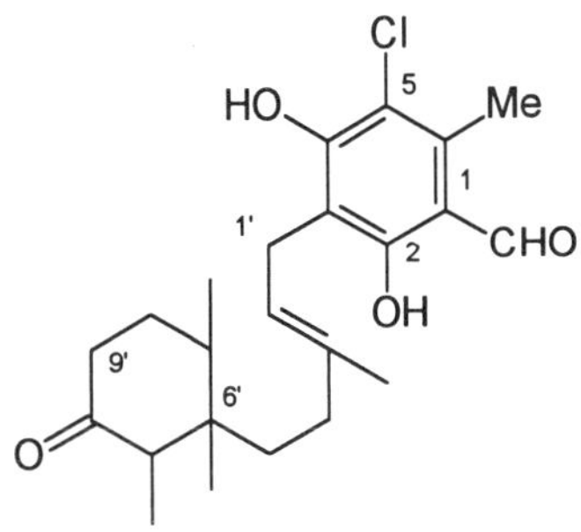

Fungal Source
Unknown *Fusarium* species designated LL-Z1272.

Isolation/Purification
Fungal culture extracted with ethyl acetate, chromatographed on a silica gel column, eluted with CH_2Cl_2, followed by a second silica gel column, eluted with a gradient of CH_2Cl_2-ethyl ether, 1:1 (v:v). Final purification was by countercurrent distribution using a MeOH-heptane system.

General Characteristics
Fractional crystallization from acetone-hexane; mp., 129.5-130.5°C; $[\alpha]_D^{25}$ +6° (*c*=1.0, in MeOH).

Biological Activity
This compound inhibited the growth of the protozoan, *Tetrahymena pyriformis.*

Spectral Data

UV:
λ_{max} 231, 293, and 346nm (ϵ = 23,000, 12,000, and 9150, respectively).

IR:
Spectrum is characterized by strong carbonyl absorption around 1630cm^{-1} in accord with a 2,4-dihydroxybenzaldehyde system.

^{1}H NMR:
757(C-2, chelated OH); 399(C-4, non-chelated OH); 604(C-1, aldehyde H); 155(C-6,

AR-CH_3); 203(C-1', AR-CH_2); 315(C-2', -CH=); 108(C-3' CH_3); 33(C-6' CH_3); 52, 53Hz (C-7' and C-11' CH_3). The non-aromatic portions which cover the residual $C_{15}H_{25}$ portion of the molecule suggested the presence of a farnesyl moiety. Spectrum exhibited an aldehydic proton signal at 600-607Hz and the chelated phenolic hydrogen signal at 757-761Hz. The remaining para phenolic hydroxyl signal was observed at 389-405Hz by exchange with CD_3OD. The 3-proton singlet at 146-155Hz was assigned to the aromatic Me group which is considerably deshielded by the adjacent carbonyl.

Mass Data:

Found: C, 67.93; H, 7.81; C1, 9.08%; $C_{23}H_{31}O_4Cl$ requires: C, 67.89; H, 7.68; Cl, 8.71 %.

Reference

G. A. Ellestad, R. H. Evans, Jr., and M. P. Kunstmann; Some New Terpenoid Metabolites from an Unidentified *Fusarium* Species; Tetrahedron. Vol. 25, pp 1323-1334(1969).

Common/Systematic Name
Antibiotic LL-Z1272ϵ

Molecular Weight/Molecular Formula
$C_{23}H_{32}O_4$; MW = 372.23006

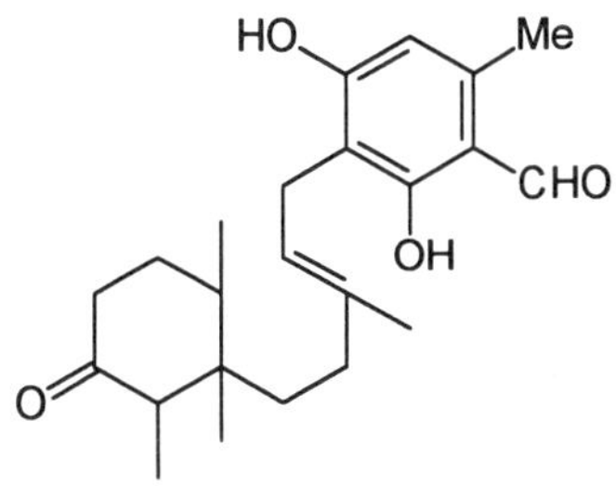

Fungal Source
Unknown *Fusarium* species designated LL-Z1272.

Isolation/Purification
Fungal culture extracted with ethyl acetate, chromatographed on a silica gel column, eluted with CH_2Cl_2, followed by a second silica gel column, eluted with a gradient of CH_2Cl_2-ethyl ether, 1:1 (v/v). Final purification was by countercurrent distribution using a MeOH-heptane system.

General Characteristics
Crystallization from AcOEt-hexane; mp., 171.5-172.5°C; $[\alpha]_D^{25}$ +6° (*c*=0.93, in MeOH).

Biological Activity
This compound inhibited the growth of the protozoan, *Tetrahymena pyriformis.*

Spectral Data

UV:
λ_{max} 223, 233 sh, and 340nm sh (ϵ = 15,600, 11,500, 16,500, and 3700, respectively).

IR:
Spectrum was characterized by strong carbonyl absorption around 1630cm^{-1} in accord with a 2,4-dihydroxybenzaldehyde system.

^{1}H NMR:
758(C-2, chelated OH); 396(C-4, non-chelated OH); 600(C-1, aldehyde H); 146(C-6, AR-CH_3); 376(C-5, ARH); 196(C-1', ARCH$_2$); 315(C-2', -CH=); 108(C-3' CH_3); 32(C-6' CH_3); 51, 51Hz (C-7' and C-11' CH_3). The non-aromatic portions which cover

the residual $C_{15}H_{25}$ portion of the molecule suggested the presence of a farnesyl moiety. Spectrum exhibited an aldehydic proton signal at 600-607Hz and the chelated phenolic hydrogen signal at 757-761Hz. The remaining para phenolic hydroxyl signal was observed at 389-405Hz by exchange with CD_3OD. The 3-proton singlet at 146-155Hz was assigned to the aromatic Me group which is considerably deshielded by the adjacent carbonyl.

Mass Data:

Exhibited a molecular ion at *m/e* of 372.228 in accord with the molecular formula $C_{23}H_{32}O_4$; found: C, 73.15; H, 8.62%; $C_{23}H_{32}O_4$ requires: C, 74.16; H, 8.66%.

Reference

G. A. Ellestad, R. H. Evans, Jr., and M. P. Kunstmann; Some New Terpenoid Metabolites from an Unidentified *Fusarium* Species; Tetrahedron, Vol. 25, pp. 1323-1334(1969).

Common/Systematic Name

Antibiotic LL-Z1272ζ

Molecular Weight/Molecular Formula

$C_{25}H_{31}O_6Cl$; MW = 462.18092

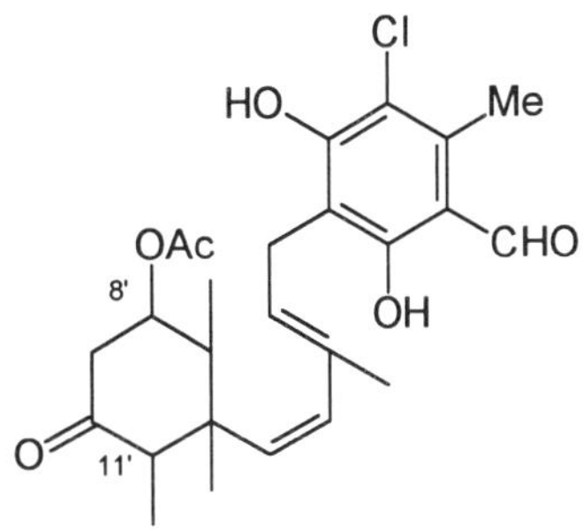

Fungal Source

Unknown *Fusarium* species designated LL-Z1272.

Isolation/Purification

Fungal culture extracted with ethyl acetate, chromatographed on a silica gel column eluted with CH_2Cl_2, followed by a second silica gel column eluted with a gradient of CH_2Cl_2-ethyl ether, 1:1 (v/v). Final purification was by countercurrent distribution using a MeOH-heptane system.

General Characteristics

Semicrystalline solid from acetone-hexane; mp., 156.5-157.0°C; $[\alpha]_D^{25}$ -1.5° (c=1.0, in MeOH).

Biological Activity

This compound inhibited the growth of the protozoan, *Tetrahymena pyriformis*.

Spectral Data

UV:

λ_{max} 229, 293, and 347nm (ϵ = 39,800, 11,700, and 9600, respectively).

IR:

Spectrum was characterized by strong carbonyl absorption around 1630cm^{-1} in accord with a 2,4-dihydroxybenzaldehyde system.

^{1}H NMR:

761(C-2, chelated OH); 400(C-4, non-chelated OH); 607(C-1, aldehyde H); 155(C-6,

AR-CH_3); 212(C-1', ARCH_2); 334(C-2', -CH=); 116(C-3' CH_3); 322(C-7' CH_3); 355(C-4'-C-5'); 44(C-6' CH_3); , 53, 53Hz (C-7' and C-11' CH_3). The non-aromatic portions which cover the residual $C_{15}H_{25}$ portion of the molecule suggested the presence of a farnesyl moiety. Spectrum exhibited an aldehydic proton signal at 600-607Hz and the chelated phenolic hydrogen signal at 757-761Hz. The remaining para phenolic hydroxyl signal was observed at 389-405Hz by exchange with CD_3OD. The 3-proton singlet at 146-155Hz was assigned to the aromatic Me group which is considerably deshielded by the adjacent carbonyl.

Mass Data:

Found: C, 65.01; H, 7.03; C1, 8.02%; $C_{25}H_{31}O_6Cl$ requires: C, 64.86; H, 6.75; Cl, 7.66%.

Reference

G. A. Ellestad, R. H. Evans, Jr., and M. P. Kunstmann; Some New Terpenoid Metabolites from an Unidentified *Fusarium* Species; Tetrahedron, Vol. 25, pp. 1323-1334(1969).

Common/Systematic Name

Ascochlorin; LL-Z1272γ

2,4-Dihydroxy-3-substituted-5-chloroorcylaldehyde

Molecular Formula/Molecular Weight

$C_{23}H_{30}O_4Cl$; MW = 404.17544

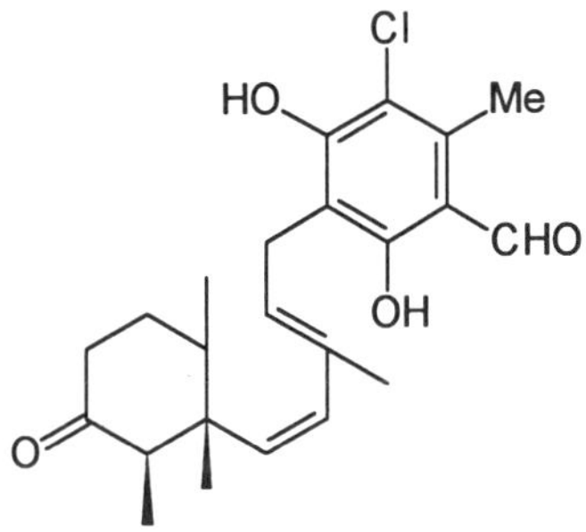

General Characteristics

Ascochlorin appeared as pale yellow crystalline needles when extract was concentrated *in vacuo*. Recrystallization from acetone-hexane; mp., 161.0-163.0°C. Analytical sample had mp., 172-173°C; $[\alpha]_D$ -1.5° (*c*=1.0, in MeOH).

Fungal Source

Ascochyta viciae, Nectria coccinea, N. lucida, Acremonium luzulae, Cylindrocladium iliciola, and *Fusarium* species.

Isolation/Purification

The fungus was grown in glucose, peptone, and ammonium chloride and the mycelia were collected by filtration. Several spots positive to $FeCl_3$ were seen on a thin-layer chromatogram of the mycelial acetone extract using *n*-hexane-acetone (4:1, v/v) as the solvent. The mother liquor after crystallization was concentrated *in vacuo* and the residue was chromatographed on a column packed with silica gel. The column was eluted with *n*-hexane-acetone (9:1, v/v). Further eluted with *n*-hexane-acetone (4:1, v/v). Each fraction was spotted on a silica gel thin-layer plate and the plate was developed with *n*-hexane-acetone (4:1, v/v). The fractions containing substances positive to $FeCl_3$ were collected and concentrated *in vacuo*. The following substances were eluted successfully and obtained as crystalline forms; LL-Z11272δ, LL-Z1272ϵ, ascochlorin, ascofuranone, ascofuranol, and 4'-hydroxy-5'-hydroascochlorin.

Biological Activity

Antibiotic against viruses and fungi *in vitro*. Some of these prenyl phenols show hypolipidemic activity in both normolipidemic and hyperlipidemic rats.

Spectral Data

UV:

λ_{max} 250(ϵ = 29,600), 292(ϵ = 10,600), and 350nm (ϵ = 11,000).

IR:

IR spectrum showed a strong chelated carbonyl band at 1630cm^{-1}, probably due to the 2,4-dihydroxybenzaldehyde system.

^{1}H NMR:

($CDCl_3$) 10.14(1H, s, C-1-aldehyde); 12.72(1H, s, C-2-chelated OH); 6.57(1H, s, C-4-OH); 2.57(3H, s, C-6-methyl); 3.50(2H, d, C-1'); 5.50(1H, t, C-2'); 1.92(3H, s, C-3 '-methyl); 5.91(1H, d, C-4'); 5.32(1H, d, C-5'); 0.70(3H, s, C-6'-methyl); 0.81(3H, d, C-7'-methyl); and 0.83ppm (3H, d, C-11-methyl).

References

G. A. Ellestad, R. H. Evans, Jr., and M. P. Kunstmann; Some New Terpenoid Metabolites from an Unidentified *Fusarium* Species; Tetrahedron, Vol. 25, pp. 1323-1334(1969).

Y. Nawata, K. Ando, G. Tamura, K. Arima, and Y. Iitaka; The Molecular Structure of Ascochlorin; J. Antibiotics, Vol. 22, pp. 511-512(1969).

H. Sasaki, T. Hosokawa, Y. Nawata, and K. Ando; Isolation and Structure of Ascochlorin and Its Analogs; Agr. Biol. Chem., (8), pp. 1463-1466(1974).

Common/Systematic Name
Ascofuranone

Molecular Formula/Molecular Weight
$C_{23}H_{29}O_5Cl$; MW = 420.17035

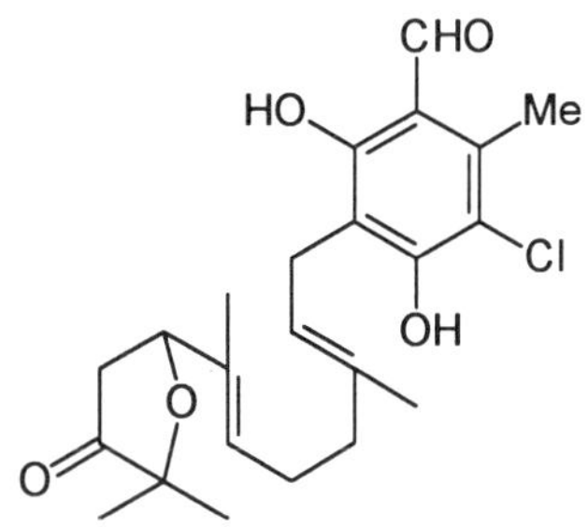

General Characteristics
Colorless needles from *n*-hexane-acetone (4:1); mp., 84-85°C; $[\alpha]_D^{21}$ -50° (*c*=1, in MeOH). Gave intense purple-brown color with $FeCl_3$. Insoluble in water, slightly soluble in *n*-hexane, readily soluble in most organic solvents.

Fungal Source
Ascochyta viciae (mutant strain No. 34).

Isolation/Purification
Culture filtrate was extracted with chloroform-methanol. The extract was chromatographed on silica gel (*n*-hexane-acetone, 95:5, v/v). The substance was crystallized from *n*-hexane-acetone, 4:1 (v/v).

Biological Activity
Less toxic to mice and rats than ascochlorin; effectively lowered serum lipid levels.

Spectral Data

UV:
λ_{max}^{MeOH} 228(ϵ=20300), 290(15000), and 350nm (8200); gave 57nm bathochromic shift of principal ET band under alkaline condition.

IR:
(KBr) 3300(hydroxyl), 1735(carbonyl), and 1630cm^{-1} (carbonyl).

^{1}H NMR:
($CDCl_3$, TMS) 10.12(s, C1-Aldehyde); 12.68(s, C2-Chelated OH); 6.55(s, C4-OH); 2.50(s, C6-Methyl); 3.38(d, 41'-H); 5.21(t, C2'-H); 1.63(s, C3'-Methyl); 2.07(s, C4',5'-

H); 5.50(m, C6'-H); 1.79(s, C7'-Methyl); 4.50 (t, C8'-H); 2.39(d, C9'-H); 1.20 and 1.26ppm (2s, C11'-Dimethyl).

Mass Data:
420(M^+, 12%), 334(30), 330(6), 267(8), 253(87), 239(25), 221(12), 199(100), 167(13), and 153*m/e* (23); found: C, 65.24; H, 6.93; Cl, 8.26%; calcd for $C_{23}H_{29}O_5Cl$, C, 65.62; H, 6.94; Cl, 8.43%.

TLC Data

Silica gel; *n*-hexane-acetone (4:1); R_f=0.42. Detection: spraying with sulfuric acid and heating at 110°C for 5 minutes.

Reference

H. Sasaki, T. Hosokawa, M. Sawada, and K. Ando; Isolation and Structure of Ascofuranone and Ascofuranol, Antibiotics with Hypolipidemic Activity; J. Antibiotics, Vol. 26, No. 11, pp. 676-680(1973).

Common/Systematic Name
Ascofuranol

Molecular Formula/Molecular Weight
$C_{23}H_{31}O_5Cl$; MW = 422.18600

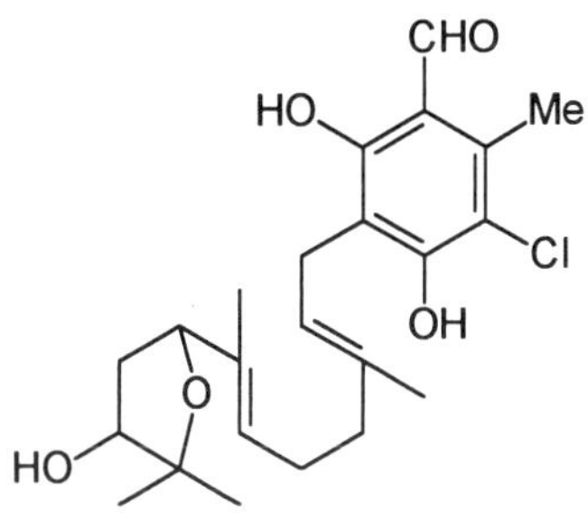

General Characteristics
Colorless needles from *n*-hexane-acetone, 4:1; mp., 75 °C; $[\alpha]_D^{21}$ -7° (*c*=1, in MeOH); gave positive reaction with $FeCl_3$; the solubility was nearly the same as that of ascofuranone.

Fungal Source
Ascochyta viciae (mutant strain No. 34)

Isolation/Purification
Culture filtrate was extracted with chloroform-methanol. The extract was chromatographed on silica gel (*n*-hexane-acetone, 4:1, v/v). The substance was crystallized from *n*-hexane-acetone, 4:1 (v/v).

Spectral Data

UV:
λ_{max}^{MeOH} 228(ε=15800), 290(8700), and 352nm (13900); gave 57nm bathochromic shift of principal ET band at alkaline condition.

IR:
(KBr) 3300(hydroxyl), 1630(carbonyl), and 1090cm^{-1} (secondary alcohol).

^{1}H NMR:
($CDCl_3$, TMS) 10.11(s, C1-aldehyde); 12.66(s, C2- chelated OH); 6.98(s, C4-OH); 2.58(s, C6-methyl); 3.37(s, C1'- H); 5.18(t, C2'-H); 1.60(s, C3'-methyl); 2.06(s, C4', 5'-H); 5.48(m, C6'-H); 1.78(s, C7'-methyl); 4.30 (t, C8' H); 2.35 (d, C9'-H); 1.20 and 1.26(2s, C11'-dimethyl); 2.38(s, C10'-OH); and 3.95(t, C10'-H).

Mass Data:
422(M^+, 11%), 253(26), 239(19), 223(9), 199(100), 183(9), 169(24), 155(21), 151(31), and 13*m/e* (636); found: C, 65.31; H, 7.35; Cl, 8.42%.

TLC Data
Silica gel; *n*-hexane-acetone (4:1); R_f=0.23; detection: spraying with sulfuric acid and heating at 110°C for 5 minutes.

Reference
H. Sasaki, T. Hosokawa, M. Sawada, and K. Ando; Isolation and Structure of Ascofuranone and Ascofuranol, Antibiotics with Hypolipidemic Activity; J. Antibiot., Vol. 26, No. 11, pp. 676-680(1973).

Common/Systematic Name

Chloronectrin

Molecular Formula/Molecular Weight

$C_{25}H_{33}O_6Cl$; MW = 464.19657

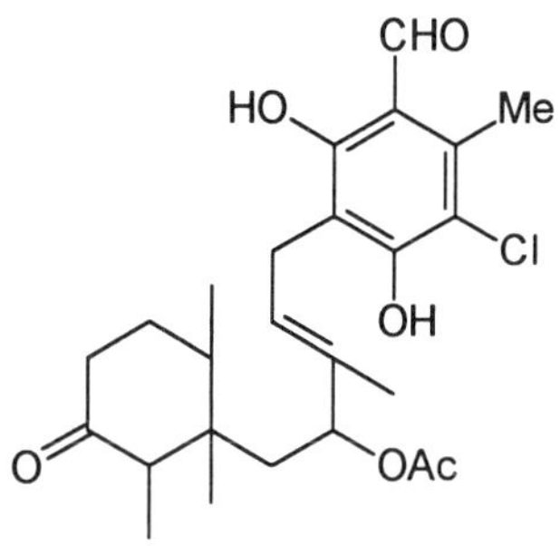

General Characteristics

A gum.

Fungal Source

Nectria coccinea (CMI 120337C).

Isolation/Purification

Mycelium extracted with chloroform, evaporated to dryness, triturated with petroleum ether to give a solid. The solid was chromatographed on a silica gel column; chloronectrin was eluted with benzene-chloroform (1:1, v/v). This fraction was subjected to repeated TLC using Merck GF silica gel to yield chloronectrin as a gum.

Biological Activity

Antibiotic.

Spectral Data

IR:

($CHCl_3$) 3515, 1728, 1712, and 1633cm^{-1}.

^{1}H NMR:

($CDCl_3$) -2.44(bonded OH); -0.11(CHO); ca 3.5 br (OH); 4.44(t, *J*=8.0Hz, CH_2CH=); 4.64(dd, *J*=4.0 and 8.0Hz, CHOAc); 6.62(d, *J*=8.0Hz), CH_2CH=); 7.43($ArCH_3$); 7.99($OCOCH_3$); 8.21(d, *J*=2.0Hz, CH=CCH_3); 9.06(d, *J*=7.0Hz, $CHCH_3$); 9.22(d, *J*=7.0Hz, $CHCH_3$); and 9.48τ (CCH_3).

Mass Spectrum:
EIMS: 464.1979(M^+), $C_{25}H_{33}O_6Cl$ requires 464.1965.

Reference

D. C. Aldridge, A Borrow, R. G. Foster, M. S. Large, H. Spencer, and W. B. Turner; Metabolites of *Nectria coccinea*; J. Chem. Soc., Perkin Trans I, pp. 2136-2141(1972).

Common/Systematic Name
Deacetylchloronectrin
4'-Hydroxy-5-hydroascochlorin

Molecular Formula/Molecular Weight
$C_{23}H_{31}O_5Cl$; MW = 422.18600

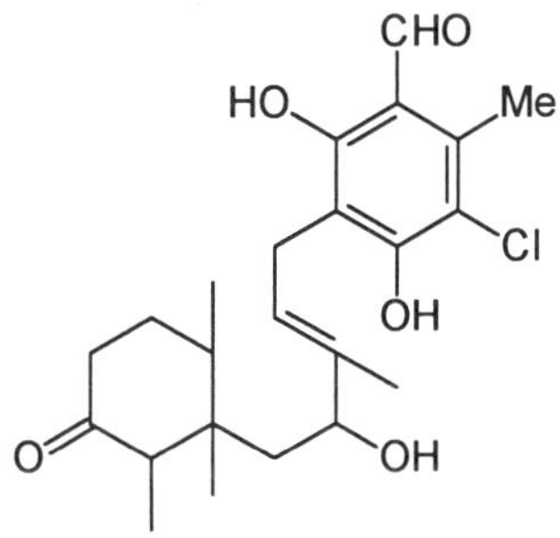

General Characteristics
4'-Hydroxy-5-hydroascochlorin was obtained as pale yellow needles. Recrystallized from *n*-hexane-acetone (4:1); formed colorless crystalline needles; mp., 138°C. A deacetylation product of chloronectin.

Fungal Source
Ascochyta viciae.

Biological Activity
Some of these prenyl phenols show hypolipidemic activity in both normolipidemic and hyperlipidemic rats.

Isolation/Purification
See Ascochlorin.

Spectral Data

UV:
Absorption maxima, 228(ϵ 16,400), 293(ϵ 9400), and 356nm (ϵ 13,500).

IR:
(KBr) 3400(OH), 1710(C=O), and 1630(chelated carbonyl)

^{1}H NMR:
($CDCl_3$) 10.12(1H, s, C1-aldehyde); 12.68(1H, s, C2-chelated OH); 6.49(1H, s, C4-OH); 2.60(3H, s, C6-methyl); 3.39(2H, d, C1'); 5.51(1H, t, C2'); 1.83(3H, s,

C3'-methyl); 4.22(1H, d, d, C4'); 1.53(1H, d, 1.60 1H, d, C5'); 0.55(3H, s, C6'-methyl); 0.82(3H, d, C7'-methyl); and 0.96ppm (3H, d, C11-methyl).

Mass Data:

The base peak, *m/e* 139, was derived from the trimethylcyclohexanone moiety. Prominent chlorine-containing peaks, *m/e* 265 and 251, are formed by fission of either C-4'-5' or 5'-6' followed by rearrangement with accompanying loss of H_2O. A chlorine-containing peak at *m/e* 199 was a tropylium ion common to ascochlorin and its analogs. Analysis: calculated for $C_{23}H_{31}O_5Cl$, C 65.32, H 7.33, Cl 8.40%: found C 65.39, H 7.41, Cl 8.36%.

Reference

H. Sasaki, T. Hosokawa, Y. Nawata, and K. Ando; Isolation and Structure of Ascochlorin and Its Analogs; Agr. Biol. Chem., (8), pp. 1463-1466(1974).

19

Helminthosporols

- Helminthosporol
- Prehelminthosporolactone
- Prehelminthosporol
- Dihydroprehelminthosporol
- Helminthosporal acid
- Helminthosporic acid
- Helminthosporal

Common/Systematic Name

Helminthosporol

Molecular Formula/Molecular Weight

$C_{15}H_{24}O_2$; MW = 236.17763

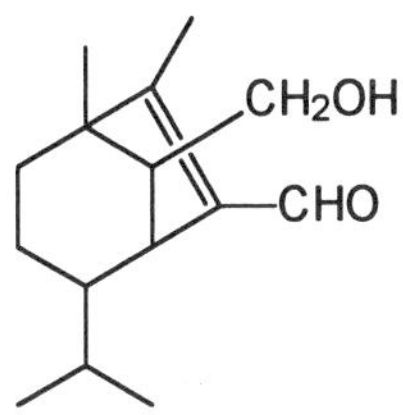

General Characteristics

Recrystallization from *n*-hexane gave colorless needles melting at 98°C; $[\alpha]_D^{23}$ -28.7° (*c*=1.93, in $CHCl_3$). It was soluble in most organic solvents such as benzene, chloroform, ethyl acetate, ether, acetone, and alcohols, but almost insoluble in water.

Fungal Source

Culture broth of *Helminthosporium sativum*, reclassified as *Bipolaris sorokiniana* which is known to cause seedling blight, foot and root rot, head blight, and leaf spot of cereals and grasses.

Isolation/Purification

The cultured broth was filtered and the filtrate treated with charcoal. The charcoal and the mycelia were separated and extracted respectively with acetone. The neutral fraction from these acetone extracts was purified by silicic acid column chromatography using benzene, benzene-2% ethyl acetate and benzene-7% ethyl acetate. Helminthosporol was obtained from the latter fraction and crystallized from *n*-hexane. The fungus was cultured on liquid CZ8 medium. The cultures were stirred at 350rpm for two days at 30°C and aerated at 10,000cm^3/min. The mycelia were separated from the culture filtrate by gravity filtration. The filtrate was extracted with EtOAc in a separatory funnel and concentrated under reduced pressure. The crude extract was suspended in a 9:1 (v/v) mixture of H_2O-MeOH and the resulting aqueous solution extracted three times with hexane, three times with EtOAc and once with *n*-BuOH (2:1, v/v). Drying with anhydrous Na_2SO_4 and evaporation yielded low, medium, and high polarity fractions, respectively. Leaf-spot assay of the three extracts indicated that the strongest biological activity was located in the low polarity fraction. The toxic, low polarity fraction was purified by preparative TLC C_6H_6-Me_2CO (80:20, v/v; R_f 0.47), CH_2Cl_2-Me_2CO (90:10, v/v ;R_f 0.50) which yielded highly purified helminthosporol.

Biological Activity

Plant growth regulator; promoted shoot growth of rice seedlings.

Spectral Data

UV:

λ_{max}^{EtOH} 267nm (ϵ = 9,700)

IR:

(Nujol) 3440, 1645, 1610, and 1025cm^{-1}; ($CHCl_3$) 3500, 2740, 1665, and 1620cm^{-1}.

^{1}H NMR:

($CDCl_3$) 10.03(1H, s, H-14); 3.63(1H, dd, *J*=11.5, 5.5Hz, H-13); 3.32(1H, dd, *J*=11, 8Hz, H-13); 2.00(3H, s, H-12); 1.66(1H, dd, *J*=8, 5Hz, H-8); 1.05(3H, d, *J*=6Hz, H-10 or H-11); 1.02(3H, s, H-15); and 0.75ppm (3H, d, *J*=5Hz, H-10 or H-11).

Mass Data:

Found: C, 76.22; H, 10.14%; MW 227 (Rast); calcd for $C_{15}H_{24}O_2$; C, 76.22; H, 10.24%; MW 236.

References

L. M. Peña-Rodriguez, N. A. Armingeon, and W. S. Chilton; Toxins from Weed Pathogens, I. Phytotoxins from A *Bipolaris* Pathogen of Johnson Grass; J. Natural Products, Vol. 51, pp. 821-828(1988).

S. Tamura, A. Sakurai, K. Kainuma, and M. Takai; Isolation of Helminthosporol as a Natural Plant Growth-Regulator and Its Chemical Structure; Agric. Biol. Chem., Vol. 29, pp. 216-221(1965).

Common/Systematic Name

Prehelminthosporolactone

Molecular Formula/Molecular Weight

$C_{15}H_{22}O_2$; MW = 234.16198

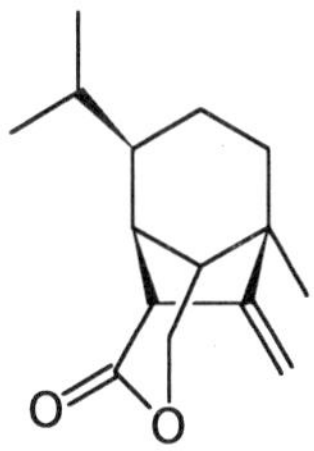

Fungal Source

Bipolaris sp. strain 36 (ATTC 64838), a fungal pathogen of Johnson grass [*Sorghum halepense* (L.) Pers].

General Characteristics

Colorless needles

Isolation/Purification

Purification of the crude fungal mixture was accomplished by flash column chromatography using Merck silica gel 60 and stepwise elution with *n*-hexane-Et_2O (90:10 to 30:70, v/v).

Biological Activity

Phytotoxic, produced lesions on Johnson grass and sorghum resembling those observed in the fungal disease in the field; a reddish brown center surrounded by a black circle at the edge of the drop containing the phytotoxin, with an outer, chlorotic zone extending beyond the drop.

Reference

L. M. Peña-Rodriguez and W. S. Chilton; Victoxinine and Prehelminthosporolactone, Two Minor Phytotoxic Metabolites Produced by *Bipolaris sp.*, A Pathogen of Johnson Grass; J. Natural Products, Vol. 52, pp. 899-901(1989).

Common/Systematic Name
Prehelminthosporol

Molecular Formula/Molecular Weight
$C_{15}H_{24}O_2$; MW = 236.17763

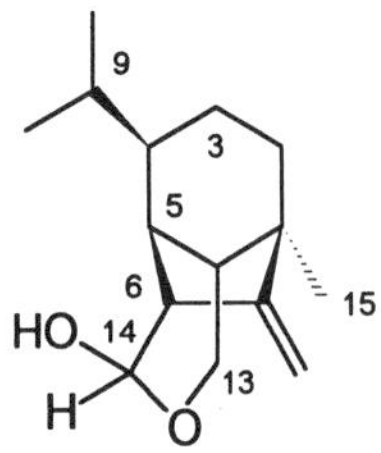

General Characteristics
A colorless oil. Prehelminthosporol occurs as a mixture of epimers at the hemiacetal carbon. It is hydrophobic, has low water solubility, and is unstable to heat.

Fungal Source
Bipolaris sp., a pathogen of Johnson grass (*Sorghum halepense* L.); *Helminthosporium sativum* (reclassified as *Bipolaris sorokiniana,* which is known to cause seedling blight, foot and root rot, head blight and leaf spot of cereals and grasses); *Cochliobolus sativus* and *C. setariae,* which are the perfect stages of *Bipolaris sorokiniana* and *B. setariae.*

Isolation/Purification
The fungus was cultured on liquid CZ8 medium. The cultures were stirred at 350RPM for two days at 30°C and aerated at 10,000 cm^3/min. The mycelia were separated from the culture filtrate by gravity filtration. The filtrate was extracted with EtOAc in a separatory funnel and concentrated under reduced pressure. The crude extract was suspended in a 9:1 (v/v) mixture of H_2O-MeOH and the resulting aqueous solution extracted three times with hexane, three times with EtOAc and once with *n*-BuOH (2:1, v/v). Drying with anhydrous Na_2SO_4 and evaporation yielded low, medium, and high polarity fractions, respectively. Leaf-spot assay of the three extracts indicated that the strongest biological activity was located in the low polarity fraction. The toxic, low polarity fraction was purified by flash column chromatography by stepwise elution with C_6H_6-Me_2CO mixtures (95:5 to 70:30, v/v) which yielded a highly active colorless oil identified as prehelminthosporol.

Biological Activity
Very phytotoxic to sorghum and Johnson grass at 25μg/5μl droplet. Toxin interacts with host membranes (plasmalemma and tonoplast) resulting in cell death and leakage of metabolites; induces both chlorosis and necrosis in plant tissues.

Spectral Data

IR:

(thin film on KBr) 3404, 3067, and 1655cm^{-1}.

^{1}H NMR:

($CDCl_3$) major epimer 4.94ppm*(1H, d, *J*=2.7Hz, H-14); 4.90(1H, s, H-12); 4.81(1H, s, H-12); 3.99(1H, dd, *J*=11.5, 2Hz, H-13); 3.57(1H, dd, *J*=11.5, 2Hz, H-13); 3.18(1H, brs, OH); 2.69(1H, br s, H-6); 2.52(1H, br s, H-5); 1.16(3H, s, H-15); 0.94(3H, d, *J*=6Hz, H-10 or H-11); and 0.88ppm (3H, d, *J*=6Hz, H-10 or H-11); minor isomer 4.84(1H, s, H-12); 4.73(1H, s, H-12); 4.67*(1H, br s, H-14); 3.93(1H, dd, *J*=11.5, 1.5Hz, H-13); 3.77(1H, dd, *J*=11.5, 2Hz, H-13); 2.73(1H, brs, OH); 2.61(1H, br s, H-6); 2.09(1H, br s, H-5); 1.19(3H, s, H-15); 0.97(3H, d, *J*=6Hz, H-10 or H-11); and 0.87ppm (3H, d, *J*=6Hz, H-10 or H-11).

* Carlson *et al.*, 1991 have assigned a different shift and multiplicity for H-14 (H-14 gave a doublet (*J*=12Hz) at 4.66ppm for the major epimer and an unresolved doublet at 4.95ppm for the minor isomer).

^{13}C NMR:

($CDCl_3$) epimeric mixture 157.83(s, C-7); 157.25(s, C-7'); 101.92(t, C-12); 101.07(t, C-12'); 97.09(d, C-14); 96.00(d, C-14'); 67.64(t, C-13); 64.71(t, C-13'); 51.64(d, C-5, C-6 or C-8); 50.48(d, C-5, C-6 or C-8 and C-5', C-6' or C-8'); 48.87(d, C-5', C-6' or C-8'); 46.54(d, C-5, C-6 or C-8); 4.78(s, C-1); 44.53(d, C-5',C-6' or C-8'); 43.60(s, C-1'); 41.95(t, C-2 and C-2'); 36.96(d, C-4 and C-4'); 30.73(d, C-9 and C-9'); 25.79(t, C-3 and C-3'); 20.91(q, C-10, C-11 or C-15 and C-10', C-11' or C-15'); 28.81(q, C-10, C-11 or C-15); 20.16(q, C-10', C-11' or C-15'); and 18.88ppm (q, C-10, C-11 or C-15 and C-10',C-11' or C-15').

Mass Spectrum:

HRMS: $[M]^+$ calcd for $C_{15}H_{24}O_2$, 236.1776, found 236.1796*m/e*; CIMS: (isobutane) 237(37%), 219(100), and 191*m/e* (77). Plasma spray mass spectroscopy 237(M^+ + 1) and 219*m/e* (loss of water).

TLC Data

R_f: 0.70(C_6H_6-Me_2CO; 80:20, v/v), 0.58(C_6H_6-Et_2O; 60:40, v/v), and 0.52(petroleum ether-Me_2CO; 85:15, v/v).

HPLC Data

Normal phase column was Nucleosil 50-5, silicic acid (Machery-Nagel, Duren, FRG). The mobile phase was hexane/ethyl acetate (80:20, v/v) at a flow rate of 3ml/min. Reverse phased column was Kromasil 100-5 (C_{18}) with a mobile phase of acetonitrile/water (70:30, v/v) at a flow rate of 1 ml/min.

References

H. Carlson, P. Nilsson, H. B. Jansson, and G. Odham; Characterization and Determination of Prehelminthosporol, a Toxin from the Plant Pathogenic Fungus *Bipolaris sorokiniana*, Using Liquid Chromatography/Mass spectrometry; J. Microbiol. Methods, Vol. 13, pp. 259-269(1991).

L. M. Peña-Rodriguez, N. A. Armingeon, and W. S. Chilton; Toxins from Weed Pathogens, I. Phytotoxins from A *Bipolaris* Pathogen of Johnson Grass; J. Natural Products, Vol. 51, pp. 821-828(1988).

Common/Systematic Name

Dihydroprehelminthosporol

Molecular Formula/Molecular Weight

$C_{15}H_{26}O_2$; MW = 238.19328

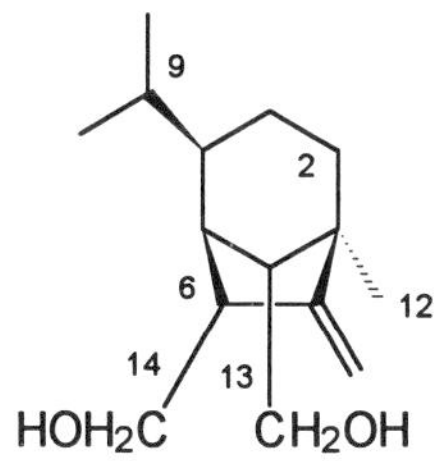

Fungal Source

Bipolaris sp., a pathogen of Johnson grass (*Sorghum halepense* L.); *Helminthosporium sativum* (reclassified as *Bipolaris sorokiniana* which is known to cause seedling blight, foot and root rot, head blight and leaf spot of cereals and grasses); *Cochliobolus sativus* and *C. setariae,* which are the perfect stages of *Bipolaris sorokiniana* and *B. setariae.*

Isolation/Purification

The fungus was cultured on liquid CZ8 medium. The cultures were stirred at 350RPM for two days at 30°C and aerated at 10,000 cm^3/min. The mycelia were separated from the culture filtrate by gravity filtration. The filtrate was extracted with EtOAc in a separatory funnel and concentrated under reduced pressure. The crude extract was suspended in a 9:1 (v/v) mixture of H_2O-MeOH and the resulting aqueous solution extracted three times with hexane, three times with EtOAc and once with *n*-BuOH (2:1, v/v). Drying with anhydrous Na_2SO_4 and evaporation yielded low, medium, and high polarity fractions, respectively. Leaf-spot assay of the three extracts indicated that the strongest biological activity was located in the low polarity fraction. The toxic, low polarity fraction was purified by preparative TLC [C_6H_6-Et_2CO (30:70, v/v; R_f ,0.43)] which yielded highly purified dihydroprehelminthosporol.

Biological Activity

Very phytotoxic to sorghum and Johnson grass at 12.5µg/5µl droplet.

Spectral Data

IR:

(thin film on KBr) 3335, 3076, and 1647cm^{-1}.

^{1}H NMR:

($CDCl_3$) 4.90(1H, d, *J*=3Hz, H-12); 4.79(1H, d, *J*=3Hz, H-12); 3.68(1H, dd, *J*=11, 6Hz, H-14); 3.66(1H, dd, *J*=11, 6.5Hz, H-13); 3.46(1H, dd, *J*=11, 9Hz, H-14);

3.75(1H, dd, *J*=11, 9Hz, H-13); 2.52(1H, br t, *J*=8Hz, H-6); 2.37(1H, s, H-5); 1.57(1H, br t, *J*=8 Hz, H-8); 0.99(3H, s, H-15); 0.96(3H, d, *J*=7Hz, H-10 or H-11); and 0.84ppm (3H, d, *J*=7Hz, H-10 or H-11).

^{13}C NMR:

($CDCl_3$) 158.69(s, C-7); 104.50(t, C-12); 66.44(t, C-13 or C-14); 63.11(t, C-13 or C-14); 57.39(d, C-5, C-6 or C-8); 49.27(d, C-5, C-6 or C-8); 47.28(s, C-1); 45.67(d, C-5, C-6 or C-8); 42.72(t, C-2); 37.71(d, C-4); 30.61(d, C-9); 24.97(t, C-3); 21.34(q, C-10, C-11 or C-15); 20.66(q, C-10, C-11 or C-15); and 20.00ppm (q, C-10, C-11 or C-15).

Mass Spectrum:

HRMS: $[M]^+$ calcd for $C_{15}H_{26}O_2$, 238.1933; found 238.1928*m/e*; CIMS: (isobutane) 239(16%), 221(100), 208(16), 203(78), and 191*m/e* (86).

Reference

L. M. Peña-Rodriguez, N. A. Armingeon, and W. S. Chilton; Toxins from Weed Pathogens, I. Phytotoxins from A *Bipolaris* Pathogen of Johnson Grass; J. Natural Products, Vol. 51, pp. 821-828(1988).

Common/Systematic Name
Helminthosporal acid

Molecular Formula/Molecular Weight
$C_{15}H_{22}O_3$; MW = 250.15689

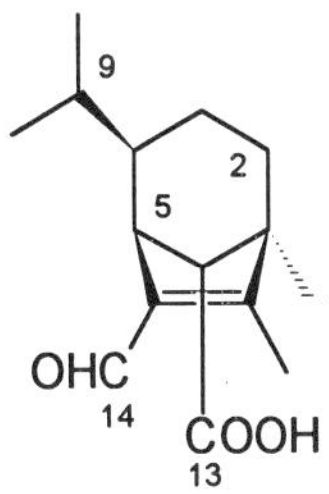

Fungal Source
Bipolaris sp., a pathogen of Johnson grass (*Sorghum halepense* L.) and *Helminthosporium sativum* (reclassified as *Bipolaris sorokiniana* which is known to cause seedling blight, foot and root rot, head blight and leaf spot of cereals and grasses.).

Isolation/Purification
The fungus was cultured on liquid CZ8 medium. The cultures were stirred at 350RPM for two days at 30°C and aerated at 10,000 cm^3/min. The mycelia were separated from the culture filtrate by gravity filtration. The filtrate was extracted with EtOAc in a separatory funnel and concentrated under reduced pressure. The crude extract was suspended in a 9:1 (v/v) mixture of H_2O-MeOH and the resulting aqueous solution extracted three times with hexane, three times with EtOAc and once with *n*-BuOH (2:1, v/v). Drying with anhydrous Na_2SO_4 and evaporation yielded low, medium, and high polarity fractions, respectively. Leaf-spot assay of the three extracts indicated that the strongest biological activity was located in the low polarity fraction. The toxic, low polarity fraction was purified by preparative TLC (C_6H_6-Et_2CO, 70:30, v/v; R_f, 0.62) which yielded highly purified helminthosporal acid.

Spectral Data

UV:
λ_{max}^{MeOH} 267nm.

IR:
(thin film on KBr) 1702, 1670, and 1631cm^{-1}.

^{1}H NMR:
($CDCl_3$) 9.99(1H, s, H-14); 3.39(1H, s, H-5); 2.35(1H, s, H-8); 2.02 (3H, s, H-12); and 1.17ppm (3H, s, H-15).

Mass Spectrum:
HRMS $[M]^+$ calcd for $C_{15}H_{22}O_3$, 250.1542; found 250.1563*m/e*; CIMS: (isobutane) 251(100%) and 233*m/e* (1).

Reference

L. M. Peña-Rodriguez, N. A. Armingeon, and W. S. Chilton; Toxins from Weed Pathogens, I. Phytotoxins from A *Bipolaris* Pathogen of Johnson Grass; J. Natural Products, Vol. 51, pp. 821-828(1988).

Common/Systematic Name
Helminthosporic acid

Molecular Formula/Molecular Weight
$C_{15}H_{24}O_3$; MW = 252.17254

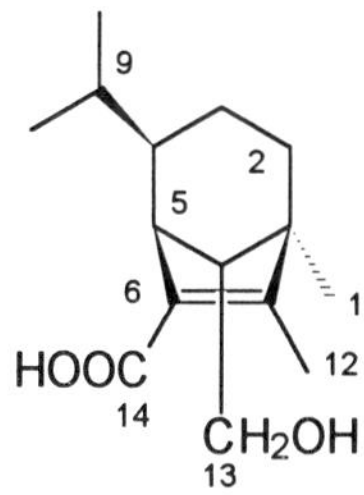

Fungal Source
Bipolaris sp., a pathogen of Johnson grass (*Sorghum halepense* L.), *Helminthosporium sativum* (reclassified as *Bipolaris sorokiniana* which is known to cause seedling blight, foot and root rot, head blight and leaf spot of cereals and grasses.).

Isolation/Purification
The fungus was cultured on liquid CZ8 medium. The cultures were stirred at 350RPM for two days at 30°C and aerated at 10,000 cm^3/min. The mycelia were separated from the culture filtrate by gravity filtration. The filtrate was extracted with EtOAc in a separatory funnel and concentrated under reduced pressure. The crude extract was suspended in a 9:1 (v/v) mixture of H_2O-MeOH and the resulting aqueous solution extracted three times with hexane, three times with EtOAc and once with *n*-BuOH (2:1, v/v). Drying with anhydrous Na_2SO_4 and evaporation yielded low, medium, and high polarity fractions, respectively. Leaf-spot assay of the three extracts indicated that the strongest biological activity was located in the low polarity fraction. The toxic, low polarity fraction was purified by preparative TLC (C_6H_6-Me_2CO, 80:20, v/v; R_f 0.24) which yielded highly purified helminthosporic acid.

Spectral Data

IR:
(thin film on KBr) 3343, 1670, and 1623cm^{-1}.

^{1}H NMR:
($CDCl_3$) 3.66(1H, dd, *J*=11, 4Hz, H-13); 3.39(1H, dd, *J*=11, 9Hz, H-13); 3.15(1H, s, H-5); 2.00(3H, s, H-12); 1.63(1H, dd, *J*=9, 4Hz, H-8); 1.02(3H, d, *J*=6Hz, H-10 or H-11); 0.96(3H, s, H-15); and 0.77ppm (3H, br s, H-10 or H-11)

Mass Spectrum:
HRMS: $[M]^+$ calcd for $C_{15}H_{24}O_3$, 252.1725; found 252.1725*m/e*; CIMS: (isobutane) 253(5%) and 235*m/e* (100).

Reference

L. M. Peña-Rodriguez, N. A. Armingeon, and W. S. Chilton; Toxins from Weed Pathogens, I. Phytotoxins from A *Bipolaris* Pathogen of Johnson Grass; J. Natural Products, Vol. 51, pp. 821-828(1988).

Common/Systematic Name

Helminthosporal

Molecular Formula/Molecular Weight

$C_{15}H_{22}O_2$; MW = 234.16198

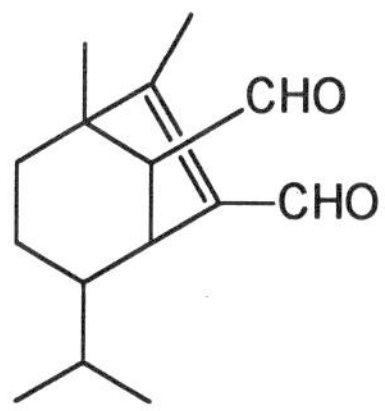

General Characteristics

Crystallized twice from light petroleum to give helminthosporal; mp., 56-59°C; $[\alpha]_D$ -19° (c=1.18, in $CHCl_3$). Unstable to air, even at low temperature.

Fungal Source

Helminthosporium sativum, a fungus which produces a seedling blight, foot and root rot, head blight, and leaf spot of cereals and grasses. The disease is widespread in North America and elsewhere and in particular affects wheat and barley crops.

Isolation/Purification

The culture filtrate was adjusted to pH 2 and activated charcoal was added and stirred. The charcoal with the absorbed toxin was separated from the liquid in a Tolhurst centrifuge and suspended in ethanol. After filtering on a Buchner funnel, the charcoal cake was suspended in chloroform, stirred, and filtered. This process was repeated three times. The chloroform extracts were combined and the solvent removed *in vacuo* to yield a crude oil extract. The oil was heated under nitrogen at 120°C for 12 hours. During this time small amounts of water, ethanol, and chloroform were lost, leaving a residue which was dissolved in light petroleum (bp, 60-80°C) and applied to a column of alumina (Merck, acid-washed). Elution with the same solvent gave a trace of oil. Elution with ether gave, after evaporation, a brown oil. Distillation of this oil (bp, 115-120/0.015mm) afforded a yellow oil which slowly set. A portion was crystallized twice from light petroleum to give helminthosporal.

Biological Activity

Phytotoxic, produces seedling blight, foot and root rot,and leaf spot of cereals in North America. The action of the fungus in producing disease is largely dependent on the action of the toxin.

Spectral Data

UV:

λ_{max}^{EtOH} 266nm (ϵ=11,000).

IR:
(CCl_4) 1715, 1685, and 1618cm^{-1}.

Mass Data:
Found: C, 76.72; H, 9.27; C-Me, 12.56%; MW. (isopiestic, av.) 232, (Rast, av.) 247; calcd for $C_{15}H_{22}O_2$: C, 76.88; H, 9.46; C-Me, (2), 12.83%; MW 234.

References

P. De Mayo. E. Y. Spencer, and R. W. White; Helminthosporal, The Toxin from *Helminthosporium sativum*: Isolation and Characterization; Canadian Journal of Chemistry, Vol.39, pp. 1608-1612(1961).

P. De Mayo. E. Y. Spencer, and R. W. White; The Constitution of Helminthosporal; J. Am. Chem. Soc., Vol. 84, pp. 494-495(1962).

P. De Mayo. E. Y. Spencer, and R. W. White; IV. The Structure and Stereochemistry of Helminthosporal; Canadian Journal of Chemistry, Vol.41, pp. 2996-3004(1963).

20

Hebevinosides and Hebelomic Acids

Hebevinoside I
Hebevinoside II
Hebevinoside III
Hebevinoside IV
Hebevinoside V
Hebevinoside VI
Hebevinoside VII
Hebevinoside VIII
Hebevinoside IX
Hebevinoside X
Hebevinoside XI
Hebevinoside XII
Hebevinoside XIII
Hebevinoside XIV
Hebelomic acid A (HS-A)
Hebelomic acid B
Hebelomic acid E
Hebelomic acid F
HS-B
HS-C

Common/Systematic Name

Hebevinoside I

3β,16β-Dihydroxy-7β-methoxycucurbita-5,24-diene-3-*O*-β-D-xylopyranoside-16-*O*-(6-*O*-acetyl)-β-D-glucopyranoside

Molecular Formula/Molecular Weight

$C_{44}H_{72}O_{13}$; MW = 808.49729

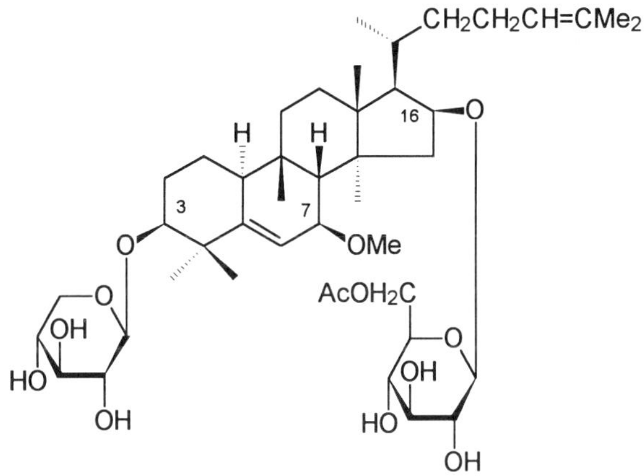

General Characteristics

Amorphous; dihydrohexaacetate: colorless needles from methanol; m.p. 176-177°, $[\alpha]_D^{22}$ + 16.° (*c*=0.51, in MeOH).

Fungal Source

Hebeloma vinosophyllum (a toxic mushroom).

Isolation/Purification

The *n*-butanol soluble fraction was isolated from the 90% MeOH extract of dried fruit bodies of cultivated *H. vinosophyllum*. Repeated column chromatography of the *n*-BuOH-soluble fraction on silica gel with C_6H_6-acetone and on Sephadex LH-20 with $CHCl_3$-MeOH had afforded several hebevinosides.

Spectral Data

UV:

$\lambda_{max}^{Acetonitrile}$ End absorption.

IR:

(KBr) 3400, 1735(ester), 1635, 1075, and 1040cm^{-1}.

^{1}H NMR
($CDCl_3$) 0.88(3H, d, *J*=6.3Hz); 0.89(6H, d, *J*=6.6Hz); 1.99, 2.01, 2.0, 2.04, 2.06(each 3H, s, Ac x 5); and 2.02ppm (6H, s, Ac x 2).

^{13}C NMR:
(C_5D_5N) C-3, 87.3; C-5, 148.0; C-6, 119.7; C-7, 77.9; C-16, 82.3; C-24, 126.9; C-25, 130.1; OMe, 56.3; xyl-1, 107.3; xyl-2, 74.8; xyl-3, 78.5; xyl-4, 71.1; xyl-5, 66.8; Glu-1, 106.6; glu-2, 75.1; glu-3, 78.0; glu-4, 71.7; glu-5, 75.5; glu-6, 64.7; $CO\underline{C}H_3$, 20.8; and $\underline{C}OCH_3$, 170.6ppm.

Mass Spectrum:
(FD-MS of dihydrohexaacetate) 1062*m/e* (M^+).

Reference

H. Fujimoto, K. Suzuki, H. Hagiwara, and M. Yamazaki; New Toxic Metabolites from a Mushroom, *Hebeloma vinosophyllum*. I. Isolation and Structures of Hebevinosides I, II, III, IV and V, Chem. Pharm. Bull., Vol. 34, pp. 88-89(1986).

Common/Systematic Name

Hebevinoside II

3β,7β,16β-Trihydroxycucurbita-5,24-diene-3-*O*-β-D-xylopyranoside-16-*O*-(6-*O*-acetyl)-β-D-glucopyranoside

Molecular Formula/Molecular Weight

$C_{45}H_{72}O_{14}$; MW = 836.49221

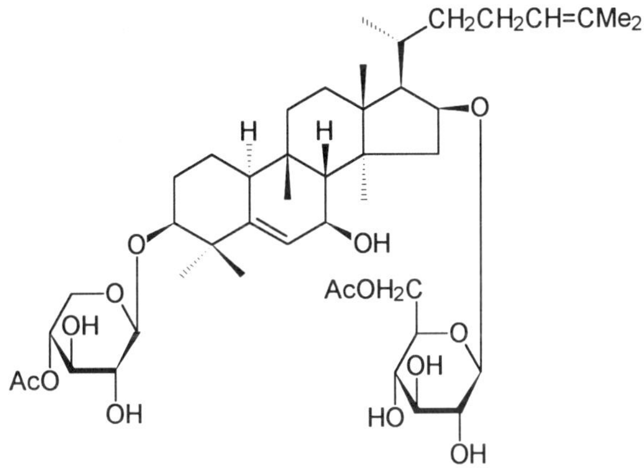

General Characteristics

Amorphous; $[\alpha]_D^{20}$ +30° (*c*=0.20, in acetone).

Fungal Source

Hebeloma vinosophyllum (a toxic mushroom).

Isolation/Purification

The *n*-butanol soluble fraction was isolated from the 90% MeOH extract of dried fruit bodies of cultivated *H. vinosophyllum*. Repeated column chromatography of the *n*-BuOH-soluble fraction on silica gel with C_6H_6-acetone and on Sephadex LH-20 with $CHCl_3$-MeOH had afforded several hebevinosides.

Spectral Data

UV:

$\lambda_{max}^{Acetonitrile}$ End absorption.

IR:

(KBr) 3400, 1735(ester), 1640, 1075, and 1035cm^{-1}.

^{1}H NMR:

($CDCl_3$) 0.88, 1.12, 1.17, 1.38, 1.61(each 3H, s); 1.04(3H, d, *J*=6.8Hz, H_3-21); 1.70, 1.73(each 3H, s, H_3-26 and H_3-27); 1.96, 2.04(each 3H, s, Ac); 4.42(1H, brd, *J*=5.1Hz, H-7); 4.67, 4.82(each 1H, d, *J*=7.7Hz, H-xyl-1 and H-glc-1); 5.53(1H, brt, *J*=7.7Hz, H-24); and 6.12ppm (1H, d, *J*=5.1Hz, H-6); acetate ($CDCl_3$) 1.99, 2.01, 2.04, 2.10(each 3H, s, Ac X 4); 2.00, 2.02ppm (each 6H, s, Ac X 4).

^{13}C NMR:

(C_5D_5N) C-3, 87.4; C-5, 146.2; C-6, 122.5; C-7, 67.5; C-16, 82.2; C-24, 126.8; C-25, 130.0; xyl-1, 107.3; xyl-2, 74.9; xyl-3, 74.9; xyl-4, 73.1; xyl-5, 63.1; glu-1, 106.5; glu-2, 75.2; glu-3, 78.3; glu-4, 71.7; glu-5, 75.4; glu-6, 64.7; $COCH_3$, 20.8, 20.8; and $COCH_3$, 170.5, 170.7ppm.

Reference

H. Fujimoto, K. Suzuki, H. Hagiwara, and M. Yamazaki; New Toxic Metabolites from Mushroom, *Hebeloma vinosophyllum*. I. Isolation and Structure of Hebevinosides I, II, III, IV, and V; Chem. Pharm. Bull., Vol. 34, pp. 88-89(1986).

Common/Systematic Name

Hebevinoside III

3β,7β,16β-Trihydroxycucurbita-5,24-diene-3-*O*-β-D-xylopyranoside-16-*O*-(6-*O*-acetyl)-β-D-glucopyranoside

Molecular Formula/Molecular Weight

$C_{43}H_{70}O_{13}$; MW = 794.48164

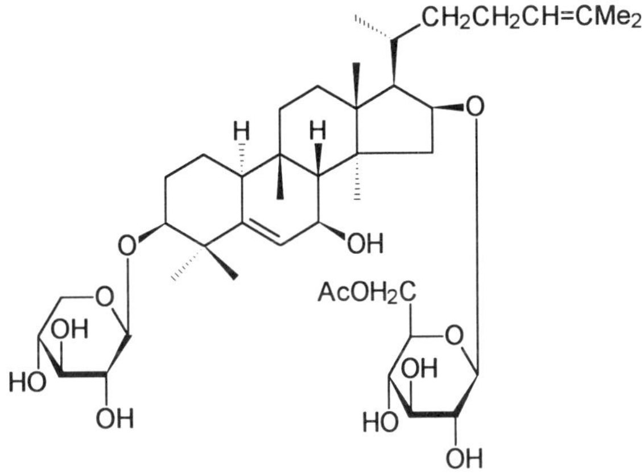

General Characteristics

Amorphous.

Fungal Source

Hebeloma vinosophyllum, a toxic mushroom.

Isolation/Purification

The *n*-butanol soluble fraction was isolated from the 90% MeOH extract of dried fruit bodies of cultivated *H. vinosophyllum*. Repeated column chromatography of the *n*-BuOH-soluble fraction on silica gel with C_6H_6-acetone and on Sephadex LH_{20} with $CHCl_3$-MeOH had afforded several hebevinosides.

Spectral Data

UV:

$\lambda_{max}^{Acetonitrile}$ End absorption.

IR:

(KBr) 3400, 1735 (ester), 1640, 1080, and 1040cm^{-1}.

^{1}H NMR:

($CDCl_3$) 0.88, 1.12, 1.18, 1.36, 1.61(each 3H, s); 1.04(3H, d, *J*=6.0Hz, H_3-21); 1.69, 1.72(each 3H, s, H_3-26 and H_3-27); 2.05(3H, s, Ac); 4.41(1H, brd, *J*=5.6Hz, H-7); 4.67, 4.8(each 1H, d, *J*=7.7Hz, H-xyl-l and H-glc-l); 5.52(1H, brt, *J*=6.0Hz, H-24) and 6.12ppm (1H, d, *J*=5.6Hz, H-6).

^{13}C NMR:

(C_5D_5N) C-3, 87.4; C-5, 146.3; C-6, 122.4; C-7, 67.5; C-16, 82.2; C-24, 126.8; C-25, 129.9; xyl-1, 107.7; xyl-2, 75.1; xyl-3, 78.4; xyl-4, 71.1; xyl-5, 66.8; glu-1, 106.6; glu-2, 75.0; glu-3, 78.6; glu-4, 71.9; glu-5, 75.6; glu-6, 64.8; CO<u>C</u>H_3, 20.8; and <u>C</u>OCH_3, 170.8ppm.

Reference

H. Fujimoto, K. Suzuki, H. Hagiwara, and M. Yamazaki; New Toxic Metabolites from a Mushroom, *Hebeloma vinosophyllum*. I. Isolation and Structures of Hebevinosides I, II, III, IV and V, Chem. Pharm. Bull., Vol. 34, pp. 88-89(1986).

Common/Systematic Name

Hebevinoside IV

3β,16β-Dihydroxy-7β-methoxycucurbita-5,24-diene-3-*O*-β-D-xylopyranoside

Molecular Formula/Molecular Weight

$C_{36}H_{60}O_7$; MW = 604.43390

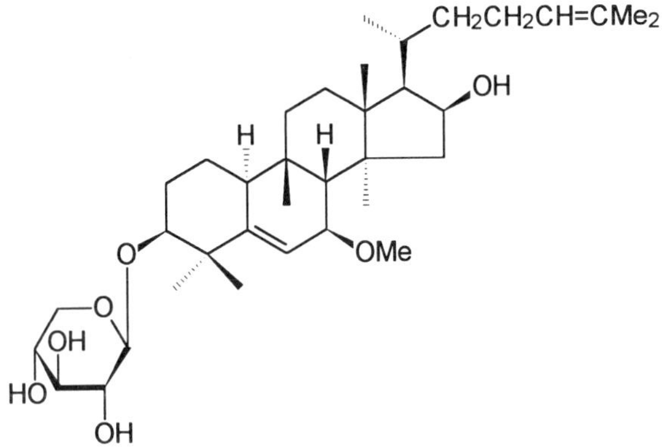

General Characteristics

Colorless needles from methanol; mp., 172-173°C; positive to both the Liebermann-Burchard and Molisch reactions. On catalytic hydrogenation afforded a dihydro derivative and on acetylation, a tetraacetate. Catalytic hydrogenation of acetate gave dihydrotetraacetate; mp., 122-125°C, $[\alpha]_D^{17}$ +64° (*c*=0.081, in MeOH). Tetraacetate; mp., 129-130°C.

Fungal Source

Hebeloma vinosophyllum (a toxic mushroom).

Isolation/Purification

The *n*-butanol soluble fraction was isolated from the 90% MeOH extract of dried fruit bodies of cultivated *H. vinosophyllum*. Repeated column chromatography of the *n*-BuOH-soluble fraction on silica gel with C_6H_6-acetone and on Sephadex LH_{20} with $CHCl_3$-MeOH had afforded several hebevinosides.

Spectral Data

UV:

$\lambda_{max}^{Acetonitrile}$ End absorption.

IR:

(KBr) 3410, 1638(ester), 1075, and 1038cm^{-1}.

^{1}H NMR:

(CDCl$_3$) 0.69, 0.97, 1.00, 1.12, 1.22(each 3H, s); 0.99(3H, d, *J*=8.1Hz, H3-21); 1.62, 1.70(each 3H, s, H3-26 and H3-27); 3.34(3H, s, OCH$_3$); 3.42(1H, brd, *J*=5.6Hz, H-7); 3.98(1H, dd, *J*1=11.8, *J*2=4.5Hz, H-3); 4.36(1H, d, *J*=6.0Hz, H-xyl- l); 4.44(1H, dd, *J*1=7.3, *J*2=13.7Hz, H-16); 5.19(1H, br t, *J*=7.7Hz, H-24); and 5.74ppm (1H, d, *J*=5.6Hz, H-6). dihydro derivative (CDCl$_3$) 0.88(6H, d, *J*=7Hz); 0.97ppm (3H, d, *J*=5Hz). acetate derivative (CDCl$_3$) 2.03(6H, s, Ac x 2); 2.04, and 2.05ppm (each 3H, s, Ac x 2).

^{13}C NMR:

(C$_5$D$_5$N) C-3, 87.4; C-5, 147.9; C-6, 119.3; C-7, 77.8; C-16, 71.1; C-24, 126.2; C-25, 130.5; OMe, 56.2; xyl-1, 107.8; xyl-2, 74.9; xyl-3, 78.5; xyl-4, 71.2; and xyl-5, 67.1ppm.

Mass Spectrum:

EIMS: 572*m/e* (19%, M^+ - CH_3OH); FDMS: (acetate derivative) 772*m/e* (M^+).

Reference

H. Fujimoto, K. Suzuki, H. Hagiwara, and M. Yamazaki; New Toxic Metabolites from a Mushroom, *Hebeloma vinosophyllum*. I. Isolation and Structures of Hebevinosides I, II, III, IV and V, Chem. Pharm. Bull., Vol. 34, pp. 88-89(1986).

Common/Systematic Name

Hebevinoside V

3β,16β-Dihydroxy-7β-methoxycucurbita-5,24-diene-3-*O*-(4-*O*-acetyl)-β-D-xylopyranoside-16-*O*-(6-*O*-acetyl)-β-D-glucopyranoside

Molecular Formula/Molecular Weight

$C_{46}H_{74}O_{14}$; MW = 850.50786

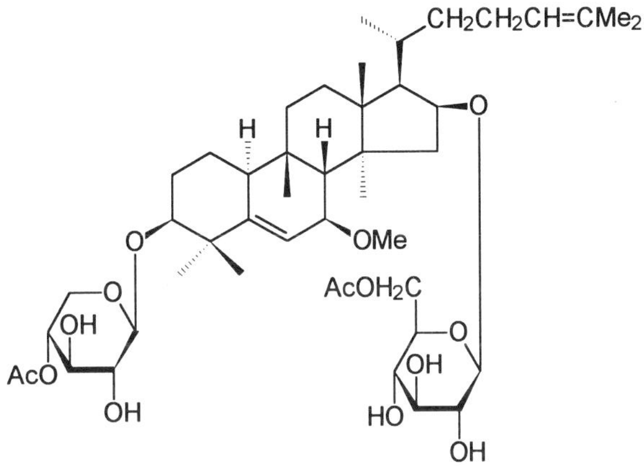

General Characteristics

Obtained as colorless needles from methanol; mp., 165-166°C; $[\alpha]_D^{20}$ +83° (c=0.053, in $CHCl_3$).

Fungal Source

Hebeloma vinosophyllum (a toxic mushroom).

Isolation/Purification

The *n*-butanol soluble fraction was isolated from the 90% MeOH extract of dried fruit bodies of cultivated *H. vinosophyllum*. Repeated column chromatography of the *n*-BuOH-soluble fraction on silica gel with C_6H_6-acetone and on Sephadex LH_{20} with $CHCl_3$-MeOH had afforded several hebevinosides.

Spectral Data

UV:

$\lambda_{max}^{Acetonitrile}$ End absorption.

IR:

(KBr) 3420, 1743, 1725, 1640, 1087, and 1032cm^{-1}.

^{1}H NMR:

($CDCl_3$) 0.68, 0.97, 1.01, 1.12, 1.22(each 3H, s); 0.96(3H, d, J=5.6Hz, H_3-21); 1.62, 1.69(each 3H, s, H_3-26 and H_3-27); 2.10(6H, s, Ac x 2); 3.34(3H, s, OMe); 3.42(1H, br d, J=6.2Hz, H-7); 4.07(1H, dd, J_1=12.0, J_2=4.3 Hz, H-glc-6); 4.16(1H, dd, partly overlapped with other signals, H_b-glc. 6); 4.42(2H, d, J=6.4Hz, H-xyl-l and H-glc-l); 4.81(1H, dt, J_1=J_2 =7.7, J_3=5.0Hz, H-xyl-4); 5.14(1H, br t, J_1=J_2=6.9Hz, H-24); and 5.75ppm (1H, d, J=6.2Hz, H-6).

^{13}C NMR:

(C_5D_5N) C-3, 87.4; C-5, 147.8; C-6, 119.5; C-7, 77.8; C-16, 82.2; C-24, 126.9; C-25, 130.0; OMe, 56.3; xyl-1, 107.5; xyl-2, 75.0; xyl-3, 78.4; xyl-4, 73.1; xyl-5, 63.2; glu-1, 106.8; glu-2, 75.0; glu-3, 78.3; glu-4, 71.7; glu-5, 75.4; glu-6, 64.5; CO$\underline{C}H_3$, 20.9, 20.9; and $\underline{C}OCH_3$, 170.5, 170.6ppm.

Reference

H. Fujimoto, K. Suzuki, H. Hagiwara, and M. Yamazaki; New Toxic Metabolites from a Mushroom, *Hebeloma vinosophyllum*. I. Isolation and Structures of Hebevinosides I, II, III, IV and V, Chem. Pharm. Bull., Vol. 34, pp. 88-89(1986).

Common/Systematic Name

Hebevinoside VI

6-3-*O*-β-D-Xylopyranoside-16-*O*-β-D-glucopyranoside

Molecular Formula/Molecular Weight

$C_{41}H_{68}O_{12}$; MW = 752.47108

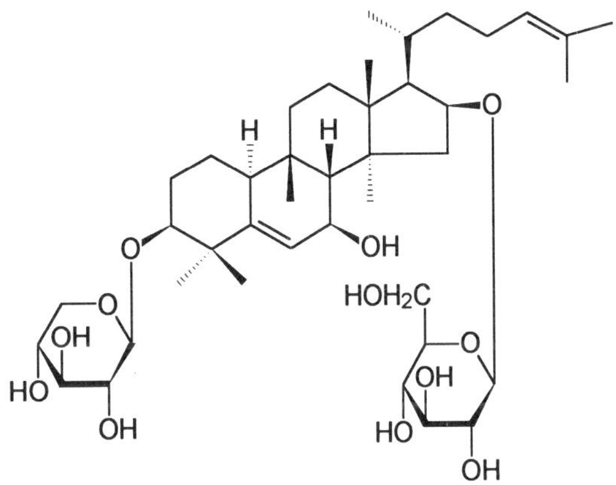

General Characteristics

Amorphous; $[\alpha]_D^{23}$ + 56°(*c*=0.054, in pyridine).

Fungal Source

Hebeloma vinosophyllum.

Isolation/Purification

Dried fruit bodies obtained by cultivation of *H. vinosophyllum* were cut into fine pieces and shaken in *n*-hexane at room temperature. The defatted fruit bodies were extracted with acetone whose pH was kept at 7 by occasional addition of small amounts of pyridine at room temperature to afford the acetone extract. The acetone extract was separated by centrifugal TLC with $CHCl_3$-acetone (1:1, v/v), $CHCl_3$-acetone-tetrahydrofuran (THF) (1:1:0.1), (1:1:0.2), (1:1:0.5, v/v), and THF. A mixture of the latter half of the fraction eluted with $CHCl_3$-acetone (1:1, v/v) and the first half of the fraction eluted with $CHCl_3$-acetone-THF (1:1:0.1, v/v/v) was further chromatographed on a Sephadex LH_{20} column with $CHCl_3$ and $CHCl_3$-THF (20:1, v:v) to afford hebevinoside VIII. A mixture of the latter half of the fraction eluted with $CHCl_3$-acetone-THF (1:1:0.1, v/v/v) and the fraction eluted with $CHCl_3$-acetone-THF (1:1:0.2, v/v/v) was further chromatographed on a Sephadex LH_{20} column with $CHCl_3$-THF (10:1, v/v) to afford hebevinoside II, hebevinoside VII, and hebevinoside IX. A mixture of the fraction eluted with $CHCl_3$-acetone-THF (1:1:0.5, v/v/v) and the first half of the fraction eluted with THF was further chromatographed on a Sephadex LH_{20} column with $CHCl_3$-THF (10:1, v/v) and THF to afford hebevinosides III and VI.

Biological Activity

IP administration of 100mg/kg to mice was LD_{100} ; mice died after paralysis; LD_{50} was 66mg/kg (IP dose).

Spectral Data

UV:

End absorption.

IR:

(KBr) 3380, 1635, 1075, and 1035cm^{-1}.

^{1}H NMR:

(C_5D_5N) 0.78; 1.12; 1.17; 1.35; 1.63(3H each, s); 1.04(3H, d, *J*=6.4Hz, H-21); 1.68, 1.72(3H each, s, H-26 and -27); 4.27(1H, d, *J*=4.8Hz, H-7); 4.74, 4.81(1H each, d, *J*=7.9Hz, H-glc-1 and H-xyl-1); 5.52(1H, t, *J*=7.1Hz, H-24); and 6.03ppm (1H, d, *J*=4.8Hz, H-6).

^{13}C NMR:

(C_5D_5N) C-3, 87.4; C-5, 146.2; C-6, 122.6; C-7, 67.3; C-16, 81.9; C-24, 126.8; C-25, 130.0; xyl-1, 107.6; xyl-2, 75.0; xyl-3, 78.5; xyl-4, 71.1; xyl-5, 67.1; glu-1, 106.6; glut, 75.7; glu-3, 78.6; glu-4, 72.1; glu-5, 78.1; and glu-6, 63.2ppm.

Reference

H. Fujimoto, H. Hagiwara, K. Suzuki, and M. Yamazaki; New Toxic Metabolites from a Mushroom, *Hebeloma vinosophyllum*. II. Isolation and Structures of Hebevinosides VI, VII, VIII, IX, X, and XI; Chem. Pharm. Bull., Vol. 35, pp. 2254-2269(1987).

Common/Systematic Name

Hebevinoside VII

6-3-*O*-β-D-Xylopyranoside-16-*O*-(4,6-di-*O*-acetyl)-β-D-glucopyranoside

Molecular Formula/Molecular Weight

$C_{43}H_{70}O_{13}$; MW = 794.48164

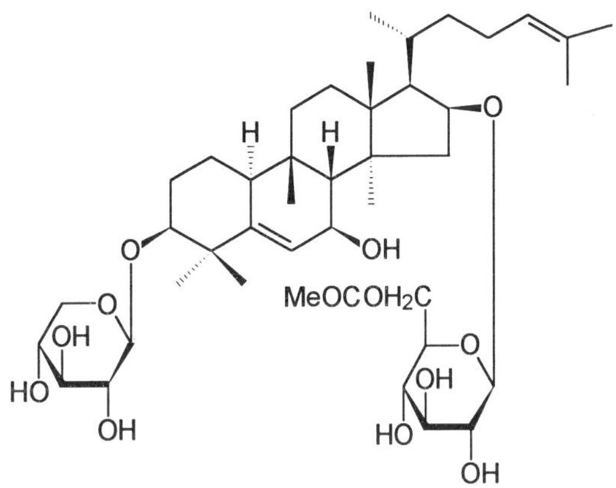

General Characteristics

Amorphous; $[\alpha]_D^{20}$ +82° (*c*=0.22, in acetone).

Fungal Source

Hebeloma vinosophyllum.

Isolation/Purification

Dried fruit bodies obtained by cultivation of *H. vinosophyllum* were cut into fine pieces and shaken in *n*-hexane at room temperature. The defatted fruit bodies were extracted with acetone whose pH was kept at 7 by occasional addition of small amounts of pyridine at room temperature to afford the acetone extract. The acetone extract was separated by centrifugal TLC with $CHCl_3$-acetone (1:1, v/v), $CHCl_3$-acetone-tetrahydrofuran (THF) (1:1:0.1), (1:1:0.2), (1:1:0.5, v/v/v), and THF. A mixture of the latter half of the fraction eluted with $CHCl_3$-acetone (1:1, v/v) and the first half of the fraction eluted with $CHCl_3$-acetone-THF (1:1:0.1, v/v/v) was further chromatographed on a Sephadex LH_{20} column with $CHCl_3$ and $CHCl_3$ -THF (20:1, v/v) to afford hebevinoside VIII. A mixture of the latter half of the fraction eluted with $CHCl_3$-acetone-THF (1:1:0.1, v/v/v) and the fraction eluted with $CHCl_3$-acetone-THF (1:1:0.2, v/v/v) was further chromatographed on a Sephadex LH_{20} column with $CHCl_3$-THF (10:1, v/v) to afford hebevinoside II, hebevinoside VII and hebevinoside IX. A mixture of the fraction eluted with $CHCl_3$-acetone-THF (1:1:0.5, v/v/v) and the first half of the fraction eluted with THF was further chromatographed on a Sephadex LH_{20} column with $CHCl_3$-THF (10:1, v/v) and THF to afford hebevinosides III and VI.

Spectral Data

UV:

$\lambda_{max}^{Acetonitrile}$ End absorption.

IR:

(KBr) 3450, 1735(ester), 1630, 1070, and 1035cm^{-1}.

^{1}H NMR:

(C_5D_5N) 0.86; 1.08; 1.19; 1.38; 1.63(each 3H, s); 1.03(3H, d, J=6.0Hz, 21-H_3); 1.71, 1.75(each 3H, s, 26 and 27H_3); 2.03, 2.09(each 3H, s AC); 4.43(1H, partly overlapped with other signals, H-7); 4.66(1H, d, J=7.7Hz, H-glc-1); 4.82(1H, d, J=7.7Hz, H-xyl-1); 5.51(1H, t, J=9.8Hz, H-24); and 6.12ppm (1H, d, J=4.9Hz, H-6).

^{13}C NMR:

(C_5D_5N) C-3, 87.4; C-5, 146.4; C-6, 122.5; C-7, 67.4; C-16, 82.5; C-24, 126.9; C-25, 130.1; xyl-1, 107.6; xyl-2, 75.1; xyl-3, 78.5; xyl-4, 71.1; xyl-5, 67.1; glu-1, 106.5; glu-2, 75.6*; glu-3, 75.5*; glu-4, 72.4; glu-5, 72.4; glu-6, 63.6; $COCH_3$, 20.8, 20.9; $COCH_3$, 170.5 and 170.7ppm.

* Assignments may be reversed.

Mass Spectrum:

LREIMS: 818(M^+ - H_2O), 686(18.5%, M^- - $C_5H_{10}O_5$[xyl.]), 668(10.2%, M^+ - H_2O, - $C_5H_{10}O_5$[xyl.]), and 653*m/e* (21.3%, M^+ - H_2O - $C_5H_{10}O_5$ - CH_3).

Reference

H. Fujimoto, H. Hagiwara, K. Suzuki, and M. Yamazaki; New Toxic Metabolites from a Mushroom, *Hebeloma vinosophyllum*. II. Isolation and Structures of Hebevinosides VI, VII, VIII, IX, X, and XI; Chem. Pharm. Bull., Vol. 35, pp. 2254-2269(1987).

Common/Systematic Name

Hebevinoside VIII

6-3-*O*-(4-*O*-Acetyl)-β-D-xylopyranoside-16-*O*-(4,6-di-*O*-acetyl)-β-D-glucopyranoside

Molecular Formula/Molecular Weight

$C_{47}H_{74}O_{15}$; MW = 878.50277

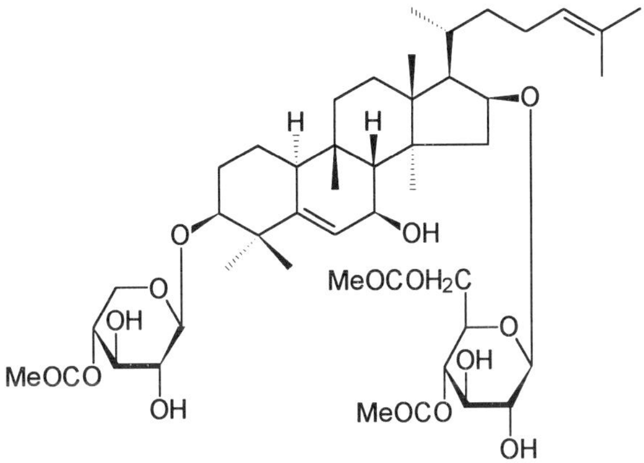

General Characteristics

Amorphous; $[\alpha]_D^{20}$ = + 36.5° (*c*=0.52, in $CHCl_3$).

Fungal Source

Hebeloma vinosophyllum.

Isolation/Purification

Dried fruit bodies obtained by cultivation of *H. vinosophyllum* were cut into fine pieces and shaken in *n*-hexane at room temperature. The defatted fruit bodies were extracted with acetone whose pH was kept at 7 by occasional addition of small amounts of pyridine at room temperature to afford the acetone extract. The acetone extract was separated by centrifugal TLC with $CHCl_3$-acetone (1:1, v/v), $CHCl_3$-acetone-tetrahydrofuran (THF) (1:1:0.1), (1:1:0.2), (1:1:0.5, v/v), and THF. A mixture of the latter half of the fraction eluted with $CHCl_3$-acetone (1:1, v/v) and the first half of the fraction eluted with $CHCl_3$-acetone-THF (1:1:0.1, v/v/v) was further chromatographed on a Sephadex LH_{20} column with $CHCl_3$ and $CHCl_3$-THF (20:1, v/v) to afford hebevinoside VIII. A mixture of the latter half of the fraction eluted with $CHCl_3$-acetone-THF (1:1:0.1, v/v/v) and the fraction eluted with $CHCl_3$-acetone-THF (1:1:0.2, v/v/v) was further chromatographed on a Sephadex LH_{20} column with $CHCl_3$-THF (10:1, v/v) to afford hebevinoside II, hebevinoside VII, and hebevinoside IX. A mixture of the fraction eluted with $CHCl_3$-acetone-THF (1:1:0.5, v/v/v) and the first half of the fraction eluted with THF was further chromatographed on a Sephadex LH_{20} column with $CHCl_3$-THF (10:1, v/v) and THF to afford hebevinosides III and VI.

Spectral Data

UV:

$\lambda_{max}^{Acetonitrile}$ End absorption.

IR:

(KBr) 3420, 1740(ester), 1640, 1075, and 1025cm^{-1}.

^1H NMR:

($CDCl_3$) 0.66; 1.00; 1.08; 1.22; 1.26(each 3H, s); 0.95(3H, d, J=6.6Hz, 21-H_3); 1.61, 1.68(each 3H, s, 26 and 27H_3); 2.10, 2.12, 2.12(each 3H, s, Ac); 3.29(1H, dd, J_1=11.7, J_2=8.1Hz, H-xyl-5a); 3.55(1H, m, H-glc-5); 3.62(1H, t, J=9.5Hz, H-xyl-3); 3.70(1H, t, J=8.1Hz, H-xyl-3); 3.93(1H, d, J=5.5Hz, H-7); 3.95(1H, dd, partly overlapped with the signal at 3.93ppm, H-xyl-5b); 4.07(1H, dd, J_1=12.1, J_2=4.8Hz, H-glc-6a); 4.16(1H, dd, partly overlapped with other signals, H-glc-6b); 4.22(1H, d, J=5.6Hz, H-glc-l); 4.40(1H, d, J=5.9Hz, H-xyl-1); 4.81(1H, dt, J_1=J_2=8.1, J_3=5.1Hz, H-xyl-4); 4.91(1H, t, J=9.5Hz, H-glc-4); 5.14(1H, t, J=6.8Hz, H-24); and 5.74ppm (1H, d, J=5.5Hz, H-6).

^{13}C NMR:

(C_5D_5N) C-3, 87.4; C-5, 146.3; C-6, 122.6; C-7, 67.5; C-16, 82.5; C-24, 126.9; C-25, 130.1; xyl-1, 107.4; xyl-2, 75.2; xyl-3, 74.9; xyl-4, 73.1; xyl-5, 63.2; glu-1, 106.5; glu-2, 75.6*; glu-3, 75.5*; glu-4, 72.4; glu-5, 72.4; glu-6, 63.6; CO<u>C</u>H_3, 20.8, 20.9, 20.9; <u>C</u>OCH_3, 170.5, 170.6, and 170.7ppm.

* Assignments may be reversed.

Mass Spectrum:

LREIMS: 860(47.1%, M^+ - H_2O), 686(27.4%, M^- - $C_7H_{12}O_6$[xyl.]), 668(83.9%, M^+ - H_2O - $C_7H_{12}O_6$ [xyl-1]), and 653m/e (36.5%, M^+ - H_2O - $C_7H_{12}O_6$ - CH_3).

Reference

H. Fujimoto, H. Hagiwara, K. Suzuki, and M. Yamazaki; New Toxic Metabolites from a Mushroom, *Hebeloma vinosophyllum*. II. Isolation and Structures of Hebevinosides VI, VII, VIII, IX, X, and XI; Chem. Pharm. Bull., Vol. 35, pp. 2254-2269(1987).

Common/Systematic Name

Hebevinoside IX

6-3-*O*-*p*-D-Xylopyranoside

Molecular Formula/Molecular Weight

$C_{35}H_{58}O_7$; MW = 590.41825

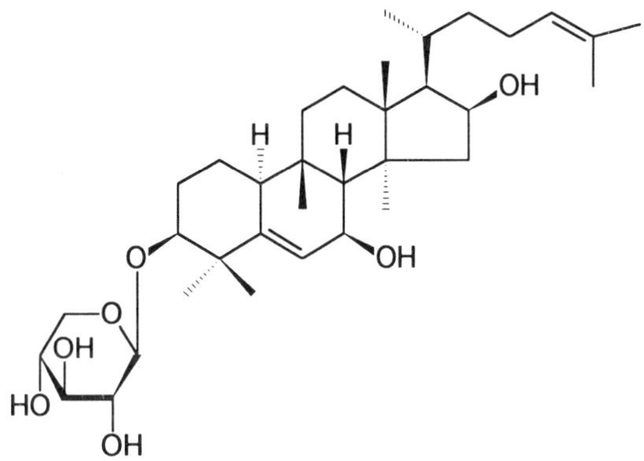

General Characteristics

Amorphous.

Fungal Source

Hebeloma vinosophyllum.

Isolation/Purification

Dried fruit bodies obtained by cultivation of *H. vinosophyllum* were cut into fine pieces and shaken in *n*-hexane at room temperature. The defatted fruit bodies were extracted with acetone whose pH was kept at 7 by occasional addition of a small amounts of pyridine at room temperature to afford the acetone extract. The acetone extract was separated by centrifugal TLC with $CHCl_3$-acetone (1:1, v/v), $CHCl_3$-acetone-tetrahydrofuran (THF) (1:1:0.1), (1:1:0.2), (1:1:0.5, v/v), and THF only. A mixture of the latter half of the fraction eluted with $CHCl_3$-acetone (1:1, v/v) and the first half of the fraction eluted with $CHCl_3$-acetone-THF (1:1:0.1, v/v/v) was further chromatographed on a Sephadex LH_{20} column with $CHCl_3$ and $CHCl_3$-THF (20:1, v/v) to afford hebevinoside VIII. A mixture of the latter half of the fraction eluted with $CHCl_3$-acetone-THF (1:1:0.1, v/v/v) and the fraction eluted with $CHCl_3$-acetone-THF (1:1:0.2, v/v/v) was further chromatographed on a Sephadex LH_{20} column with $CHCl_3$-THF (10:1, v/v) to afford hebevinoside II, hebevinoside VII, and hebevinoside IX. A mixture of the fraction eluted with $CHCl_3$-acetone-THF (1:1:0.5, v/v/v) and the first half of the fraction eluted with THF was further chromatographed on a Sephadex LH_{20} column with $CHCl_3$-THF (10:1, v/v) and THF to afford hebevinosides III and VI.

Spectral Data

UV:

$\lambda_{max}^{Acetonitrile}$ End absorption.

^{1}H NMR:

($CDCl_3$) 0.61; 0.93; 0.97; 1.02; 1.15(each 3H, s); 0.90(3H, d, *J*=6.2Hz, H_3-21); 1.54, 1.62(each 3H, s, H_3-26 and H_3-27); 3.89(1H, d, *J*=4.9Hz, H-7); 4.33(1H, d, *J*=5.9Hz, H-xyl-l); 5.10(1H, t, *J*=6.2Hz, H-24); and 5.68ppm (1H, d, *J*=4.9Hz, H-6).

Reference

H. Fujimoto, H. Hagiwara, K. Suzuki, and M. Yamazaki; New Toxic Metabolites from a Mushroom, *Hebeloma vinosophyllum*. II. Isolation and Structures of Hebevinosides VI, VII, VIII, IX, X, and XI; Chem. Pharm. Bull., Vol. 35, pp. 2254-2269(1987).

Common/Systematic Name

Hebevinoside X

7-3-*O*-β-D-Xylopyranoside-16-*O*-β-D-glucopyranoside

Molecular Formula/Molecular Weight

$C_{42}H_{70}O_{12}$; MW = 766.48673

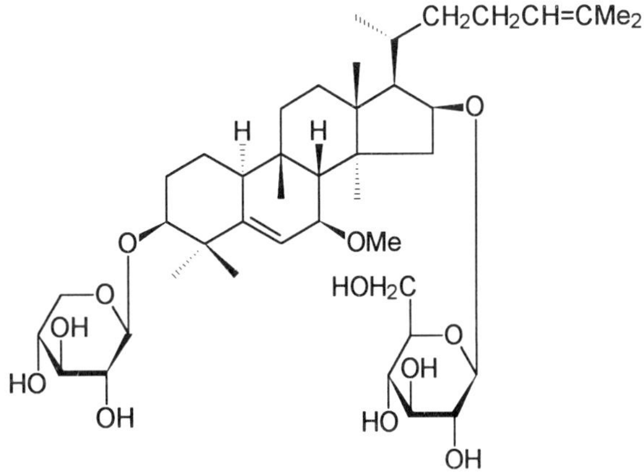

General Characteristics

Amorphous; $[\alpha]_D^{20}$ +63° (*c*=0.49, in MeOH).

Fungal Source

Hebeloma vinosophyllum; however, it was shown to be an artifact produced during purification of other natural hebevinosides.

Isolation/Purification

The *n*-butanol-soluble fraction was isolated from the 90% MeOH extract of dried fruit bodies of cultivated *H. vinosophyllum*. Repeated column chromatography of the *n*-BuOH-soluble fraction on silica gel with C_6H_6-acetone and on Sephadex LH_{20} with $CHCl_3$-MeOH had afforded several previously reported hebevinosides. A residual chromatographic fraction was further chromatographed on a column of silica gel with C_6H_6-acetone (5:1) and (2:1, v/v) to afford hebevinoside XI; C_6H_6-acetone (1:1, v/v) afforded hebevinoside X.

Spectral Data

UV:

λ_{max}^{MeOH} End absorption.

IR:
(KBr) 3400, 1740(ester), 1075, and 1035cm^{-1}.

^{1}H NMR:
($CDCl_3$) 0.74; 1.10; 1.16; 1.24; 1.63(each 3H, s); 1.08(3H, d, *J*=7.3Hz, 21-H_3); 1.69, 1.73(each 3H, s, 26 and 27H_3); 3.21(3H, s, OMe); 3.40(1H, d, *J*=5.3Hz, H-7); 4.79(2H, d, *J*=7.6Hz, H-glc-1 and H-xyl-1); 5.55(1H, t, *J*=6.8Hz, H-24); and 5.88ppm (1H, d, *J*=5.3Hz, H-6).

^{13}C NMR:
(C_5D_5N) C-3, 87.4; C-5, 147.7; C-6, 119.4; C-7, 77.6; C-16, 82.1; C-24, 126.8; C-25, 130.0; OMe, 56.2; xyl-1, 107.7; xyl-2, 74.9; xyl-3, 78.5; xyl-4, 71.1; xyl-5, 67.0; glu-1, 106.8; glu-2, 75.6; glu-3, 78.7; glu-4, 72.0; glu-5, 78.2; and glu-6, 63.2ppm.

Reference

H. Fujimoto, H. Hagiwara, K. Suzuki, and M. Yamazaki; New Toxic Metabolites from a Mushroom, *Hebeloma vinosophyllum*. II. Isolation and Structures of Hebevinosides VI, VII, VIII, IX, X, and XI; Chem. Pharm. Bull., Vol. 35, pp. 2254-2269(1987).

Common/Systematic Name

Hebevinoside XI

7-3-*O*-β-D-Xylopyranoside-16-*O*-(4,6-di-*O*-acetyl)-β-D-glucopyranoside

Molecular Formula/Molecular Weight

$C_{44}H_{72}O_{13}$; MW = 808.49729

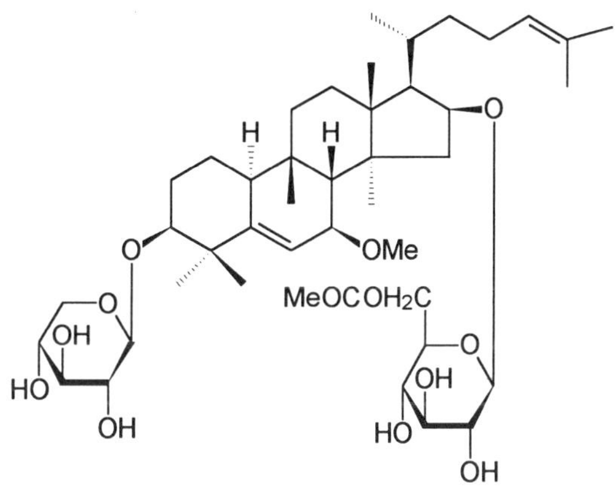

General Characteristics

Amorphous; $[\alpha]_D^{20}$ +65° (*c*=0.17, in MeOH).

Fungal Source

Hebeloma vinosophyllum; however, it was shown to be an artifact produced during purification of other natural hebevinosides.

Isolation/Purification

The *n*-butanol-soluble fraction was isolated from the 90% MeOH extract of dried fruit bodies of cultivated *H. vinosophyllum*. Repeated column chromatography of the *n*-BuOH-soluble fraction on silica gel with C_6H_6-acetone and on Sephadex LH_{20} with $CHCl_3$-MeOH had afforded several previously reported hebevinosides. A residual chromatographic fraction was further chromatographed on a column of silica gel with C_6H_6-acetone (5:1) and (2:1, v/v) to afford hebevinoside XI; C_6H_6-acetone (1:1, v/v) afforded hebevinoside X.

Spectral Data

UV:

λ_{max}^{MeOH} End absorption.

IR:

(KBr) 3450, 1747(ester), 1645, 1085 and 1040cm^{-1}.

^{1}H NMR:

($CDCl_3$) 0.68; 0.97; 1.01; 1.12; 1.21(each 3H, s); 0.96(3H, d, J=5.0Hz, 21-H_3); 1.69, 1.60(each 3H, s, 26 and 27H_3); 2.07, 2.12(3H, s, Ac); 3.33(1H, d, J=5.3Hz, H-7); 4.79(2H, d, J=7.6Hz, H-glc-1 and H-xyl-1); 5.55(1H, t, J=6.8Hz, H-24); and 5.88ppm (1H, d, J=5.3Hz, H-6).

^{13}C NMR:

(C_5D_5N) C-3, 87.4; C-5, 147.7; C-6, 119.4; C-7, 77.6; C-16, 82.1; C-24, 126.8; C-25, 130.0; OMe, 56.2; xyl-l, 107.7; xyl-2, 74.9; xyl-3, 78.5; xyl-4, 71.1; xyl-5, 67.0; glu-l, 106.8; glu-2, 75.6; glu-3, 78.7; glu-4, 72.0; glu-5, 78.2; and glu-6, 63.2ppm.

Reference

H. Fujimoto, H. Hagiwara, K. Suzuki, and M. Yamazaki; New Toxic Metabolites from a Mushroom, *Hebeloma vinosophyllum*. II. Isolation and Structures of Hebevinosides VI, VII, VIII, IX, X, and XI; Chem. Pharm. Bull., Vol. 35, pp. 2254-2269(1987).

Common/Systematic Name

Hebevinoside XII

16-3-*O*-ß-D-Xylopyranoside-16-*O*-(4-*O*-acetyl)-ß-D-glucopuranoside

Molecular Formula/Molecular Weight

$C_{45}H_{74}O_{11}$; MW = 790.52311

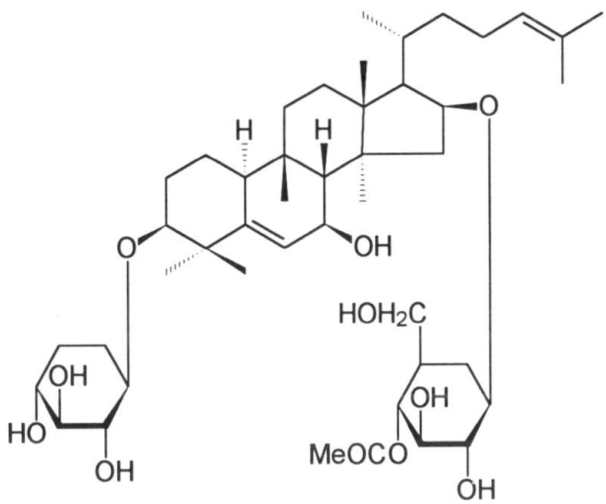

General Characteristics

Amorphous solid; $[\alpha]_D^{22}$ +60° (*c*=0.36, in pyridine).

Fungal Source

Hebeloma vinosophyllum (poisonous mushroom).

Isolation/Purification

Dried mycelia with primordia at growth stage II were cut into fine pieces and extracted 4 times with acetone at room temperature for 12h. The combined acetone extracts were divided with *n*-hexane into hexane soluble and hexane insoluble fractions. The hexane insoluble fraction was successively chromatographed on an ODS column in the HPLC system with 20% aqueous acetone (flow rate: 4.0ml/min) and an ODS column in the HPLC system with 75% aqueous CH_3CN (flow rate: 2.0ml/min).

Biological Activity

Neurotoxic.

Spectral Data

UV:

$\lambda_{max}^{Acetonitrile}$ End absorption.

IR:
(KBr) 3400, 1730, 1640, 1075, and 1035cm^{-1}.

^{1}H NMR:
(C_5D_5N) 0.78; 1.11; 1.20; 1.38; 1.67(each 3H, s); 1.05(3H, d, *J*=6.4Hz, H_3-21); 1.73; 1.77(each 3H, s, H_3-26, -27); 2.06(3H, s, Ac); 4.36 (1H, d, *J*=4.9Hz, H-7); 4.75 (1H, d, *J*=7.6Hz, H-glc-1); 4.84 (1H, d, *J*=7.6Hz, H-xyl-1); 5.56 (lH, t, *J*=7.7Hz, H-24); 5.65 (1H, t, *J*=9.6Hz, H-glc-4); and 6.04ppm (1H, d, *J*=4.9Hz, H-6).

^{13}C NMR:
(C_5D_5N) C-3, 87.5; C-5, 146.2; C-6, 122.6; C-7, 67.3; C-16, 82.1; C-24, 126.9; C-25, 130.0; xylose-1, 107.7; xylose-2, xylose-3, 78.5; xylose-4, 71.1; xylose-5, 67.1; glucose-1, 106.5; glucose-2, 75.7*; glucose-3, 75.9*; glucose-4, 73.1; glucose-5, 75.9*; glucose-6, 62.5; CO$\underline{C}H_3$, 21.1; and $\underline{C}OCH_3$, 170.6ppm.

* Assignments may be interchanged.

Reference

H. Fujimoto, K. Maeda, and M. Yamazaki, New Toxic Metabolites from a Mushroom, *Hebeloma vinosophyllum* III. Isolation and Structures of Three New Glycosides, Hebevinosides XII, XIII and XIV and Productivity of the Hebevinosides at Three Growth Stages of the Mushroom; Chem. Pharm. Bull., Vol. 39, pp. 1958-1961(1991).

Common/Systematic Name

Hebevinoside XIII

16-3-*O*-(3,4-Di-*O*-acetyl)-β-D-xylopyranoside-16-*O*-(4,6-di-*O*-acetyl)-β-D-glucopyranoside

Molecular Formula/Molecular Weight

$C_{51}H_{80}O_{14}$; MW = 916.55481

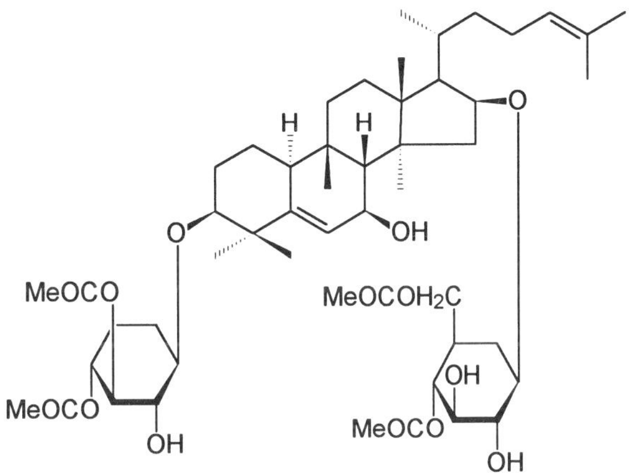

General Characteristics

Amorphous solid; $[\alpha]_D^{22}$ +35° (c=0.42, in $CHCl_3$).

Fungal Source

Hebeloma vinosophyllum (poisonous mushroom).

Isolation/Purification

Dried mycelia with primordia at growth stage II were cut into fine pieces and extracted 4 times with acetone at room temperature for 12h. The combined acetone extracts were divided with *n*-hexane into hexane soluble and hexane insoluble fractions. The hexane insoluble fraction was successively chromatographed on an ODS column in the HPLC system with 20% aqueous acetone (flow rate: 4.0ml/min) and an ODS column in the HPLC system with 75% aqueous CH_3CN (flow rate: 2.0ml/min).

Biological Activity

Neurotoxic.

Spectral Data

UV:

$\lambda_{max}^{Acetonitrile}$ End absorption.

IR:

(KBr) 3430, 1750, 1645, 1070, and 1040cm^{-1}.

^{1}H NMR:

($CDCl_3$) 0.66; 1.00; 1.05; 1.08; 1.21(each 3H, s); 0.95(3H, d, J=6.6Hz, H_3-21); 1.61; 1.68(each 3H, s, H_3-26, -27); 2.02; 2.08; 2.11; 2.12(each 3H, s, Ac); 3.31(1H, dd, J_1=11.6, J_2=8.7 Hz, H-xyl-5a); 3.64(1H, m, H-glc-5); 3.67(1H, t-like, H-glc-3); 4.06(1H, d, J=5.2Hz, H-7); 4.08(1H, dd, J_1=11.8, J_2=5.0 Hz, H-glc-6a); 4.22(1H, dd, H-glc-6b); 4.45(1H, d, J=7.6Hz, H-glc-1); 4.51(1H, d, J=6.7Hz, H-xyl-1); 4.90(1H, t, J=7.8Hz, H-glc-4); 5.00(1H, brt, J=8.7Hz, H-xyl-4); 5.08(1H, t, J=9.3Hz, H-xyl-3); 5.13(1H, t, J=8.5Hz, H-24); and 5.61ppm (1H, d, J=5.2Hz, H-6).

^{13}C NMR:

(C_5D_5N) C-3, 87.6; C-5, 146.1; C-6, 122.7; C-7, 67.4; C-16, 82.5; C-24, 126.9; C-25, 130.0; xylose-1, 106.9; xylose-2,72.1; xylose-3, 75.4; xylose-4, 70.6; xylose-5, 62.8; glucose-1, 106.5; glucose-2, 75.6; glucose-3, 75.4; glucose-4, 72.4; glucose-5, 72.4; glucose-6, 63.5; $CO\underline{C}H_3$, 20.6, 20.8, 20.9, 20.9; and $\underline{C}OCH_3$, 170.2, 170.4, 170.4, 170.6ppm.

Reference

H. Fujimoto, K. Maeda, and M. Yamazaki; New Toxic Metabolites from a Mushroom, *Hebeloma vinosophyllum* III. Isolation and Structures of Three New Glycosides, Hebevinosides XII, XIII and XIV and Productivity of the Hebevinosides at Three Growth Stages of the Mushroom; Chem. Pharm. Bull., Vol. 39, pp. 1958-1961(1991).

Common/Systematic Name
Hebevinoside XIV
16-3-*O*-(3,4-Di-*O*-acetyl)-β-D-xylopyranoside-16-*O*-(6-*O*-acetyl)-β-D-glucopyranoside

Molecular Formula/Molecular Weight
$C_{49}H_{78}O_{13}$; MW = 874.54424

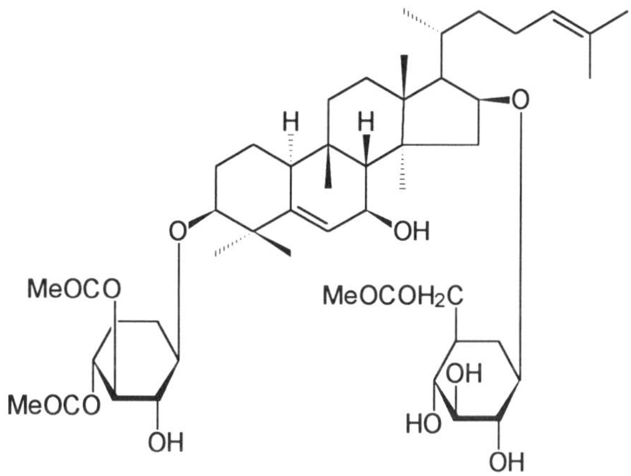

General Characteristics
Amorphous solid; $[\alpha]_D^{24}$ +15 (*c*=0.40, in $CHCl_3$).

Fungal Source
Hebeloma vinosophyllum (poisonous mushroom).

Isolation/Purification
Dried mycelia with primordia at growth stage II were cut into fine pieces and extracted 4 times with acetone at room temperature for 12h. The combined acetone extracts were divided with *n*-hexane into hexane soluble and hexane insoluble fractions. The hexane insoluble fraction was successively chromatographed on an ODS column in the HPLC system with 20% aqueous acetone (flow rate: 4.0ml/min) and an ODS column in the HPLC system with 75% aqueous CH_3CN (flow rate: 2.0ml/min).

Biological Activity
Neurotoxic.

Spectral Data

UV:
$\lambda_{max}^{Acetonitrile}$ End absorption.

IR:

(KBr) 3450, 1740, 1635, 1065, and 1030cm^{-1}.

^{1}H NMR:

($CDCl_3$) 0.66; 1.00; 1.04; 1.07; 1.21(each 3H, s); 0.94(3H, d, *J*=6.4Hz, H_3-21); 1.61; 1.68(each 3H, s, H_3-26, -27); 2.03; 2.09; 2.13(each 3H, s, Ac); 3.31(1H, dd, J_1=12.2, J_2=9.0Hz, H-xyl-5a); 3.52(1H, m, H-glc-5); 3.94(1H, d, *J*=4.8Hz, H-7); 4.03(1H, dd, J_1=11.6, J_2=5.2Hz, H-glc-6a); 4.16(1H, dd, H-glc-6b); 4.18(1H, d, *J*=7.6Hz, H-glc-1); 4.34(1H, d, *J*=7.3Hz, H-xyl-1); 4.89(1H, dt, J_1=J_2=9.0Hz, J_3=5.2Hz, H-xyl-4); 5.09(1H, t, *J*=9.0Hz, H-xyl-3); 5.13(1H, t, *J*=7.3Hz, H-24); and 5.71ppm (1H, d, *J*=4.8Hz, H-6).

^{13}C NMR:

(C_5D_5N) C-3, 87.5; C-5, 146.1; C-6, 122.7; C-7, 67.5; C-16, 82.2; C-24, 126.8; C-25, 130.0; xylose-1, 106.8; xylose-2, 72.2; xylose-3, 75.5; xylose-4, 70.6; xylose-5, 62.7; glucose-1, 106.5; glucose-2, 75.9; glucose-3, 78.3; glucose-4, 71.8; glucose-5, 75.5; glucose-6, 64.7; $CO\underline{C}H_3$, 20.6, 20.8, 20.9; and $\underline{C}OCH_3$, 170.2, 170.4, 170.8ppm.

Reference

H. Fujimoto, K. Maeda, and M. Yamazaki; New Toxic Metabolites from a Mushroom, *Hebeloma vinosophyllum* III. Isolation and Structures of Three New Glycosides, Hebevinosides XII, XIII and XIV and Productivity of the Hebevinosides at Three Growth Stages of the Mushroom; Chem. Pharm. Bull., Vol. 39, pp. 1958-1961(1991).

Common/Systematic Name

Hebelomic acid A

3β-Acetyl-2α-(3'-hydroxy-3'-methyl)glutarylcrustulinol

Molecular Formula/Molecular Weight

$C_{38}H_{60}O_{11}$; MW = 692.41356

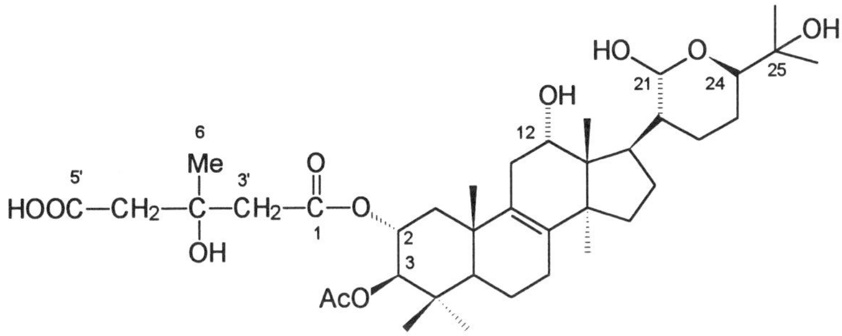

General Characteristics

A colorless powder; mp., 201-202°C; $[\alpha]_D^{17}$ -7 9° (*c*=1.0, in MeOH).

Fungal Source

Hebeloma spoliatum, H. crustuliniforme, and *H. sinapizans.*

Isolation/Purification

Dried fruit-bodies were cut into small pieces and extracted for 3 hours with MeOH at room temperature. The methanolic extract was divided with 60% (v/v) aqueous MeOH-acetone (3:5, v/v) into a supernatant and a precipitate. The supernatant was chromatographed on a silica gel column eluted with $CHCl_3$-MeOH (100:1), (100:1), (50:1), (15:1) and (5:1, v/v) to afford five fractions. Fraction IV was subjected to MPLC with *n*-hexane-acetone (2:1) and (1:1, v/v) at a flow rate of 12ml/min and HPLC with MeOH-H_2O (40:1, v/v) at a flow rate of 4.5 ml/min successively, to afford HS-A, which was treated with *n*-hexane-acetone to give HS-A in a pure state. Fraction III was subjected to HPLC with CH_3CN-acetone (3:1, v/v) at a flow rate of 2.5ml/min and MPLC with $CHCl_3$-MeOH (70:1, v/v) to afford HS-B and HS-C.

Biological Activity

Toxic to mice; IP dose of 100mg/kg produced death after paralysis of the limbs. It appeared to exhibit a paraverine-like relaxation of smooth muscle.

Spectral Data

UV:

λ_{max}^{MeOH} End absorption.

IR:
3425(0-H), 2930(C-H), 1735(C=O), and 1375cm^{-1} (C-O).

^{1}H NMR:
($CDCl_3$) 0.61; 0.91; 0.94; 1.05; 1.11; 1.15; 1.18; 1.37(3H each, s, H-18, -30, -29, -19, -26, -27, -28, -6'); 2.06(3H, s, CH_3CO); 2.62, 2.66(1H each, AB-type d, *J*=14.8Hz); 2.65, 2.71(each 1H, AB-type d, *J*=15.5Hz, H-2', -4'); 3.72(1H, dd, J_1=11.6, J_2=2.0Hz, H-24); 3.87(1H, d, *J*=8.2Hz, H-12); 4.79(1H, d, *J*=10.3Hz, H-3); 5.18(1H, td, J_1=10.3, J_2=5.0Hz, H-2); and 5.46ppm (1H, brs, H-21).

^{13}C NMR:
(C_5D_5N) C-1, 40.9; C-2, 70.3; C-3, 80.2; C-4, 39.5; C-5, 50.2; C-6, 18.3; C-8, 135.6; C-9, 132.9; C-10, 38.2; C-12, 72.7; C-13, 50.5; C-14, 50.0; C-17, 39.9; C-18, 17.9; C-19, 19.9; C-20, 44.0; C-21, 93.3; C-24, 75.0, C-25, 71.2; C-26, 26.1; C-27, 26.8; C-28, 24.2; C-29, 28.3; C-30, 17.2; C-1', 171.3; C-2', 46.5; C-3', 69.9; C-4', 46.6; C-5', 174.8; C-6', 28.4; <u>C</u>OCH$_3$, 170.8; and CO<u>C</u>H$_3$, 21.1ppm.

Mass Data:
FAB-MS positive ion 715*m/e* [$(M - Na)^+$]; positive ion (plus KI) 731*m/e* [$(M + K)^+$]; negative ion 691*m/e* [$(M-H)^-$]. Anal. calcd for $C_{38}H_{60}O_{11}$: C, 65.87; H, 8.73%; found: C, 65.45; H, 8.67%.

<u>Reference</u>

L. Garlaschelli, G. Vidari, M. Virtuani, P. Vita-Finzi, and G. Mellerio; The Structures of New Lanostane Triterpenes from the Fruiting Bodies of *Hebeloma senescens*; J. Nat. Prod., Vol. 58, pp. 992-1002(1995).

H. Fujimoto, Y. Takano, and M. Yamaazaki; Isolation, Identification and Pharmacological Studies on Three Toxic Metabolites from a Mushroom, *Hebeloma spoliatum*; Chem. Pharm. Bull., Vol. 40, pp. 869-872(1992).

Common/Systematic Name

Hebelomic acid B

Molecular Formula/Molecular Weight

$C_{40}H_{62}O_{12}$; MW = 734.42413

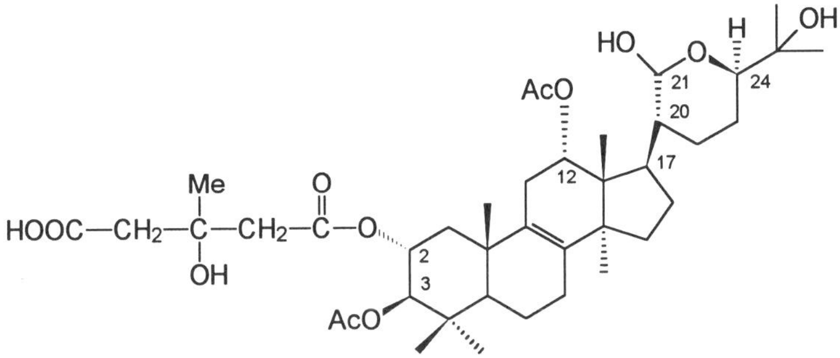

General Characteristics

Amorphous powder; mp., 160-163 °C; $[\alpha]_D^{21}$ +27.7° (*c*=0.7, in $CHCl_3$).

Fungal Source

Hebeloma senescens, an inedible mushroom.

Isolation/Purification

The more polar fractions from silica gel column chromatography containing metabolites showing a free carboxylic group (yellow spots with bromocresol green solution on TLC plates) were further purified. Concentration of fraction 14 by rotary evaporation gave a white precipitate, which was crystallized from acetonitrile and found to be identical to hebelomic acid A. Concentration of fraction 12 gave a precipitate consisting of a 1:1 (v/v) mixture of hebelomic acids A and B, which were then separated from one another. Fraction 11 was further separated by silica gel column chromatography, using a methylene chloride-acetone-acetic acid gradient, to provide, in order of elution, hebelomic acids D and C, along with additional quantities of hebelomic acid B and hebelomic acid A. Fraction 10 was further separated by silica gel column chromatography with methylene chloride-acetone-acetic acid (75:25:2.5 and 70:30:2.5, v/v/v) to afford fractions A, B, C, D, E, F, G, and H. Fractions were monitored by silica gel TLC developed with methylene chloride-acetone-acetic acid (73:24.5:2.5, v/v/v). Crystallization of fraction B from acetone/hexane afforded hebelomic acid F. Fractions A, F, and H contained almost pure hebelomic acids D, B, and A, respectively. Fractions C, D, and E were pooled and separated on different silica gel columns, eluted with hexane-acetone-acetic acid (48.7:48.7:2.5), benzene-acetone-methanol (85:10:5), benzene-methanol-acetic acid (93:5:2), benzene-methylene chloride-methanol-acetic acid (26:59:3:2, v/v/v/v), to afford hebelomic acid E.

Biological Activity

Showed moderate antibacterial activity in the Kirby-Bauer test against *Bacillus subtilis* and *Staphylococcus oxford*, but no activity against *Escherichia coli* or *Candida albicans*. The LD_{50} value in the brine shrimp (*Artemia salina*) lethality assay was 386μg/ml.

Spectral Data

IR:

(KBr) 3450(OH), 1725(C=O), 1375, 1235, 1073, 1030, 1000, 950, and 897cm^{-1}.

^{1}H NMR:

($CDCl_3$) 0.75; 0.91; 0.94; 1.11; 1.12; 1.15; 1.37(24H overall, 7 x s, H_3-18, -30, -29, -19, -28, -26, -27, -6'); 2.04(3H, s, MeCOO-12); 2.06(3H, s, MeCOO-3); 2.52; 2.69 (1H each, Abq, J=16.0Hz, H_2-2' or H_2-4'); 2.64, 2.69(1H each, Abq, J=15.5Hz, H_2-4' or H_2-2'); 3.69(1H, dd, J_1=11.5Hz, J_2=2.0Hz, H-24); 4.78(1H, d, $J_{2,3}$=10.0Hz, H-3); 4.93(1H, br d, J=7.3Hz, H-12); 5.18(1H, td, $J_{2,3}$=$J_{2,1a}$=11.0Hz, $J_{2,1\beta}$=4.5Hz, H-2); and 5.28ppm (1H, br s, H-21). (C_5D_5N) 0.75; 0.95; 1.02; 1.31; 1.34; 1.74(24H overall, 6 x s, H_3-18, -30, -29, -19, -28, -26, -27, -6'); 2.10; 2.16(3H each, 2 X s, 2 X MeCOO-); 3.11(2H, ABq, H_2-2 or H_2-4); 3.18(2H, ABq, H_2-4 or H_2-2); 4.20(1H, dd, J_1=11.5 Hz, J_2=2.0Hz, H-24); 5.08(1H, d, $J_{2,3}$=10.0Hz, H-3); 5.18(1H, br d, J=7.5Hz, H-12); 5.51(1H, td, $J_{2\beta,3\alpha}$=$J_{1\alpha,2\beta}$=11.0Hz, $J_{2,1\beta}$=4.5Hz, H-2); and 5.70ppm (1H, br s, H-21).

^{13}C NMR:

($CDCl_3$) C-1, 40.9(t); C-2, 69.8(d); C-3, 80.1(d); C-4, 39.5(s); C-5, 50.1(d); C-6, 18.2(t); C-7, 27.0*(t); C-8, 135.6(s); C-9, 131.8(s); C-10, 38.1(s); C-11, 26.6*(t); C-12, 73.8(d); C-13, 49.7(s); C-14, 48.2(s); C-15, 32.2(t); C-16, 31.4(t); C-17, 38.9(d); C-18, 17.6**(q); C-19, 19.9(q); C-20, 44.2(d); C-21, 93.6(d); C-22, 26.2(t); C-23, 23.6(t); C-24, 74.5(d); C-25, 71.3(s); C-26, 26.2***(q); C-27, 26.6***(q); C-28, 25.4(q); C-29, 28.3(q); C-30, 17.1**(q); C-1', 170.8(s); C-2', 46.4****(t); C-3', 69.8(s); C-4', 46.3****(t); C-5', 174.6(s); C-6', 28.3(q); <u>C</u>OMe, 171.2(s), 170.5(s); and CO<u>Me</u>, 21.1(q), 21.7ppm (q).

*, **, ***, **** Assignments may be reversed.

Mass Data:

CIMS: (NH_3) 734(M^+ + NH_4-H_2O), 674(734 - HOAc), and 632m/e (M^+ + NH_4 - 2 X HOAc); DCIMS: (NH_3) 752(M^+ + NH_3) and 734m/e (M^+ + NH_4-H_2O). EIMS: 656(86%) (M^+ - H_2O - HOAc), 641m/e (656 - Me), 641(77), 554(25), 514(20), 512(46), 497 (24), 479 (32), 436(21), 435(43), 419(28), 376(23), 353(22), 295(30), 277(23), 253(31), 251(25), 237(23), 226(20), 211(29), 208(33), 197(20), 181(100), 173(25), 171(30), 168(60), 159(33), 157(29), 145(44), 133(35), 124(42), 123(38), 121(30), 109(24), 107(33), 95(41), 93(21), 81(21), and 43m/e (29); anal. calcd. for $C_{40}H_{62}O_{12}$: C, 65.37; H, 8.50%; found: C, 65.57, H, 8.38%.

TLC Data

Silica gel GF_{254} plates developed with acetone-methylene chloride-acetic acid (24:73.5:2.5, v/v/v), R_f = 0.22.

Reference

L. Garlaschelli, G. Vidari, M. Virtuani, P. Vita-Finzi, and G. Mellerio; The Structures of New Lanostane Triterpenes from the Fruiting Bodies of *Hebeloma senescens*; J. Nat. Prod., Vol. 58, pp. 992-1002(1995).

Common/Systematic Name

Hebelomic acid E

Molecular Formula/Molecular Weight

$C_{38}H_{60}O_{10}$; MW = 676.41865

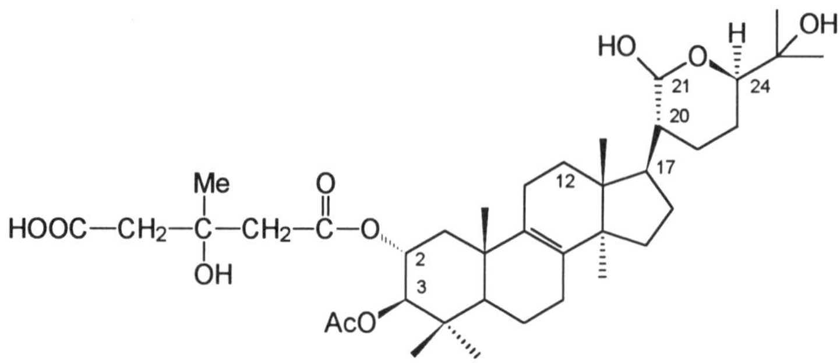

General Characteristics

Amorphous powder; mp., 170-172°C; $[\alpha]_D^{21}$ - 9.6° (*c*=0.6, in CH_2Cl_2).

Fungal Source

Hebeloma senescens, an inedible mushroom.

Isolation/Purification

The more polar fractions from silica gel column chromatography containing metabolites showing a free carboxylic group (yellow spots with bromocresol green solution on TLC plates) were further purified. Concentration of fraction 14 by rotary evaporation gave a white precipitate, which was crystallized from acetonitrile and found to be identical to hebelomic acid A. Concentration of fraction 12 gave a precipitate consisting of a 1:1 (v/v) mixture of hebelomic acids A and B, which were then separated from one another. Fraction 11 was further separated by silica gel column chromatography, using a methylene chloride-acetone-acetic acid gradient, to provide, in order of elution, hebelomic acids D and C, along with additional quantities of hebelomic acid B and hebelomic acid A. Fraction 10 was further separated by silica gel column chromatography with methylene chloride-acetone-acetic acid (75:25:2.5 and 70:30:2.5, v/v/v) to afford fractions A, B, C, D, E, F, G, and H. Fractions were monitored by silica gel TLC developed with methylene chloride-acetone-acetic acid (73:24.5:2.5, v/v/v). Crystallization of fraction B from acetone/hexane afforded hebelomic acid F. Fractions A, F, and H contained almost pure hebelomic acids D, B, and A, respectively. Fractions C, D, and E were pooled and separated on different silica gel columns, eluted with hexane-acetone-acetic acid (48.7:48.7:2.5), benzene-acetone-methanol (85:10:5), benzene-methanol-acetic acid (93:5:2), benzene-methylene chloride-methanol-acetic acid (26:59:3:2, v/v/v), to afford hebelomic acid E.

Biological Activity

Showed moderate antibacterial activity in the Kirby-Bauer test against *Bacillus subtilis* and *Staphylococcus oxford*, but no activity against *Escherichia coli* or *Candida albicans*.

Spectral Data

IR:

(KBr) 3424(OH), 2948, 1742(C=O), 1456, 1376, 1237, 1094, 1076, 1032, 951, 902, 803, 752, and 730cm^{-1}.

^{1}H NMR:

($CDCl_3$) 0.72; 0.88; 0.90; 0.94; 1.13; 1.15; 1.18; 1.36(3H each, 8 X s, H_3-18, -30, -29, -19, -28, -26, -27, -6'); 2.10(3H, s, MeCOO-); 2.40-2.80(4H, m, H_2-2' and H_2-4'); 3.78(1H, br d, *J*=11.0Hz, H-24); 4.77(1H, d, *J*=10.0Hz, H-3); 5.17(1H, br t, *J*=11.0Hz, H-2); and 5.38ppm (1H, br s, H-21).

^{13}C NMR:

($CDCl_3$) C-1, 41.1(t); C-2, 71.1(d); C-3, 80.3(d); C-4, 39.4(s); C-5, 49.9(d); C-6, 18.0(t); C-7, 27.0(t); C-8, 135.0(s); C-9, 133.2(s); C-10, 38.1(s); C-11, 21.3*(t); C-12, 26.2*(t); C-13, 44.1(s); C-14, 49.7(s); C-15, 30.6**(t); C-16, 30.5**(t); C-17, 45.0***(d); C-18, 16.6(q); C-19, 20.1(q); C-20, 42.8***(d); C-21, 93.0(d); C-22, 26.0*(t); C-23, 23.2(t); C-24, 74.2(d); C-25, 72.2(s); C-26, 26.4****(q); C-27, 24.6****(q); C-28, 23.8(q); C-29, 27.2(q); C-30, 17.5(q); C-1', 171.5(s); C-2', 45.1(t); C-3', 69.9(s); C-4', 45.1(t); C-5', 174.7(s); C-6', 28.2(q); <u>C</u>OMe, 171.3(s); and CO<u>Me</u>, 21.1ppm (q).

*, **, ***, **** Assignments may be reversed.

Mass Data:

CIMS: (NH_3 + NH_4Cl) 676(M^++ NH_4-H_2O), 658(676 - H_2O), and 574*m/e* (M^+ + NH_4 - 2 X HOAc); DCIMS: (NH_3) 694(M^+ + NH_4^+) and 676*m/e* (M^+ + NH_4-H_2O); anal. calcd. for $C_{38}H_{60}O_{10}$: C, 67.43; H, 8.93%; found: C, 67.66, H, 8.76%.

TLC Data

Silica gel GF_{254} plates developed with acetone-methylene chloride-acetic acid (24:73.5:2.5, v/v/v), R_f = 0.29. Spots were visualized under UV light or by spraying plates with 0.5% vanillin solution in sulfuric acid-ethanol (4:1, v/v) and heating at 120°C for 1 min. or with 0.04% bromocresol green solution in water in the cold.

Reference

L. Garlaschelli, G. Vidari, M. Virtuani, P. Vita-Finzi, and G. Mellerio; The Structures of New Lanostane Triterpenes from the Fruiting Bodies of *Hebeloma senescens*; J. Nat. Prod., Vol. 58, pp. 992-1002(1995).

Common/Systematic Name

Hebelomic acid F

Molecular Formula/Molecular Weight

$C_{40}H_{62}O_{11}$; MW = 718.42921

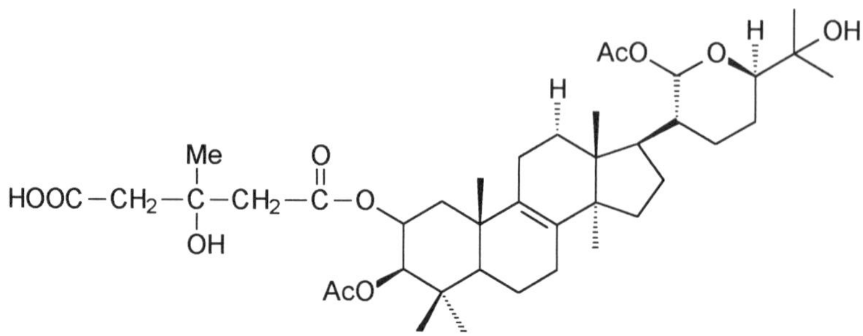

General Characteristics

Amorphous powder; $[\alpha]_D^{21}$ - 32.5° (*c*=0.8, in methylene chloride).

Fungal Source

Hebeloma senescens, an inedible mushroom.

Isolation/Purification

The more polar fractions from silica gel column chromatography containing metabolites showing a free carboxylic group (yellow spots with bromocresol green solution on TLC plates) were further purified. Concentration of fraction 14 by rotary evaporation gave a white precipitate, which was crystallized from acetonitrile and found to be identical to hebelomic acid A. Concentration of fraction 12 gave a precipitate consisting of a 1:1 (v/v) mixture of hebelomic acids A and B, which were then separated from one another. Fraction 11 was further separated by silica gel column chromatography, using a methylene chloride-acetone-acetic acid gradient, to provide, in order of elution, hebelomic acids D and C, along with additional quantities of hebelomic acid B and hebelomic acid A. Fraction 10 was further separated by silica gel column chromatography with methylene chloride-acetone-acetic acid (75:25:2.5 and 70:30:2.5, v/v/v) to afford fractions A, B, C, D, E, F, G, and H. Fractions were monitored by silica gel TLC developed with methylene chloride-acetone-acetic acid (73:24.5:2.5, v/v/v). Crystallization of fraction B from acetone/hexane afforded hebelomic acid F. Fractions A, F, and H contained almost pure hebelomic acids D, B, and A, respectively. Fractions C, D, and E were pooled and separated on different silica gel columns, eluted with hexane-acetone-acetic acid (48.7:48.7:2.5), benzene-acetone-methanol (85:10:5), benzene-methanol-acetic acid (93:5:2), benzene-methylene chloride-methanol-acetic acid (26:59:3:2, v/v/v), to afford hebelomic acid E.

Biological Activity

Showed moderate antibacterial activity in the Kirby-Bauer test against *Bacillus subtilis* and *Staphylococcus oxford*, but no activity against *Escherichia coli* or *Candida albicans*. The LD_{50} value in the brine shrimp (*Anemia saliva*) lethality assay was 19.7μg/ml.

Spectral Data

IR:

(KBr) 3464(OH), 2953, 1743(C=O), 1455, 1373, 1337, 1238, 1162, 1118, 1078, 1036, 1012, 959, 941, 928, 910, and 866cm^{-1}.

^{1}H NMR:

($CDCl_3$) 0.73; 0.86; 0.92; 0.95; 1.12; 1.13; 1.15; 1.38(3H each, 8 x s, H_3-18, -30, -29, -19, -28, -26, -27, -6'); 2.07, 2.10(3H each, 2 x s, 2 x MeCOO-); 2.54, 2.64(2H, ABq, *J*=15Hz, H_2-2' and HZ-4'); 2.61, 2.72(2H, ABq, *J*=15.0Hz, H_2-4' or H_2-2'); 3.48(1H, br d, *J*=15.0Hz, H-24); 4.78(1H, d, *J*=10Hz, H-3); 5.17(1H, td, $J_{2\beta,3\alpha}$=$J_{2\beta1\beta}$=4.2Hz, H_2); and 6.10ppm (1H, br s, H-21).

^{13}C NMR:

($CDCl_3$) C-1, 41.1(t); C-2, 71.2(d); C-3, 80.3(d); C-4, 39.3(s); C-5, 50.0(d); C-6, 18.0(t); C-7, 27.1(t); C-8, 134.7(s); C-9, 133.4(s); C-10, 38.1(s); C-11, 21.2(t); C-12, 26.2*(t); C-13, 43.9(s); C-14, 49.7(s); C-15, 30.6**(t); C-16, 29.7**(t); C-17, 45.0***(d); C-18, 16.5(q); C-19, 20.1(q); C-20, 41.2***(d); C-21, 93.0(d); C-22, 25.3*(t); C-23, 23.8(t); C-24, 76.6(d); C-25, 71.6(s); C-26, 25.6****(q); C-27, 24.4****(q); C-28, 23.8(q); C-29, 27.1(q); C-30, 17.5(q); C-1', 171.4(s); C-2', 45.0(t); C-3', 69.6(s); C-4', 44.9(t); C-5', 174.0(s); C-6', 28.2(q); <u>C</u>OMe, 171.2(s), 170.1(s); and CO<u>Me</u>, 21.0(q), 21.1ppm (q).

*, **, ***, **** Assignments may be reversed.

Mass Data:

CIMS: (NH_3) 676(M^+ + NH_4 -HOAc), 658, 614(676 - H_2O), and 574*m/e* (M^+ + NH_4 - HMGA); DCIMS: (NH_3) 676(M^+ + NH_4^+ + HOAc). EIMS: 658(M^+ - HOAc)(2%), 556(M^+ - HMGA) (20), 541(556 - Me)$^+$ (13), 496(556 - HOAc) (10), 481(496 - Me)$^+$ (3), 421(481 - HOAc) (20), 294(29), 168(100), 125(20), 119(16), 107(21), 95(20), 69(19), and 43*m/e* (68); Anal. calcd for $C_{40}H_{62}O_{11}$: C, 66.83; H, 8.69%; found: C, 66.71, H, 8.82%.

TLC Data

Silica gel GF_{254} plates developed with acetone-methylene chloride-acetic acid (24:73.5:2.5, v/v/v), R_f = 0.47. Spots were visualized under UV light or by spraying plates with 0.5% vanillin solution in sulfuric acid-ethanol (4:1, v/v) and heating at 120°C for 1 min. or with 0.04% bromocresol green solution in water in the cold.

Reference

L. Garlaschelli, G. Vidari, M. Virtuani, P. Vita-Finzi, and G. Mellerio; The Structures of New Lanostane Triterpenes from the Fruiting Bodies of *Hebeloma senescens*; J. Nat. Prod., Vol. 58, pp. 992-1002(1995).

Common/Systematic Name

HS-B

3β,21-Diacetyl-2-(3'-hydroxy-3'-methyl)glutarylcrustulinol

Molecular Formula/Molecular Weight

$C_{40}H_{62}O_{12}$; MW = 734.42413

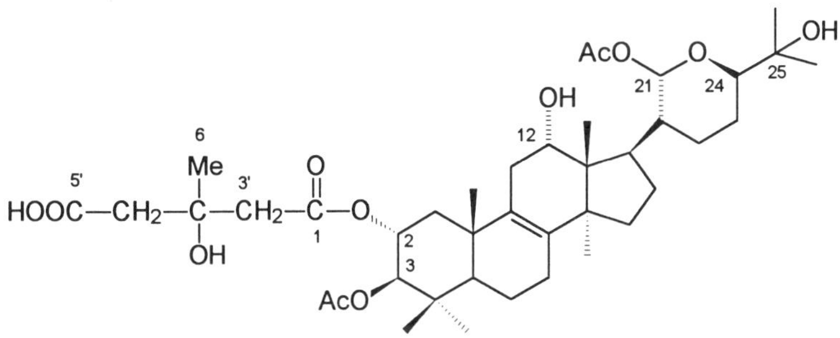

General Characteristics

A colorless amorphous solid; $[\alpha]_D^{25}$ -21.4° (*c*=0.70, in MeOH).

Fungal Source

Hebeloma spoliatum.

Isolation/Purification

Dried fruit bodies were cut into small pieces and extracted for 3 hours with MeOH at room temperature. The methanolic extract was divided with 60% (v/v) aqueous MeOH-acetone (3:5, v/v) into a supernatant and a precipitate. The supernatant was chromatographed on a silica gel column eluted with $CHCl_3$-MeOH (100:1), (100:1), (50:1), (15:1) and (5:1, v/v) to afford five fractions. Fraction IV was subjected to MPLC with *n*-hexane-acetone (2:1) and (1:1, v/v) at a flow rate of 12ml/min and HPLC with MeOH-H_2O (40:1, v/v) at a flow rate of 4.5 ml/min successively, to afford HS-A, which was treated with *n*-hexane-acetone to give HS-A in a pure state. Fraction III was subjected to HPLC with CH_3CN-acetone (3:1, v/v) at a flow rate of 2.5ml/min and MPLC with $CHCl_3$-MeOH (70:1, v/v) to afford HS-B and HS-C.

Biological Activity

Toxic to mice; IP dose of 100mg/kg produced death after paralysis of the limbs. It appeared to exhibit a paraverine-like relaxation of smooth muscle.

Spectral Data

UV:

λ_{max}^{MeOH} End absorption.

IR:

3450(O-H), 2950(C-H), 1740(C=O), and 1370cm^{-1} (C-O).

^{1}H NMR:

(CD_3COCD_3) 0.70; 0.91; 0.94; 1.07; 1.07; 1.14; 1.16; 1.35(each 3H, s. H3-18, -30, -29, -19, -26, -27, -28, -6'); 2.02, 2.03(each 3H, s, CH_3CO); 2.61, 2.64(each 1H, AB-type d, *J*=15.1Hz); 2.65, 2.70(each 1H, AB-type d, *J*=15.7Hz) (H2-2', -4'); 3.51(1H, dd, J_1=10.8, J_2=1.9Hz, H-24); 3.77(1H, d, *J*=7.2Hz, H-12); 4.76(1H, d, *J*=10.8Hz, H-3); 5.16(1H, td, J_1=10.8, J_2=4.4Hz, H-2); and 6.14ppm (1H, br s, H-21).

^{13}C NMR:

(C_5D_5N) C-1, 41.2; C-2, 70.3; C-3, 80.1; C-4, 39.5; C-5, 50.1; C-6, 18.3; C-8, 135.6; C-9, 132.3; C-10, 38.1; C-12, 71.4; C-13, 49.8; C-14, 49.4; C-17, 38.4; C-18, 17.6; C-19, 19.9; C-20, 42.4; C-21, 93.0; C-24, 77.2, C-25, 70.9; C-26, 25.6; C-27, 26.9; C-28, 25.7; C-29, 28.3; C-30, 17.3; C-1', 170.2; C-2', 46.5; C-3', 69.9; C-4', 46.4; C-5', 174.6; C-6', 28.4; $\underline{C}OCH_3$, 170.8; and $CO\underline{C}H_3$, 21.0ppm.

Mass Data:

FAB-MS positive ion 757*m/e* [(M + Na)$^+$]; positive ion (plus KI) 773*m/e* [(M + K)$^+$]; negative ion 733*m/e* [(M - H)$^-$]; HRFAB-MS positive ion (plus KI) calcd for $C_{40}H_{62}O_{12}$ [(M + K)$^+$] 773.3879; found: 773.7859; negative ion calcd for $C_{40}H_{62}O_{12}$ [(M - H)$^-$] 733.4163*m/e*.

Reference

H. Fujimoto, Y. Takano, and M. Yamaazaki; Isolation, Identification and Pharmacological Studies on Three Toxic Metabolites from a Mushroom, *Hebeloma spoliatum*; Chem. Pharm. Bull., Vol. 40, pp. 869-872(1992).

Common/Systematic Name

HS-C

3β-Acetyl-2-(3'-hydroxy-3-methyl)glutarylanhydrocrustulinol

Molecular Formula/Molecular Weight

$C_{38}H_{58}O_{10}$; MW = 674.40300

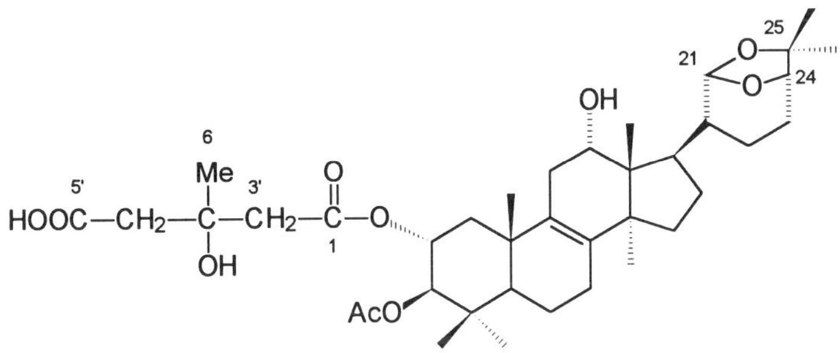

General Characteristics

A colorless powder; $[\alpha]_D^{25}$ -3.1° (*c*=0.38, in MeOH).

Fungal Source

Hebeloma spoliatum.

Isolation/Purification

Dried fruit bodies were cut into small pieces and extracted for 3 hours with MeOH at room temperature. The methanolic extract was divided with 60% (v/v) aqueous MeOH-acetone (3:5, v/v) into a supernatant and a precipitate. The supernatant was chromatographed on a silica gel column eluted with $CHCl_3$-MeOH (100:1), (100:1), (50:1), (15:1) and (5:1, v/v) to afford five fractions. Fraction IV was subjected to MPLC with *n*-hexane-acetone (2:1) and (1:1, v/v) at a flow rate of 12ml/min and HPLC with MeOH-H_2O (40:1, v/v) at a flow rate of 4.5 ml/min successively, to afford HS-A, which was treated with *n*-hexane-acetone to give HS-A in a pure state. Fraction III was subjected to HPLC with CH_3CN-acetone (3:1, v/v) at a flow rate of 2.5ml/min and MPLC with $CHCl_3$-MeOH (70:1, v/v) to afford HS-B and HS-C.

Biological Activity

Toxic to mice; IP dose of 100mg/kg produced death after paralysis of the limbs. It appeared to exhibit a paraverine-like relaxation of smooth muscle

Spectral Data

UV:

λ_{max}^{MeOH} End absorption.

21

Siccanochromenes and Grifolins

Siccanochromene A
Siccanochromene B
Siccanochromene C
Siccanochromene E
Siccanin
Grifolin
Grifolin Monomethyl Ether
Grifolin Dimethyl Ether
Neogrifolin
Neogrifolin Dimethyl Ether
Grifolin-1-acetoxy-3-methyl Ether
Scutigeral

Common/Systematic Name

Siccanochromene A

Molecular Formula/Molecular Weight

$C_{22}H_{30}O_2$; MW = 326.22458

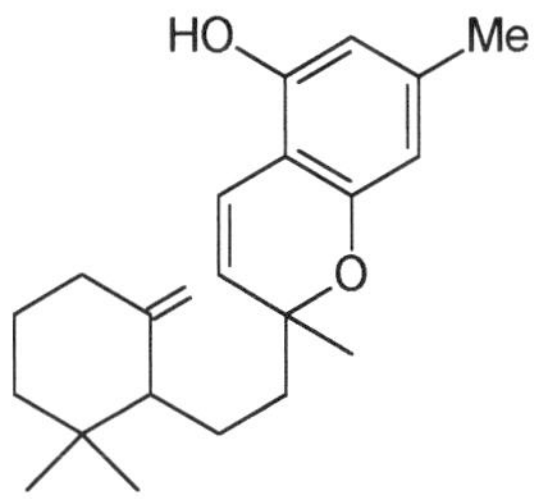

General Characteristics

Noncrystalline solid; $[\alpha]_D^9$ + 68.8° (EtOH).

Fungal Source

Helminthosporium siccans, the plant pathogenic fungus which is parasitic on rye-grass, *Lolium multifolium* Lam.

Isolation/Purification

The mycelia were separated from the fermented broth by filtration, extracted with hot acetone, and most of the acetone evaporated *in vacuo* and combined with the filtrated broth. The combined filtrates were extracted with ether or EtOAc, dried (Na_2SO_4) and evaporated to dryness. The crude metabolites were directly or after hydrolysis (in 5% KOH-EtOH at room temperature overnight) fractionated on silica gel columns and/or Florisil columns repeatedly. Changing eluting solvents from 0-100% ether in benzene on silica gel columns, siccanochromene A and H, siccanin, siccanochromene B, C, D, E, F, and G were eluted out successively; each fraction was further purified by changing solvent systems (to 0-100% MeOH in $CHCl_3$ and/or packing material to Florisil) until no impurity was detected by TLC, GC, NMR, and/or MS. The purity was also confirmed on acetylation.

Biological Activity

Some chromenes exhibit antibacterial activity against *Staphylococcus aureus.*

Spectral Data

UV:

λ_{max}^{EtOH} 229.5(ϵ=25,700), 278.5(9,320), and 286.5nm (8,700).

IR:

($CHCl_3$) 1630 and 1580cm^{-1}.

^{1}H NMR:

The NMR spectrum of siccanochromene A indicated the presence of a geminal dimethyl group (0.84 and 0.92, 6H), a methyl attached to an oxygen atom (1.26, 3H), an aromatic methyl group (2.17, 3H), an exocyclic methylene group 4.71, 4.48, 2H, d, J=2 cps), protons on a phenyl-conjugated disubstituted double bond (AB pattern at 6.55, 5.31, 2H, J=11 cps) and two aromatic protons (6.10, 6.96, 2H, J=1.5 cps); C14-Methyl, C15-methyl, 0.92, 0.82 (6H, s); C12-methyl, 1.26 (3H, s); C7'-methyl, 2.17 (3H, s); C13-H, 4.48, 4.71(2H, J=2.0Hz); C10-H, 5.31(2H, J_{AB}=11Hz); C11-H, 6.55; and C3'-H, C5'-H, 5.96, 6.10ppm (2H, s).

Mass Spectrum:

326(M^+), 311(M^+ -150, 247(M^+ -79), and 175*m/e* (base peak).

References

S. Nozoe and K. T. Suzuki; The Isolation and Structure of Siccanochromenes; Tetrahedron, Vol. 27, pp. 6063-6071(1971).

S. Nozoe, K. T. Suzuki, and S. Okuda; The Isolation and the Structure of Siccanochromenes; Tet. Lett., pp. 3643-3646(1968).

Common/Systematic Name
Siccanochromene B

Molecular Formula/Molecular Weight
$C_{22}H_{30}O_3$; MW = 342.21950

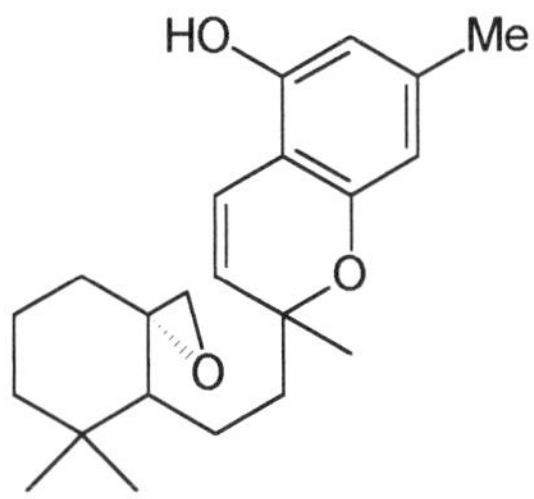

General Characteristics
Noncrystalline solid; $[\alpha]_D^{8.5}$ +121° (EtOH).

Fungal Source
Helminthosporium siccans, the plant pathogenic fungus which is parasitic on rye-grass, *Lolium multifolium*.

Isolation/Purification
The mycelia were separated from the fermented broth by filtration, extracted with hot acetone, and most of the acetone evaporated *in vacuo* and combined with the filtrated broth. The combined filtrates were extracted with ether or EtOAc, dried (Na_2SO_4) and evaporated to dryness. The crude metabolites were directly or after hydrolysis (in 5% KOH-EtOH at room temp. overnight) fractionated on silica gel columns and/or Florisil columns repeatedly. Changing eluting solvents from 0-100% ether in benzene on silica gel columns, siccanochromene A and H, siccanin, siccanochromene B, C, D, E, F, and G were eluted out successively; each fraction was further purified by changing solvent systems (0-100% MeOH in $CHCl_3$ and/or packing material to Florisil) until no impurity was detected by TLC, GC, NMR, and/or MS. The purity was also confirmed on acetylation

Biological Activity
Some chromenes exhibit antibacterial activity against *Staphylococcus aureus.*

Spectral Data

UV:
λ_{max}^{EtOH} 229.5(ϵ=25,700), 278.5(9,320), and 286.5nm (8,700).

^{1}H NMR:

The NMR spectrum of siccanochromene B showed singlets at 2.36 and 2.53(2H, AB pattern, *J*=5.0Hz) due to the methylene protons of the oxirane ring instead of the exocyclic methylene of siccanochromene A. Additional chemical shifts were at 0.82; 1.04(geminal dimethyl, 6H); 1.29(methyl attached to an oxygen atom, 3H); 2.16(aromatic methyl, 3H); 6.55; 5.33(protons on a disubstituted double bond, AB pattern, 2H, *J*=11Hz); 5.97; 6.08(aromatic protons, 2H, *J*=1.5Hz); C14-methyl, C15-methyl, 1.04, 0.82(6H, s); C12-methyl, 1.28(3H, s); C7'-methyl, 2.16(3H, s); C13-H, 2.36, 2.53 (2H, AB, *J*=5.0Hz); C10-H, 5.33(2H, J_{AB}=11Hz); C11-H, 6.55; and C3'-H, C5'-H, 5.97, and 6.08ppm (2H, s).

Mass Spectrum:

342(M^+), 327(M^+ -15), 324(M^+ - 18), 309(M^+ -33), and 175*m/e* (base peak).

References

S. Nozoe and K. T. Suzuki; The Isolation and Structure of Siccanochromenes; Tetrahedron, Vol. 27, pp. 6063-6071(1971).

S. Nozoe, K. T. Suzuki, and S. Okuda; The Isolation and the Structure of Siccanochromenes; Tet. Lett., pp. 3643-3646(1968).

Common/Systematic Name
Siccanochromene C

Molecular Formula/Molecular Weight
$C_{22}H_{30}O_3$; MW = 342.21950

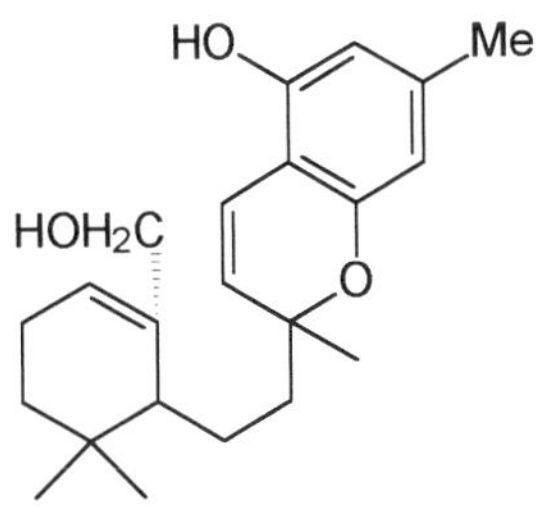

General Characteristics
Noncrystalline solid.

Fungal Source
Helminthosporium siccans, the plant pathogenic fungus which is parasitic on rye-grass, *Lolium multifolium* Lam.

Isolation/Purification
See siccanochromene A.

Biological Activity
Some chromenes exhibit antibacterial activity against *Staphylococcus aureus.*

Spectral Data

UV:
Essentially the same as siccanochromene A.

IR:
Similar to siccanochromene A.

^{1}H NMR:
The NMR spectrum indicated the presence of an unsaturated primary alcohol (3.95, 2H, s) which changed to an AB type quartet at 4.29 and 4.43, J=12Hz, upon acetylation and an olefinic proton (5.54,1H, multiplet); C14-methyl, C15-methyl, 0.91, 0.82(6H, s); C12-methyl, 1.27(3H, s); C7'-methyl, 2.09(3H, s); C13-H, 3.92(2H, s); C10-H, 5.28(2H, J_{AB}=11Hz); C11-H, 6.54; C3'-H, C5'-H, 5.97, 6.06(2H, s); and C4-H, 5.54ppm (1H, m).

Mass Spectrum:
EIMS: 342*m/e* (M^+).

Reference

S. Nozoe and K. T. Suzuki; The Isolation and Structure of Siccanochromenes; Tetrahedron, Vol. 27, pp. 6063-6071(1971).

Common/Systematic Name
Siccanochromene E

Molecular Formula/Molecular Weight
$C_{22}H_{30}O_3$; MW = 342.21950

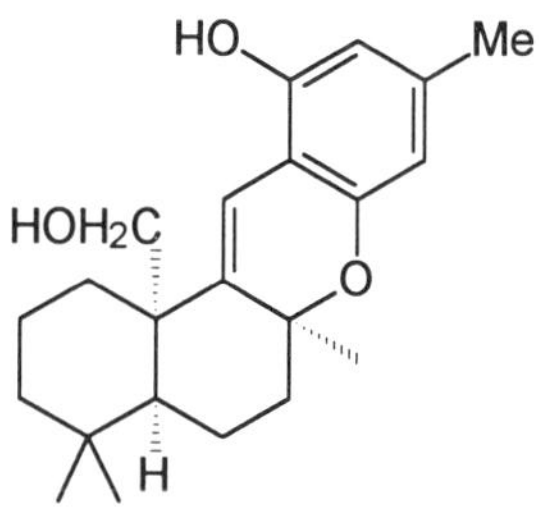

General Characteristics
Crystals; mp., 191-192°C; $[\alpha]_D^{20}$ -86° (EtOH).

Fungal Source
Helminthosporium siccans Drechsler, a parasitic fungus of rye grass (*Lothium multifolium* Lam.).

Isolation/Purification
Siccanochromene E was isolated from the non-saponifiable fraction of the fermentation broth of *H. siccans* after repeated silica gel or Florisil column chromatography.

Biological Activity
Possible antifungal and antibacterial activity.

Spectral Data

UV:
λ_{max}^{EtOH} 234(ϵ=26,000), 287(13,900), and 293nm(14,100).

IR:
($CHCl_3$) 1628 and 1580cm^{-1}.

^{1}H NMR:
($CDCl_3$) 0.94, 1.01, 1.39, and 2.18(4 X methyl groups); 3.80(2H, AB quartet, *J*=11Hz); 6.21 and 6.25(2H, s, aromatic H's), and 6.59ppm (1H, s, olefinic proton).

Mass Spectrum:
EIMS: 342(M$^+$), 327(M$^+$ -CH_3), 311(M$^+$ -31), and 175*m/e* (base peak).

Reference

K. Hirai, K. T. Suzuki, and S. Nozoe; The Structure and the Chemistry of Siccanin and Related Compounds; Tetrahedron, Vol. 27, pp. 6057-6061(1971).

Common/Systematic Name
Siccanin

Molecular Formula/Molecular Weight
$C_{22}H_{30}O_3$; MW = 342.21450

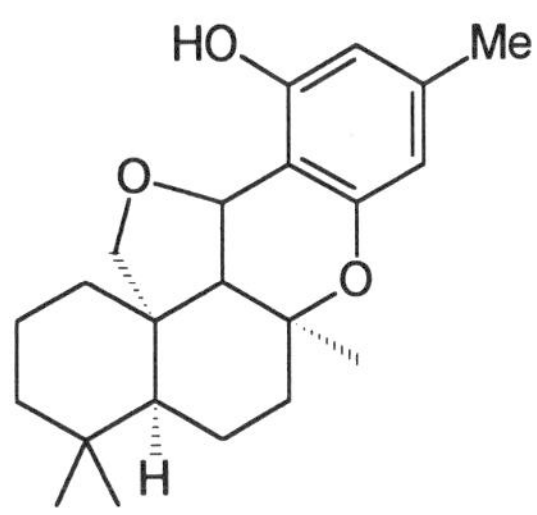

General Characteristics
Crystals; mp., 138°C; $[\alpha]_D^{16}$ -150° (*c*=7.75, in $CHCl_3$).

Fungal Source
Helminthosporium siccans Drechsler, the plant pathogenic fungus which is parasitic on rye-grass, *Lolium multifolium* Lam.

Biological Activity
Siccanin exhibits inhibitory activity against a variety of fungi, especially strong activity against *Trichophyton interdigitale* and *T. asteroide* at 0.1μg/ml.

Spectral Data

UV:
λ_{max}^{EtOH} 278(sh) and 285nm (ε=1,800).

IR:
($CHCl_3$) 3500, 1633, and 1575cm^{-1}.

^{1}H NMR:
($CDCl_3$) The NMR spectrum showed the signals at 0.80, 0.84(6H, *gem*-dimethyls), 1.25(-OCCH$_3$), 2.20(aromatic methyl), 6.15 and 6.30(aromatic protons), 3.46, 4.24(2H, AB quartet, -CH$_2$O-), 5.16(1H, d, benzylic methine), 1.94(methine), and 6.57ppm (-OH).

Mass Spectrum:
342*m/e* (M$^+$).

Reference
K. Hirai, S. Nozoe, K. Tsuda, Y. Iitaka, K. Ishibashi, and M. Shirasaka; The Structure of Siccanin; Tet. Lett., pp. 2177-2179(1967).

Common/Systematic Name

Grifolin

E,E-5-Methyl-(3,7,11-trimethyl-2,6,10-dodecatrienyl)-1,3-benzenediol

Molecular Formula/Molecular Weight

$C_{22}H_{32}O_2$; MW = 328.24023

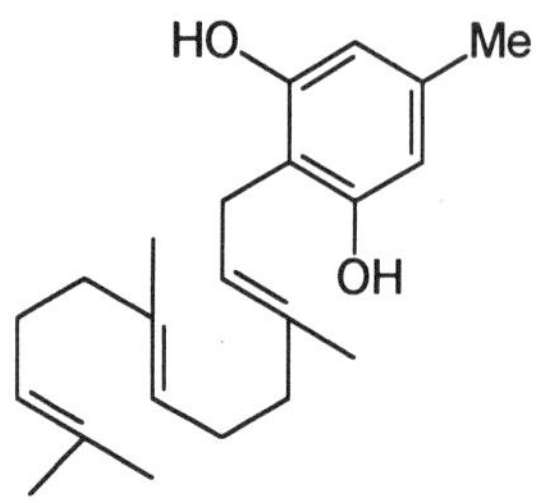

General Characteristics

Colorless needles from petroleum ether; mp., 43°C.

Fungal Source

Grifola confluens (shiromaitake), *Albatrellus confluens*, *A. ovinus* = *Polyporus ovinus*, an edible mushroom.

Isolation/Purification

Fresh *Albatrellus ovinus* mushrooms were extracted with acetone, concentrated and partitioned between 60% aqueous EtOH and benzene. The benzene layer was dried and repeatedly chromatographed on silica gel using petrol-benzene (3:1, v/v), benzene and benzene-ethyl ether (9:1, v/v). Grifolin was recrystallized from petroleum ether to give colorless needles.

Biological Activity

Antibiotic activity.

Spectral Data

UV:

275(log ϵ = 2.97) and 281nm (2.96).

IR:

(KBr) 3430, 1631, 1590, 1515, and 1043cm^{-1}; (CCl_4) 3620, 3460, 2920, 1635, 1585,1450, 1170, and 1044cm^{-1}.

^{1}H NMR:

($CDCl_3$) 1-H, 6.24(s); 5-H, 6.24(s); 8-H, 2.21 (s);1'-H, 3.38(d); 2'-H, 5.20(m);

4',5',8',9'-H, 1.92, 2.10(m); 6', 10'-H, 5.08(m); 12'-H, 1.67(s); 13'-H, 1.58(s); 14'-H, 1.59(s); and 15'-H, 1.81ppm (s).

Mass Spectrum:
EIMS: 328(M^+, 3%), 201(18), 191(21), 175(100), 137(100), and 69*m/e* (74).

References

H. Besl, G. Höfle, B. Jendrny, E. Jägers, and W. Steglich; Farnesylphenole aus *Albatrellus*-Arten (*Basidiomycetes*); Chem. Ber., Vol. 110, pp. 3370-3376(1977).

T. Goto, H. Kakisawa and Y. Hirata, The Structure of Grifolin, an Antibiotic from a Basidiomycete, Tetrahedron, Vol. 19, pp. 2079-2083(1963).

J. Vrkoč, M. Budešinsky, and L. Dolejš; Phenolic Meroterpenoids from the Basidiomycete *Albatrellus ovinus*; Phytochemistry, Vol. 16, pp. 1409-1411(1977).

Common/Systematic Name

Grifolin monomethyl ether

E,E-5-Methyl-(3,7,11-trimethyl-2,6,10-dodecatrienyl)-1-hydroxy-3-methoxybenzene

Molecular Formula/Molecular Weight

$C_{23}H_{34}O_2$; MW = 342.25588

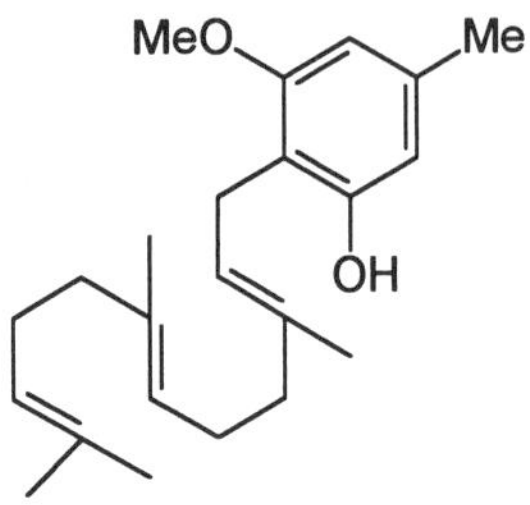

Fungal Source

Albatrellus ovinus and *Polyporus ovinus*, an edible mushroom.

Isolation/Purification

Fresh *Albatrellus ovinus* mushrooms were extracted with acetone, concentrated and partitioned between 60% aqueous EtOH and benzene. The benzene layer was dried and repeatedly chromatographed on silica gel using petrol-benzene (3:1, v/v), benzene and benzene-ethyl ether (9:1, v/v). Purity of chromatographic fractions was monitored by silica gel TLC (benzene-ethyl ether, 9:1, v/v) (R_f=0.85) and GLC [Gas Chrom-Q coated with 3% QF-1 (220°C) and 3% OV-17(240°C)].

Biological Activity

Antibiotic activity

Spectral Data

IR:

(KBr) 3450, 1615, 1585, and 1510cm^{-1}.

^{1}H NMR:

($CDCl_3$) 3.77, 3H; 2.26, Ar-methyl; 6.30, Ar-H(2H); 3.37 Ar-CH_2; 1.58(6H); 1.67(3H); 4.95-5.35(3H, -CH=); and 2.03ppm, (-CH_2-CH_2-).

Mass Spectrum:

EIMS: 342(M$^+$, 7%), 327(3), 299(3), 286(5), 273(1), 260(1), 231(3), 229(5), 217(2), 205(21), 204(3), 203(6), 191(12), 190(7), 189(11), 177(5), 163(5), 151(100), 121(15), 69(28), 43(4), and 41*m/e* (43).

Reference
J. Vrkoč, M. Budešinsky, and L. Dolejš; Phenolic Meroterpenoids from the Basidiomycete *Albatrellus ovinus*; Phytochemistry, Vol. 16, pp. 1409-1411(1977).

Common/Systematic Name

Grifolin dimethyl ether

E,E-5-Methyl-2-(3,7,11-trimethyl-2,6,10-dodecatrienyl)-1,3-dimethoxybenzene

Molecular Formula/Molecular Weight

$C_{24}H_{36}O_2$; MW = 356.27153

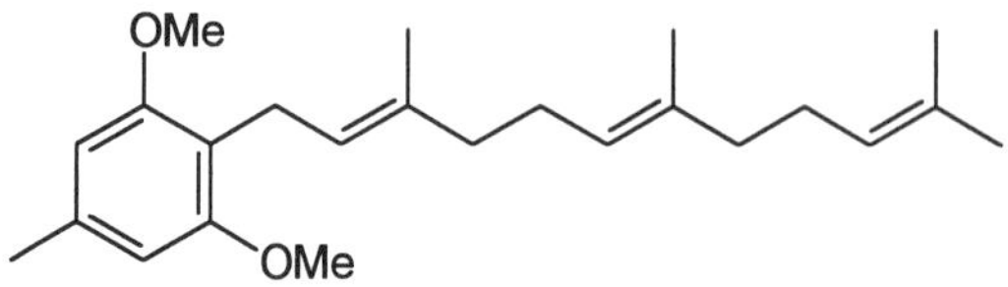

Fungal Source

Albatrellus ovinus and *Polyporus ovinus,* an edible mushroom.

Isolation/Purification

Fresh *Albatrellus ovinus* mushrooms were extracted with acetone, concentrated and partitioned between 60% aqueous EtOH and benzene. The benzene layer was dried and repeatedly chromatographed on silica gel using petrol-benzene (3:1, v/v), benzene and benzene-ethyl ether (9:1, v/v). Purity of chromatographic fractions was monitored by silica gel TLC (benzene-ethyl ether, 9:1, v/v)(R_f=0.85) and GLC (Gas Chrom-Q coated with 3% QF-1 (220° C) and 3% OV-17 (240°C).

Biological Activity

Antibiotic activity.

Spectral Data

IR:

(KBr) 1608, 1566, 1492, and 1118cm^{-1}.

^{1}H NMR:

($CDCl_3$) 3.78, 3H; 3.78, 3H; 2.32, Ar-methyl; 6.36, Ar-H; 3.32 Ar-CH_2; 1.58(6H); 1.68(3H); 4.95-5.30(3H, -CH=); and 1.99ppm (3H, -CH_2-CH_2-).

Mass Spectrum:

EIMS: 356(M$^+$, 13%), 219(46), 165(100), 151(8), 150(8), and 69*m/e* (17).

Reference

J. Vrkoč, M. Budešinsky, and L. Dolejš; Phenolic Meroterpenoids from the Basidiomycete *Albatrellus ovinus*; Phytochemistry, Vol. 16, pp. 1409-1411(1977).

Common/Systematic Name
Neogrifolin
E,E-5-Methyl-4-(3,7,11-trimethyl-2,6,10-dodecatrienyl)-1,3-benzenediol

Molecular Formula/Molecular Weight
$C_{22}H_{32}O_2$; MW = 328.24023

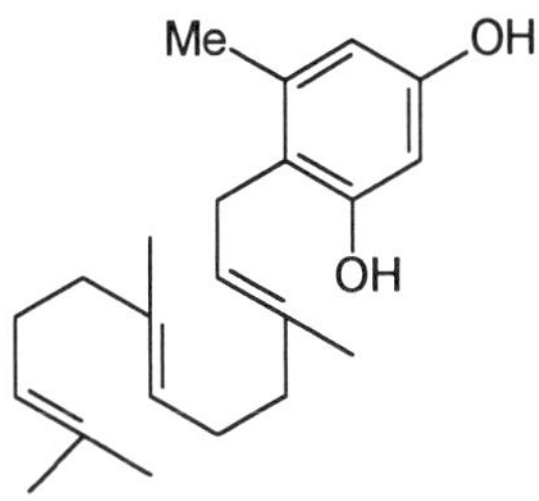

General Characteristics
Crystals; mp., 43-45°C.

Fungal Source
Grifola confluens (shiromaitake), *Albatrellus confluens*, *A. ovinus*, and *Polyporus ovinus*, an edible mushroom.

Isolation/Purification
Fresh *Albatrellus ovinus* mushrooms were extracted with acetone, concentrated and partitioned between 60% aqueous EtOH and benzene. The benzene layer was dried and repeatedly chromatographed on silica gel using petrol-benzene (3:1, v/v), benzene and benzene-ethyl ether (9:1, v/v). Purity of chromatographic fractions was monitored by silica gel TLC (benzene-ethyl ether, 9:1, v/v) (R_f=0.85) and GLC [Gas Chrom-Q coated with 3% QF-1 (220°C) and 3% OV-17 (240°C)].

Biological Activity
Antibiotic activity.

Spectral Data

UV:
$\lambda_{max}^{Acetonitrile}$ 280(log ϵ=3.54) and 227nm (3.83).

IR:
(KBr) 3400, 1600, 1516, 1496, and 1110cm^{-1}; ($CHCl_3$) 3600, 3550-3150, 3030, 3005, 2975, 2925, 2860, 1720, 1610, 1590, 1485, 1455, 1435, 1370, 1325, 1220, 1130, 1040, 975, and 885cm^{-1}.

^{1}H NMR:
(CDCl$_3$) 2.22, Ar-methyl; 6.19(1H),6.24(1H), Ar-H; 3.29 Ar-CH$_2$; 1.58(6H); 1.67(3H); 5.10(3H, -CH=); and 2.03ppm, (-CH$_2$-CH$_2$-).

^{13}C NMR:
(CDCl$_3$) C-1, 109.9; C-2, 154.2; C-3, 101.1; C-4, 155.3; C-5, 118.1; C-6, 138.6; C-8, 20.0; C-1', 26.7; C-2', 124.4; C-3', 131.3; C-4', 25.9; C-5', 17.7, C-6', 16.2; and C-7', 16.0ppm.

Mass Spectrum:
EIMS: 328.2407(M$^+$, 20%), 191(42), 177(10), 175(27), and 137*m/e* (100).

References
H. Besl, G. Höfle, B. Jendrny, E. Jägers, and W. Steglich; Farnesylphenole aus *Albatrellus*-Arten (*Basidiomycetes*); Chem. Ber., Vol. 110, pp. 3370-3376(1977).

J. Vrkoč, M. Budešinsky, and L. Dolejš; Phenolic Meroterpenoids from the Basidiomycete *Albatrellus ovinus*; Phytochemistry, Vol. 16, pp. 1409-1411(1977).

Common/Systematic Name

Neogrifolin dimethyl ether

E,E-5-Methyl-4-(3,7,11-trimethyl-2,6,10-dodecatrienyl)-1,3-dimethoxybenzene

Molecular Formula/Molecular Weight

$C_{24}H_{36}O_2$; MW = 356.27153

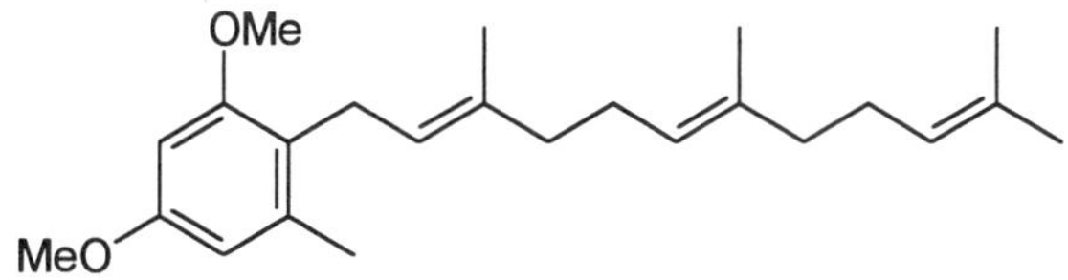

Fungal Source

Grifola confluens (shiromaitake), *Albatrellus confluens*, *A. ovinus*, and *Polyporus ovinus*, an edible mushroom.

Isolation/Purification

Fresh *Albatrellus ovinus* mushrooms were extracted with acetone, concentrated and partitioned between 60% aqueous EtOH and benzene. The benzene layer was dried and repeatedly chromatographed on silica gel using petrol-benzene (3:1, v/v), benzene and benzene-ethyl ether (9:1, v/v). Purity of chromatographic fractions was monitored by silica gel TLC (benzene-ethyl ether, 9:1, v/v) (R_f=0.85) and GLC [Gas Chrom-Q coated with 3% QF-1 (220°C) and 3% OV-17 (240°C)].

Biological Activity

Antibiotic activity.

Spectral Data

IR:

(KBr) 1608, 1586, 1492, and 1118cm^{-1}.

^{1}H NMR:

($CDCl_3$) 3.77(OCH_3); 3.77(OCH_3); 2.25, Ar-methyl; 6.32(2H), Ar-H; 3.28 Ar-CH_2; 1.58(6H); 1.67(3H); 5.07(3H, -CH=); and 1.99ppm, (-CH_2-CH_2-).

Mass Spectrum:

Identical with that of grifolin dimethyl ether except for the intensity of some peaks.

Reference

J. Vrkoč, M. Budešinsky, and L. Dolejš; Phenolic Meroterpenoids from the Basidiomycete *Albatrellus ovinus*; Phytochemistry, Vol. 16, pp. 1409-1411(1977).

Common/Systematic Name

Grifolin-1-acetoxy-3-methyl ether

E,E-5-Methyl-(3,7,11-trimethyl-2,6,10-dodecatrienyl)-1-acetoxy-3-methoxybenzene

Molecular Formula/Molecular Weight

$C_{25}H_{36}O_3$; MW = 384.26645

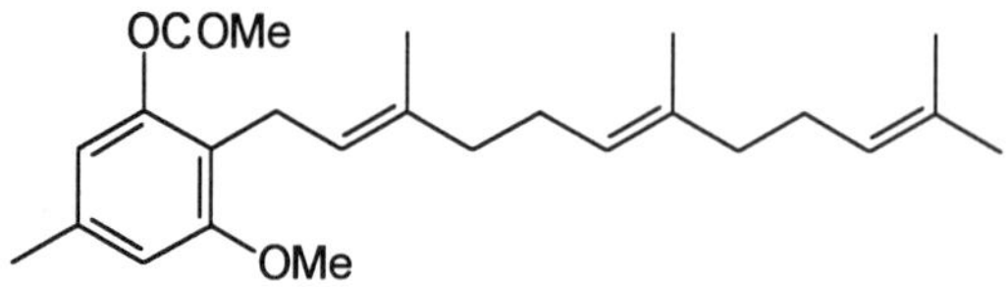

General Characteristics

Liquid.

Fungal Source

Albatrellus ovinus= Polyporus ovinus, an edible mushroom.

Isolation/Purification

Fresh *Albatrellus ovinus* mushrooms were extracted with acetone, concentrated and partitioned between 60% aqueous EtOH and benzene. The benzene layer was dried and repeatedly chromatographed on silica gel using petrol-benzene (3:1, v/v), benzene and benzene-ethyl ether (9:1, v/v). Grifolin-1-acetoxy-3-methyl ether was obtained as a liquid.

Biological Activity

Antibiotic activity.

Spectral Data

IR:

(KBr) 1768, 1595, 1485, and 1212cm^{-1}.

^{1}H NMR:

($CDCl_3$) 2.26, CH_3C=O-; 3.77(OCH_3); 2.25, Ar-methyl; 6.47(2H), 6.51 Ar-H; 3.32 Ar-CH_2; 1.58(6H); 1.68(3H); 5.07(3H, -CH=); and 2.00ppm (-CH_2-CH_2-).

Mass Spectrum:

EIMS: 384(M^+, 8%), 342(8), 273(3), 260(8), 248(11), 247(5), 246(5), 231(5), 218(6), 217(10), 205(100), 204(6), 193(33), 191, 175(10), 163(7), 151(95), 137(13), 121(23), 109(7), 107(8), 95(8), 93(6), 91(7), 81(19), 69(61), 67(8), 57(2), 55(9), 53(5), 43(14), and 41*m/e* (47).

Reference

J. Vrkoč, M. Budešinsky, and L. Dolejš; Phenolic Meroterpenoids from the Basidiomycete *Albatrellus ovinus*; Phytochemistry, Vol. 16, pp. 1409-1411(1977).

Common/Systematic Name
Scutigeral

Molecular Formula/Molecular Weight
$C_{23}H_{32}O_4$; MW = 372.23006

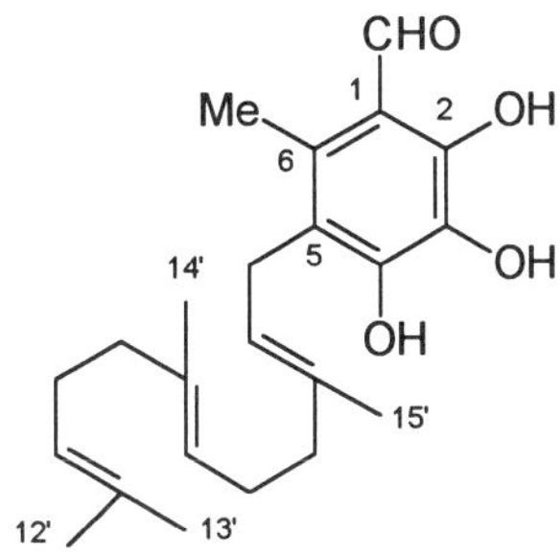

General Characteristics
Crystals from *n*-hexane; mp., 81°C.

Fungal Source
Albatrellus ovinus = *Polyporus ovinus* and *Albatrellus subrubescens* (edible mushrooms).

Biological Activity
Antibiotic activity.

Spectral Data

UV:
λ_{max}^{MeOH} 310nm (log ϵ=3.74); $\lambda_{max}^{MeOH+NaOH}$ 425 and 295nm (3.33 and 3.64 respectively).

IR:
(KBr) 3500-3200, 3440, 2970, 1230, 1195, 1150, 1120, 1090, 1075, 915, 890, 840, and 715cm^{-1}.

^{1}H NMR:
($CDCl_3$) 7-H, 10.17(s); 8-H, 2.43(s); 1'-H, 3.35(d); 2'-H, 5.07(m); 4'-, 5'-, 8'-, 9'-H, 1.95, 2.04(m); 6', 10'-H, 5.07(m); 12'-H, 1.67(s); 13'-H, 1.58(s); 14'-H, 1.58(s); and 15'-H, 1.77ppm (s).

^{13}C NMR:
($CDCl_3$) C-1, 113.1; C-2, 149.8; C-3, 128.6; C-4, 150.2; C-5, 120.1; C-6, 133.1; C-7, 194.3; C-8, 12.9; C-1', 24.5; C-2', 122.0; C-3', 135.1; C-4', 39.8; C-5', 26.6; C-6', 124.1; C-7', 135.7; and C-8', 39.8ppm.

Mass Spectrum:
EIMS: 372.2319(M^+, 18%), 236(11), 235(11), 221(13), 219(24), 191(12), 182(11), 181(100), 153(11), 152(20), 151(24), 136(14), and 125*m/e* (12).

Reference

H. Besl, G. Höfle, B. Jendrny, E. Jägers, and W. Steglich; Farnesylphenole aus *Albatrellus*-Arten (*Basidiomycetes*); Chem. Ber., Vol. 110, pp. 3370-3376(1977).

22

Fusarochromanones

Fusarochromanone
Diacetylfusarochromanone
C-3'-*N*-Acetylfusarochromanone
Formylfusarochromanone

Common/Systematic Name
Fusarochromanone

Molecular Formula/Molecular Weight
$C_{15}H_{20}N_2O_4$; MW = 292.14231

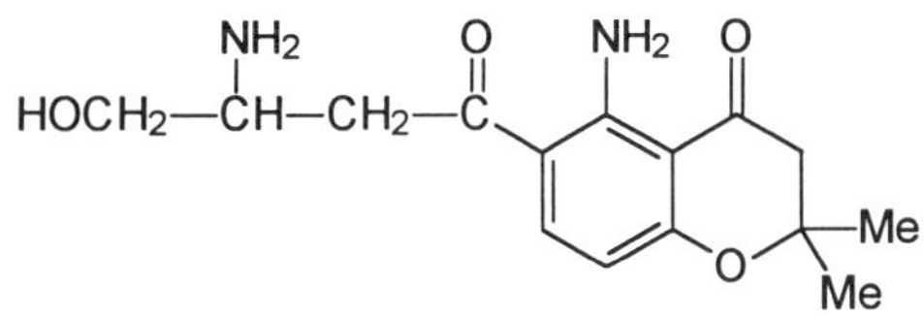

General Characteristics
Crystals from methylene chloride-hexane; mp., 132-134°C.

Fungal Source
Fusarium equiseti (Corda) Sacc. (Previously identified as *F. roseum graminearum*) collected from overwintered oats in Fairbanks, Alaska.

Isolation/Purification
Five hundred grams of crude culture material was extracted 5 times with $CHCl_3$-MeOH-NH_4OH (90:10:1, v/v/v). The extracts were pooled and concentrated *in vacuo* at 40°C to give an oily residue which was dissolved in 50mL methanol in 200mL of distilled water. This mixture was further concentrated to approximately 150mL of solution and applied to an XAD-2 column rinsed with 1L of distilled water, and eluted with 1L of 90% aqueous methanol. The methanol fraction was concentrated *in vacuo* at 40°C and reconstituted in 50ml $CHCl_3$-MeOH (9:1, v/v). This solution was mixed with 50g of anhydrous sodium sulfate and the mixture was dried at room temperature. The dry yellow powder was then packed on the top of a column containing 150g of Florisil and eluted with 2L of $CHCl_3$-MeOH (9:1, v/v) followed by 2L of $CHCl_3$-MeOH-NH_4OH (90:10:1, v/v/v). Fusarochromanone was detected by irradiating the column with long-wave UV (365nm) light since the toxin produced an intense blue fluorescence under the UV light. The fractions containing fusarochromanone were concentrated (40°C) and subsequently purified by TLC using $CHCl_3$-MeOH-NH_4OH (80:20:2, v/v/v). Fusarochromanone was crystallized from methylene chloride-hexane.

Biological Activity
Caused tibial dyschondroplasia, a bone malformation in poultry which is characterized by the presence of a cone of cartilage which extends distally from the proximal tibiotarsal physis; also causes low hatchability of fertile eggs.

Spectral Data

UV:
λ_{max}^{MeOH} 248(ϵ = 21,800), 277(8,800), and 383nm (11,700).

IR:
(KBr) 3373, 3267, 3160(NH_2, OH), 1655(CO), and 1562cm^{-1} (CO, -H bonded).

^{1}H NMR:
($CDCl_3$) 2.7, H-3; 7.86, H-7; 6.08, H-8; 3.03, 3.88, H-2'; 3.50, H-3'; 3.60, H-4'; 1.26, CMe_2; and 9.46, 9.60ppm , $ArNH_2$.

^{13}C NMR:
($CDCl_3$) 79.32, C-2; 48.96, C-3; 193.70, C-4; 154.71, C-5; 111.71, C-6; 140.16, C-7; 104.10, C-8; 165.99, C-9; 104.10, C-10; 198.80, C-1'; 42.44, C-2'; 49.71, C-3'; 65.82, C-4'; 26.28ppm, C-11 and C-12.

Mass Spectrum:
LREIMS yielded an ion at nominal mass of 292. Fast atom bombardment mass spectrometry indicated a protonated species at 293*m/e*, which was peak matched to 293.1524 to fit a composition $C_{15}H_{20}N_2O_4$ within 7.8ppm. The FAB also showed an ion 558.2960*m/e* which fitted to within 6.3ppm for a composition of $C_{30}H_{41}N_4O_8$, presumably due to a dimer.

Reference

S. V. Pathre, W. B. Gleason, Y. Lee, and C. J. Mirocha; The Structure of Fusarochromanone: New Mycotoxin from *Fusarium roseum, "Graminearum"*; Can. J. Chem., Vol. 64, pp. 1308-1311(1986).

Common/Systematic Name

Diacetylfusarochromanone

2,2-Dimethyl-5-amino-6-(3'-*N*-acetyl-4'-*O*-acetylbutyryl)-4-chromone

Molecular Formula/Molecular Weight

$C_{19}H_{24}N_2O_6$; MW = 376.16344

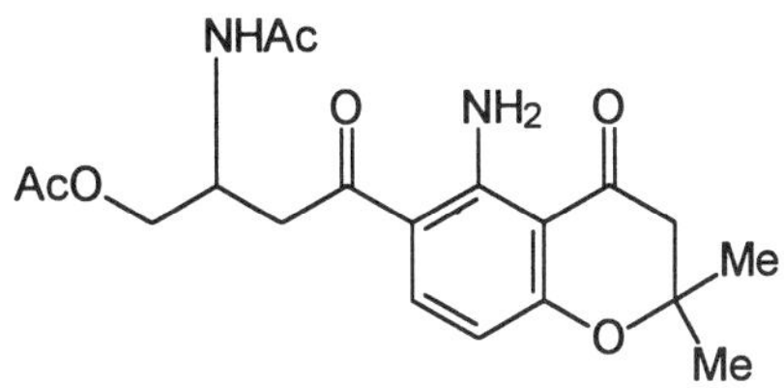

General Characteristics

Obtained as needle-shaped colorless crystals in Me_2CO.

Fungal Source

Fusarium equiseti (previously identified as *Fusarium roseum graminearum*) was isolated from an oat kernel collected in Fairbanks, Alaska and deposited as Alaska 2-2.

Isolation/Purification

The *F. equiseti* isolate was grown on autoclaved rice with 60% water for 4 weeks at room temperature. The dried, powdered rice culture was extracted with 90% aqueous MeOH. The extract was concentrated under reduced pressure, defatted with petroleum ether, and partitioned with CH_2Cl_2. The CH_2Cl_2 layer was concentrated and fractionated on a Florisil column. Elution was with $CHCl_3$-MeOH (19:1, v/v) followed by $CHCl_3$-MeOH (9:1, v/v). Fractions containing 2,2-dimethyl-5-amino-6-(3'-*N*-formyl-4'-*O*-hydroxylbutyryl)-4-chromone by TLC were combined and further purified by preparative TLC in $CHCl_3$-MeOH (9:1, v/v); combined fractions containing 2,2-dimethyl-5-amino-6-(3'-*N*-acetyl-4'-*O*-acetylbutyryl)-4-chromone were purified by TLC in EtOAc-Me_2CO (9:1, v/v). The TLC spots and bands exhibited blue fluorescence under long wavelength UV.

Biological Activity

Chemically related to fusarochromanone (which caused tibia dyschondroplasia, a common bone disease in chickens), and reduced hatchability in fertile chicken eggs.

Spectral Data

UV:

λ_{max}^{MeOH} 278, 248, and 214nm.

IR:

(film) 3395, 3285, 2998, 2932, 2856, 1734, 1662, 1653, 1595, 1560, 1498, 1466, 1375, 1319, 1278, 991, and 893cm^{-1}.

^{1}H NMR:

(CDCl$_3$) 1.44(6H, s, H-11 and H-12); 1.97(3H, s, NHAc); 2.04(3H, s, OAc); 2.69(2H, s, H-3); 3.01(1H, dd, *J*=6.2, 16Hz, Ha-2'); 3.25(1H, dd, *J*=5.7, 16Hz, Hb-2'); 4.14(1H, dd, *J*=5.3, 11Hz, Ha-4'); 4.32(1H, dd, *J*=6.1, 11Hz, Hb-4'); 4.48(1H, m, H-3'); 6.02(1H, d, *J*=8.9Hz, H-8); 6.39(1H, d, *J*=8.3Hz, NHAc); 7.82(1H, d, *J*=8.9Hz, H-7); and 9.34-9.60ppm (2H, b, ArNH).

Mass Spectrum:

EIMS yielded an M^+ at *m/e* 376. FABMS yielded an $[M + H]^+$ at *m/e* 377 and $[M + Na]^+$ at *m/e* 399, and negative CIMS yielded an $[M]^-$ at *m/e* 376. The HREIMS indicated a molecular ion at *m/e* 376.1661 (calcd 376.1634 for $C_{19}H_{24}N_2O_6$).

Reference

W. Xie, C. J. Mirocha, and Y. Wen; Formyl Fusarochromanone and Diacetyl Fusarochromanone, Two New Metabolites of *Fusarium equiseti*; J. Nat. Prod., Vol. 54, pp. 1165-1167(1991).

Common/Systematic Name

C-3'-*N*-Acetylfusarochromanone

Molecular Formula/Molecular Weight

$C_{17}H_{22}N_2O_5$; MW = 334.15287

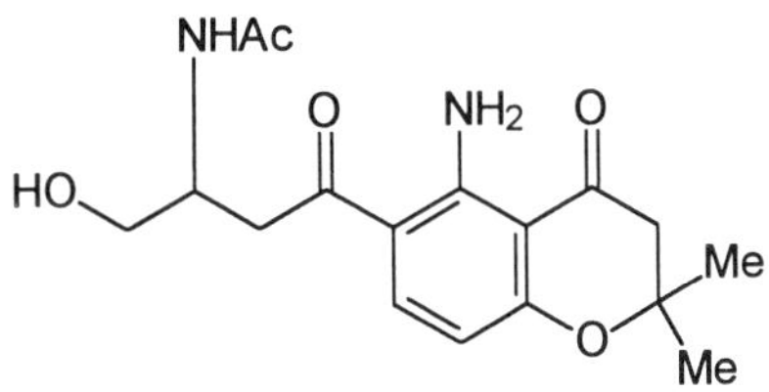

General Characteristics

Yellow-green, needle-shaped crystals from acetone-petroleum ether.

Fungal Source

Fusarium equiseti (previously identified as *F. roseum graminearum*) collected from overwintered oats in Fairbanks, Alaska.

Isolation/Purification

Dried rice culture material was extracted with 90% aqueous methanol three times, filtered and concentrated *in vacuo*. The filtrate was defatted by partitioning with petroleum ether. The aqueous methanol fraction was partitioned three times with dichloromethane, evaporated to dryness *in vacuo*, and redissolved in a small volume of chloroform. This extract was applied to a Florisil column eluted with chloroform-methanol (9:1, v/v). Fractions were monitored by TLC (chloroform-methanol, 4:1 v/v). Fractions containing mainly C-3'-*N*-acetylfusarochromanone were combined, evaporated to dryness, redissolved in acetone; petroleum ether was added to the solution dropwise until the solution became cloudy. The test tube was placed in a freezer overnight. The mother liquor was pipetted out and the yellow-green, needle-shaped crystals were vacuum dried and collected.

Biological Activity

Chemically related to fusarochromanone, a metabolite that caused tibial dyschondroplasia and reduced hatchability of fertile eggs.

Spectral Data

^{1}H NMR:

($CDCl_3$) 2.68(2H, s, H-3); 7.92(1H, d, *J*=9.1Hz, H-7); 6.06(1H, d, *J*=4.1Hz, H-8); 3.06, 3.28(2H, m, H-2'); 4.4(1H, d, *J*=7.4Hz, , H-3'); 3.70(2H, m, H-4'); 1.49(6H, s, CMe_2); 9.43, 950(2H, b, $ArNH_2$); 1.89(3H, NAc); and 6.50ppm (1H, d, *J*=7.5Hz, NHAc).

^{13}C NMR:

($CDCl_3$) 79.52(C-2, s); 48.95(C-3, t); 193.81(C-4, s); 154.97(C-5, s); 111.48(C-6, s); 140.62(C-7, d); 164.55(C-8,d); 166.39(C-9, s); 104.55(C10, s); 26.53(C-11, C-12, q, 26.50, q); 198.82(C-1', s);39.58(C-2', t); 49.70(C-3', d); 64.27(C-4', t); 23.48(CH_3C=O, q); and 170.73ppm (CH_3C=O, s).

Mass Spectrum:

Spectrum displayed a base peak at 218*m/e* and molecular ion peak at 334*m/e*. The molecular ion of 334 was confirmed by the EI and CI mass spectra of the TMS derivative obtained from GC-MS. The EI spectrum yielded a molecular ion of 406*m/e*, which was 334 plus a TMS (72) group. The CI spectrum in which ion 407(M^+ + 41) were found confirmed the molecular ion of the TMS derivative as 406*m/e*.

HPLC Data

Reversed-phase μBondpak C_{18} column, mobile phase, methanol-water-acetic acid (120:20:1, v/v/v) eluted at 1ml/minute. Fluorescence detection at 450 (emission) and 384 (excitation).

References

S. V. Pathre, W. B. Gleason, Y. Lee, and C. J. Mirocha; The Structure of fusarochromanone: New Mycotoxin from *Fusarium roseum, "Graminearum"*; Can. J. Chem., Vol. 64, pp. 1308-1311(1986).

W. Xie, C. J. Mirocha, R. J. Pawlosky, Y. Wen, and X. Xu; Biosynthesis of Fusarochromanone and Its Monoacetyl Derivative by *Fusarium equiseti*; Appl. Environ. Microbiol., Vol. 55, pp. 794-797(1989).

Common/Systematic Name

Formylfusarochromanone

2,2-Dimethyl-5-amino-6-(3'-*N*-formyl-4'-*O*-hydroxylbutyryl)-4-chromone

Molecular Formula/Molecular Weight

$C_{16}H_{20}O_5N_2$; MW = 320.13722

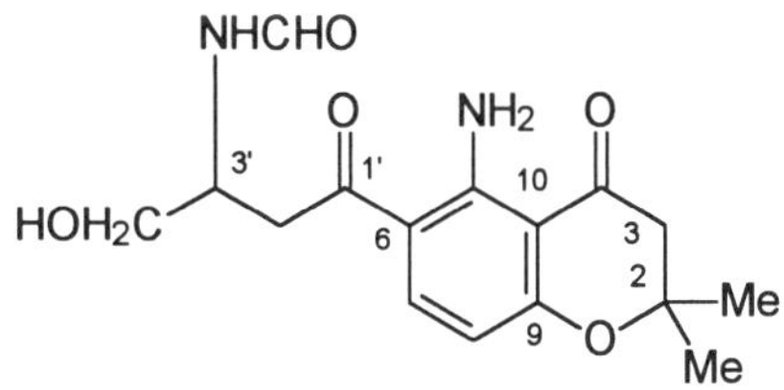

General Characteristics

Obtained as a yellow oil.

Fungal Source

Fusarium equiseti (previously identified as *Fusarium roseum graminearum*) was isolated from an oat kernel collected in Fairbanks, Alaska and deposited as Alaska 2-2.

Isolation/Purification

The *F. equiseti* isolate was grown on autoclaved rice with 60% water for 4 weeks at room temperature. The dried, powdered rice culture was extracted with 90% aqueous MeOH. The extract was concentrated under reduced pressure, defatted with petroleum ether, and partitioned with CH_2Cl_2. The CH_2Cl_2 layer was concentrated and fractionated on a Florisil column. Elution was with $CHCl_3$-MeOH (19:1, v/v) followed by $CHCl_3$-MeOH (9:1, v/v). Fractions containing 2,2-dimethyl-5-amino-6-(3'-*N*-formyl-4'-*O*-hydroxylbutyryl)-4-chromone by TLC were combined and further purified by preparative TLC in $CHCl_3$-MeOH (9:1, v/v); combined fractions containing 2,2-dimethyl-5-amino-6-(3'-*N*-formyl-4'-*O*-acetylbutyryl)-4-chromone were purified by TLC in EtOAc-Me_2CO (9:1, v/v). The TLC spots and bands exhibited blue fluorescence under long wavelength UV.

Biological Activity

Chemically related to fusarochromanone (which caused tibia dyschondroplasia, a common bone disease in chickens), and reduced hatchability in fertile chicken eggs.

Spectral Data

UV:

λ_{max}^{MeOH} 278, 248, and 214nm.

IR:

(film) 3399, 3287, 2932, 2864, 1724, 1682, 1657, 1599, 1568, 1560, 1496, 1466, 1388, 1373, 1277, 1174, 1159, 1109, and 893cm^{-1}.

^{1}H NMR:
($CDCl_3$) 1.45(6H, s, H-11 and H-12); 2.70(2H, s, H-3); 3.12(1H, dd, *J*=5.9, 16.3Hz, Ha-2'); 3.34(1H, dd, *J*=6.4, 16.3Hz, Hb-2'); 3.79(2H, d, *J*=4.2Hz, H-4'); 4.26-4.39(1H, m, H-3'); 6.08(1H, d, *J*=9.1Hz, H-8); 6.57(2H, d, *J*=6.0Hz, NH-3'); 7.89(1H, d, *J*=9.1Hz, H-7); 8.14(1H, d, *J*=1.6Hz, CHO); and 9.50ppm (2H, s, ArNH).

Mass Spectrum:
EIMS: 320(M^+) and at 302*m/e* (M^+-H_2O) indicating the presence of an -OH group. Acetylation yielded a monoacetate with a molecular ion at 362*m/e* (M^+). FABMS yielded an $[M + H]^+$ at 321 and $\{M + Na\}^+$ at 343*m/e* ; negative CIMS yielded M^- at *m/e* 320*m/e* (M^+). HREIMS: calcd 320.1372 for a molecular formula of $C_{16}H_{20}O_5N_2$, which was compatible with 8 double bond equivalents.

Reference

W. Xie, C. J. Mirocha, and Y. Wen; Formyl Fusarochromanone and Diacetyl Fusarochromanone, Two New Metabolites of *Fusarium equiseti*; J. Nat. Prod., Vol. 54, pp. 1165-1167(1991).

23

Aphidicolins

Aphidicolin
3-Deoxyaphidicolin
Aphidicolin 17-monoacetate
Aphidicolin 3,18-orthoacetate

Common/Systematic Name

Aphidicolin

Aphidicolane-3α,16,17,18-tetraol

Molecular Formula/Molecular Weight

$C_{20}H_{34}O_4$; MW = 338.24571

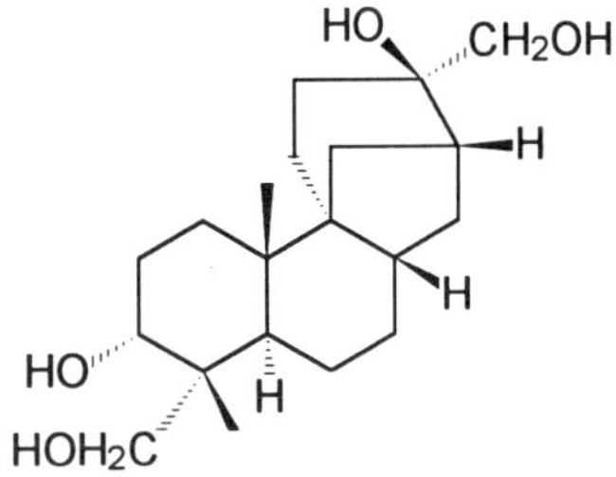

General Characteristics

Needles from ethyl acetate; mp., 227-233°C; $[\alpha]_D^{27}$ +12° (*c*=1.0, in MeOH).

Fungal Source

Cephalosporium aphidicola Petch, *Nigrospora sphaerica,* and *Harziella entomophilla.*

Isolation/Purification

The culture medium was filtered, adjusted to pH 6.6 and extracted with chloroform. The extract was concentrated and aphidicolin crystallized over time; it was recrystallized from ethyl acetate to give needles.

Biological Activity

Aphidicolin reduced the mitotic rate of mouse "L" cells growing in tissue culture; inhibited the growth of *Herpes simplex* type 1 both in cultures of human embryonic lung cells and in the rabbit cornea; and showed marked inhibitory activity in lettuce and rice seedlings (root growth).

Spectral Data

IR:

(Nujol) 3520, 3440, 3360, 1077, 1047, and 1027cm^{-1}.

^{1}H NMR:

(pyridine-d_5) 0.76(3H, s); 0.98(3H, s); 1.2-3.0(ca 20H, br m); 3.70(3H, br); 5.60(1H, br); 5.84(1H, br); and 6.08ppm (1H, br).

^{13}C NMR:

(CD_3CO_2D) C-1 or 7, 27.4(t); C-3, 76.9(d); C-4, 41.2(s); C-5, 41.7(d); C-7 or -1, 27.1(t); C-8, 34.6(d); C-9, 50.0(s); C-10, 40.6(s); C-12, 41.0(d); C-15, 28.3(t); C-16, 77.3(s); C-17, 68.1(t); C-18, 17.8(qt); C-19, 71.5(t); C-20, 15.5(qt); C-2, 33.5(t); C-6, 32.1(t); C-11, 27.6(t); C-13, 25.3(t); and C-14, 23.9ppm (t).

Mass Spectrum:

EIMS: 307*m/e* (M^+ - CH_2OH)

References

K. M. Brundret, W. Dalziel, B. Hesp, J. A. J. Jarvis, and S. Neidle; X-Ray Crystallographic Determination of the Structure of the Antibiotic Aphidicolin: a Tetracyclic Diterpenoid Containing a New Ring system; J. C. S. Chem. Commun., pp. 1027-1028(1972).

W. Dalziel, B. Hesp, K. M. Stevenson, and J. A. J. Jarvis; The Structure and Absolute Configuration of the Antibiotic Aphidicolin: a Tetracyclic Diterpenoid Containing a New Ring System; J. Chem. Soc., Perkin Trans. I, pp. 2841-2851(1973).

K. Kawada, Y. Kimura, K. Katagiri, A. Suzuki, and S. Tamura, Isolation of Aphidicolin as a Root Growth Inhibitor from *Harziella entomophilla*, Agric. Biol. Chem., Vol. 42, pp. 1611-1612(1978).

A. N. Starratt and S. R. Loschiavo; The Production of Aphidicolin by *Nigrospora sphaerica*; Canad. J. Microbiol., Vol. 20, pp. 416-417(1974).

Common/Systematic Name
3-Deoxyaphidicolin

Molecular Formula/Molecular Weight
$C_{20}H_{34}O_3$; MW = 322.25080

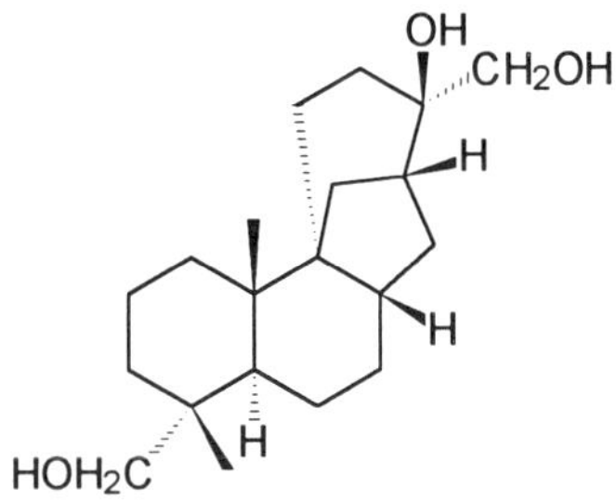

General Characteristics
Crystals; mp., 138.0-140.5°C; $[\alpha]_D^{27}$ +22.6° (*c*=0.98, in EtOH).

Fungal Source
Phoma betae Fr., which causes a leaf spot disease on sugar beet.

Isolation/Purification
The culture broth was separated by filtration into mycelium and filtrate, and the mycelium was extracted with acetone. To the extracts from the mycelium was added water and the mixture was again extracted with ether to give brown extracts. The residual aqueous layer was further extracted with ethyl acetate and the extracts were combined with the ether extracts. After removal of the solvent, the dark brown residue was chromatographed on a silica gel column which was eluted with chloroform-methanol (9:1, v/v) to give active fractions for a bioassay using lettuce seedlings. The active fractions which inhibited the root elongation of lettuce seedlings were further separated by repeated chromatography. The aphidicolin derivatives appeared as pinkish spots on TLC after anisaldehyde reagent spray and gave in the order of elution aphidicolin 3,18-orthoacetate, 3-deoxyaphidicolin, aphidicolin 17-monoacetate and aphidicolin.

Biological Activity
Aphidicolin 3,18-orthoacetate, 3-deoxyaphidicolin, aphidicolin 17-monoacetate, and aphidicolin inhibited the root growth of lettuce seedlings by 54.5, 50.9, 58.1 and 74.3%, respectively.

Spectral Data

IR:
(KBr) 3360, 2930, 2850, 1440, and 1030cm^{-1}; diacetate: 3470, 2930, 1730, 1230, and 1030cm^{-1}

^{1}H NMR:

(pyridine-d_5) 0.88(3H, s); 1.03(3H, s); 3.2(1H, dd, J=10.8, 3.4Hz); 3.68(1H, dd, J=10.8, 3.4Hz); 3.72(1H, d, J=10.8Hz); 3.81(1H, d, J=10.8 Hz). Diacetate ($CDCl_3$) 0.87(3H, s); 0.93(1H, dd, J=13.3, 5.9Hz); 0.98(3H, s); 2.07(3H, s); 2.10(3H, s); 3.63(1H, d, J=10.8Hz); 3.82(1H, d, J=10.8Hz); 3.95(1H, d, J=11.5Hz); and 3.99(1H, d, J=11.5Hz).

Mass Spectrum:

3-Deoxyaphidicolin has the molecular formula $C_{20}H_{34}O_3$ from the high resolution mass spectrum, 291.2319(M^+ -CH_2OH, $C_{19}H_{31}O_2$, calcd 291.2321) and the field desorption mass spectrum 322(M^+), and 304(M^+ -H_2O).

<u>Reference</u>

K. Hayashi, M. Hashimoto, S. Sakamura, and R. Sakai; 3-Deoxyaphidicolin and Aphidicolin Analogues as Phytotoxins from *Phoma betae*; Agric. Biol. Chem., Vol. 48, pp. 1687-1689(1984).

Common/Systematic Name

Aphidicolin 17-monoacetate

Molecular Formula/Molecular Weight

$C_{22}H_{36}O_5$; MW = 380.25627

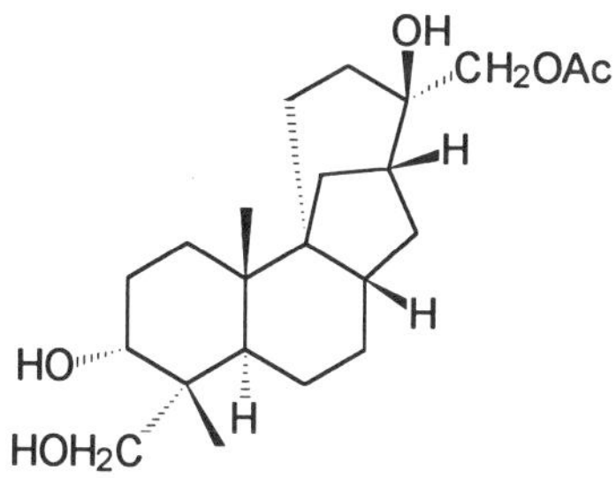

General Characteristics

Crystals; mp., 196.5-199.0°C; 193.5-196°C; $[\alpha]_D^{27}$ + 5.0°.

Fungal Source

Phoma betae Fr., which causes a leaf spot disease on sugar beet.

Isolation/Purification

See 3-deoxyaphidicolin.

Biological Activity

Aphidicolin 3,18-orthoacetate, 3-deoxyaphidicolin, aphidicolin 17-monoacetate, and aphidicolin inhibited the root growth of lettuce seedlings by 54.5, 50.9, 58.1, and 74.3%, respectively.

Spectral Data

IR:

(KBr) 3300 and 1750cm^{-1}.

^{1}H NMR:

(pyridine-d_5) 0.79(3H, s); 1.03(3H, s); 1.95(3H, s); 3.65(1H, br, d, J=10Hz); 3.82(1H, dd, J=10.7, 4.3Hz); 3.92(1H, br, s); and 4.29ppm (2H, s).

Mass Spectrum:

Aphidicolin 17-monoacetate had the molecular formula $C_{22}H_{36}O_5$ from the high resolution mass spectrum, 362.2451 (M^+ -H_2O, $C_{22}H_{34}O_4$; calcd 362.2454).

Reference

K. Hayashi, M. Hashimoto, S. Sakamura, and R. Sakai; 3-Deoxyaphidicolin and Aphidicolin Analogues as Phytotoxins from *Phoma betae*; Agric. Biol. Chem., Vol. 48, pp. 1687-1689(1984).

Common/Systematic Name

Aphidicolin 3,18-orthoacetate

Molecular Formula/Molecular Weight

$C_{22}H_{36}O_5$; MW = 380.25627

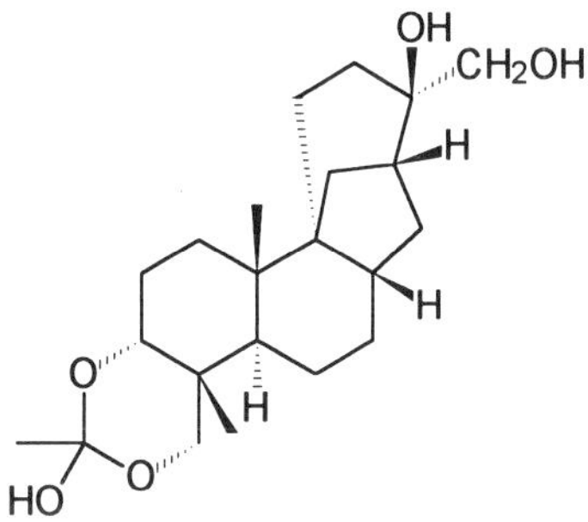

General Characteristics

Oily compound; $[\alpha]_D^{25}$ -5.4° (*c*=1.6, in $CHCl_3$).

Fungal Source

Phoma betae Fr., which causes a leaf spot disease on sugar beet.

Isolation/Purification

See 3-deoxyaphidicolin.

Biological Activity

Aphidicolin 3,18-orthoacetate, 3-deoxyaphidicolin, aphidicolin 17-monoacetate, and aphidicolin inhibited the root growth of lettuce seedlings by 54.5, 50.9, 58.1, and 74.3%, respectively.

Spectral Data

IR:

(neat) 3400cm^{-1}.

^1H NMR:

($CDCl_3$) 0.73(3H, s); 0.99(3H, s); 1.41(3H, s); 3.23(1H, d, *J*=12.2Hz); 3.37(1H, d, *J*=10.7Hz); 3.40(1H, d, *J*=10.7Hz); 3.62(1H, d, *J*=12.2Hz); and 3.62-3.65ppm (1H, m).

Mass Spectrum:

$C_{22}H_{36}O_5$ from HRMS 363.2572 (M^+ - H_2O + H^+, $C_{22}H_{35}O_4$, calcd. 363.2536), and FD-MS 380(M^+).

Reference

K. Hayashi, M. Hashimoto, S. Sakamura, and R. Sakai; 3-Deoxyaphidicolin and Aphidicolin Analogues as Phytotoxins from *Phoma betae*; Agric. Biol. Chem., Vol. 48, pp. 1687-1689(1984).

Neovasinins

Neovasinin
Neovasinone
Neovasipyridone A
Neovasipyridone B
Neovasipyridone C

Common/Systematic Name

Neovasinin

Molecular Formula/Molecular Weight

$C_{17}H_{24}O_5$; MW = 308.16237

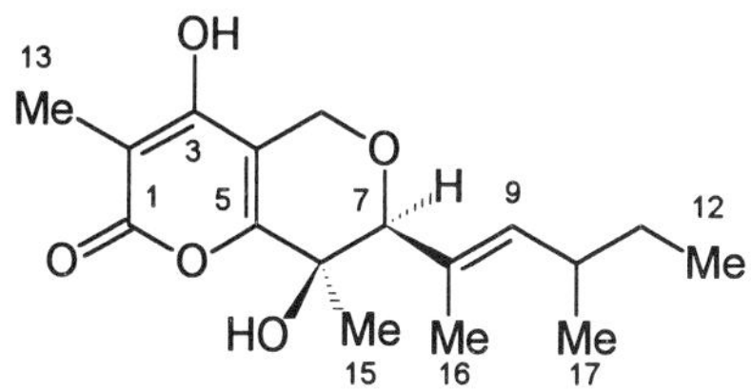

General Characteristics

Colorless plates; mp., 204-206°C; monoacetyl derivative; mp., 127-129°C.

Fungal Source

Neocosmospora vasinfecta (strain NHL 2298), a pathogen that causes root- and fruit-rot and seedling damping off in the Malvaceae, Leguminosae, Piperaceae, Cucurbitacea, etc., including pepper, peanuts, soybeans, beans, coconuts, and *Albizziac*.

Isolation/Purification

Fungal culture filtrate was treated with active charcoal at pH 2, extracted with acetone and the extract concentrated. The concentrated extract was chromatographed on a silica gel column eluting with benzene and acetone. Neovasinone eluted with 5% acetone in benzene; neovasinin eluted with 10% acetone in benzene. Neovasinin was further purified by Sephadex LH_{20} column chromatography eluting with methanol. It was crystallized from *n*-hexane-ethyl acetate solution as colorless plates.

Biological Activity

The metabolite inhibited root growth of lettuce seedings 50% at 300mg/liter.

Spectral Data

UV:

λ_{max}^{EtOH} 212(ϵ = 36,800), 240 (3,900), and 289nm (11,300).

IR:

(KBr) 3560, 3120, 1670, 1590, 1240, and 1110cm^{-1}.

^{1}H NMR:

(DMSO-d_6) 3.70(s, H-7); 5.07(dd, *J*=9.3, 1.2Hz, H-9); 2.33(m, H-10); 1.25(m, H-11); 0.84(t, *J*=7.3Hz, H-12); 1.85(s, H-13); 4.29, 4.55(Abq, *J*=15.1Hz, H-14); 1.22(s, H-15); 1.70(d, *J*=1.2Hz, H-16); 0.91(d, *J*=6.3Hz, H-17); 5.29(br, s, 3-OH); and 3.40ppm (br, s, 6-OH).

^{13}C NMR:

(DMSO-d_6) 164.3(C-1, s); 98.9(C-2, s); 161.8(C-3, s); 107.6(C-4, s); 155.7(C-5, s); 68.2(C-6, s); 87.7(C-7, d); 131.1(C-8, s); 136.2(C-9, d); 33.2(C-10, d); 29.9(C-11, t); 11.9(C-12, q); 9.1(C-13, q); 61.7(C-14, d); 20.6(C-15, q); 13.3(C-16, q); and 20.5ppm (C-17, q).

Mass Spectrum:

LREIMS: 308(M^+, 6%), 182(80), 139(100), 126(15), and 111*m/e* (15).

Reference

H. Nakajima, K. Nishimura, T. Hamasaka, Y. Kimura, and S. Udagawa; Structure of Neovasinin, A New Metabolite Produced by the Fungus, *Neocosmospora vasinfecta* E.F. Smith, and Its Biological Activity to Lettuce Seedlings; Agric. Biol. Chem., Vol. 51, pp. 2831-2833(1987).

Common/Systematic Name

Neovasinone

Molecular Formula/Molecular Weight

$C_{17}H_{22}O_6$; MW = 322.14164

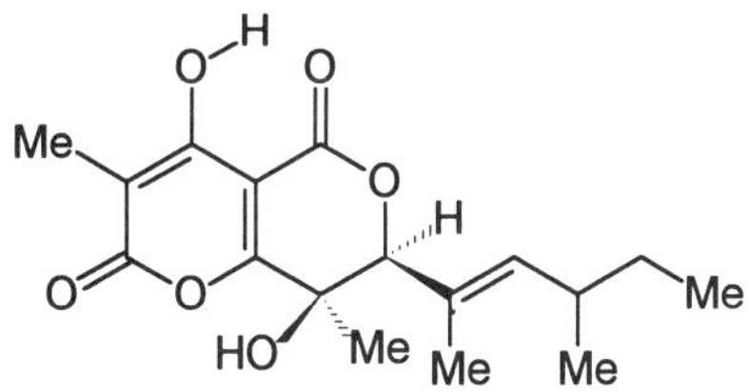

General Characteristics

Colorless plates from ethyl acetate-*n*-hexane; mp., 193-195°C; $[\alpha]_D^{25}$ - 90° (*c*=0.20, in MeOH).

Fungal Source

Neocosmospora vasinfecta var *africana.*

Isolation/Purification

The culture filtrate was treated at pH 2.0 with active charcoal and the active charcoal was extracted with acetone. The extract was concentrated and the residue chromatographed on a silica gel column with 5% acetone in benzene. The combined neovasinone fractions were further purified by silica gel column chromatography using 30% ethyl acetate in *n*-hexane followed by crystallization from ethyl acetate-*n*-hexane to afford colorless plates.

Biological Activity

Plant growth regulator activity; the administration of 10 and 100mg/L of neovasinone accelerated the root elongation of lettuce seedlings.

Spectral Data

UV:

λ_{max}^{MeOH} 223(ε=19,000), 271(9900), and 300nm (2400).

IR:

(KBr) 3450, 1700, 1680, 1600, and 1460cm^{-1}.

^{1}H NMR:

($CDCl_3$) 4.68(s, 1H, H-7); 5.32(dd, 1H, *J*=9.8, 1.3 Hz, H-9); 2.43(dddd, 1H, *J*=9.9, 7.5, 6.8, 7.2Hz, H-10); 1.35(m, 2H); 0.89(t, *J*=7.7Hz, 3H, H-12); 1.94(s, 3H, H-13); 1.60(s, 3H, H-15); 1.87(d, *J*=1.3Hz, 3H, H-16); and 1.00ppm (d, *J*=7.2 Hz, 3H, H-17).

^{13}C NMR:

($CDCl_3$) 162.8(Sq), C-1; 101.4(Sdq), C-2; 162.3(Sqd), C-3; 98.1(Sd), C-4; 168.8(Sbr), C-5; 69.4(Sbr,q), C-6; 90.54(Dm), C-7; 127.3(Sm), C-8; 142.2(Dm), C-9; 34.2(Dm), C-10; 30.0(Tm), C-11; 12.1(Qdt), C-12; 8.3(Qs), C-13; 167.1(Sbr s), C-14; 20.5(Qbr s), C-15; 13.2(Qdd), C-16; and 201ppm (Qm), C-17.

Mass Data:

LREIMS: 322*m/e* (M^+); found C, 63.18; H, 6.81%; calcd for $C_{17}H_{22}O_6$: C, 63.34; H, 6.88%.

Reference

H. Nakajima, K. Nishimura, T. Hamasaki, Y. Kimura, T. Dyokota, and S. Udagawa; Structure of Neovasinone, A New α-Pyrone Plant Growth Regulator Produced by the Fungus, *Neocosmospora vasinfecta* E.F. Smith; Agric. Biol. Chem., Vol.51, pp. 1221-1224(1987).

Common/Systematic Name

Neovasipyridone A

Molecular Formula/Molecular Weight

$C_{21}H_{35}NO_3$; MW = 349.26169

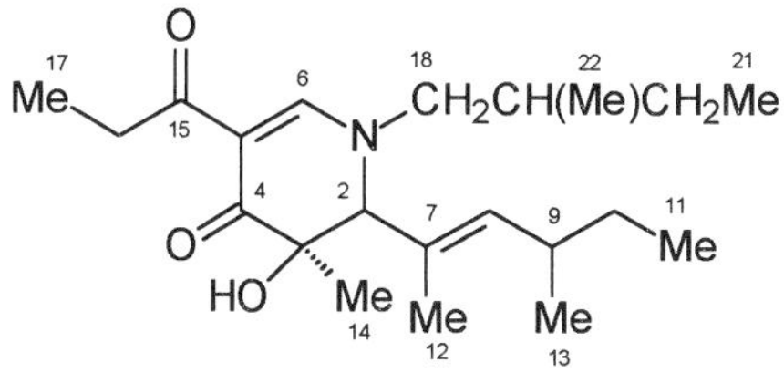

General Characteristics

An oil; $[\alpha]_D^{20}$ + 366° (*c*=1.0, in EtOH).

Fungal Source

Neocosmospora vasinfecta (NHL 2298), a pathogen which causes root and fruit rot and seedling damping-off in the Malvaceae, Leguminosae, Piperaceae, and Cucurbitaceae.

Isolation/Purification

See Neovasipyridone B for isolation.

Spectral Data

UV:

λ_{max}^{EtOH} 202 (log ε = 3.79), 262 (4.10), and 322nm (4.14).

IR:

(thin film) 3398, 2964, 2930, 2874, 2858, 1647, 1580, 1462, 1375, 1350, 1315,1267, 1241, 1172, and 1114cm^{-1}.

^{1}H NMR:

($CDCl_3$) 3.82(s, H-2); 8.21(s, H-6); 5.12(dq, *J*=9.5, 0.9Hz, H-8); 2.28(m, H-9); 1.10-1.40(m, H-10); 0.83(t, *J*=7.6Hz, H-11); 1.43(d, *J*=0.9Hz, H-12); 0.93(d, *J*=6.8Hz, H-13); 1.38(s, H-14); 2.86(dq, *J*=17.1,7.2Hz, H-16); 2.98(dq, *J*=17.1, 7.2Hz, H-16); 1.10(t, *J*=7.2Hz, H-17); 3.10(dd, *J*=13.5, 8.6Hz, H-18); 3.23(dd, *J*=13.5, 8.6Hz, H-18); 1.80(m, H-19); 1.25(m, H-20); 0.95(t, *J*=7.2Hz, H-21); and 0.98ppm (d, *J*=6.3Hz, H-22).

^{13}C NMR:
(CDCl$_3$) 76.7, C-2; 71.1, C-3; 191.7, C-4; 104.9, C-5; 158.8, C-6; 127.6, C-7; 140.1, C-8; 34.0, C-9; 30.1, C-10; 11.8, C-11; 13.1, C-12; 20.6, C-13; 28.8, C-14; 197.6, C-15; 34.9, C-16; 8.5, C-17; 61.2, C-18; 33.7, C-19; 26.4, C-20; 10.8, C-21; and 16.8ppm, C-22.

Mass Spectrum:
FABMS: 350$[M+H]^+$ (100%), 332(10), 306(14), 236(30) and 196*m/e* (12); exact mass calcd for $C_{21}H_{36}NO_3$ 349.26169; found 349.2655*m/e*.

Reference
H. Nakajima, K. Shimomum, T. Furumoto, and T. Hamasaki; Neovasipyridones A, B and C: Metabolites Related to Neovasinin, A Phytotoxin of the Fungus, *Neocosmospora vasinfecta*; Phytochemistry; Vol. 40, pp. 1643-1647(1995).

Common/Systematic Name

Neovasipyridone B

Molecular FormulaJMolecular Weight

$C_{21}H_{35}NO_3$; MW = 349.26169

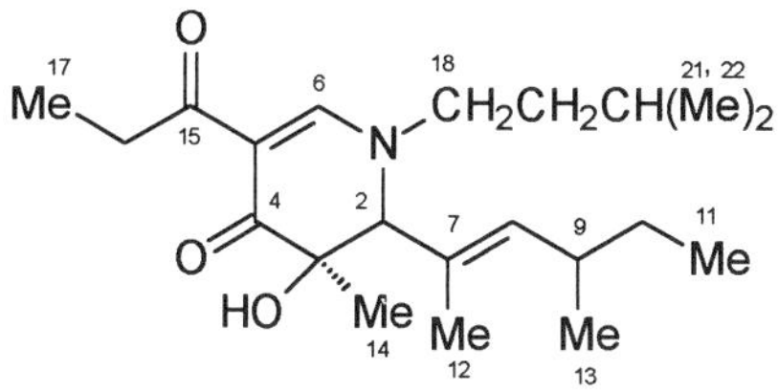

General Characteristics

An oil; $[\alpha]_D^{20}$ + 392° (*c*=1.0, in EtOH).

Fungal Source

Neocosmospora vasinfecta (NHI, 2298),a pathogen which causes root and fruit rot and seedling damping-off in the Malvaceae, Leguminosae, Piperaceae, and Cucurbitaceae.

Isolation/Purification

The dried mycelial mats were soaked in acetone overnight after which the acetone extract was obtained by filtration. The acetone extract was evaporated to dryness, the residue dissolved in $NaHCO_3$ and treated with ethyl acetate. The ethyl acetate was placed over Na_2SO_4 and evaporated to dryness. The residue was partitioned with MeOH-H_20 (9:1, v/v) and *n*-hexane. The *n*-hexane layer was extracted with MeOH-H_2O (9:1, v/v). The combined MeOH solutions were evaporated to dryness and the residue was subjected to silica gel column chromatography. The column was eluted successively with each of 0%, 2%, 5%, 10% and 20% acetone in *n*-hexane. Fractions 1-3 eluted with 10% acetone in *n*-hexane contained the neovasipyridones. The first fraction was purified by Sephadex LH_{20} column chromatography with MeOH as the solvent. Each 7ml of eluate constituted one fraction and fractions 41-44 were combined and evaporated to dryness. The residue was further purified by preparatory TLC with a solvent system of acetone-chloroform (5:95, v/v) to afford neovasipyridone A. The second and third fractions were combined and evaporated to dryness. The residue was subjected to Sephadex LH_{20} column chromatography with MeOH as the solvent. Each 10ml of eluate constituted one fraction, and fractions 56-61 were combined and evaporated to dryness. The residue contained neovasipyridones A, B, and C. Removal of neovasipyridone A from the mixture was achieved by preparatory TLC with ethyl acetate-benzene (7:93, v/v) (quadruple development), giving neovasipyridone A as a constituent of a higher R_f band on TLC. The lower R_f band contained neovasipyridones B and C. Neovasipyridone B was separated from C by repeated HPLC [column: DAISOPAK SP-120-5-ODS-A, solvent: MeOH-H_2O (9:1, v/v), flow rate: 1.0ml min^{-1}, detector: 220nm), giving neovasipyridones B (Rt, 28.0min) and C (Rt, 33.0min].

Spectral Data

UV:

λ_{max}^{EtOH} 200(log ϵ = 3.71), 263(4.15), and 322nm (4.19).

IR:

(thin film) 3362, 2960, 2874, 1638, 1574, 1460, 1392, 1377, 1334, 1292, 1257, 1201, 1156, 1116, 1083, and 1029cm^{-1}.

^{1}H NMR:

($CDCl_3$) 3.80(s, H-2); 8.25(s, H-6); 5.12(dq, *J*=10.8, 1.2Hz, H-8); 2.25(m, H-9); 1.10-1.42(m, H-10); 0.82 (t, *J*=7.5Hz, H-11); 1.41(d, *J*=1.2Hz, H-12); 0.91(d, *J*= 6.8Hz, H-13); 1.34(s, H-14); 2.84(dq, *J*=16.5, 7.5 Hz, H-16); 2.97(dq, *J*=16.5, 7.5Hz, H-16); 1.09(t, *J*=7.5Hz, H-17); 3.29(dd, *J*=13.5, 7.5Hz, H-18); 3.39(dd, *J*=13.5, 7.0Hz, H-18); 1.60(m, H-19); 1.60(m, H-20); 0.95(t, *J*=6.5Hz, H-21); and 0.96ppm (d, *J*=6.5Hz, H-22).

^{13}C NMR:

($CDCl_3$) 76.1, C-2; 70.9, C-3; 191.6, C-4; 104.8, C-5; 158.7, C-6; 127.7, C-7; 140.0, C-8; 34.0, C-9; 30.1, C-10; 11.8, C-11; 13.0, C-12; 20.6, C-13; 28.6, C-14; 197.6, C-15; 34.8, C-16; 8.6, C-17; 53.6, C-18; 37.6, C-19; 25.7, C-20; 22.0, C-21; and 22.5ppm, C-22.

Mass Spectrum:

FABMS: 350[M+H]$^{+}$(100%), 306 (15), 196 (10),154 (39), 149(27), 136 (33), and 107*m/e* (15); exact mass calcd for $C_{21}H_{36}NO_3$ 350.2695; found 350.2666*m/e*.

Reference

H. Nakajima, K. Shimomum, T. Furumoto, and T. Hamasaki; Neovasipyridones A, B and C: Metabolites Related to Neovasinin, A Phytotoxin of the Fungus, *Neocosmospora vasinfecta*; Phytochemistry, Vol. 40, pp. 1643-1647(1995).

Common/Systematic Name

Neovasipyridone C

Molecular Formula/Molecular Weight

$C_{20}H_{33}NO_3$; MW = 335.24604

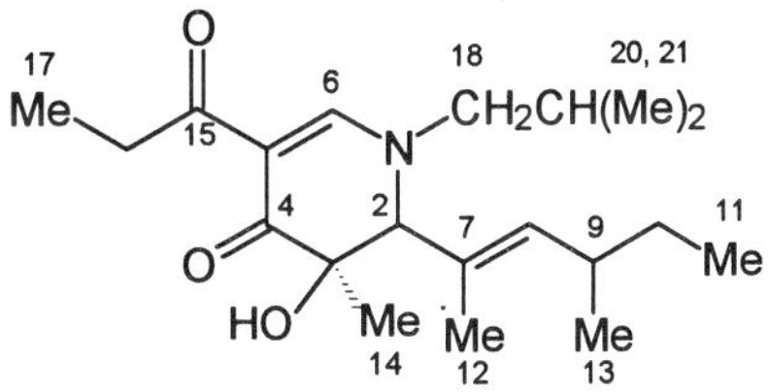

General Characteristics

An oil; $[\alpha]_D^{20}$ + 383° (c= 1.0, in EtOH).

Fungal Source

Neocosmospora vasinfecta (NHL 2298), a pathogen which causes root and fruit rot and seedling damping-off in the Malvaceae, Leguminosae, Piperaceae and Cucurbitaceae.

Isolation/Purification

The dried mycelial mats were soaked in acetone overnight after which the acetone extract was obtained by filtration. The acetone extract was evaporated to dryness, the residue dissolved in $NaHCO_3$ and treated with ethyl acetate. The ethyl acetate was placed over Na_2SO_4 and evaporated to dryness. The residue was partitioned with MeOH-H_2O (9:1, v/v) and *n*-hexane. The *n*-hexane layer was extracted with MeOH-H_2O (9:1). The combined MeOH solutions were evaporated to dryness and the residue was subjected to silica gel column chromatography. The column was eluted successively with each of 0%, 2%, 5%, 10% and 20% acetone in *n*-hexane. Fractions 1-3 eluted with 10% acetone in *n*-hexane contained the neovasipyridones. The first fraction was purified by Sephadex LH_{20} column chromatography with MeOH as the solvent. Each 7ml of eluate constituted one fraction, and fractions 41-44 were combined and evaporated to dryness. The residue was further purified by preparatory TLC with a solvent system of acetone-chloroform (5:95, v/v) to afford neovasipyridone A. The second and third fractions were combined and evaporated to dryness. The residue was subjected to Sephadex LH_{20} column chromatography with MeOH as the solvent. Each 10ml of eluate constituted one fraction, and fractions 56-61 were combined and evaporated to dryness. The residue contained neovasipyridones A, B and C. Removal of neovasipyridone A from the mixture was achieved by preparatory TLC with ethyl acetate-benzene (7:93, v/v) (quadruple development), giving neovasipyridone A as a constituent of a higher R_f band on TLC. The lower R_f band contained neovasipyridones B and C. Neovasipyridone B was separated from C by repeated HPLC [column: DAISOPAK SP-120-5-ODS-A, solvent: MeOH-H_2O (9:1, v/v), flow rate: 1.0 ml min^{-1}, detector: 220 nm, giving neovasipyridones B (Rt, 28.0 min) and C (Rt, 33.0 min)].

Spectral Data

UV:

203 (log ε = 3.98), 262 (4.11), and 323nm (4.14).

IR:

(thin film) 3406, 2967, 2926, 2874, 1639, 1579, 1460, 1369, 1315, 1261, 1157, and 1113cm^{-1}.

^{1}H NMR:

($CDCl_3$) 3.80(s, H-2); 8.20(s, H-6); 5.13(dq, *J*=9.5, 1.0Hz, H-8); 2.28(m, H-9); 1.07-1.40 (m, H-10); 0.81(t, *J*=7.5Hz, H-11); 1.41(d, *J*=1.0Hz, H-12); 0.91(d, *J*=6.8Hz, H-13); 1.37(s, H-14); 2.84(dq, *J*=17.0, 7.0 Hz, H-16); 2.97(dq, *J*=17.0, 7.0Hz, H-16); 1.08(t, *J*=7.0Hz, H-17); 2.94 (dd, *J*=14.0, 9.0Hz, H-18); 3.28(dd, *J*=14.0, 5.2Hz, H-18); 2.03(m, H-19); 0.97(d, *J*=6.8Hz, H-20); and 0.98ppm (d, *J*= 6.8Hz, H-21).

^{13}C NMR:

($CDCl_3$) 76.3, C-2; 71.0, C-3; 191.7, C-4; 104.8, C-5; 158.1, C-6; 127.5, C-7; 140.1, C-8; 34.0, C-9; 30.1, C-10; 11.8, C-11; 13.1, C-12; 20.6, C-13; 28.9, C-14; 197.6, C-15; 34.9, C-16; 8.5, C-17; 62.7, C-18; 27.2, C-19; 19.8, C-20; and 20.1ppm, C-21.

Mass Spectrum:

EIMS 335 (M^+, 55%), 292 (100), 278 (23), 222 (15), 182 (41), and 125*m/e* (19); exact mass calcd for $C_{20}H_{33}NO_3$ 335.2461; found 335.2488*m/e*.

Reference

H. Nakajima, K. Shimomum, T. Furumoto, and T. Hamasaki; Neovasipyridones A, B and C: Metabolites Related to Neovasinin, A Phytotoxin of the Fungus, *Neocosmospora vasinfecta;* Phytochemistry, Vol. 40, pp. 1643-1647 (1995).

25

Koninginins

- Koninginin A
- Koninginin B
- Koninginin C
- Koninginin D
- Koninginin E

Common/Systematic Name

Koninginin A

Molecular Formula/Molecular Weight

$C_{16}H_{28}O_4$; MW = 284.19876

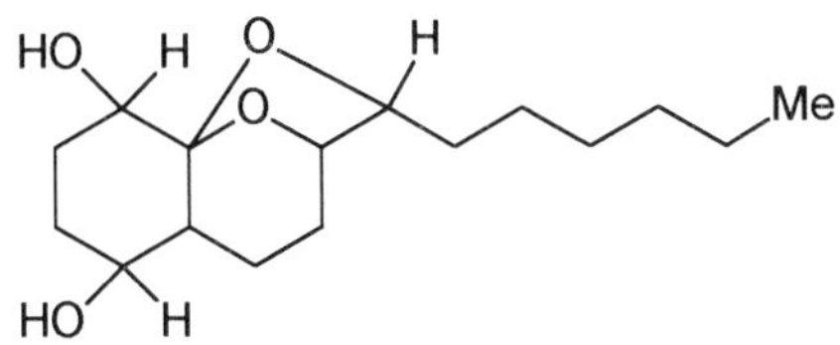

General Characteristics

Crystallized from isooctane; mp., 80-84°C; $[\alpha]_D$ -22° (c=0.7, in $CHCl_3$).

Fungal Source

Trichoderma koningii was isolated from *Diffenbachia* sp. (ATCC 46314) and *T. harzianum* was isolated from wheat roots.

Isolation/Purification

Fungal cultures were extracted with acetone, reduced to an aqueous phase under vacuum, and partitioned twice with ethyl acetate. The ethyl acetate extract was chromatographed on a silica gel column eluted with benzene, ethyl ether, ethyl acetate, acetone, and acetonitrile. The ethyl ether and ethyl acetate fractions were combined, reduced in volume and further purified by silica gel column chromatography eluted with a linear gradient from benzene to acetone. The koninginin A-containing fractions were further purified with a silica gel C_{18} reversed-phase column eluted with different proportions of acetonitrile-water. The koninginin A containing fractions were combined, frozen and the supernatant liquid removed. The supernatant fluid was reduced in volume and chromatographed on a column containing hydrated silica gel eluted with benzene-ethyl acetate (5:4, v/v). Final purification was performed on a silica gel-silver nitrate column eluted with methylene chloride-acetone (9:1, v/v) followed by crystallization as fine needles.

Biological Activity

The metabolite inhibited the growth of etiolated wheat coleoptiles at 10^{-3} M. It was not active against several bacteria and fungi.

Spectral Data

UV:

λ_{max}^{EtOH} End absorption.

IR:

(KBr) 3500, 3460, 2950, 2925, 2855, 1450, 1410, 1340, 1255, 1242, 1216, 1202, 1175, 1132, 1102, 1080, 1068, 1052, 1042, 1002, 955, 938, 915, 855, 820, 770, 722, 665, and 640cm^{-1}.

^{13}C NMR:

($CDCl_3$) 72.66(C-1, d); 30.85(C-2, t); 25.38(C-3, t); 69.78(C-4, d); 109.20(C-5, s); 41.43(C-6, d); 20.60(C-7, t); 27.26(C-8, t); 79.00(C-9, d); 79.30(C-10, d); 35.12(C-11, t); 25.54(C-12, t); 29.08(C-13, t); 31.69(C-14, t); 25.53(C-15, t); and 14.02ppm (C-16, q).

^{1}H NMR:

0.88(3H, *J*=6.6Hz, H-a); 1.30(8H, m, H-b's); 1.50(2H, *J*=6.7Hz, H-c); 1.57(3H, *J*=3.2, 7.2Hz, H-d); 1.72(1H, *J*=6.3, 13.9Hz, H-e); 1.85(2H, *J*=3.6, 11.6, H-f); 1.96(1H, *J*=6.8, 12.7Hz, H-g); 2.11(1H, *J*=4.7, 9.5Hz, H-h); 2.26(1H, *J*=3.6, 6.1, 13.1, H-i); 3.61(1H, *J*=5.4, 11.6Hz, H-j); 3.89(1H, *J*=2.5Hz, H-k); 4.03(1H, *J*=6.6Hz, H-l); and 4.32ppm (1H, *J*=2.1Hz, H-m).

TLC Data

Silica gel 60 thin-layer plates developed with toluene-ethyl acetate-formic acid (5:4:1, v/v/v), R_f = 0.56-0.60; detected as a salmon-pink colored spot when treated with anisaldehyde followed by heating.

References

F. Almassi, E. L. Ghisalberti, M. J. Narbey, and K. Sivasithamparam; New Antibiotics from Strains of *Trichoderma harzianum*; J. Natural Products, Vol. 54, pp. 396-402(1991).

H. G. Cutler, D. S. Himmelsbach, R. F. Arrendale, P. D. Cole, and R. H. Cox; Koninginin A: A Novel Plant Growth Regulator from *Trichoderma koningii*; Agric. Biol. Chem., Vol. 53, pp. 2605-2611(1989).

Common/Systematic Name
Koninginin B

Molecular Formula/Molecular Weight
$C_{16}H_{26}O_4$; MW = 282.18311

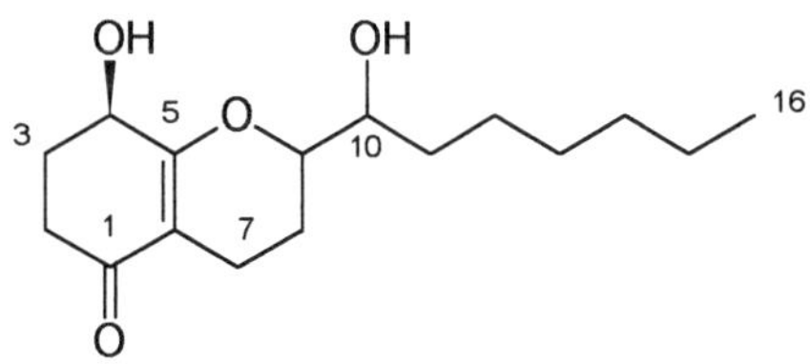

General Characteristics
Colorless crystalline solid; mp., 72-74°C.

Fungal Source
Trichoderma koningii was isolated from *Diffenbachia* sp. (ATCC 46314).

Isolation/Purification
Fungal cultures were extracted with acetone, reduced to an aqueous phase under vacuum, and partitioned twice with ethyl acetate. The ethyl acetate extract was chromatographed on a silica gel column eluted with benzene, ethyl ether, ethyl acetate, acetone, and acetonitrile. The ethyl ether and ethyl acetate fractions were combined, reduced in volume and further purified by silica gel column chromatography eluted with a linear gradient from benzene to acetone. The koninginin B fractions were reduced in volume and chromatographed on a column containing hydrated silica gel eluted with benzene-ethyl acetate (5:4, v/v). Biologically active fractions were pooled and chromatographed on C_{18} reversed-phase silica gel with acetonitrile-water (1:1, v/v). Final purification was performed on a silica gel-silver nitrate column eluted with methylene chloride-acetone (9:1, v/v) to yield koninginin B.

Biological Activity
The metabolite inhibited the growth of etiolated wheat coleoptiles at 10^{-3} M. It was marginally active against gram-positive bacteria and inactive against gram-negative bacteria.

Spectral Data

UV:
λ_{max}^{EtOH} 262nm (log ϵ = 4.24).

IR:

(KBr) 3450, 2960, 2930, 2860, 1650, 1612, 1453, 1402, 1381, 1287, 1242, 1196, 1152, 1078, 1048, 981, 860, 728, and 692cm^{-1}.

^{1}H NMR:

($CDCl_3$) 2.50($J_{2a,2b}$=-17Hz, H-2a); 2.56($J_{2a,3a}$=14.4, $J_{2a,3b}$=2.3, $J_{2b,3a}$=5.1, $J_{2b,3b}$=5.2Hz, H-2b); 1.79($J_{3a,3b}$=12.9Hz, H-3a); 2.34($J_{3a,4}$=-12.1, $J_{3b,4}$=4.8Hz, H-3a); 4.05(H-4); 2.09($J_{7a,7b}$=-15.9Hz, H-7a); 2.57($J_{7a,8a}$=11.6, $J_{7a,8b}$=6.1, $J_{7b,8a}$=13.2, $J_{7b,8b}$=2.1Hz, H-7b); 1.68($J_{8a,8b}$=-13.0Hz, H-8a); 1.96($J_{8a,9}$=5.2, $J_{8b,9}$=1.9Hz, H-8b); 3.80($J_{9,10}$=11.2Hz); 3.68($J_{10,11}$=5.9Hz, H-10); 1.55(H-11); 1.30, 1.55(H-12); 1.30(H-13); 1.30(H-14); 1.30($J_{15,16}$=6.8Hz, H-15); and 0.89ppm (H-16).

^{13}C NMR:

($CDCl_3$) 198.10(C-1); 2714(C-2); 20.20(C-3); 71.02(C-4); 171.58(C-5); 109.9(C-6); 17.70(C-7); 22.59(C-8); 80.80(C-9); 72.89(C-10); 32.76(C-11); 25.44(C-12); 29.25(C-13); 31.75(C-14); 22.59(C-15); and 14.06ppm (C-16).

Reference

H. G. Cutler, D. S. Himmelsbach, B. Yagen, R. F. Arrendale, J. M. Jacyno, P. D. Cole, and R. H. Cox; Koninginin B: A Biologically Active Congener of Koninginin A from *Trichoderma koningii*; J. Agric. Food Chem., Vol. 39, pp. 977-980(1991).

Common/Systematic Name

Koninginin C

Molecular Formula/Molecular Weight

$C_{16}H_{28}O_4$; MW = 284.19876

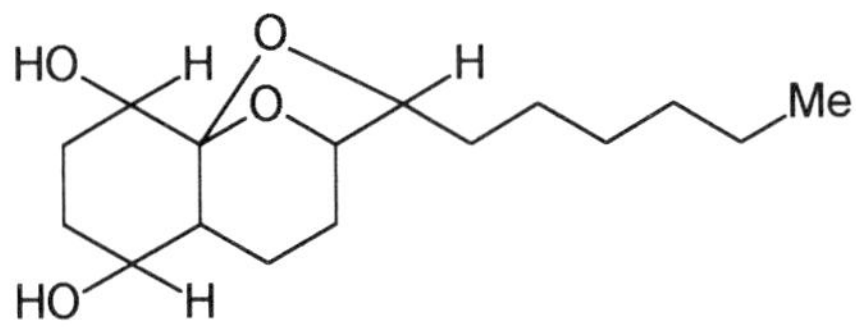

General Characteristics

Fine white crystals from isooctane; mp., 70-72°C.

Fungal Source

Trichoderma koningii was isolated from *Diffenbachia* sp. (ATCC 46314) and *T. harzianum* was isolated from wheat roots.

Isolation/Purification

Fungal cultures were extracted with acetone, reduced to an aqueous phase under vacuum, and partitioned twice with ethyl acetate. The ethyl acetate extract was chromatographed on a silica gel 60 column eluted with benzene, ethyl ether, ethyl acetate, acetone, and acetonitrile. The koninginin C containing fractions were further purified with a silica gel C_{18} reversed-phase column eluted with different proportions of acetonitrile-water; hydrated silica gel eluted with benzene-ethyl acetate (5:4, v/v). The koninginin C was crystallized as fine white crystals from isooctane.

Biological Activity

The metabolite inhibited the growth of etiolated wheat coleoptiles at 10^{-3} M.

Spectral Data

UV:

λ_{max}^{EtOH} End absorption.

^{13}C NMR:

($CDCl_3$) 25.15(C-1, t); 76.89(C-2, d); 29.01(C-3, t); 69.30(C-4, d); 108.47(C-5, s); 36.15(C-6, d); 20.42(C-7, t); 27.07(C-8, t); 78.92(C-9, d); 79.17(C-10, d); 35.65(C-11, t); 23.32(C-12, t); 29.17(C-13, t); 31.71(C-14, t); 22.54(C-15, t); and 14.04ppm (C-16, q).

Mass Spectrum:

HREIMS: 284.198929*m/e* (M^+); calcd for $C_{16}H_{28}O_4$, 284.198760.

TLC Data

Silica gel 60 thin-layer plates developed with toluene-ethyl acetate-formic acid (5:4:1, v/v/v); R_f, 0.56-0.60; detected as a salmon-pink colored spot when treated with anisaldehyde followed by heating.

Reference

S. R. Parker, H. G. Cutler, and P. R. Schreiner; A Biologically Active Natural Product from *Trichoderma koningii*; Bioscience, Biotech., Biochem., Vol. 59, pp. 1126-1127 (1995).

Common/Systematic Name

Koninginin D

4,8-Dihydroxy-2-(1-hydroxyheptyl)-3,4,5,6,7,8-hexahydro-*2H*-1-benzopyran-5-one

Molecular Formula/Molecular Weight

$C_{16}H_{26}O_5$; MW = 298.17802

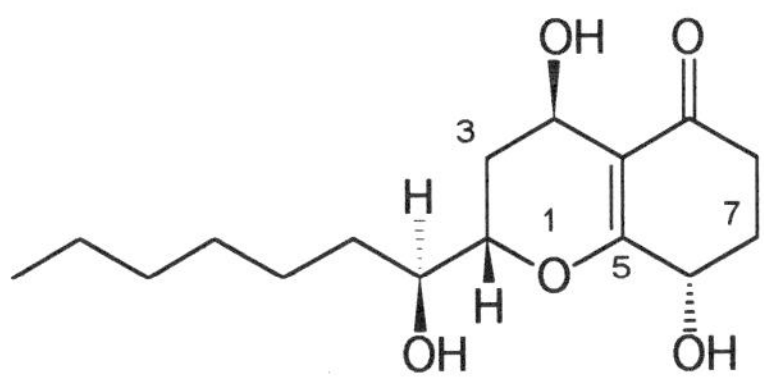

General Characteristics

Crystallized from $CHCl_3$/pentane as solvated needles which collapsed to a powdery solid on recovery. The triol had mp of 122-123°C; $[\alpha]_D$ + 166.9° (*c*=0.3, in $CHCl_3$).

Fungal Source

Trichoderma koningii (IMI 308477) isolated from soil.

Isolation/Purification

The culture broth was extracted with ethyl acetate, dried over anhydrous sodium sulfate and evaporated to dryness. Column chromatography on neutral alumina (activity I) using ethyl acetate as eluting solvent gave 4,8-dihydroxy-2-(1-hydroxyheptyl)-3,4,5,6,7,8-hexahydro-*2H*-1-benzopyran-5-one which was crystallized from chloroform-pentane.

Biological Activity

The metabolite inhibited the growth of the take all fungus, *Gaeumannomyces graminis*, and inhibited the growth of several other soil fungi.

Spectral Data

UV:

λ_{max}^{EtOH} 260nm (log ϵ = 3.9).

IR:

($CHCl_3$) 3600, 3380, 1655, and 1620cm^{-1}.

^{1}H NMR:

($CDCl_3$) 2.38($J_{2a,2b}$=16.9Hz, $J_{2a,3a}$=6.3, $J_{2a,3b}$=4.9, H-2a); 2.64($J_{2b,3a}$=9.4, $J_{2b,3b}$=5.7Hz, H-2b); 1.6($J_{3a,3b}$=14, $J_{3a,4}$=4.7, H-3a); 2.2($J_{3b,4}$=4.7, $J_{4,7}$=1.0Hz, H-3b); 5.81($J_{7,8a}$=3.5, $J_{7,8b}$=2.1Hz), 1.75($J_{8a,8b}$=15, $J_{8a,9}$=3.5Hz, H-8a); 1.99($J_{8b,9}$=2.2, H-8b); 4.15($J_{9,10}$=3.9Hz); 5.08($J_{10,11}$=6.7Hz); 1.3, 1.6; 1.25, 1.25, 1.25, 1.25; and 0.9ppm ($J_{15,16}$=6.5Hz).

^{13}C NMR:

(CDCl$_3$) 194.8(C-1); 31.6(C-2); 26.4(C-3); 66.5(C-4); 170.4(C-5); 112.1(C-6); 59.6(C-7); 29.1(C-8); 74.2(C-9); 72.9(C-10); 29.9(C-11); 25.1(C-12); 29.2(C-13); 32.8(C-14); 22.5(C-15); and 14.0pp (C-16).

Mass Spectrum:

LREIMS: 298(M^+,3%), 280(5), 262(4), 244(8), 236(6), 209(13), 191(7), 181(12), 170(18), and 165*m/e* (39); HREIMS: 298.1759*m/e*, $C_{16}H_{26}O_5$ requires 298.1780.

TLC Data

Silica gel thin-layer plates developed with ethyl acetate; R_f, 0.13; detected under UV light.

Reference

R. W. Dunlop, A. Simon, K. Sivasithamparam, and E. L. Ghisalberti; An Antibiotic From *Trichoderma koningii* Active Against Soilborne Plant Pathogens; J. Natural Products, Vol. 52, pp. 67-74(1989).

Common/Systematic Name

Koninginin E

Molecular Formula/Molecular Weight

$C_{16}H_{26}O_4$; MW = 282.18311

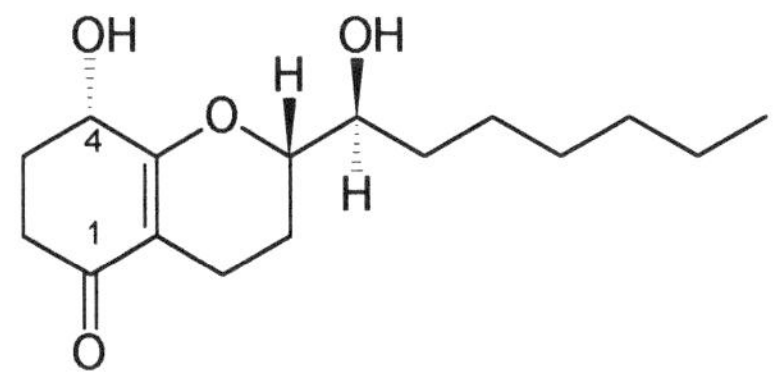

General Characteristics

Off-white amorphous solid from isooctane.

Fungal Source

Trichoderma koningii isolated from *Diffenbachia* sp. (ATCC 46314).

Isolation/Purification

Following incubation, cultures were extracted with acetone. The extract was filtered under vacuum and the filtrate reduced to a non-volatile, aqueous volume. The aqueous volume was extracted three times with an equal volume of ethyl acetate and the combined extract washed with an equal volume of distilled water. After drying over anhydrous sodium sulfate, the ethyl acetate extract was reduced to an oily residue and thoroughly azeotroped with benzene to remove any remaining water. The residue was mixed 1:1 (w/w) with Celite and used to fill a 10ml stainless steel extraction thimble. The contents were made compact using a blunt ended glass rod and the sealed thimble placed in the extraction chamber of an SFX 2-10 extraction module. A ten minute static extraction was followed by a dynamic extraction, at constant pressure, with supercritical carbon dioxide. SFE fractions were dissolved in acetone and analyzed directly by GC-MS. Koninginin E was observed in the methanol amended fractions. The combined koninginin E containing SFE fractions were reduced under nitrogen and applied to a C_{18} column prepared in 1:1 (v/v) acetonitrile-water. The column was eluted stepwise with each of 1:1 (v/v) followed by 3:1 (v/v) acetonitrile-water. Final purification was achieved by precipitation from isooctane solution.

Biological Activity

The metabolite significantly inhibited (60%) the growth of etiolated wheat coleoptiles at 10^{-3} M.

Spectral Data

UV:

λ_{max}^{EtOH} 262nm.

IR:

(KBr) 3450, 3366, 2955, 2925, 2856, 1650, 1620, 1592, 1402, 1379, 1295, 1250, and 1082cm^{-1}.

^{13}C NMR:

($CDCl_3$) 197.6(C-1); 33.3(C-2); 29.0(C-3); 65.9(C-4); 169.3(C-5); 111.4(C-6); 17.6(C-7); 22.8(C-8); 81.3(C-9); 73.4(C-10); 32.7(C-11); 25.1(C-12); 29.3(C-13); 31.8(C-14); 22.6(C-15); and 14.1ppm (C-16).

TLC Data

Silica gel 60 thin-layer plates developed with toluene-ethyl acetate-formic acid (5:4:1, v/v/v); R_f, 0.28-0.30; detected by spraying with acidified anisaldehyde and heating.

References

E. L. Ghisalberti, and C. Y. Rowland; Antifungal Metabolites from *Trichoderma harzianum*; J. Nat. Prod., Vol. 56, pp. 1799-1804(1993).

S. R. Parker, H. G. Cutler, and P. R. Schreiner; Koninginin E: Isolation of a Biologically Active Natural Product from *Trichoderma koningii*; Bioscience Biotech Biochem., Vol. 59, pp. 1747-1749(1995).

26

Curvularins

Curvularin
6-Chlorodehydrocurvularin
11-Acetyldehydrocurvularin
11β-Hydroxycurvularin
trans-Dehydrocurvularin
cis-Dehydrocurvularin
11α-Hydroxylcurvularin
12-Oxocurvularin
Methoxycurvularin
Citreofuran

Common/Systematic Name

Curvularin

Molecular Formula/Molecular Weight

$C_{17}H_{22}O_5$; MW = 306.14672

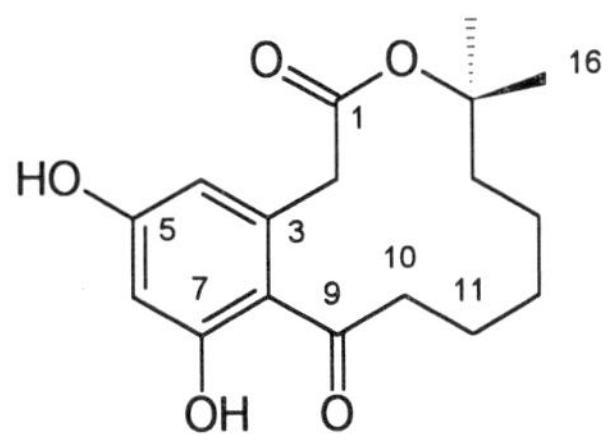

General Characteristics

Curvularin separated as plates from benzene containing a small amount of methanol; mp., 206-206.5°C; $[\alpha]_D^{18}$ -36.3° (*c*=3.8, in ethanol). Curvularin was soluble in ethanol, methanol, dioxane, acetone, pyridine, and concentrated sulfuric acid, moderately soluble in acetic acid and ether, and sparingly soluble in benzene, light petroleum, chloroform, and water. It gave yellow solutions with aqueous ammonia, sodium carbonate, and sodium hydroxide, from which it was recovered unchanged by acidification. Alkaline solutions of curvularin darken rapidly in air, eventually becoming purple. This sensitivity to oxygen in the presence of alkali is also demonstrated by the rapid decomposition of curvularin at its m.p. when heated in open soda-glass capillaries.

Fungal Source

Curvularia sp. (No. F334 in the Nobel Division culture collection; Commonwealth Mycological Institute Cat. No. I.M.I. 52,980), *Drechslera* sp. (tentatively identified as *D. rostrata*, IMI 356287), *Cercospora* spp., *Penicillium citreo-viride B* (hybrid strain derived from IFO 6200 and 4692), *P. gilmanii, Alternaria cinerariae,* and *Penicillium* sp. 511 (isolated from soil).

Isolation/Purification

The culture filtrate of *Curvularia* sp. was decanted from the mycelium and extracted with ethyl ether which, after removal of the solvent, gave crude curvularin as colorless or pale yellow-brown crystals. The crude curvularin was recrystallized from hot benzene containing a small amount of methanol to give plates. *Penicillium* sp. 511 was grown on potato-sucrose-agar for thirty days. The slant cultures were soaked in acetone for one week at room temperature. An aqueous concentrate after evaporating acetone was partitioned against EtOAc. The EtOAc extract was chromatographed on a silica gel column (Wakogel C-100). The column was eluted with 20, 30, 40, and 50% EtOAc/benzene. Purification of curvularin, 8-dehydrocurvularin, 8-hydroxycurvularin, and 8-methoxycurvularin was guided by a sea urchin egg assay.

Biological Activity

Non-specific phytotoxic action, caused stem collapse and vascular necrosis of cut zinnia (*Zinnia elegans* Jacq.) seedlings at concentrations down to 0.1 and 0.33mg/ml respectively within 24hr; caused necrosis of cuttings of Canada thistle (*Cirsium arvense* L.) Scop. within 16 hours; cytotoxic against sea urchin (*Hemicentrotus pulcherrimus*) embryogenesis where it induced barbell-shaped mitotic spindles. It was shown to act on components of the mitotic apparatus and to effectively inhibit cell division.

Spectral Data

UV:

λ_{max}^{EtOH} 223, 272, and 304.5nm (ϵ = 11,300, 6350, and 5100, respectively) and an inflection at 229nm (ϵ = 10,150), unaffected by the addition of hydrochloric acid and 0.005N aqueous sodium hydroxide at 233 and 351nm (ϵ = 10,200 and 9300) and inflections at 258 and 328nm (ϵ = 7,250 and 8,100).

IR:

(KBr) 3300, 1706, 1665, 1615, 1608, 1160, 1050, and 841cm^{-1}.

^{1}H NMR:

(acetone-d_6) 6.37(1H, d, *J*=2.2Hz); 6.33(1H, d, *J*=2.2Hz); 4.90(1H, m); 3.76(1H, d, *J*=15Hz); 3.69(1H, d, *J*=15Hz); 3.09(1H, m); 2.76(1H, m); 1.2~1.8(8H, m); and 1.10ppm (3H, d, *J*=6.3Hz).

Mass Spectrum:

HREIMS: 292.1309*m/e* M$^+$, ($C_{16}H_{20}O_5$, calcd 292.1309).

References

R. G. Coombe, J. J. Jacobs, and T. R. Watson; Constituents of Some *Curvularia* Species; Aust. J. Chem., Vol. 21, pp. 783-788(1968).

A. Kobayoshi, T. Hino, S. Yata, T. J. Itoh, H. Sato, and K. Kawazi; Unique Spindle Poisons, Curvularin and Its Derivatives, Isolated from *Penicillium* Species; Agric. Biol. Chem., Vol. 52, pp. 3119-3123(1988).

S. Lai, Y. Shizuri, S. Uamamura, K. Kawai, Y. Terada, and H. Furukawa; Novel Curvularin-Type Metabolites of a Hybrid Strain Me 0005 Derived from *Penicillium citreo-viride* B. IFO 6200 and 4692; Tet. Lett., Vol. 70, pp. 2241-2244(1989).

O. C. Musgrave; Curvularin. Part I. Isolation and Partial Characterization of a Metabolic Product from a New Species of *Curvularia*; J. Chem. Soc., pp. 4301-4305(1956).

H. Raistrick and F. A. H. Rice; 2,3-Dihydro-3,6-dihydroxy-2-methyl-4-pyrone and Curvularin from *Penicillium gilmanii*; J. Chem. Soc. (C), pp. 3069-3070(1971).

D. J. Robeson and G. A. Strobel; α,β-Dehydrocurvularin and Curvularin from *Alternaria cinerariae*; Z. Naturforsch., Vol. 36, pp. 1081-1083(1981).

Common/Systematic Name

6-Chlorodehydrocurvularin

E-6-Chloro-10,11-dehydrocurvularin

Molecular Formula/Molecular Weight

$C_{16}H_{17}O_5Cl$; MW = 324.07645

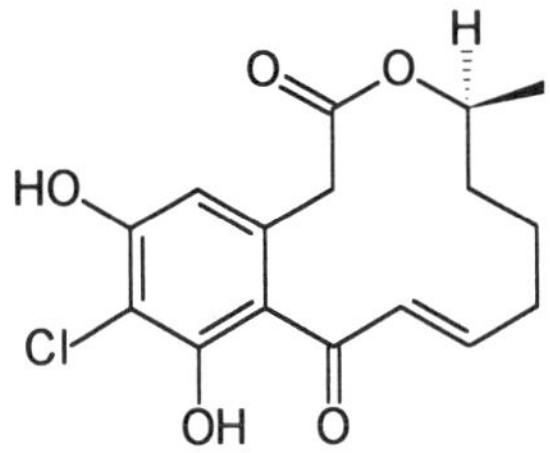

General Characteristics

Crystals from CH_2Cl_2-EtOAc as needles; mp., 189-190°C; $[\alpha]_D$ -51° (*c*=0.2, in EtOH).

Fungal Source

Drechslera species (Deuteromycotina) (tentatively classified as *D. rostrata*) was isolated from soil in which wheat seedlings were growing (culture deposited with the Commonwealth Mycological Institute; IMI 356287). *Drechslera* sp. are an important group of fungal plant pathogens which have been responsible for devastating disease epidemics, particularly on rice, corn, and sorghum.

Isolation/Purification

The fungal culture was grown on Czapek-Dox medium. The liquid medium of a 3-month-old culture of *Drechslera* sp, after filtration from the mycelial mat, was extracted repeatedly with EtOAc. The combined organic layer was dried over Na_2SO_4 and evaporated under reduced pressure to give a residue. The mycelial mat was extracted with MeOH and the extract was washed with EtOAc to give a residue. TLC analysis (Si gel; CH_2Cl_2-EtOAc, 1:1, v/v) of both extracts showed them to be similar and to contain mainly two components. Radial plate chromatography of the EtOAc-soluble portion of the mycelial extract using gradient elution (CH_2Cl_2 to EtOAc) gave two fractions. Fraction 1 appeared to be mainly linoleic acid. Fraction 2 was purified by column chromatography on silica gel to give a crystalline sample of curvularin. The fungal culture was also grown on potato-dextrose broth. A 3-month-old culture of *Drechslera* sp. grown on this medium was similarly processed to give two residues: EtOAc extract and EtOAc-soluble portion of mycelial extract. TLC analysis (Si gel; CH_2Cl_2-EtOAc, 1:1, v/v) of both fractions showed them to be similar. Radial plate chromatography of the EtOAc extract and gradient elution (CH_2Cl_2-EtOAc, 1:1, v/v, to EtOAc) yielded a fraction, corresponding to the major component, 6-chlorodehydrocurvularin, as a solid.

Biological Activity

Biological activity of 6-chlorodehydrocurvularin is not known but the close chemical relationship to the phytotoxic metabolite curvularin suggests it may also be phytotoxic.

Spectral Data

^{1}H NMR:

($CDCl_3$) 6.68(1H, ddd, *J*=16.0, 3.5, 3.5Hz, H-11); 6.66(1H, br d, *J*=16.0Hz, H-10); 6.48(1H, br s, H-4); 4.85(1H, ddq, *J*=2.0, 8.0, 6.5Hz, H-15); 4.02(1H, br d, *J*=17.8Hz, H-2a); 3.54(1H, br d, *J*=17.8, H-2b); and 1.25(3H, d, *J*=6.5Hz, H-16). The following assignments were made with the aid of decoupling techniques: 2.5 and 2.35, H-12; 1.99 and 1.69, H-13; 1.89 and 1.65ppm, H-14; (500MHz, acetone-d_6) 6.77(1H, d, *J*=15.4, H-11); 6.62(1H, ddd, *J*=15.4, 8.9, 4.9Hz, H-10); 6.57(1H, br s, H-4); 4.74(1H, m, H-15); 4.05(1H, br d, *J*=17.6Hz, H-2a); 3.62(1H, br d, *J*=17.6, H-2b); and 1.17ppm (3H, d, *J*=6.4Hz, H-16).

^{13}C NMR:

($CDCl_3$) C-1, 170.9; C-2, 43.7; C-3, 130.9; C-4, 112.5; C-5, 156.1; C-6, 107.5; C-7, 160.8; C-8, 114.9; C-9, 198.3; C-10, 135.9; C-11, 149.7; C-12, 32.8; C-13, 24.1; C-14, 34.0; C-15, 72.8; and C-15 CH_3, 20.0ppm.

Mass Spectrum:

EIMS: 326(15), 324(36) (M^+), 239(29), 237(9), 231(6), 229(14), 226(16), 225(11), 224(24), 223(18), 213(12), 211(37), 210(31), 209(32), 203(12), 201(37), 186(15), 184(46), 160(24), 115(26), 81(60), and 55*m/e* (100).

Reference

E. L. Ghisalberti, and C. Y. Rowland; 6-Chlorodehydrocurvularin, A New Metabolite from a *Drechslera* sp.; Journal of Natural Products (Lloydia), Vol. 56, pp. 2175-2177 (1993).

Common/Systematic Name
11-Acetyldehydrocurvularin

Molecular Formula/Molecular Weight
$C_{18}H_{20}O_6$; MW = 332.12599

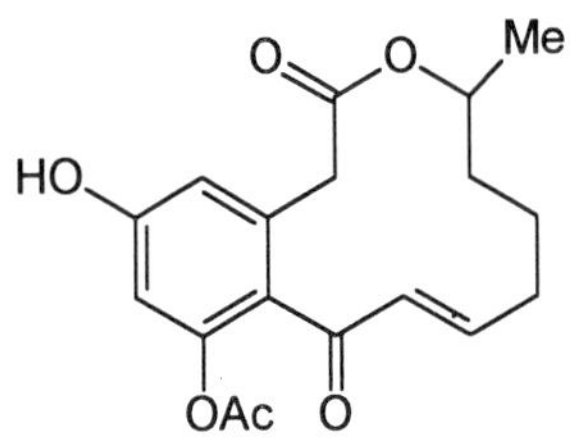

General Characteristics
Crystals; mp., 105-110°C; $[\alpha]_D^{20}$ -31.7° (*c*=0.4, in MeOH).

Fungal Source
Cercospora scirpicola (CBS 104.40).

Biological Activity
Phytotoxic.

Spectral Data

UV:
222, 232sh, 275, and 308nm (ϵ = 10,200, 8,800, 5,800, and 4,900, respectively).

^{1}H NMR:
(acetone-d_6/DMSO-d_6) 1.10(d, 3H, d, *J*=6Hz); 1.2-2.1(6 aliphatic protons); 2.09(Ac), 3.74(2H, q, Ar-$\underline{CH_2}$-CO); 4.94(1H, m, O-$\underline{CH}$-Me); and 6.26-6.48ppm (12 aromatic and 2 vinylic protons).

Reference
G. Assante, R. Locci, L. Camarda, L. Merlini, and G. Nasini; Screening of the Genus *Cercospora* for Secondary Metabolites; Phytochemistry, Vol. 16, pp. 243-247(1977).

Common/Systematic Name

11β-Hydroxycurvularin

Molecular Formula/Molecular Weight

$C_{16}H_{20}O_6$; MW = 308.12599

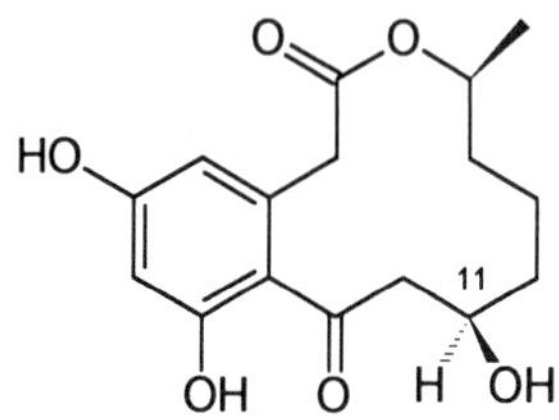

General Characteristics

β-Hydroxycurvularin was obtained as a colorless solid; mp.; 138-140°C; $[\alpha]_D^{24}$ -10.9° (*c*=0.19, in EtOH).

Fungal Source

Penicillium citreo-viride B (hybrid strain derived from IFO 6200 and 4692), *Alternaria tomato,* and *Penicillium* sp. 511 (isolated from soil).

Isolation/Purification

Penicillium sp. 511 was grown on potato-sucrose-agar for thirty days. The slant cultures were soaked in acetone for one week at room temperature. Following evaporation of acetone the resulting aqueous concentrate was partitioned against EtOAc. The EtOAc extract was chromatographed on a silica gel column (Wakogel C-100). The column was eluted with 20, 30, 40, and 50% EtOAc/benzene. Purification of curvularin, 8-dehydrocurvularin, 8-hydroxycurvularin, and 8-methoxycurvularin was guided by a sea urchin egg assay.

Biological Activity

Highly phytotoxic. Marked sporulation-suppressing activity against *Alternaria tomato.* Cytotoxic against sea urchin (*Hemicentrotus pulcherrimus*) embryogenesis where it caused miniature spindles. It was shown to act on components of the mitotic apparatus and to effectively inhibit cell division.

Spectral Data

UV:

λ_{max}^{EtOH} 223, 275, and 306nm (log ε = 3.98, 3.71, and 3.66, respectively).

IR:

(KBr) 3300br, 1700, 1605, and 1585cm^{-1}.

^{1}H NMR:

(CD$_3$OD) 1.16(3H, d, *J*=5.9Hz); 1.32(2H, m); 1.48(2H, m); 1.62(2H, m); 2.96(1H, dd, *J*=9.0, 13.7Hz); 3.62(1H, dd, *J*=5.1, 13.7Hz); 3.67(1H, d, *J*=15.6Hz); 3.98(1H, d, *J*=15.6Hz); 4.08(1H, m); 4.87(1H, m); 6.24(1H, d, *J*=2.2Hz; and 6.28ppm (1H, d, *J*=2.2Hz).

Mass Spectrum:

HREIMS: 308.12230*m/e*, M^+, ($C_{16}H_{20}O_6$); prominent M^+ - 18 ion for loss of H_2O.

References

S. Hyeon, A. Ozaki, A. Suzuki, and S. Tamura; Isolation of α,β-Dehydrocurvularin and β-Hydroxycurvularin from *Alternaria tomato* as Sporulation Suppressing Factors; Agric. Biol. Chem., Vol. 40, pp. 1663-1664(1976).

A. Kobayoshi, T. Hino, S. Yata, T. J. Itoh, H. Sato, and K. Kawazi, Unique Spindle Poisons, Curvularin and Its Derivatives, Isolated from *Penicillium* Species; Agric. Biol. Chem., Vol. 52, pp. 3119-3123(1988).

S. Lai, Y. Shizuri, S. Uamamura, K. Kawai, Y. Terada, and H. Furukawa; Novel Curvularin-Type Metabolites of a Hybrid Strain Me 0005 Derived from *Penicillium citreo-viride* B. IFO 6200 and 4692; Tet. Lett., Vol. 70, pp. 2241-2244(1989).

Common/Systematic Name

α,β-Dehydrocurvularin; *trans*-Dehydrocurvularin; 10,11-Dehydrocurvularin

Molecular Formula/Molecular Weight

$C_{16}H_{18}O_5$; MW = 290.11542

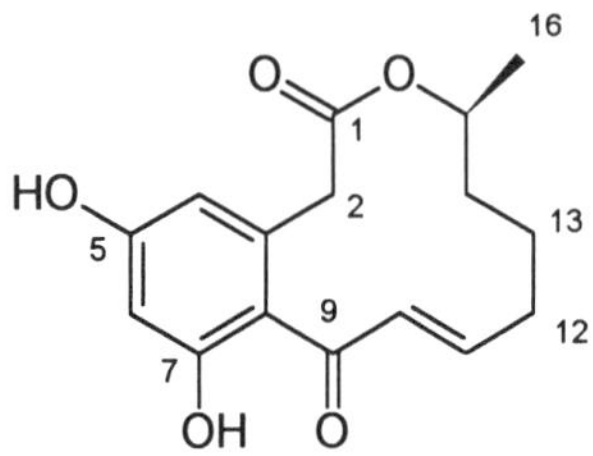

General Characteristics

Pale yellow crystals from benzene; mp., 223-224°C; $[\alpha]_D$ -82.4° (*c*=0.4, in MeOH); $[\alpha]_D$ -82.4° (*c*=0.9, in EtOH).

Fungal Source

Curvularia scirpicola, *Penicillium* sp. 511 (isolated from soil), *Alternaria cinerariae* (syn. *Alternaria senecionis*) a pathogen of the ornamental *Cinerariae* (ATCC 11784), and *Alternaria tomato*.

Isolation/Purification

The fungus was cultured in shaking flasks containing a medium composed of 1% potato starch (Difco) and 2% sucrose at 25°C for 6 days in darkness. The culture filtrate was extracted with ethyl acetate at pH 6.4. After evaporation of the solvent *in vacuo*, the residue was applied to a silica gel column (Wacogel C-100), and the column was eluted with benzene and benzene-ethyl acetate, successively. Marked sporulation-suppressing activity was observed in eluates with benzene containing 10% and 50% ethyl acetate. After evaporation of the solvent from 10% ethyl acetate eluate, the residue was recrystallized from benzene to give an active principle as pale yellow crystals.

Biological Activity

Exhibited antifungal activity against *Cladosporium herbarium*; highly phytotoxic. Marked sporulation-suppressing activity against *A. tomato*.

Spectral Data

UV:

227(ϵ = 14,700), 299(5,710), and 330nm (5,180).

IR:

(KBr) 3428, 3312, 1712, and 1636cm^{-1}.

^{1}H NMR:
1.21, 4-Me; 1.70, 1.90, 6-H_2; 2.37, 2.43, 7-H_2; 1.67, 1.90, 5-H_2; 3.62, 4.07, 1-H_2; 4.70, 4-H; 6.30, 12-H; 6.34, 14-H; 6.74, 9-H; and 6.54ppm, 8-H.

^{13}C NMR:
20.23(C-4Me); 24.90(C-6); 33.15(C-7); 34.83(C-5); 43.59(C-1); 73.09(C-4); 103.16(C-12); 113.90(C-14); 116.21(C-16); 133.11(C-9); 139.63(C-15); 150.35(C-8); 163.51(C-13); 165.92(C-11); 172.47(C-2); and 198.11ppm (C-10).

Mass Spectrum:
HREIMS: 290.1177*m/e* M^+, calcd 290.1157.

References

K. Arai, B. J. Rawlings, Y. Yoshizawa, and J. C. Vederas; Biosyntheses of Antibiotic A26771 B by *Penicillium turbatum* and Dehydrocurvularin by *Alternaria cinerariae*: Comparison of Stereochemistry of Polyketide and Fatty Acid Enoyl Thiol Ester Reductases; J. Am. Chem. Soc., Vol. 111, pp. 3391-3399(1989).

G. Assante, R. Locci, L. Camarda, L. Merlini, and G. Nasini; Screening of the Genus *Cercospora* for Secondary Metabolites; Phytochemistry, Vol. 16, pp. 243-247(1977).

S. Hyeon, A. Ozaki, A. Suzuki, and S. Tamura; Isolation of α,β-Dehydrocurvularin and β-Hydroxycurvularin from *Alternaria tomato* as Sporulation Suppressing Factors; Agric. Biol. Chem., Vol. 40, pp. 1663-1664(1976).

D. J. Robeson, and G. A. Strobel; α,β-Dehydrocurvularin and Curvularin from *Alternaria cinerariae*; Z. Naturforsch., Teil C, Vol. 36, pp. 1081-1083(1981).

Common/Systematic Name

cis-Dehydrocurvularin

Molecular Formula/Molecular Weight

$C_{16}H_{18}O_5$; MW = 290.11542

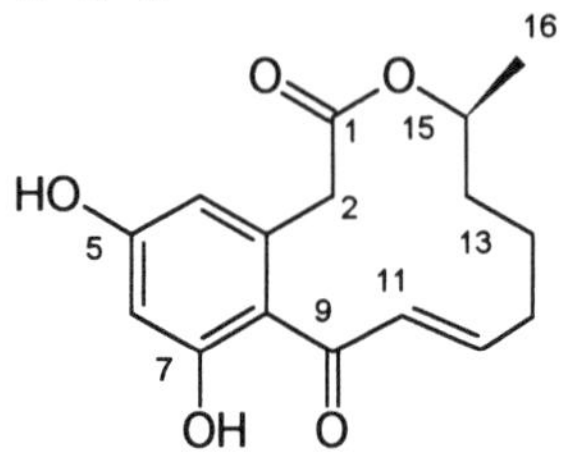

General Characteristics

cis-Dehydrocurvularin was obtained as an amorphous powder; $[\alpha]_D^{22}$ +7.3° (c=0.781, in EtOH).

Fungal Source

Penicillium citreo-viride B (IFO 6200 and 4692).

Isolation/Purification

Polished rice was inoculated with a suspension of mycelium of the hybrid strain ME 0005, incubated stationarily at 25°C for 23 days, and extracted with acetone and then with EtOAc. The combined extracts were partitioned between EtOAc and water. The EtOAc extract was chromatographed on silica gel using a gradient solvent of MeOH-$CHCl_3$ (1-50%). The dark brown oil eluted with $CHCl_3$-MeOH (100:3-4, v/v) was separated by repeated preparative TLC (Kieselgel PF_{254}) using hexane-EtOAc (5:1, v/v), hexane-EtOAc (2:1, v/v), benzene-acetone (5:1, v/v), chloroform-MeOH (15:1, v/v), chloroform-acetone (2:1, v/v), and then chloroform-MeOH (10:1, v/v) to give *cis*-dehydrocurvularin and citreofuran.

Spectral Data

UV:

λ_{max}^{EtOH} 210, 220, 272, 256, and 287nm (log ϵ = 4.21, 4.20, 3.91, and 3.74, respectively).

IR:

(KBr) 3300br, 1705, 1620, and 1590cm^{-1}.

^1H NMR:

(CD_3OD) 1.23(3H, d, J=6.4Hz); 1.77(1H, m); 2.03(1H, m); 2.66(1H, ddd, J=2.9, 11.5, 15.0Hz); 2.84(1H, ddd, J=2.9, 5.9, 15Hz); 3.10(1H, d, J=14.4Hz); 3.20(1H, d, J=14.4Hz); 5.16(1H, m); 6.09(1H, d, J=3.2Hz); 6.21(1H, d, J=3.2Hz); 6.30(1H, d, J=2.4Hz); and 6.34ppm (1H, d, J=2.4Hz).

^{13}C NMR:
(CD_3OD) 21.2(q); 26.5(t); 37.2(t); 42.5(t); 73.9(d); 102.5(d); 107.5(d); 111.0(d); 111.8(s); 112.2(d); 139.7(s); 148.7 (s); 156.0(s); 157.3(s); 159.5(s); and 174.1ppm (s).

Mass Spectrum:
HREIMS: 290.1161*m/e* (M^+, $C_{16}H_{18}O_5$).

Reference

S. Lai, Y. Shizuri, S. Uamamura, K. Kawai, Y. Terada, and H. Furukawa; Novel Curvularin-Type Metabolites of a Hybrid Strain Me 0005 Derived from *Penicillium citreo-viride* B IFO 6200 and 4692; Tet. Lett., Vol. 70, pp. 2241-2244(1989).

Common/Systematic Name

11α-Hydroxycurvularin

Molecular Formula/Molecular Weight

$C_{16}H_{20}O_6$; MW = 308.12599

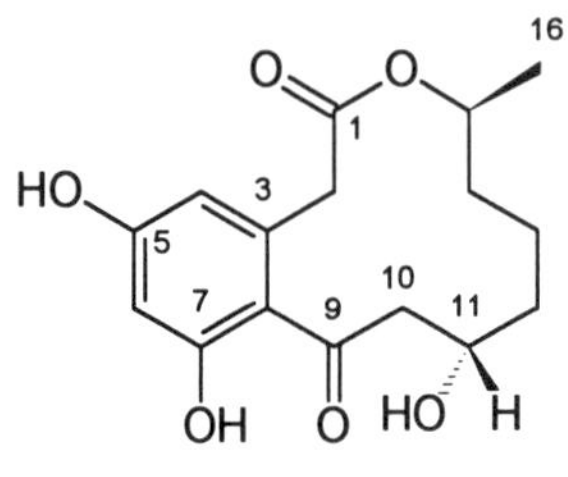

General Characteristics

α-Hydroxycurvularin; mp., 150-152°C; $[\alpha]_D^{26}$ -29.4° (*c*=0.33, in EtOH).

Fungal Source

Penicillium citreo-viride B (hybrid strain derived from IFO 6200 and 4692).

Isolation/Purification

Polished rice was inoculated with a suspension of mycelium of the hybrid strain ME 0005, incubated stationarily at 25°C for 23 days, and extracted with acetone and then with EtOAc. The combined extracts were partitioned between EtOAc and water. The EtOAc extract was chromatographed on silica gel using a gradient solvent of MeOH-$CHCl_3$ (1-50%). The dark brown oil eluted with $CHCl_3$-MeOH (100:9-10, v/v) was separated by repeated preparative TLC (Kieselgel PF_{254}) using $CHCl_3$-MeOH (10:1, v/v) then hexane-methylene chloride-EtOAc-MeOH (3:3:3:1, v/v/v/v) to afford a pale yellow solid, which was further separated by preparative HPLC (Develosil ODS-5) using CH_3CN-H_2O then MeOH-H_2O (1:1, v/v) to afford α-hydroxycurvularin.

Spectral Data

UV:

λ_{max}^{EtOH} 222, 274, and 307nm (log ϵ = 3.98, 3.78, and 3.69, respectively).

IR:

(KBr) 3300br, 1700, 1605, and 1585cm^{-1}.

^1H NMR:

(CD_3OD) 1.13(3H, d, *J*=6.3Hz); 1.45(4H, m); 1.61(1H, m); 1.68(1H, m); 3.23(2H, d, *J*=6.8Hz); 3.59(1H, d, *J*=15.6Hz); 3.97(1H, d, *J*=15.6Hz); 4.09(1H, m); 4.98(1H, m); 6.21(1H, d, *J*=2.2Hz); and 6.28ppm (1H, d, *J*=2.2Hz).

Mass Spectrum:
HREIMS: 308.1242*m/e* M^+, ($C_{16}H_{20}O_6$); prominent M^+ - 18 for loss of H_2O.

Reference

S. Lai, Y. Shizuri, S. Uamamura, K. Kawai, Y. Terada, and H. Furukawa; Novel Curvularin-Type Metabolites of a Hybrid Strain Me 0005 Derived from *Penicillium citreo-viride* B. IFO 6200 and 4692; Tet. Lett., Vol. 70, pp. 2241-2244(1989).

Common/Systematic Name

12-Oxocurvularin

Molecular Formula/Molecular Weight

$C_{16}H_{18}O_6$; MW = 306.11034

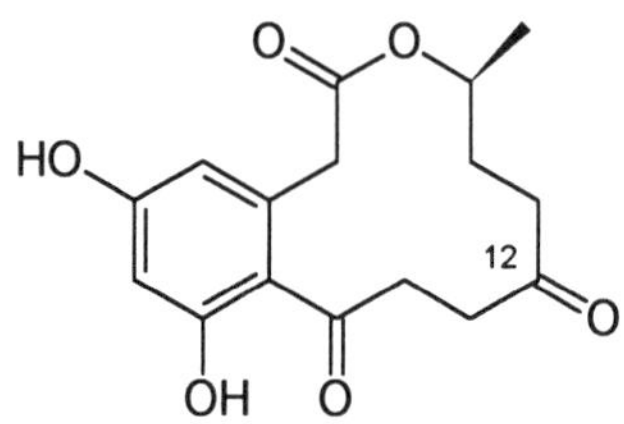

General Characteristics

12-Oxocurvularin was an amorphous powder; $[\alpha]_D^{29}$ -43.5° (*c*=0.47, in EtOH).

Fungal Source

Penicillium citreo-viride B (hybrid strain derived from IFO 6200 and 4692)

Isolation/Purification

Polished rice was inoculated with a suspension of mycelium of the hybrid strain ME 0005, incubated stationarily at 25°C for 23 days, and extracted with acetone and then with EtOAc. The combined extracts were partitioned between EtOAc and water. The EtOAc extract was chromatographed on silica gel using a gradient solvent of MeOH-$CHCl_3$ (1-50%). The dark brown oil eluted with $CHCl_3$-MeOH (100:5-6, v/v) was separated by repeated preparative TLC (Kieselgel PF_{254}) using $CHCl_3$-MeOH (10:1, v/v), hexane-acetone (1:1, v/v), chloroform-MeOH (15:1, v/v), hexane-acetone (2:10, v/v), hexane-EtOAc-MeOH (7:2:1,v/v/v), then benzene-MeOH (4:1, v/v) to give 12-oxocurvularin.

Spectral Data

UV:

λ_{max}^{EtOH} 222, 272, and 299nm (log ϵ = 3.99, 3.69, and 3.60, respectively).

IR:

(KBr) 3300br, 1710, 1608, and 1590cm^{-1}.

^{1}H NMR:

(CD_3OD) 1.13(3H, d, *J*=6.4Hz); 1.72(1H, m); 2.04(1H, m); 2.25(1H, ddd, *J*=2.9, 7.8, 16.6Hz); 2.66(1H, m); 2.73(1H, ddd, *J*=2.3, 9.8, 16.6Hz); 3.01(2H, m); 3.42(1H, m); 3.54(1H, d, *J*=14.9Hz); 3.64(1H, d, *J*=14.9Hz); 4.94(1H, m); 6.20(1H, d, *J*=2.4Hz); and 6.28ppm (1H, d, *J*=2.4Hz).

^{13}C NMR:
(CD_3OD) 19.8(q); 30.5(t); 38.3(t); 38.8(t); 39.7(t); 40.4(t); 71.8(d); 102.4(d); 111.4(d); 120.6(s); 137.1(s); 158.7(s); 161.2 (s); 172.4(s); 208.3(s); and 212.5ppm (s).

Mass Spectrum:
HREIMS: 306.1082*m/e* M^+, ($C_{16}H_{18}O_6$).

Reference

S. Lai, Y. Shizuri, S. Uamamura, K. Kawai, Y. Terada and H. Furukawa, Novel Curvularin-Type Metabolites of a Hybrid Strain Me 0005 Derived from *Penicillium citreo-viride* B. IFO 6200 and 4692; Tet. Lett., Vol. 70, pp. 2241-2244(1989).

Common/Systematic Name

Methoxycurvularin

Molecular Formula/Molecular Weight

$C_{17}H_{22}O_6$; MW = 322.14164

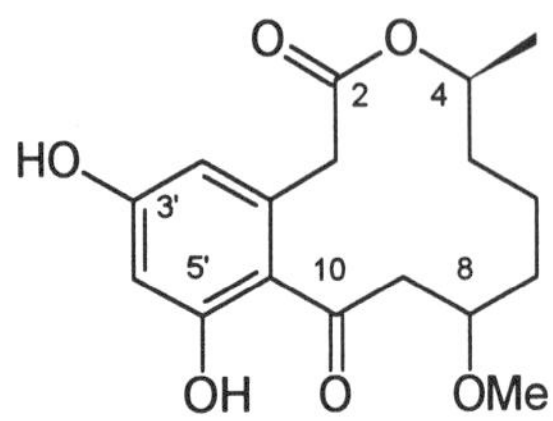

General Characteristics

Methoxycurvularin was obtained as a colorless oil from 50% EtOAc eluate; $[\alpha]_D^{20}$ -13.9° (*c*=3.3, in EtOH).

Fungal Source

Penicillium sp. 511 (isolated from soil).

Isolation/Purification

Penicillium sp. 511 was grown on potato-sucrose-agar for thirty days. The slant cultures were soaked in acetone for one week at room temperature. After evaporation of acetone the resulting aqueous concentrate was partitioned against EtOAc. The EtOAc extract was chromatographed on a silica gel column (Wakogel C-100). The column was eluted with 20, 30, 40 and 50% EtOAc/benzene. Purification of curvularin, 8-dehydrocurvularin, 8-hydroxycurvularin and 8-methoxycurvularin was guided by a sea urchin egg assay.

Biological Activity

Cytotoxic against sea urchin (*Hemicentrotus pulcherrimus*) embryogenesis. It was shown to act on components of the mitotic apparatus and to effectively inhibit cell division.

Spectral Data

UV:

λ_{max}^{EtOH} 223, 237, 275, and 308nm (ϵ = 9,800, 8,200, 6,100, and 15,400, respectively).

IR:

(KBr) 3300br, 1710, 1615, 1595, and 1460cm^{-1}.

^{1}H NMR:

($CDCl_3$) 6.10(1H, d, *J*=3Hz); 6.00(1H, d, *J*=3Hz); 4.70(1H, m); 4.04-3.60(1H, m);

3.92(1H, d, J=16Hz); 3.62(1H, d, J=16Hz); 3.26(3H, s); 3.20(1H, dd, J=3, 15Hz); 2.94(1H, dd, J=2.5, 15Hz); 1.44(6H, m); and 1.08ppm (3H, d, J=6.5Hz).

^{13}C NMR:
(CD_3OD) 203.7, s; 171.3, s; 160.4, s; 137.1, s; 136.1, s; 111.5, d; 101.7, d; 77.0, d; 74.3, d; 56.1, q; 40.0, t; 33.8, t; 32.1, t; 23.3, t; 21.3, t; and 18.9ppm, q.

Mass Spectrum:
EIMS: 322(M^+, 6%), 290(100), 203(57), 177(38), 167(39), 150(59), 121(35), and 69*m/e* (67).

Reference

A. Kobayoshi, T. Hino, S. Yata, T. J. Itoh, H. Sato, and K. Kawazi; Unique Spindle Poisons, Curvularin and Its Derivatives, Isolated from *Penicillium* Species; Agric. Biol. Chem., Vol. 52, pp. 3119-3123(1988).

Common/Systematic Name
Citreofuran

Molecular Formula/Molecular Weight
$C_{16}H_{16}O_5$; MW = 288.09977

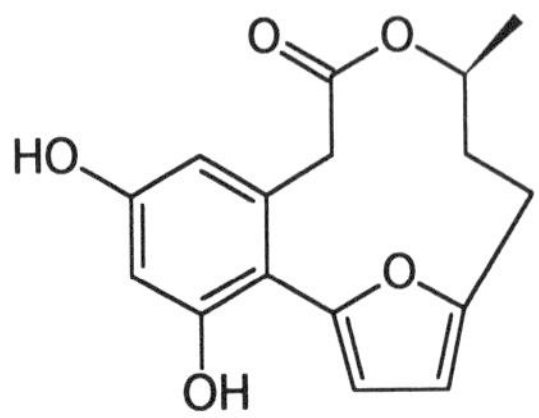

General Characteristics
Melting point, 203-205°C; $[\alpha]_D^{27}$ + 112° (*c*=0.18, in EtOH).

Fungal Source
Penicillium citreo-viride B (hybrid strain derived from IFO 6200 and 4692).

Isolation/Purification
Polished rice was inoculated with a suspension of mycelium of the hybrid strain ME 0005, incubated stationarily at 25°C for 23 days, and extracted with acetone and then with EtOAc. The combined extracts were partitioned between EtOAc and water. The EtOAc extract was chromatographed on silica gel using a gradient solvent of MeOH-$CHCl_3$ (1-50%). The dark brown oil eluted with $CHCl_3$-MeOH (100:3-4, v/v) was separated by repeated preparative TLC (Kieselgel PF_{254}) using hexane-EtOAc (5:1, v/v), hexane-EtOAc (2:1, v/v), benzene-acetone (5:1, v/v), chloroform-MeOH (15:1, v/v), chloroform-acetone (2:1, v/v), and then chloroform-MeOH (10:1, v/v) to give *cis*-dehydrocurvularin and citreofuran.

Spectral Data

UV:
λ_{max}^{EtOH} 210, 220, 272, 256, and 287nm (log ϵ = 4.21, 4.20, 3.91, and 3.74, respectively).

IR:
(KBr) 3300br, 1705, 1620, and 1590 cm^{-1}.

^1H NMR:
(CD_3OD) 1.23(3H, d, *J*=6.4Hz); 1.77(1H, m); 2.03(1H, m); 2.66 (1H, ddd, *J*=2.9, 11.5, 15.0Hz); 2.84(1H, ddd, *J*=2.9, 5.9, 15Hz); 3.10(1H, d, *J*=14.4Hz); 3.20(1H, d, *J*=14.4 Hz); 5.16(1H, m); 6.09(1H, d, *J*=3.2Hz); 6.21(1H, d, *J*=3.2Hz); 6.30(1H, d, *J*=2.4Hz); and 6.34ppm (1H, d, *J*=2.4Hz).

^{13}C NMR:

(CD_3OD) 21.2(q); 26.5(t); 37.2(t); 42.5(t); 73.9(d); 102.5(d); 107.5(d); 111.0(d); 111.8(s); 112.2(d); 139.7(s); 148.7 (s); 156.0(s); 157.3(s); 159.5(s); and 174.1ppm (s).

Mass Spectrum

HREIMS: 288.10150*m/e* (M^+, $C_{16}H_{16}O_5$).

Reference

S. Lai, Y. Shizuri, S. Uamamura, K. Kawai, Y. Terada, and H. Furukawa; Novel Curvularin-Type Metabolites of a Hybrid Strain Me 0005 Derived from *Penicillium citreo-viride* B. IFO 6200 and 4692, Tet. Lett., Vol. 70, pp. 2241-2244(1989).

27

Arcyroxepin A and Related Bisindoles

Arcyroxepin A
Arcyriaflavin B
Arcyriaflavin C
Arcyriarubin B
Arcyriarubin C

Common/Systematic Name

Arcyroxepin A

Molecular Formula/Molecular Weight

$C_{20}H_{11}N_3O_3$; MW = 341.08004

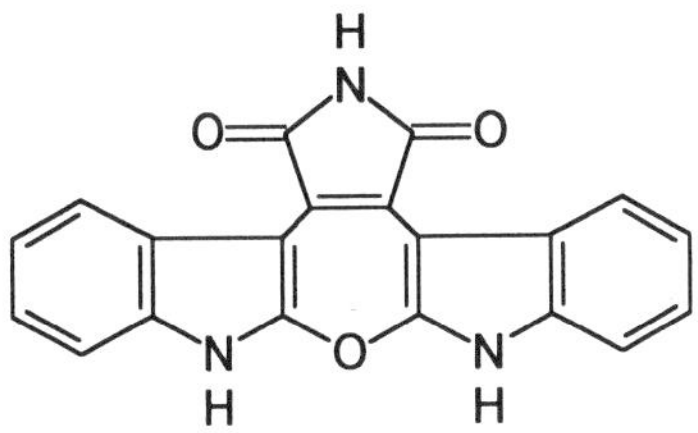

General Characteristics

Red crystals; mp., 268-270°C.

Fungal Source

Arcyria denudata (fruiting bodies).

Isolation/Purification

Extracted with methanol, separated by chromatography on Sephadex LH-20 (eluent methanol).

Biological Activity

In the platelet diffusion test, arcyroxepin A exhibited medium inhibiting action against *Bacillus brevis* and *B. subtilis.*

Spectral Data

UV:

λ_{max}^{MeOH} 471(log ε=3.68), 362(3.51), 283(3.83), 273(3.82), and 226nm (4.44).

IR:

(KBr) 3420 br, s; 1760 w; 1710, ss; 1625, w; 1510, m; 735cm^{-1} s.

^{1}H NMR:

(CD_3COCD_3, TMS) 7.0-7.40(m); 9.82(br. s, 1,1'-NH); 111.22ppm (very br. s, 10-NH).

^{13}C NMR:

(CD_3OD, TMS) 92*(C-3); 112.4(CH); 116.1(CH); 123*; 124.5*; 125.9(CH); 133*; 143.2(C-7a); 153.2*(C-2); and 177.0ppm (C-9).

* Signals are broadened by coalescence; assignment of the signals of coupled carbon by selective decoupling and increment calculation.

TLC Data

Silica gel F_{254}; benzene-ethyl formate-formic acid (10:5:3, v/v/v); R_f=0.42. Reversible color change to violet over NH_3 vapor.

Reference

W. Steglich, B. Steffan, L. Kopanski, and G. Eckhardt; Indole Pigments from the Bodies of the Slime Mold *Arcyria denudata*; Angew. Chem. Int. Ed. Engl., Vol. 19, pp. 459-460(1980).

Common/Systematic Name
Arcyriaflavin B

Molecular Formula/Molecular Weight
$C_{20}H_{11}N_3O_3$; MW = 341.08004

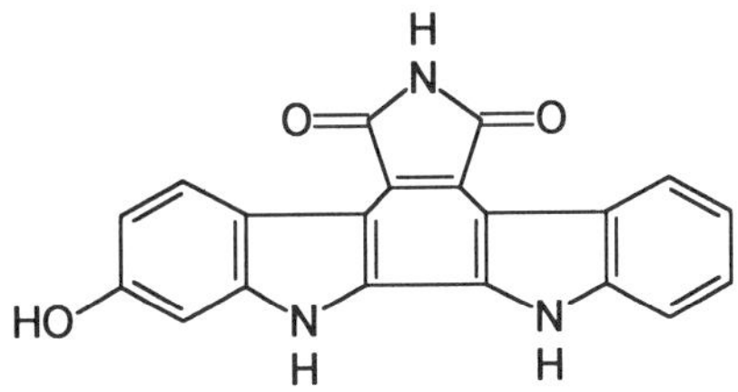

General Characteristics
Pale yellow crystals; mp., 350°C.

Fungal Source
Arcyria denudata (fruiting bodies).

Isolation/Purification
Extracted with methanol, separated by chromatography on Sephadex LH-20 (eluent methanol).

Spectral Data

UV:
λ_{max}^{MeOH} 414(log ϵ=3.19), 323(4.11), 280(3.75), 271(3.70), and 229nm (4.07).

IR:
(KBr) 3420 br, s; 2955 s; 2880 m; 1750 w; 1705 s; and 1620cm^{-1}. br, m.

^{1}H NMR:
(CD_3COCD_3, TMS) 6.93(dd, *J*=8.6 + 2.1Hz, 5-H); 7.16(d, *J*=2.1Hz 7-H); 7.24-7.80(m, 5',6',7'-H); 8.96(d, *J*=8.6Hz, 4-H); 9.15 (broadened d, *J*=7.9Hz, 4'-H); 9.64(br. s, 10-NH); and 11.65ppm (br. s, 1,1'-NH).

TLC Data
Silica gel F_{254}; benzene-ethyl formate-formic acid (10:5:3, v/v/v); R_f=0.33. Detection: strong bright yellow fluorescence at 366nm.

Reference
W. Steglich, B. Steffan, L. Kopanski, and G. Eckhardt; Indole Pigments from the Bodies of the Slime Mold *Arcyria denudata*; Angew. Chem. Int. Ed. Engl., Vol. 19, pp. 459-460(1980).

Common/Systematic Name
Arcyriaflavin C

Molecular Formula/Molecular Weight
$C_{20}H_{11}N_3O_4$; MW = 357.07496

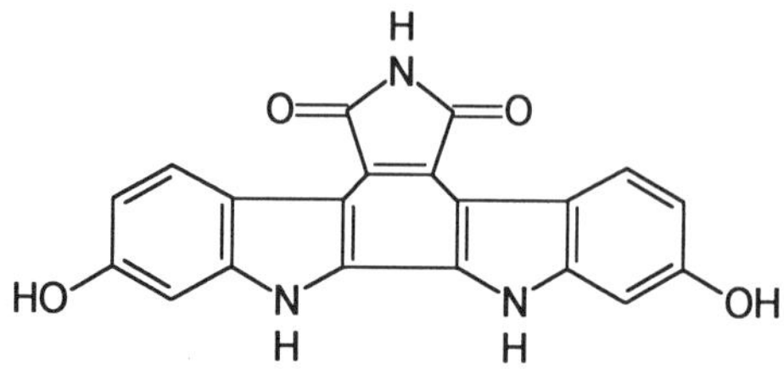

General Characteristics
Pale yellow crystals; mp., 350°C.

Fungal Source
Arcyria denudata (fruiting bodies).

Isolation/Purification
Extracted with methanol, separated by chromatography on Sephadex LH-20 (eluent methanol).

Spectral Data

UV:
λ_{max}^{MeOH} 422(log ε = 3.34), 330.5(4.29), 318(sh, 4.09), 280(3.70), 270(3.66), 255(sh, 3.77), and 229nm (4.14).

IR:
(KBr) 3420 br, s; 2960 s; 2880 m; 1740 w; 1710 s; and 1635cm^{-1} br, s.

^{1}H NMR:
(CD_3COCD_3, TMS) 6.89(dd, *J*=8.4 + 2.2Hz, 5,5'-H); 7.13(d, *J*=2.2Hz, 7,7'-H); 8.93(d, *J*=8.4Hz, 4,4'-H); 9.56(br. s, 10-NH); 11.92ppm (br. s, 1,1'-NH).

TLC Data
Silica gel F_{254}; benzene-ethyl formate-formic acid (10:5:3, v/v/v); R_f=0.21; detection: strong bright yellow fluorescence at 366nm.

Reference
W. Steglich, B. Steffan, L. Kopanski, and G. Eckhardt; Indole Pigments from the Bodies of the Slime Mold *Arcyria denudata*; Angew. Chem. Int. Ed. Engl., Vol. 19, pp. 459-460(1980).

Common/Systematic Name
Arcyriarubin B

Molecular Formula/Molecular Weight
$C_{20}H_{13}N_3O_3$; MW = 343.09569

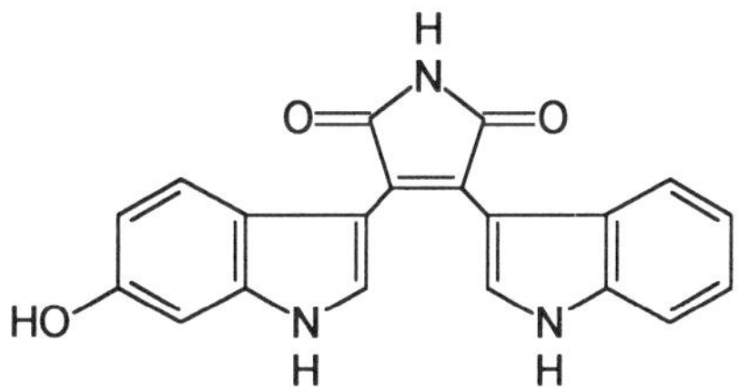

General Characteristics
Red crystals; mp., 154-155°C.

Fungal Source
Arcyria denudata (fruiting bodies).

Isolation/Purification
Extracted with methanol, separated by chromatography on Sephadex LH-20 (eluent methanol).

Biological Activity
In the platelet diffusion test, arcyriarubin B exhibited medium inhibiting action against *Bacillus brevis* and *B. subtilis*.

Spectral Data

UV:
λ_{max}^{MeOH} 465(log ε=3.77), 392(sh, 3.57), and 281nm (3.93).

IR:
(KBr) 3440 br, s; 2960 m; 1760 m; 1715 s; 1630 w; 1538cm^{-1} s.

^{1}H NMR:
(CD_3COCD_3, TMS) 6.24(dd, *J*=8.7 + 2.1Hz, 5-H); 6.72(d, *J*=8.7Hz, 4-H); 6.85 (d, *J*=2.4Hz, 7-H); 6.72-7.05 (m, 5', 6', 7'-H); 7.42("d", *J*=8Hz, 4'H); 7.73 (d, *J*=2.7Hz, 2-H); 7.82(d, *J*=2.9Hz, 2'-H); 9.63(br. s, 10-NH); 10.52(br. s, 1'-NH); 10.78ppm (br. s, 1-NH).

^{13}C NMR:
(CD_3OD, TMS) 97.7(C-7); 107.7(C-3/3'); 110.8(C-5); 112.3(C-7'); 120.7(CH); 120.9(C-3a); 122.7(CH); 123.0(CH); 123.2(C4); 127.3(C-3a'); 128.9(C-2/8');

129.7(C-2'); 129.9(C-8); 137.8(C-7a'); 138.9(C-7a); 154.5(C-6); 175.3ppm (C-9/9').

TLC Data

Silica gel F_{254}; benzene-ethyl formate-formic acid (10:5:3, v/v/v); R_f=0.33.

Reference

W. Steglich, B. Steffan, L. Kopanski, and G. Eckhardt; Indole Pigments from the Bodies of the Slime Mold *Arcyria denudata*; Angew. Chem. Int. Ed. Engl., Vol. 19, pp. 459-460(1980).

Common/Systematic Name
Arcyriarubin C

Molecular Formula/Molecular Weight
$C_{20}H_{13}N_3O_4$; MW = 359.09061

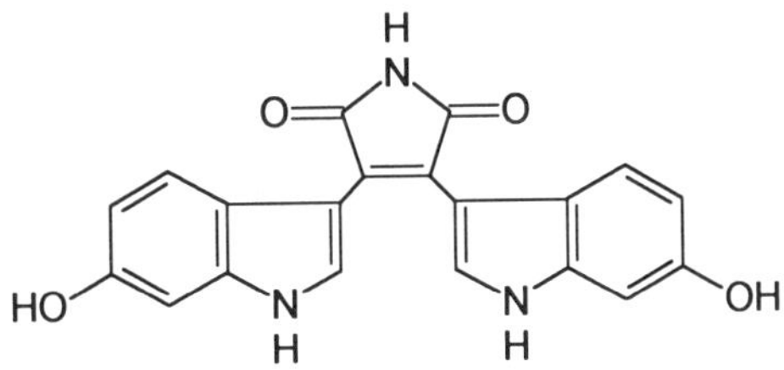

General Characteristics
Red crystals; mp., 205-206°C.

Fungal Source
Arcyria denudata (fruiting bodies).

Isolation/Purification
Extracted with methanol, separated by chromatography on Sephadex LH-20 (eluent methanol).

Biological Activity
In the platelet diffusion test, arcyriarubin C exhibited medium inhibiting action against *Bacillus brevis* and *B. subtilis.*

Spectral Data

UV:
λ_{max}^{MeOH} 474(log ϵ = 3.76) and 283nm (3.93).

IR:
(KBr) 3430 br, s; 1755 w; 1705 s; 1625 m; 1535cm^{-1} s.

^{1}H NMR:
(CD_3COCD_3, TMS) 6.27(dd, *J*=8.4 + 2.2Hz, 5,5'-H); 6.79(d, *J*=8.4Hz, -4,4'-H); 6.85(d, *J*=2.2Hz, 7,7'-H); 7.66(d, *J*=2.5Hz, 2,2'-H); 9.51(br. s, 10-NH); 10.42ppm (br. s, 1,1'-H).

^{13}C NMR:
(CD_3OD, TMS) 97.7(C-7); 107.8(C-3); 110.9(C-5); 121.0(C-3a); 123.3(C-4); 128.7(*J*=190Hz, C-2); 129.2(C-8); 138.8(C-7a); 154.4(C-6); and 175.3ppm (C-9).

TLC Data

Silica gel F_{254}; benzene-ethyl formate-formic acid (10:5:3); R_f=0.21.

Reference

W. Steglich, B. Steffan, L. Kopanski, and G. Eckhardt; Indole Pigments from the Bodies of the Slime Mold *Arcyria denudata*; Angew. Chem. Int. Ed. Engl., Vol. 19, pp. 459-460(1980).

28

Arugosins

Arugosin A
Arugosin B
Arugosin C
Arugosin D
25-*O*-Methylarugosin A

Common/Systematic Name

Arugosin A

Molecular Formula/Molecular Weight

$C_{25}H_{28}O_6$; MW = 424.18859

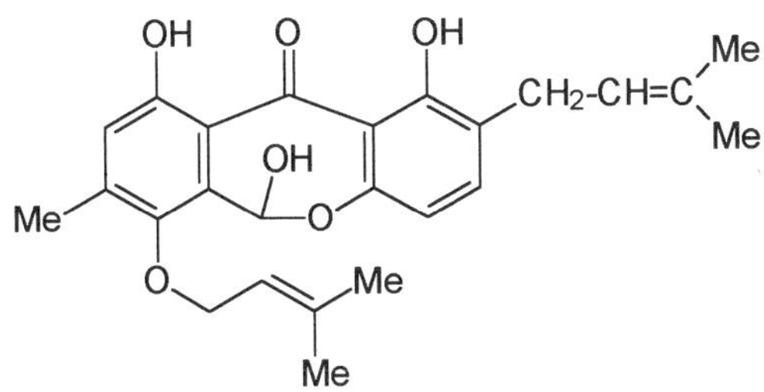

General Characteristics

Optically inactive yellow oil.

Fungal Source

Aspergillus rugulosus (I.M.I. strain 84338)

Isolation/Purification

A. rugulosus was grown for 15 days at 25 °C on a low-nitrogen medium in stationary culture. The mycelium, after separation from the culture fluid, was macerated in a blender with distilled water and extracted with ether. The extract was concentrated *in vacuo* to yield a dark brown viscous residue. Portions of the residue were distributed equally between the first six tubes of a 200-tube automatic counter-current distribution apparatus and subjected to 200 transfers with light petroleum-methanol-water (8:10:1, v/v/v). Tubes 50-80 yielded impure arugosin A as a yellow syrup, which was subjected to preparative thin-layer chromatography on Kieselgel G plates (light petroleum-ethyl acetate, 72:28, v/v). A light yellow band at R_f 0.75 was collected and the metabolite eluted with ethyl acetate. The metabolite was obtained as a viscous yellow oil after concentration *in vacuo.*

Spectral Data

UV:

λ_{max}^{EtOH} 227, 280, and 365nm (log ϵ 4.32, 3.97, 3.86, 4.01, and 395, respectively);

$\lambda_{max}^{EtOH\text{-}NaOH}$ 241, 272, 384, and 427nm (log ϵ 4.58, 4.27, 3.94, and 3.93 respectively).

IR:

($CHCl_3$) 3395(OH) and 1735cm^{-1} (C=O).

^{1}H NMR:

(CDCl$_3$) ArOH, -3.37(1H, s); ArOH, -1.09(1H, s); ArH, 2.76(1H, d, *J*=8.0Hz); ArH, 3.19(1H, s); ArCH(OH)-O, 3.46(1H, d, *J*=5.0Hz); ArH, 3.56(1H, d, *J*=8.0Hz); =CH-CH$_2$-O-, 4.48(1H, m); ArCH$_2$-CH=, 4.73(1H, m); ArCH(OH)-O-, 5.0(1H, d, *J*=5.0Hz); O-CH$_2$-CH, 5.66(2H, m, *J*=4.0, 9.0Hz); ArCH$_3$, 6.74(2H, d, *J*=8.0Hz); ArCH$_3$, 7.68(3H, s); and (CH$_3$)$_2$C=, 8.25τ (6H, d, *J*=7.0Hz).

^{13}C NMR:

(CDCl$_3$) C-1, 194.4; C-3, 6, 8, 12, = 160.1, 155.4, 150.6, 143.6; C-21, 139.0; C-20, 137.4(d); C-15, 135.8; C-16, 130.8; C-7, 128.5; C-5, 121.9; C-10, 120.7; C-2 or 13, 118.9; C-4, 118.8(d); C-2 or 13, 117.3; C-11, 112.5; C-9, 110.0(d); C-25, 92.4; C-19, 70.8(t); and C-14, 17, 18, 22, 23, 24, = 30.2, 27.3, 25.4, 17.3, 17.4, 16.5ppm (m).

Mass Spectrum:

HREIMS: 424.1886*m/e* (M$^+$); $C_{25}H_{28}O_6$ requires 424.1886.

References

J. A. Ballantine, V. Ferrito, C. H. Hassall, and M. L. Jenkins; The Biosynthesis of Phenols. Part XXIV. Arugosin C, a Metabolite of a Mutant Strain of *Aspergillus rugulosus*; J. Chem. Soc., Perkin I, pp. 1825-1830(1973).

J. A. Ballentine, D. J. Francis, C. H. Hassall, and J. L. C. Wright; The Biosynthesis of Phenols. Part XXI. The Molecular Structure of Arugosin, a Metabolite of a Wild-type Strain of *Aspergillus rugulosus*; J. Chem. Soc. (C), pp. 1175-1182(1970).

W. B. Turner and D. C. Aldridge; Fungal Metabolites II; Academic Press, New York, New York, 631 pp. (1983).

Common/Systematic Name

Arugosin B

Molecular Formula/Molecular Weight

$C_{25}H_{28}O_6$; MW = 424.18859

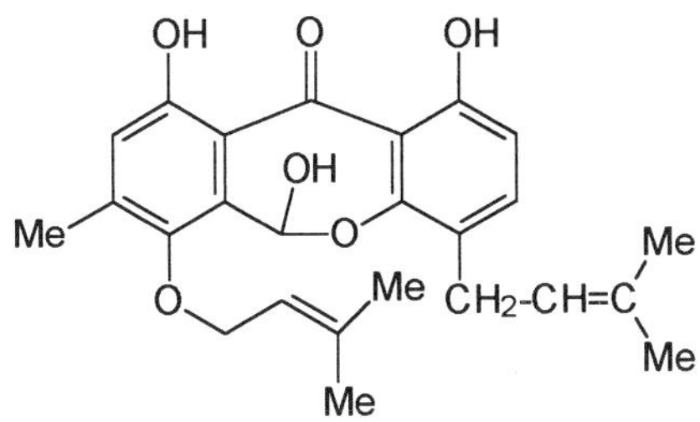

General Characteristics

Optically inactive yellow oil.

Fungal Source

Aspergillus rugulosus (I.M.I. strain 84338)

Isolation/Purification

A. rugulosus was grown for 15 days at 25°C on a low-nitrogen medium in stationary culture. The mycelium, after separation from the culture fluid, was macerated in a blender with distilled water and extracted with ether. The extract was concentrated *in vacuo* to yield a dark brown viscous residue. Portions of the residue were distributed equally between the first six tubes of a 200-tube automatic counter-current distribution apparatus and subjected to 200 transfers with light petroleum-methanol-water (8:10:1, v/v/v). Tubes 50-80 yielded impure arugosin B as a yellow syrup, which was subjected to preparative thin-layer chromatography on Kieselgel G plates (light petroleum-ethyl acetate, 72:28, v/v). A light yellow band at R_f 0.60 was collected and the metabolite eluted with ethyl acetate. The metabolite was obtained as a viscous yellow oil after concentration *in vacuo*.

Spectral Data

UV:

λ_{max}^{EtOH} 227, 280, and 365nm (log ϵ 4.32, 3.97, 3.86, 4.01, and 395, respectively;

$\lambda_{max}^{EtOH\text{-}NaOH}$ 241, 272, 384, and 427nm (log ϵ = 4.58, 4.27, 3.94, and 3.93).

IR:

($CHCl_3$) 3395(OH) and 1735cm^{-1} (C=O).

^{1}H NMR:

(CDCl$_3$) ArOH, -2.78(1H, s); ArOH, -0.87(1H, s); ArH, 2.76(1H, d, *J*=8.0Hz); ArH, 3.19(1H, s); ArCH(OH)-O, 3.46(1H, d, *J*=5.0Hz); ArH, 3.56(1H, d, *J*=8.0Hz); =CH-CH$_2$-O-, 4.48(1H, m); ArCH$_2$-CH=, 4.73(1H, m); ArCH(OH)-O-, 5.0(1H, d, *J*=5.0Hz); O-CH$_2$-CH, 5.66(2H, m, *J*=4.0, 9.0Hz); ArCH$_3$, 6.74(2H, d, *J*=8.0Hz); ArCH$_3$, 7.68(3H, s); and (CH$_3$)$_2$C=, 8.25τ (6H, d, *J*=7.0Hz).

^{13}C NMR:

(CDCl$_3$) C-1, 194.4; C-3, 6, 8, 12, = 159.4, 154.8, 149.9, 143.5; C-21, 138.8; C-20, 137.4(d); C-15, 135.2; C-16, 130.6; C-5, 121.0; C-10, 121.1; C-2 or, 13, 118.9; C-4, 118.8(d); C-2 or 13, 116.3; C-11, 116.0; C-9, 108.4(d); C-25, 91.6; C-19, 70.1(t); and C-14, 17, 18, 22, 23, 24, = 28.2, 27.3, 25.0, 17.3, 17.4, 16.5ppm (m).

Mass Spectrum:

HREIMS: 424.1886*m/e* (M$^+$); $C_{25}H_{28}O_6$ requires 424.1886.

References

J. A. Ballantine, V. Ferrito, C. H. Hassall, and M. L. Jenkins; The Biosynthesis of Phenols. Part XXIV. Arugosin C, a Metabolite of a Mutant Strain of *Aspergillus rugulosus*; J. Chem. Soc., Perkin I,, pp. 1825-1830(1973).

J. A. Ballentine, D. J. Francis, C. H. Hassall, and J. L. C. Wright; The Biosynthesis of Phenols. Part XXI. The Molecular Structure of Arugosin, a Metabolite of a Wild-type Strain of *Aspergillus rugulosus*; J. Chem. Soc. (C), pp. 1175-1182(1970).

W. B. Turner and D. C. Aldridge; Fungal Metabolites II; Academic Press, New York, New York, 631 pp. (1983).

Common/Systematic Name

Arugosin C

1,12*a*-Dihydro-6,8-dihydroxy-1-(1-hydroxy-1-methylethyl)-4-methyl-9-(3-methylbut-2-enyl[1]benzopyrano[4,5-*bc*][1]benzoxepin-7(2*H*)-one

Molecular Formula/Molecular Weight

$C_{25}H_{28}O_6$; MW = 424.18859

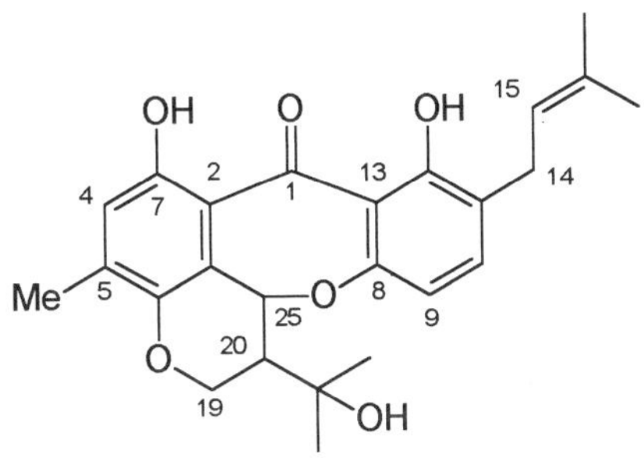

General Characteristics

Obtained as a golden yellow oil.

Fungal Source

A mutant strain of *Aspergillus rugulosus* (A.R.M. 325).

Isolation/Purification

A mutant strain of *A. rugulosus* was grown for 14 days at 25°C on a low-nitrogen medium in stationary culture. The mycelium, after separation from the culture fluid, was macerated in a blender with distilled water and extracted with ether. The extract was concentrated *in vacuo* to yield a dark brown viscous residue. Portions of the residue were distributed equally between the first six tubes of a 200-tube automatic counter-current distribution apparatus and subjected to 200 transfers with light petroleum-methanol-water (10:10:1, v/v/v). Tubes 50-80 yielded impure arugosin C as a yellow syrup, which was subjected to preparative TLC on Kieselgel G plates. A light yellow band at R_f 0.55 was collected and the metabolite eluted with ethyl acetate. The metabolite was further purified by gel filtration through Sephadex LH_{20} in methanol to yield pure arugosin C.

Spectral Data

UV:

λ_{max}^{EtOH} 227, 271, 293, 309, and 407nm (log ϵ 4.34, 3.94, 3.97, 4.01, and 395, respectively).

IR:

($CHCl_3$) 3600(OH), 3440br (OH), and 1620cm^{-1} (C=O).

CD:
(MeOH, *c*= 8.4) 409, 307, 262, and 223nm (mol. ellipticity -12,640, +12,060, +10,200, and +44,220).

ORD:
(*c*=8.4, in MeOH) 446, 371, 333, 295, 274 and 236nm (molecular rotation -4,950, -16,800, -25,500, -9,100, +10,900, and +23,600°).

^{1}H NMR:
($CDCl_3$) ArOH, -3.85(s); ArOH, -0.69(s); ArH, 2.76(d, *J*=8.0Hz); ArH, 3.23(s); ArH, 3.64(d, *J*=8.0Hz); ArCH_2-CH=, 4.72(m); ArCHO, 4.96(d, *J*=4.0Hz); O-CH_2-CH, 5.76(m, *J*=4.0, 9.0Hz); ArCH_2, 6.72(d, *J*=8.0Hz); O-CH_2-CH-CH, 7.67(m, *J*=4.0, 9.0Hz); ArCH_3, 7.78(s); $(CH_3)_2C$=, 8.28(d, *J*= 7.0Hz); C-OH, 8.44(s); $(CH_3)_2C$-O, 8.70 (s); and $(CH_3)_2C$-O, 8.75τ (s).

^{13}C NMR:
($CDCl_3$) C-1, 194.0; C-3, 6,8, 12, = 160.8, 156.6, 153.3, 143.0; C-15, 135.3; C-7, 134.2; C-16, 130.7; C-5, 122.1; C-10, 120.1; C-2 or, 13, 118.7; C-4, 118.2(d); C-2 or 13, 117.8; C-11, 110.8; C-9, 107.2(d); C-25, 73.1; C-21, 69.8; C-19, 64.2(t); C-20, 48.7(d); and C-14, 17, 18, 22, 23, 24, = 28.3, 27.7, 27.3, 25.3, 17.4, 16.1ppm.

Mass Spectrum:
HREIMS: 424.1886*m/e* (M^+) $C_{25}H_{28}O_6$ requires 424.1886; LREIMS: 424(31%), 366(10), 349(7), 293(11), 205(14), 203(17), 189(54), 188 (100), 163(12), 162(58), 161(21), 149(25), and 59*m/e* (26).

References

J. A. Ballantine, V. Ferrito, C. H. Hassall, and M. L. Jenkins; The Biosynthesis of Phenols. Part XXIV. Arugosin C, a Metabolite of a Mutant Strain of *Aspergillus rugulosus*; J. Chem. Soc., Perkin I, pp. 1825-1830(1973).

W. B. Turner and D. C. Aldridge; Fungal Metabolites II; Academic Press, New York, New York, 631 pp. (1983).

Common/Systematic Name

Arugosin D

1,12*a*-Dihydro-6,8-dihydroxy-1-(1-hydroxy-1-methylethyl)-4-methyl-9-(2-hydroxy-3-methylbut-3-enyl)[1]benzopyrano[4,5-*bc*][1]benzoxepin-7(2*H*)-one

Molecular Formula/Molecular Weight

$C_{25}H_{28}O_7$; MW = 440.18350

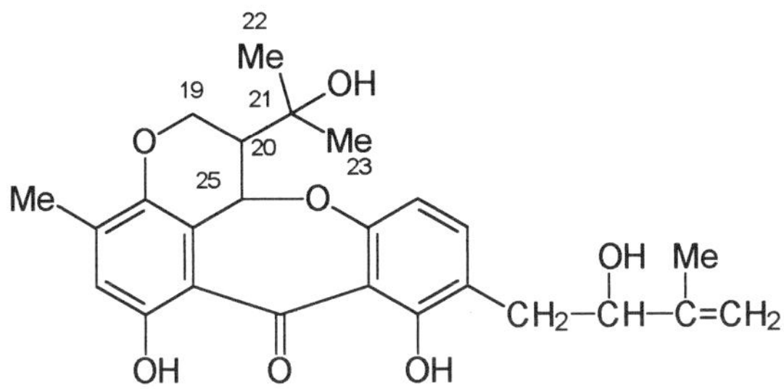

General Characteristics

Isolated as a viscous yellow oil; R_f 0.40 [ether-benzene (1:1)]; $[\alpha]_D$ -31.2° (*c*=0.95, in $CHCl_3$).

Fungal Source

Aspergillus variecolor

Isolation/Purification

The cultures of the individual strains of *A. variecolor* were extracted with light petroleum. The isolations of the mycelial metabolites from the methanol-soluble fractions of the light petroleum extracts were carried out as for shamixanthone and tajixanthone. Arugosin D was isolated from the mycelium by preparative TLC on silica gel GF (Merck).

Spectral Data

^1H NMR:

($CDCl_3$) 1-OH, -0.60 or -3.88; 3-H, 2.70; 5-H, 3.20; 14-CH_2, ca. 7.10; 15-H, ca. 5.65; 17 and 18-CH_3, 8.20(5.00, 5.16); 19-CH_2, ca. 5.70; 20-H, ca. 7.58; 23-CH_2, 24-CH_3, 8.26, 8.31; 24-CH_3, 7.80; 25-CH, 4.90; 4-H, 3.58; 11-OH, -3.88 or -0.60τ.

Mass Spectrum:

EIMS: 440.182(M^+) $C_{25}H_{28}O_7$ requires 440.182.

References

K. K. Chexal, J. S. E. Holker, and T. J. Simpson; The Biosynthesis of Fungal Metabolites. Part VI. Structures and Biosynthesis of Some Minor Metabolites from Variant Strains of *Aspergillus variecolor*; J. C. S. Perkin I, pp. 549-554(1975).

J. S. E. Holker, R. D. Lapper, and T. J. Simpson; The Biosynthesis of Fungal Metabolites. Part IV Tajixanthone: ^{13}C NMR Spectrum and Feedings with [1-^{13}C]- and [2-^{13}C]-Acetate; J. C. S. Perkin I, 2135-2142(1974).

Common/Systematic Name

25-*O*-Methylarugosin A

1,10-Dihydroxy-6-methoxy-8-methyl-2-(3-methylbut-2-enyl)-7-(3-methylbut-2-enyloxy)-dibenz[*b*,*e*]oxepin-11(6*H*)-one

Molecular Formula/Molecular Weight

$C_{26}H_{30}O_6$; MW = 438.20424

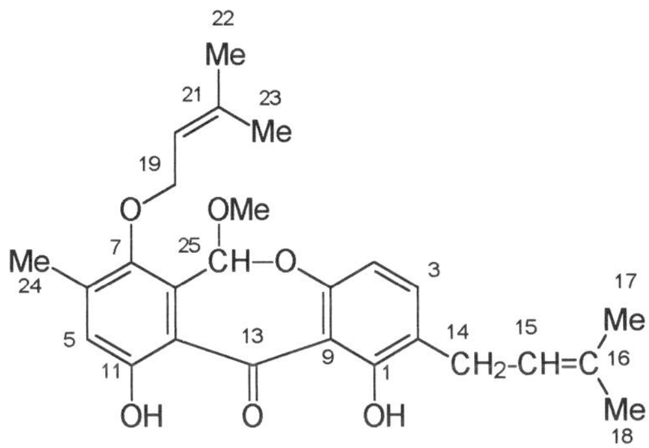

General Characteristics

Isolated as a viscous yellow oil; R_f 0.50 (benzene).

Fungal Source

Aspergillus variecolor

Isolation/Purification

The cultures of the individual strains of *A. variecolor* were extracted with light petroleum. The isolations of the mycelial metabolites from the methanol-soluble fractions of the light petroleum extracts were carried out as for shamixanthone and tajixanthone. 25-*O*-Methylarugosin A was isolated from the mycelium by preparative TLC on silica gel GF (Merck).

Spectral Data

^{1}H NMR:

($CDCl_3$) 1-OH, -0.83 or -2.43; 3-H, 2.71; 5-H, 3.14; 14-CH_2, 6.65; 15-H, 4.68; 17 and 18-CH_3, 8.12, 8.18; 19-CH_2, 5.67, 5.74; 20-H, 4.48; 23-CH_2, 24-CH_3, 8.26, 8.31; 24-CH_3, 7.66; 25-CH, 3.63; 4-H, 3.50; 11-OH, -2.43 or -0.83; and 25-OMe, 5.45τ.

Mass Spectrum:

EIMS: 438.207(M^+) $C_{26}H_{30}O_6$ requires 438.204.

References

K. K. Chexal, J. S. E. Holker, and T. J. Simpson; The Biosynthesis of Fungal Metabolites. Part VI. Structures and Biosynthesis of Some Minor Metabolites from Variant Strains of *Aspergillus variecolor*; J. C. S. Perkin I, pp. 549-554(1975).

J. S. E. Holker, R. D. Lapper, and T. J. Simpson; The Biosynthesis of Fungal Metabolites. Part IV Tajixanthone: ^{13}C NMR Spectrum and Feedings with [1-^{13}C]- and [2-^{13}C]-Acetate; J. C. S. Perkin I, 2135-2412(1974).

29

Hericenes and Hericenones

- Hericene A
- Hericene B
- Hericene C
- Hericenone A
- Hericenone B
- Hericenone E
- Hericenone F
- Hericenone G

Common/Systematic Name

Hericene A

Palmitic acid ester of 4(3',7'-dimethyl-2',6'-octadienyl)-2-formyl-3-hydroxyl-5-methoxylbenzyl alcohol

Molecular Formula/Molecular Weight

$C_{35}H_{56}O_5$; MW = 556.41278

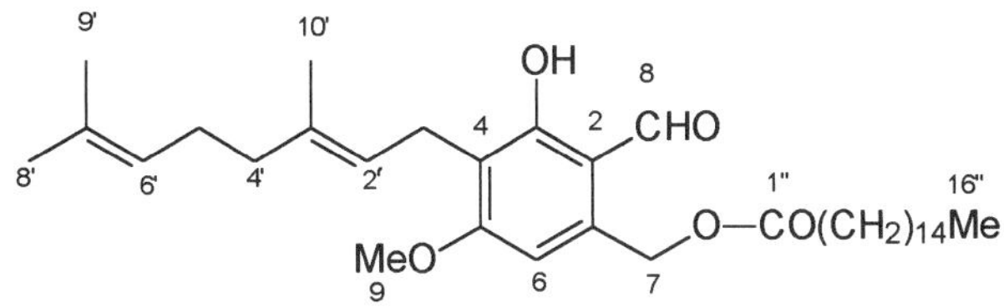

General Characteristics

Hericene A was isolated as an oil.

Fungal Source

Hericium erinaceum (strain CBS 233.87)

Isolation/Purification

Fungal cultures were extracted with EtOAc containing 1% MeOH and the extracts were evaporated to give a mixture of crude metabolites. The mixture was chromatographed on a silica gel column using hexane-EtOAc (1:1, v/v) to yield hericenes A-C, fatty acids (palmitic, linoleic, and oleic acids); the residual material was then eluted with EtOAc containing MeOH (1%) to give erinapyrone C. Further purification of the above mixture of compounds by preparative TLC (RP_{18} plates Me_2CO-H_2O, 8:1, v/v) and HPLC with a Si-60 column [250 X 50mm, 6ml/min, hexane-EtOAc (98:2)] with detection at 298nm gave hericenes A-C in a 5:3:2 ratio.

Spectral Data

IR:

(thin film) 3450, 1730, 1630, and 1570cm^{-1}.

^{1}H NMR:

($CDCl_3$) H-6, 6.52(s); H-7, 5.32(s); H-8, 10.10(s); H-9, 3.91(s); 3-OH, 12.37(s); H-1', 33.4(br d, *J*=7.0Hz); H-2', 5.17(br t, *J*=7.0Hz); H-4', 1.97(m); H-5', 2.03 (m); H-6', 5.05(br t, *J*=6.8Hz); H-8', 1.63(br s); H-9', 1.60(br s); H-10', 1.77(br s); H-2", 2.33(t, *J*=7.5Hz); H-3", 1.60(m); and H-16", 0.88ppm (t, *J*=6.5Hz).

^{13}C NMR:
($CDCl_3$) C-1, 138.41; C-2, 112.87; C-3, 162.88; C-4, 118.06; C-5, 163.47; C-6, 105.60; C-7, 62.96; C-8, 193.09; C-9, 55.90; C-1', 21.35; C-2', 121.19; C-3', 131.23*; C-4', 39.77; C-5', 26.67; C-6', 124.36; C-7', 135.75*; C-8', 25.67; C-9', 17.65; C-10', 16. 10; C-1", 173.21; C-2", 34.23; and C-16", 14.14ppm.

* Assignments may be reversed.

Mass Spectrum:
EIMS: 556(M^+), 300, and 257*m/e*; CIMS: 557(M^+ + 1)

Reference

A. Arnone, R. Cardillo, G. Nasini, and O. Vajna De Pava; Secondary Mold Metabolites: Part 46. Hericenes A-C and Erinapyrone C, New Metabolites Produced by the Fungus *Hericium erinaceum*; J. Nat. Prod., Vol. 57, pp. 602-606(1994).

Common/Systematic Name

Hericene B

Oleic acid ester of 4(3',7'-dimethyl-2',6'-octadienyl)-2-formyl-3-hydroxyl-5-methoxylbenzyl alcohol

Molecular Formula/Molecular Weight

$C_{37}H_{58}O_5$; MW = 582.42843

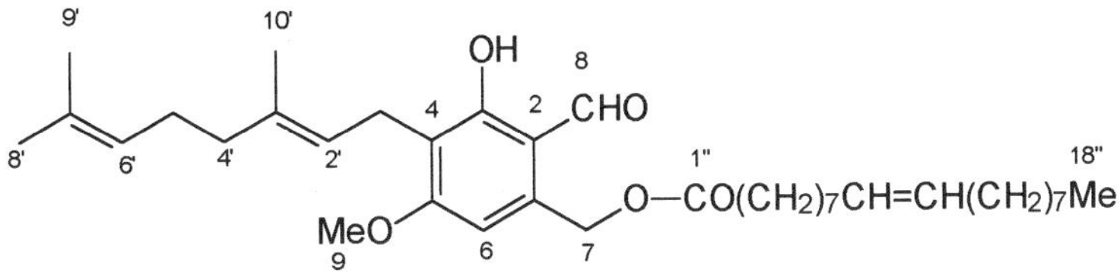

General Characteristics

Hericene B was isolated as an oil.

Fungal Source

Hericium erinaceum (strain CBS 233.87).

Isolation/Purification

Fungal cultures were extracted with EtOAc containing 1% MeOH and the extracts were evaporated to give a mixture of crude metabolites. The mixture was chromatographed on a silica gel column using hexane-EtOAc (4:1, v/v) to yield hericenes A-C, fatty acids (palmitic, linoleic, and oleic acids); the residual material was then eluted with EtOAc containing MeOH (1%) to give erinapyrone C. Further purification of the above mixture of compounds by preparative TLC (RP_{18} plates Me_2CO-H_2O, 8:1, v/v) and HPLC with a Si-60 column [250 X 50mm, 6ml/min, hexane-EtOAc (98:2, v/v)] with detection at 298nm gave hericenes A-C in a 5:3:2 ratio.

Spectral Data

^1H NMR:

($CDCl_3$) The data of the alcoholic portion of hericene B paralleled those of hericene A; the vinylic protons of the oleic portion resonated at 5.34ppm, while the remaining protons resonated between 2.4 and 0.8ppm.

^{13}C NMR:

($CDCl_3$) The data of the alcoholic portion of hericene B paralleled those of hericene A; the vinylic carbons of the oleic portion resonated at 130.02 and 129.71ppm, while the remaining carbons resonated between 32 and 14ppm.

Mass Spectrum:

EIMS: 582(M^+), 300, and 283*m/e*.

Reference

A. Arnone, R. Cardillo, G. Nasini, and O. Vajna De Pava; Secondary Mold Metabolites: Part 46. Hericenes A-C and Erinapyrone C, New Metabolites Produced by The Fungus *Hericium erinaceum*; J. Nat. Prod., Vol. 57, pp. 602-606(1994).

Common/Systematic Name

Hericene C

Steric acid ester of 4(3',7'-dimethyl-2',6'-octadienyl)-2-formyl-3-hydroxyl-5-methoxylbenzyl alcohol

Molecular Formula/Molecular Weight

$C_{37}H_{60}O_5$; MW = 584.44408

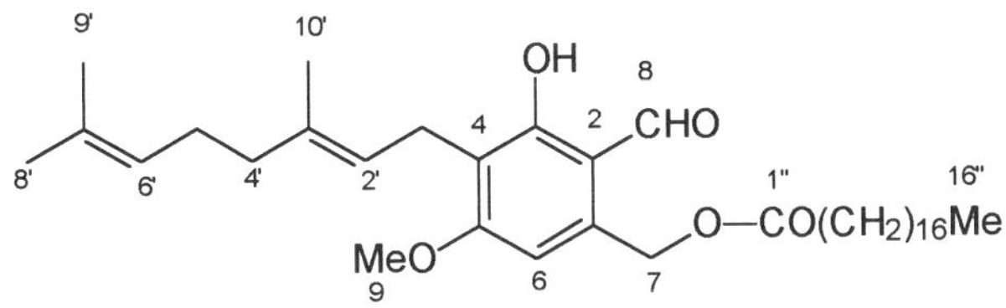

General Characteristics

Hericene C was isolated as an oil.

Fungal Source

Hericium erinaceum (strain CBS 233.87)

Isolation/Purification

Fungal cultures were extracted with EtOAc containing 1% MeOH and the extracts were evaporated to give a mixture of crude metabolites. The mixture was chromatographed on a silica gel column using hexane-EtOAc (4:1, v/v) to yield hericenes A-C, fatty acids (palmitic, linoleic, and oleic acids); the residual material was then eluted with EtOAc containing MeOH (1%) to give erinapyrone C. Further purification of the above mixture of compounds by preparative TLC (RP_{18} plates Me_2CO-H_2O, 8:1, v/v) and HPLC with a Si-60 column [250 X 50mm, 6ml/min, hexane-EtOAc (98:2, v/v)] with detection at 298nm gave hericenes A-C in a 5:3:2 ratio.

Spectral Data

^{1}H NMR:

($CDCl_3$) The data of the alcoholic portion of hericene B and C paralleled those of hericene A; the vinylic protons of the oleic portion of B resonated at 5.34ppm, while the remaining protons of B and C resonated between 2.4 and 0.8ppm.

^{13}C NMR:

($CDCl_3$) The data of the alcoholic portion of hericene B and C paralleled those of hericene A; the vinylic carbons of the oleic portion of B resonated at 130.02 and 129.71ppm, while the remaining carbons of hericenes A and C resonated between 32 and 14ppm.

Mass Spectrum:
EIMS: 584(M^+), 300, and 285*m/e*.

Reference

A. Arnone, R. Cardillo, G. Nasini, and O. Vajna De Pava; Secondary Mold Metabolites: Part 46. Hericenes A-C and Erinapyrone C, New Metabolites Produced by the Fungus *Hericium erinaceum*; J. Nat. Prod., Vol. 57, pp. 602-606(1994).

Common/Systematic Name

Hericenone A

Molecular Formula/Molecular Weight

$C_{19}H_{22}O_5$; MW = 330.14672

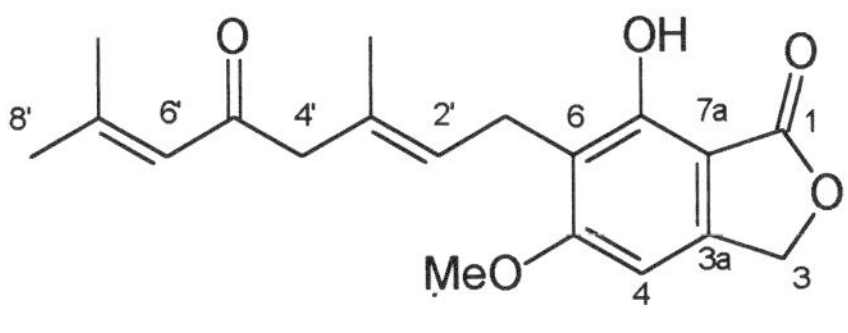

General Characteristics

Colorless crystals from chloroform; mp., 100-102°C; reacted with Folin-Ciocalteu reagent.

Fungal Source

Hericium erinaceum, an edible mushroom.

Isolation/Purification

Fresh fruiting bodies of *H. erinaceum* were extracted with acetone and the extract concentrated and fractionated by solvent partitions (chloroform then ethyl acetate). Repeated silicic acid column chromatography followed by recrystallization of the chloroform extract gave purified hericenone A as colorless crystals.

Biological Activity

Cytotoxic to HeLa cells; the minimum concentration completely inhibiting growth of HeLa cells was 100μg/ml.

Spectral Data

IR:

1760, 1660 and 3300-2600cm^{-1}.

^{1}H NMR:

($CDCl_3$) 5.25(s, H-3); 6.97(s, H-4); 3.59(d, *J*=6.41Hz, H-1'); 5.30(t, *J*=6.41Hz, H-2'); 3.18(s, H-4'); 6.09(s, H-6'); 1.91(s, H-8'); 1.81(s, 3'-CH_3); 2.17(s, 7'- CH_3); and 3.89ppm (s, OCH_3).

^{13}C NMR:

($CDCl_3$) 199.08(C-5'); 171.86(C-1); 159.18(C-5 or C-7); 157.48(C-7 or C-5); 150.61(C-7'); 133.63(C-3a); 128.22(C-3'); 125.80(C-7a or 2'); 125.02(C-2' or 7a); 123.05(C-6'); 121.35(C-6); 98.45(C-4); 68.31(C-3); 56.21(OCH_3); 54.40(C-4'); 27.82(C-8'); 23.32(C-1'); 21.06(C-7'-CH_3); and 17.18ppm (C-3'-CH_3).

Mass Spectrum:
FAB-MS exhibited MH^+ ion at 331*m/e*; HREIMS: 330.1494*m/e* (M^+, $C_{19}H_{22}O_5$).

Reference

H. Kawagishi, M. Ando, and T. Mizuno; Hericenone A and B as Cytotoxic Principles from The Mushroom *Hericium erinaceum*; Tetrahedron Letters, Vol. 31, pp. 373-376(1990).

Common/Systematic Name

Hericenone B

Molecular Formula/Molecular Weight

$C_{27}H_{31}NO_4$; MW = 433.22531

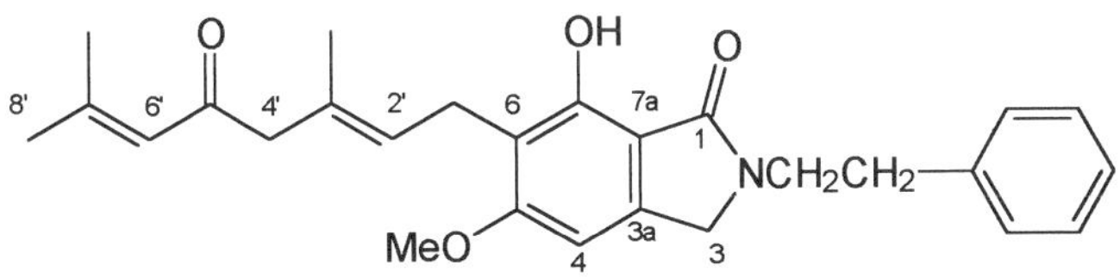

General Characteristics

Colorless crystals from chloroform, mp. 136-138°C.

Fungal Source

Hericium erinaceum, an edible mushroom.

Isolation/Purification

Fresh fruiting bodies of *H. erinaceum* were extracted with acetone and the extract concentrated and fractionated by solvent partitions(chloroform then ethyl acetate). Repeated silicic acid column chromatography followed by recrystallization of the chloroform extract gave purified hericenone B as colorless crystals.

Biological Activity

Cytotoxic to HeLa cells; the minimum concentration completely inhibiting growth of HeLa cells was 6.1μg/ml.

Spectral Data

IR:

1680cm^{-1}

^{1}H NMR:

($CDCl_3$) 4.20(s, H-3); 6.96(s, H-4); 3.56(d, *J*=6.74Hz, H-1'); 5.30(t, *J*=6.74Hz, H-2'); 3.14(s, H-4'); 6.08(s, H-6'); 1.88(s, H-8'); 1.81(s, 3'-CH_3); 2.16(s, 7'- CH_3); 3.84(t, *J*=7.33Hz, H-1'); 2.97(t, *J*=7.33Hz, H-2"); 7.20, 7.26(m, H-2''' -6'''); and 3.84ppm (s, OCH_3).

^{13}C NMR:

($CDCl_3$) 198.98(C-5'); 168.94(C-1); 158.48(C-5 or C-7); 156.81(C-7 or C-5); 150.70(C-7'); 138.76(C-1'''); 132.85(C-3a or 4'''); 132.20(C-4''' or C-3a); 128.70(C-3'); 128.61(C-2''' or C-6'''); 126.51(C-2'); 123.01(C-6'); 122.00(C-7a); 118.48 (C-6); 97.69(C-4); 56.21(OCH_3); 54.65(C-4'); 48.28 (C-3); 44.21(C-1"); 34.90(C-2"); 27.76(C-8'), 23.09(C-1'), 20.94(C-7'-CII_3); and 17.00ppm (C-3'-CII_3).

Mass Spectrum:

FAB-MS: 434(MH^+) and 456*m/e* (M^+Na^+); HREIMS: 433.2243*m/e*; LREIMS: 433(M^+, 12.9%), 342(64.6), 205(3.3), 137(4.5), 105(15.5), 83(100), and 55*m/e* (36.3).

Reference

H. Kawagishi, M. Ando, and T. Mizuno; Hericenone A and B as Cytotoxic Principles from The Mushroom *Hericium erinaceum*; Tetrahedron Letters, Vol. 31, pp. 373-376(1990).

Common/Systematic Name

Hericenone E

4-(3',7'-Dimethyl-5'-oxo-2',6'-octadienyl)-2-formyl-3-hydroxy-5-methoxybenzyl linoleate

Molecular Formula/Molecular Weight

$C_{37}H_{54}O_6$; MW = 594.39204

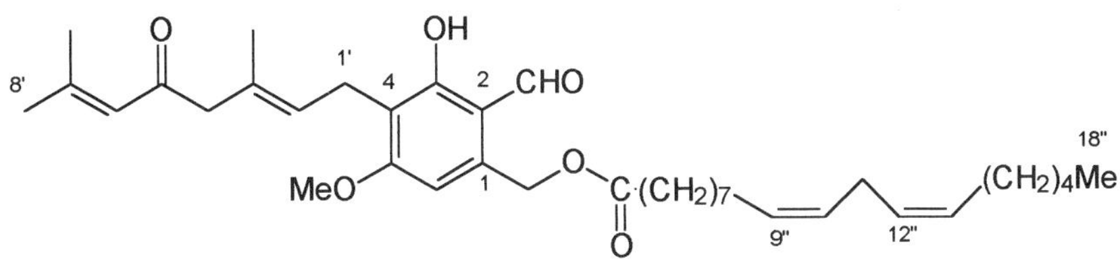

General Characteristics

Colorless oil.

Fungal Source

Hericium erinaceum, an edible mushroom.

Isolation/Purification

Fresh fruiting bodies of *H. erinaceum* were extracted with acetone, and the extract was concentrated and fractionated by solvent partitions (chloroform then ethyl acetate). Repeated silicic acid column chromatography followed by HPLC with an ODS column using methanol-water(95:5, v/v) as mobile phase gave purified hericenone E as a colorless oil.

Biological Activity

Hericenone E was non-cytotoxic to HeLa cells; however, it exhibited stimulating activity to the synthesis of nerve growth factor *in vitro*. In the presence of 33μg/ml, mouse astroglial cells secreted 13.9pg/ml NGF into culture medium.

Spectral Data

IR:

1660, 1720, and 1730cm^{-1}.

^1H NMR:

($CDCl_3$) 6.53(s, H-6); 5.32(s, H-1-CH_2); 0.11(s, CHO); 12.38(s, OH); 3.91(s, OCH_3); 3.40(d, *J*=7.33Hz, H-1'); 5.32(t, *J*=7.33Hz, H-2'); 3.01(s, H-4'); 6.09(s, H-6'); 1.84(s, H-8'); 1.78(s, H-7'-CH_3); 2.12(s, H-7'-CH_3); fatty acid, 2.33(dd, *J*=7.70, 7.32Hz, H-2"); 1.61(m, H-3"); 0.88(t, *J*=6.96Hz, terminal CH_3); and 1.25ppm (m, H-7", 15").

^{13}C NMR:

($CDCl_3$) 199.53(C-5'); 1.93.11(CHO); 173.19 (C-1"); 163.47(C-5); 162.93(C-3); 155.45(C-7); 138.68(C-1); 130.34(C-3'); 126.25(C-2'); 122.81(C-6'); 117.29(C-4); 112.88(C-2); 105 55(C-6); 62.90(C-1-CH_2); 55.93(OCH_3); 55.56(C-4'); 34.23(C-2"); 27.67(C-8'); 21.61(C-1'); 20.67(C-7'-CH_3); 16.40(C-3'-CH_3); 14.32(C-16"); 31.93; 29.70; 29.68; 29.66; 29.65; 29.44; 29.36; 29.23; 29.12; 24.88; and 22.70ppm (C-3"-15").

Mass Spectrum:

FAB-MS: 595*m/e* MH^+; HR-FAB-MS: 595.4016*m/e* for $C_{37}H_{54}O_6$.

Reference

H. Kawagishi, M. Ando, H. Sakamoto, S. Yoshida, F. Ojima, Y. Ishiguro, N. Ukai, and S. Furukawa; Hericenones C, D, and E, Stimulators of Nerve Growth Factor (NGF) Synthesis, from the Mushroom *Hericium erinaceum*, Tetrahedron Letters, Vol. 32, pp. 4561-4564(1991).

Common/Systematic Name

Hericenone F

8-Formyl-5-methoxy-2-methyl-2-(4'-methyl-2'-oxo-3'-pentenyl-)-7-chromanylmethyl palmitate

Molecular Formula/Molecular Weight

$C_{35}H_{54}O_6$; MW = 570.39204

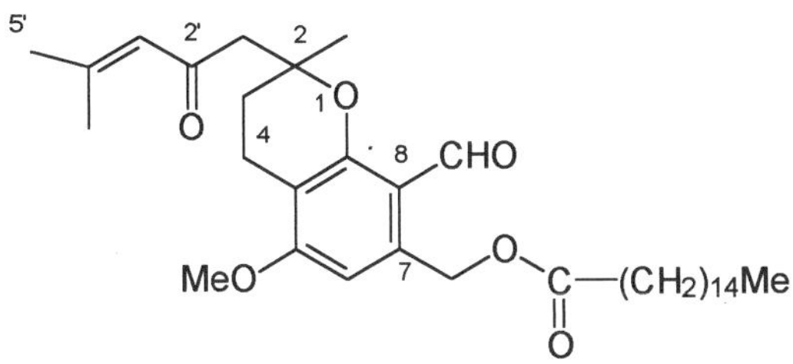

General Characteristics

Pale yellow oil.

Fungal Source

Hericium erinaceum, an edible mushroom.

Isolation/Purification

Fresh fruiting bodies of *H. erinaceum* were extracted with acetone, and the extract was concentrated and fractionated by solvent partitions between chloroform and water, then ethyl acetate. The chloroform layer was concentrated and chromatographed on a silicic acid column eluted with benzene and chloroform-acetone mixtures (9:1, 7:3, and 2:3, v/v) followed by repeated silica gel column chromatography, preparative TLC (chloroform-acetone, 49:1) and HPLC with an ODS column using methanol-water, 19:1, v/v as mobile phase gave purified hericenone F as a pale yellow oil.

Biological Activity

Hericenone F exhibited stimulating activity to the synthesis of nerve growth factor in quiescent mouse astroglial cells.

Spectral Data

IR:

1740, 1670, and 1570cm^{-1}.

^{1}H NMR:

($CDCl_3$) 2.01(ddd, *J*=13.94, 6.61, 6.61, H-3) and 1.91(13.94, 6.60, 6.60, H-3); 2.64(dd, *J*=6.61, 6.60, H-4); 6.55(s, H-6); 10.41(s, CHO); 1.44(s, 2-CH_3); 5.51(s, 7-CH_2); 3.88(s, OCH_3); 2.82(d, *J*=13.94Hz, H-1'), 2.66(d, *J*=13.94Hz, H-1'); 6.06(s, H-3'); 2.14(s, 4'-CH_3); 1.86(s, H-5'); fatty acid, 2.41(t, *J*=7.33Hz, H-2"); 1.69(m, H-3");

0.88(t, *J*=6.60Hz, terminal CH_3); and 1.25ppm (H-11', -8", -14", -7", -15", -9", -10", -12", -13", others).

^{13}C NMR:
($CDCl_3$) 76.9(C-2); 30.1(C-3); 16.5(C-4); 109.2(C-4a); 161.9(C-5); 100.8(C-6); 139.7(C-7); 115.8C-8); 158.3(C-8a); 24.5(C-2 methyl); 52.5(C-1'); 197.9(C-2'); 125.0(C-3'); 156.4(C-C-4'); 20.8(C-4' methyl); 27.8(C-5'); 190.4(CHO); 55.6(OCH_3); 64.5(C-7'-CH_2); fatty acid, 173.3(C-1"); 34.5(C-2"); 25.1(C-3"): 14.1(terminal methyl); and 31.9; 29.7; 29.7; 29.6; 29.5; 29.4; 29.3; 29.3; 22.7ppm (others).

Mass Data:
FAB-MS: 571*m/e* MH^+; HR-FAB-MS: 571.3981*m/e* calcd for $C_{35}H_{54}O_6$ 571.3999

Reference

H. Kawagishi, M. Ando, K. Shinba, H. Sakamoto, S. Yoshida, F. Ojima, Y. Ishiguro, N. Ukai, and S. Furukawa; Chromans, Hericenones F, G and H, from the Mushroom *Hericium erinaceum*; Phytochemistry, Vol. 32, pp. 175-178(1993).

Common/Systematic Name

Hericenone G

8-Formyl-5-methoxy-2-methyl-2-(4'-methyl-2'-oxo-3'-pentenyl)-7-chromanylmethyl-sterate

Molecular Formula/Molecular Weight

$C_{37}H_{58}O_6$; MW = 598.42334

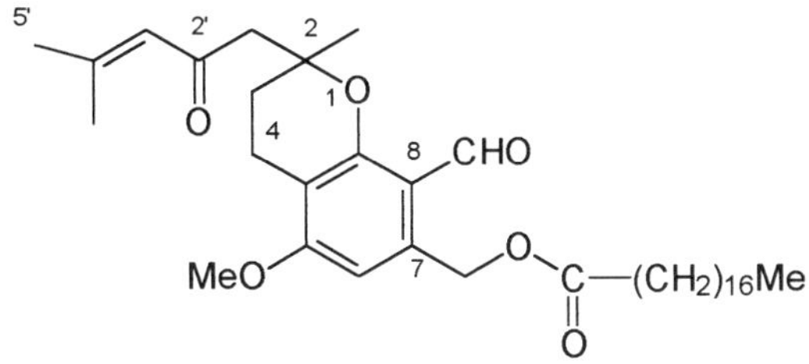

General Characteristics

Crystallized as yellow plates from chloroform-ethanol; mp., 56-58 °C.

Fungal Source

Hericium erinaceum, an edible mushroom.

Isolation/Purification

Fresh fruiting bodies of *H. erinaceum* were extracted with acetone, and the extract was concentrated and fractionated by solvent partitions between chloroform and water, then ethyl acetate. The chloroform layer was concentrated and chromatographed on a silicic acid column eluted with benzene and chloroform-acetone mixtures (9:1, 7:3, and 2:3, v/v) followed by repeated silica gel column chromatography, preparative TLC (chloroform-acetone, 49:1, v/v) and HPLC with an ODS column using methanol-water, 19:1, as mobile phase gave purified hericenone G as a yellow plate from chloroform-ethanol solution.

Biological Activity

Hericenone G exhibited stimulating activity to the synthesis of nerve growth factor in quiescent mouse astroglial cells.

Spectral Data

IR:

1740, 1680, and 1570cm^{-1}.

^{1}H NMR:

($CDCl_3$) 2.00(ddd, J=13.92, 6.60, 6.60, H-3) and 1.90(J=13.94, 6.60, 6.60, H-3); 2.63(dd, J=6.96, 6.60, H-4); 6.45(s, H-6); 10.41(s, CHO); 1.44(s, 2-CH_3); 5.51(s, 7-CH_2); 3.88(s, OCH_3); 2.82(d, J=14.29Hz, H-1'), 2.66(d, J=14.29Hz, H-1'); 6.06(s, H-

3'); 2.14(s, 4'-CH_3); 1.86(s, H-5'); fatty acid, 2.41(t, *J*=7.33Hz, H-2"); 1.69(m, H-3"); 0.88(t, *J*=6.59Hz, terminal CH_3); and 1.25ppm (H-11', -8", -14", -7", -15", -9", -10", -12", -13", others).

^{13}C NMR:

(CDCl$_3$) 76.9(C-2); 30.1(C-3); 16.5(C-4); 109.2(C-4a); 161.9(C-5); 100.8(C-6); 139.7(C-7); 115.8(C-8); 158.3(C-8a); 24.5(C-2 methyl); 52.5(C-1'); 197.9(C-2'); 125.0(C-3'); 156.4(C-C-4'); 20.8(C-4' methyl); 27.8(C-5'); 190.4(CHO); 55.6(OCH_3); 64.5(C-7'-CH_2); fatty acid, 173.3(C-1"); 34.5(C-2"); 25.1(C-3"): 14.1(terminal methyl); and 31.9; 29.7; 29.7; 29.7; 29.6; 29.5; 29.4; 29.3; 29.3; 22.7ppm (others).

Mass Data:

FAB-MS: 599*m/e* MH^+; HR-FAB-MS: 599.4299*m/e* calcd for $C_{37}H_{59}O_6$, 599.4312.

Reference

H. Kawagishi, M. Ando, K. Shinba, H. Sakamoto, S. Yoshida, F. Ojima, Y. Ishiguro, N. Ukai, and S. Furukawa; Chromans, Hericenones F, G, and H from the Mushroom *Hericium erinaceum*; Phytochemistry, Vol. 32, pp. 175-178(1993).

Miscellaneous Metabolites

30

Variecoxanthone A
Variecoxanthone B
Variecoxanthone C
Norlichexanthone
γ-Decalactone
5-Decanolide
Jasmine lactone; *cis*-7-Decen-5-olide
5-Hydroxy-3-methoxy-6-oxo-2-decanoic acid δ-lactone
6-(1-Hydroxypentyl)-4-methoxypyran-2-one
Pestalotin; LL-P880α
LL-P880β; (6*S*,1'*S*,2'*R*)-4-Methoxy-5,6-dihydro-6-(1',2'-dihydroxypentyl)-2*H*-pyranone
LL-P880γ; 4-Methoxy-6-(1',2'-dihydroxypentenyl)-2*H*-pyran-2-one
3,5-Dihydroxydecanoic acid δ-lactone
Trichoacorenol; (1*S*,4*S*,5*S*,7*R*)-1-Isopropyl-4,8-dimethylspiro[4.5]dec-8-en-7-ol
Aszonapyrone A
Oxysporone
Ascladiol
Anhydrosepedonin
Sepedonin
Alterporriol B
LL-S491β
LL-S491γ
cis-Sativenediol
trans-Sativenediol; Isosativenediol
Isosativenetriol
Patulin; Claviformin; Clavitin; Clavicin; Expansion; Penicidin; Mycoin; Leucopin; Tercinin
Isopatulin
Acrostalidic acid
Acrostalic acid
Pimara-8(9),15-diene
Isoacrostalidic acid
Deoxypatulinic acid; 2,3-Dihydro-4-pyrone-5-acetic acid
Dihydropenicillic acid
(-)-Longifolene
Victoxinine
Antibiotic A26771B
IFO 6635 Aversion Factor
Penicillic acid; 3-Methoxy-5-methyl-4-oxo-2,5-hexadienoic acid
5,5-Dimethyl-9-(2-hydroxyethyl)-8-hydroxymethyl-1-methylbicyclo[4.2.1]non-7-ene
Cynodontin; 1,4,5,8-Tetrahydroxy-2-methylanthraquinone; 2,5,7-trihydroxyemodin
Cyclonerodiol
Cyperine
Erinapyrone A; (2*S*)-2,3-Dihydro-6-hydroxymethyl-2-methyl-4*H*-pyran-4-one
Erinapyrone B; (2*R*)-2,3-Dihydro-2-hydroxymethyl-6-methyl-4*H*-pyran-4-one

Erinapyrone C
Asperflavin; (+)-3,4-Dihydro-3,6,9-trihydroxy-8-methoxy-3-methylanthracen-1(2*H*)-one
Diplodiol; Diplosporin
5-Deoxydiplosporin
Betulin (Lup-20(29)-ene-3β,28-diol)

Common/Systematic Name
Variecoxanthone A

Molecular Formula/Molecular Weight
$C_{20}H_{20}O_5$; MW = 340.13107

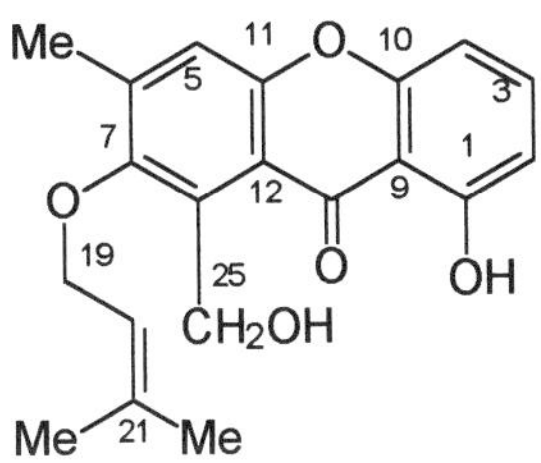

General Characteristics
Variecoxanthone A gave pale yellow needles from methanol; mp., 135°C.

Fungal Source
Aspergillus variecolor (strain CBS 135-55)

Isolation/Purification
Aspergillus variecolor was grown in static culture for 15 days at 25°C. The dried mycelium was ground and continuously extracted with light petroleum (bp., 60-80°C). The resulting semi-solid was triturated with warm methanol and the solution evaporated to give a yellow solid which was fractionated by preparative TLC (toluene-ether, 97:3, v/v). Variecoxanthone C was eluted from the band at R_f 0.3 and gave yellow needles from methanol; variecoxanthone A was eluted from the band at R_f 0.5 and gave pale yellow needles from methanol; and variecoxanthone B was eluted from the band of R_f 0.55 and afforded yellow prisms from methanol.

Spectral Data

UV:
λ_{max}^{EtOH} 374(ϵ = 5,500), 292(11,100), and 260nm (26,000).

IR:
($CHCl_3$) 3500-3100 and 1643cm^{-1}.

^{1}H NMR:
($CDCl_3$) -2.64, 1-OH; 3.22, 2-CH; 2.42, 3-CH; 3.12, 4-CH; 2.71, 5-CH; 5.58, 19-CH_2; 4.42, 20-H; 8.20, 22-CH_3; 8.29, 23-CH_3; 7.56, 24-CH_3; 4.95, 25-CH_2; and 5.60τ 25-OH.

Mass Data:

EIMS: 340.130(M^+), 272(35), 271(48), 254(100), and 242*m/e* (13%), $C_{20}H_{20}O_5$ requires 340.131; found: C, 70.5; H, 5.9%; requires C, 70.6; H, 5.9%.

Reference

K. K. Chexal, J. S. E. Holker, T. J. Simpson, and K. Young; Part V. Structure of Variecoxanthones A, B and C, Metabolites of *Aspergillus variecolor* Conversion of Variecoxanthone A into (±)-De-*C*-Prenylepishamixanthone; J. C. S. Perkin I, pp. 543-548(1975).

Common/Systematic Name

Variecoxanthone B; Emericellin

Molecular Formula/Molecular Weight

$C_{25}H_{28}O_5$; MW = 408.19367

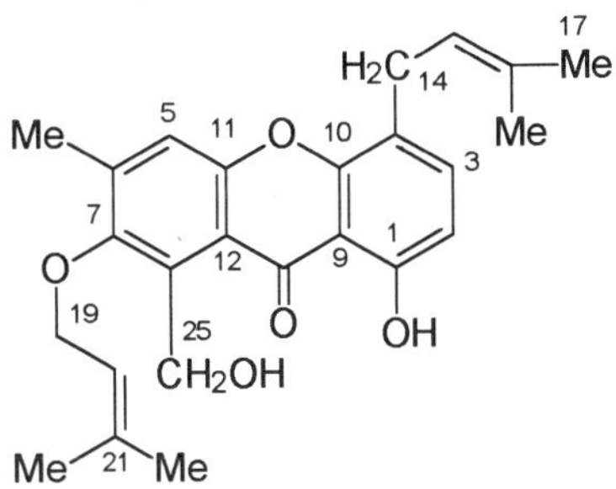

General Characteristics

Variecoxanthone B afforded yellow prisms from methanol; mp., 113-115°C; crystals from ethanol as yellow prisms; mp., 110-112°C, $[\alpha]_D$ 0°; gave a dark green color with ferric chloride.

Fungal Source

Aspergillus variecolor (strain CBS 135.55), *A. nidulans*, *A. rugulosus*, and *A. quadrilineatus*.

Isolation/Purification

See Variecoxanthone A.

Spectral Data

UV:

λ_{max}^{EtOH} 353(ε = 26,000), 258(28,400), 275(30,400), 294(28,400), and 380nm 15,400) (see Chexal, *et al.*, 1975);

λ_{max}^{EtOH} 234(24,400), 245(29,500), 256(29,800), 287(11,900), and 383nm (8,800) (characteristic of xanthone) (see Ishida, *et al.*, 1975).

IR:

($CHCl_3$) 3520-3100 and 1640cm^{-1}.

^1H NMR:

($CDCl_3$) -2.86, 1-OH; 3.23, 2-CH; 2.59, 3-CH; 2.82, 5-CH; 6.66, 14-CH_2; 4.71, 15-CH; 8.26, 17-CH_3; 8.26, 18-CH_3; 5.60, 19-CH_2; 4.43, 20-H; 8.20, 22-CH_3; 8.29, 23-CH_3; 7.59, 24-CH_3; 4.98, 25-CH_2; and 5.60τ, 25-OH.

The NMR spectrum showed two signals at 4.35ppm (1H, triplet, *J*=8Hz) and 12.50ppm (1H, singlet) disappeared by addition of deuterium oxide and assigned to the primary hydroxy and hydrogen bonded phenolic hydroxy protons. The signal at 7.28ppm (1H, singlet),which showed weak long range coupling with the aromatic methyl proton (2.45ppm, 3H, singlet),was assigned to the proton at the position 5 in xanthone nucleus. The signals at 6.68ppm (1H, doublet, *J*=8.5Hz) and 7.40ppm (1H, doublet, *J*=8.5Hz) were assigned to the protons at the position 2 and 3, or 4 and 3. The signals at 1.75-1.80ppm (3H x 4) were assigned to the protons of the four olefinic methyl groups. The signals at 4.44ppm (2H, doublet, *J*=7Hz) and 5.62ppm (1H, triplet, *J*=7Hz) were assigned to the methylene proton attached to the oxygen-bearing carbon and olefinic proton, respectively. These data indicated the presence of an *O*-isoprenyl group. The signals at 3.47ppm (2H, doublet, *J*=7Hz) and 5.31ppm (1H, triplet, *J*=7Hz) were assigned to the methylene and olefinic protons of the C-isoprenyl group.

Mass Data:

EIMS: 408.195(M^+, 5%), 340(37), 339(3500), and 322*m/e* (925), $C_{25}H_{28}O_5$ requires 408.194; found: C, 73.4; H, 7.0%; requires C, 73.5; H, 6.9%.

References

K. K. Chexal, J. S. E. Holker, T. J. Simpson, and K. Young; Part V. Structure of Variecoxanthones A, B and C, Metabolites of *Aspergillus variecolor* Conversion of Variecoxanthone A into (±)- De-*C*-Prenylepishamixanthone; J. C. S. Perkin I, pp. 543-548(1975).

M. Ishida, T. Hamasaki, Y. Hatsuda, K. Furuyama, T. Tsukihara, and Y. Katsube; Emericellin, a New Metabolite from *Aspergillus nidulans*; Agr. Biol. Chem., Vol. 39, pp. 291-292(1975).

Common/Systematic Name
Variecoxanthone C

Molecular Formula/Molecular Weight
$C_{25}H_{28}O_6$; MW = 424.18859

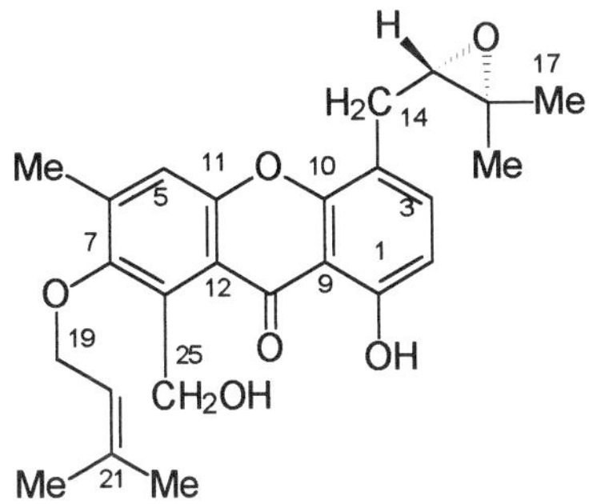

General Characteristics
Variecoxanthone C gave yellow needles from methanol; mp., 118-120°C.

Fungal Source
Aspergillus variecolor (strain CBS 135.55)

Isolation/Purification
See variecoxanthone A.

Spectral Data

UV:
λ_{max}^{EtOH} 252(ϵ=30,700), 258(24,700), 274(30,000), 294(27,800), and 376nm (15,000).

IR:
($CHCl_3$) 3500-3100 and 1641cm^{-1}.

^{1}H NMR:
($CDCl_3$) -2.62, 1-OH; 3.23, 2-CH; 2.50, 3-CH; 2.69, 5-CH; 6.97, 14-CH_2; 6.97, 15-CH; 8.56, 17-CH_3; 8.67, 18-CH_3; 5.58, 19-CH_2; 4.43, 20-H; 8.20, 22-CH_3; 8.29, 23-CH_3; 7.55, 24-CH_3; 4.94, 25-CH_2; and 5.60τ 25-OH.

Mass Spectrum:
EIMS: 424.188(M^+, 4%), 356(95), 355(96), and 338*m/e* (100); $C_{25}H_{28}O_6$ requires 424.188.

Reference

K. K. Chexal, J. S. E.Holker, T. J. Simpson, and K. Young; Part V. Structure of Variecoxanthones A, B and C, Metabolites of *Aspergillus variecolor* Conversion of Variecoxanthone A into (±)-De-*C*-prenylepishamixanthone; J. C. S. Perkin I, pp. 543-548(1975).

Common/Systematic Name

Norlichexanthone

3,6,8-Trihydroxy-1-methylxanthone

Molecular Formula/Molecular Weight

$C_{14}H_{10}O_5$; MW = 258.05282

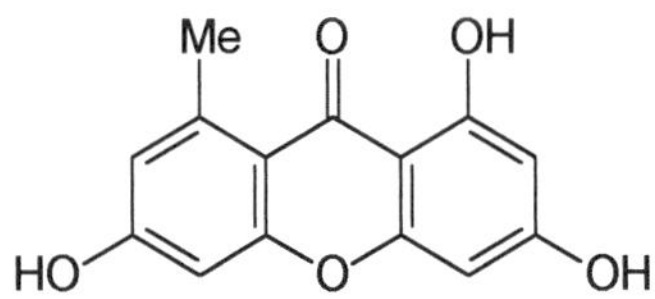

General Characteristics

Crystals from aqueous acetone; mp., 272-274°C. Gives blue-green color with $FeCl_3$.

Fungal Source

Penicillium patulum, a mutant strain.

Isolation/Purification

The filtered broth was acidified with mineral acid (pH 2), extracted with ethyl acetate, and evaporated to dryness. The residue was redissolved in methanol and evaporated onto silica gel. This was packed on top of additional silica gel and the substance was eluted with chloroform-ethyl acetate (17:3, v/v) and chloroform-ethyl acetate-ethanol (17:3:1, v/v/v).

Biological Activity

Antibacterial (MIC vs *Clostridium welchii* = 25ppm)

Spectral Data

UV:

λ_{max}^{EtOH} 241(ϵ=36500), 311nm (22500).

IR:

3500, 3050, 1650, 1620, 1500, 830, 812, and 760cm^{-1}.

^{1}H NMR:

3.4(OH); 3.4(2H, m, ArH), 3.85(AB q, J=2Hz, ArH); and 7.24τ (Ar-CH_3).

References

D. Broadbent, R. P. Mabelis, and H. Spencer; 3,6,8-Trihydroxy-1-methylxanthone - An Antibacterial Metabolite from *Penicillium patulum*; Phytochemistry, Vol. 14, pp. 2082-2083(1975).

J. Santesson; Chemical Studies on Lichens. 16. The Xanthones of *Lecanora straminea*. I. Arthothelin and Thiophanic Acid; Ark. Kemi, Vol. 30, pp. 449-454(1969).

Common/Systematic Name

γ-Decalactone

Molecular Formula/Molecular Weight

$C_{10}H_{18}O_2$; MW = 170.13068

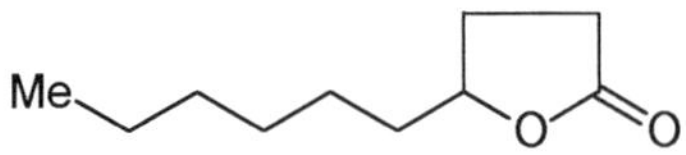

General Characteristics

A flavor component of deteriorated milk fat, butter, γ-irradiated milk fat, raspberry oil, peach, strawberry, apricot, orange, carrot seed oil, and bilberry.

Fungal Source

Sporobolomyces odorus.

Isolation/Purification

The culture broth was extracted with an equal volume of ethyl ether. The ether extract was washed with 2% sodium carbonate solution and dehydrated with anhydrous sodium sulfate. The extract was reduced in volume by fractional distillation and finally a stream of nitrogen gas. γ-Decalactone was isolated in pure form by preparative gas chromatography (10% PEG-20M as the liquid phase on Celite 545 AW DMCS at 190°C).

Spectral Data

IR:

(Thin film) 2910, 2880, 1770(γ-lactone ring), 1458, 1178, and 720cm^{-1}.

Mass Spectrum:

LREIMS: 170(M^+), 152, 134, 128, 100, 85(base peak), 55, 43, and 41*m/e*.

GLC Data

Retention time 7.2 and 4.0min. on Celite 545AW DMCS coated with 10% PEG-20M; carrier gas N_2 at 23ml/min; column temperature 180°C; injection port temperature 250°C and on 1.5% OV-17 on Chromosorb W; carrier gas N_2 at 24ml/min; column temperature 140°C, respectively.

Reference

S. Tahara, K. Fujiwara, H. Ishizaka, J. Mizutani, and Y. Obata; γ-Decalactone, One of the Constituents of Volatiles in Cultured Broth of *Sporobolomyces odorus*; Agric. Biol. Chem., Vol. 36, pp. 2585-2587(1972).

Common/Systematic Name
5-Decanolide

Molecular Formula/Molecular Weight
$C_{10}H_{18}O_2$; MW = 170.13068

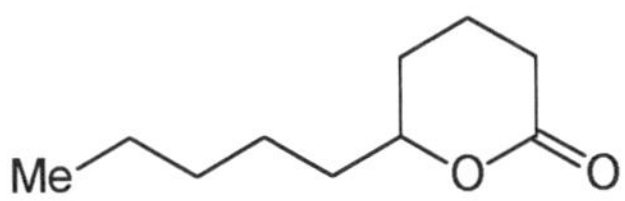

General Characteristics
5-Decanolide occurs widely in the flavor components of milk and fruits.

Fungal Source
Sporobolomyces odorus AHU 3246.

Isolation/Purification
The neutral fraction was obtained and fractionated on a silica gel column packed in isopentane and eluted with ether-isopentane (1:3, v/v) followed by ether-isopentane (1:1, v/v). The δ-lactone fractions were further purified by preparative GLC (10% PEG-20M column, column temperature 183°C, injection port 270°C, N_2 carrier at 22.5ml/min., retention time 4.72 min.).

Spectral Data

Mass Spectrum:
LREIMS: 170(M^+), 152, 141, 134, 114, 99(base peak), 71, 55, and 41*m/e*.

Reference
S. Tahara and J. Mizutani; δ-Lactones Produced by *Sporobolomyces odorus*; Agric. Biol. Chem., Vol. 39, pp. 281-282(1975).

Common/Systematic Name

Jasmine lactone

cis-7-Decen-5-olide

Molecular Formula/Molecular Weight

$C_{10}H_{16}O_2$; MW = 168.11503

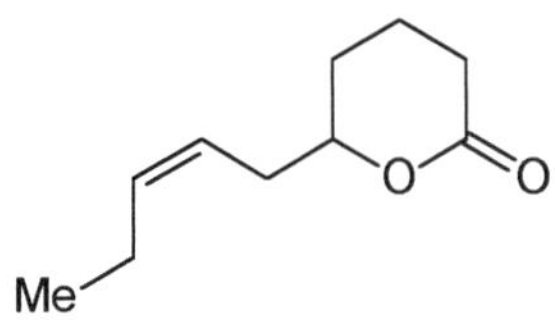

General Characteristics

Jasmine lactone was originally isolated from the oil and wax of *Jasminum grandiflorum.*

Fungal Source

Sporobolomyces odorus AHU 3246

Isolation/Purification

The neutral fraction was obtained and fractionated on a silica gel column packed in isopentane and eluted with ether-isopentane (1:3, v/v) followed by ether-isopentane (1:1, v/v). The δ-lactone fractions were further purified by preparative GLC (10% PEG-20M column, column temperature 183°C, injection port 270°C, N_2 carrier at 22.5ml/min., retention time 4.72 min.).

Spectral Data

Mass Spectrum:

LREIMS: 168(M^+), 150, 139, 108, 99, 81, 71(base peak), 55, and 41*m/e*.

Reference

S. Tahara and J. Mizutani; δ-Lactones Produced by *Sporobolomyces odorus*; Agric. Biol. Chem., Vol. 39, pp. 281-282(1975).

Common/Systematic Name

5-Hydroxy-3-methoxy-6-oxo-2-decanoic acid δ-lactone

Molecular Formula/Molecular Weight

$C_{11}H_{16}O_4$; MW = 212.10486

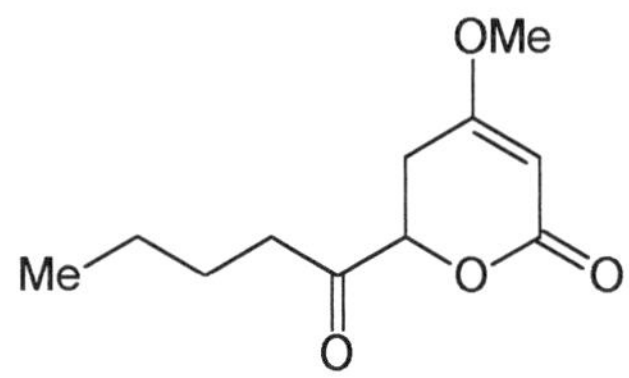

General Characteristics

Crystallized from ether-cyclohexane as colorless needles; mp., 83-84°C; $[\alpha]_D^{22}$ -4.4° (*c*=0.266, in MeOH).

Fungal Source

Penicillium sp. and an unidentified fungus.

Isolation/Purification

Filtrates from fungal liquid shake cultures were extracted thoroughly with chloroform. After removal of the solvent *in vacuo* TLC analysis of the brown residue revealed two major components. The mixture was chromatographed on a column of silica gel (Davidson Grade 923) and eluted with chloroform-methylene chloride which afforded 5-hydroxy-3-methoxy-6-oxo-2-decanoic acid-δ-lactone slightly contaminated by other products. The second major metabolite, again containing small amounts of impurities, was eluted with chloroform. The less polar compound was purified by preparative TLC on silica gel plates, using the solvent system benzene-*n*-hexane-acetone (10:9:1, v/v/v). 5-Hydroxy-3-methoxy-6-oxo-2-decanoic acid-δ-lactone crystallized from ether-cyclohexane as colorless needles.

Spectral Data

UV:

λ_{max} 233nm (ϵ = 10,600)

IR:

(KBr) 1710, 1620, 1248, and 1230cm^{-1}.

^{1}H NMR:

($CDCl_3$) 0.92(3H, t, *J*=7Hz); 1.34(2H, sextet, slightly broadened, *J*=7Hz); 1.59(2H, quintet, slightly broadened, *J*=7Hz); 2.75(4H, m, simplified by irradiation at both 1.59 and 4.80ppm); 3.78(3H, s); 4.80(1H, dd, $J_{AX}+J_{BX}$=14Hz); and 5.17ppm (1H, s).

Mass Data:

LREIMS: 212(M^+), 127(base peak, M^+ - 85, - C_5H_9O), and 85*m/e* ($C_5H_9O^+$); anal. calcd. for $C_{11}H_{16}O_4$: C, 62.25; H, 7.60; O, 30.15; found: C, 62.44; H, 7.32; O, 30.20.

Reference

G.M. Strunz, C. J. Heissner, M. Kakushima, and M. A. Stillwell; Metabolites of an Unidentified Fungus: A New 5,6-Dihydro-2-pyrone Related to Pestalotin; Can. J. Chem., Vol. 52, pp. 825-826(1974).

Common/Systematic Name

6-(1-Hydroxypentyl)-4-methoxypyran-2-one

Molecular Formula/Molecular Weight

$C_{11}H_{16}O_4$; MW = 212.10486

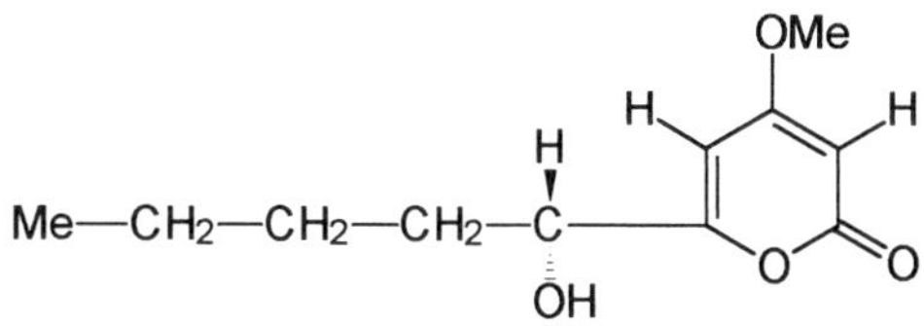

General Characteristics

Oil; $[\alpha]_D^{23}$ + 78.5° (*c*=1.0, in MeOH); monoacetate: an oil; $[\alpha]_D^{23}$ + 85.3° (*c*=0.19, in MeOH).

Fungal Source

Penicillium sp.

Isolation/Purification

The culture filtrate was extracted with ethyl acetate at pH 2, washed with sodium bicarbonate and water, and evaporated to dryness. The residue was chromatographed on a silica gel column eluted with a gradient from benzene to ethyl acetate. The 30% ethyl acetate fraction was concentrated and further purified using preparative silica gel TLC plates developed with benzene-ethyl acetate (1:1, v/v).

Spectral Data

UV:

λ_{max}^{EtOH} 227(sh, ϵ=3,650) and 289nm (9,850).

IR:

(thin film) 3460, 1720, 1700, 1650, 1570, 1410, and 1255cm^{-1}; monoacetate: (thin film) 1730, 1650, 1565, 1405, 1245, 1225, 1035, and 1020cm^{-1}.

^{1}H NMR:

($CDCl_3$) 0.89(3H, t, *J*=7.0Hz); 1.33(4H, m); 1.75(2H, m); 2.95(1H, br); 3.72(3H, s); 4.30(1H, m); 5.33(1H, d, *J*=2.0Hz); and 6.02ppm (1H, d, *J*=2.0Hz); monoacetate: 0.87(3H, t, *J*=7.0Hz); 1.25(4H, m); 1.80(2H, m); 3.72(3H, s); 5.32(2H, d and t, *J*=7.0 and 2.0Hz); and 5.82ppm (1H, d, *J*=2.0Hz).

Mass Spectrum:

Acetate: 212(M^+), 125*m/e* (base peak, $C_6H_5O_3^+$); LREIMS: 254(M^+), 211(M^+ - CH_3CO), and 125*m/e* (base peak).

TLC Data

Silica gel G_{254} plates developed with benzene-ethyl acetate (1:1, v/v); R_f 0.34.

Reference

Y. Kimura, W. J. McGahren, A. Suzuki, and S. Tamura; Structure of a New Fungal Pyrone, from an Unidentified *Penicillium* sp.; Agric. Biol. Chem., Vol. 42, pp. 1625-1626(1978).

Common/Systematic Name

Pestalotin; LL-P880α

Molecular Formula/Molecular Weight

$C_{11}H_{18}O_4$; MW = 214.12051

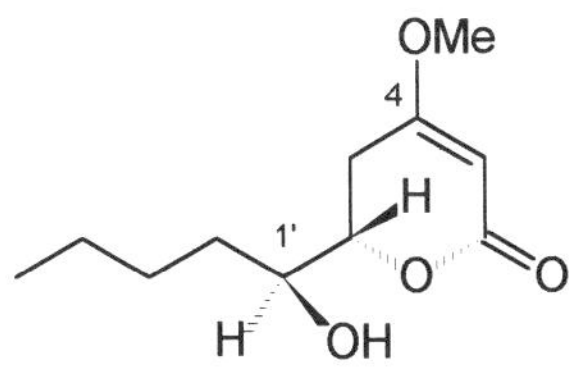

General Characteristics

Crystallization from *n*-hexane-benzene gave pestalotin as colorless plates melting at 88.0-88.5°C; $[\alpha]_D$ -86.2° (*c*=0.14, in MeOH); acetate, a colorless oil. Pestalotin was soluble in most organic solvents such as benzene, chloroform, ethyl acetate, ether, acetone, and alcohols, but sparingly in *n*-hexane.

Fungal Source

Pestalotia cryptomeriaecola, a pathogen on Japanese cedar (*Cryptomeria japonica*) and *Penicillium* sp.

Isolation/Purification

After addition of an equal volume of acetone, the culture broth was filtered to remove mycelia and was evaporated under reduced pressure. The residual aqueous solution was extracted repeatedly with ethyl acetate at pH 3. The combined extracts were washed successively with 5% sodium bicarbonate solution and water. The ethyl acetate was dried over anhydrous sodium sulfate, filtered, and evaporated under reduced pressure. The residue was dissolved in benzene and applied to a column of silicic acid eluted stepwise with 100 ml each of benzene and benzene containing 10, 20, 30, 40, and 50% ethyl acetate, respectively. The eluate with benzene containing 30% ethyl acetate afforded crude pestalotin, which was subjected to preparative TLC using silica gel GF_{254}, and benzene-ethyl acetate (1:1, v/v) as the developing solvent. The zone containing pestalotin was extracted with ethyl acetate. The extract was evaporated to dryness under reduced pressure to yield pestalotin. Final purification of pestalotin was by recrystallization from *n*-hexane-benzene.

Biological Activity

Biological activity as a gibberellin synergist. No effect observed when applied alone on rice seedlings at dosages of 30-300mg/liter. However, when applied in combination with gibberellin A_3, pestalotin enhanced the stimulative effect of gibberellin A_3 on shoot growth.

Spectral Data

UV:
$\lambda_{max}^{95\% EtOH}$ 233nm (ϵ=12,200)

IR:
(Nujol) 3480, 1710, 1635, 1255, 1090, and 1025cm^{-1}.

CD:
(0.82mg/10cc MeOH) $\Delta\epsilon_{243}$ -7.90

^{1}H NMR:
($CDCl_3$) 0.91(3H, t, *J*=6.3 Hz); 1.30~1.60(6H, broad); 2.22(1H, broad); 2.25(1H, q, *J*=4.0, 17.0Hz); 2.79(1H, *J*=2.0, 12.7, 17.0Hz); 3.62(1H, broad); 3.75(3H, s); 4.28(1H, m, *J*=4.0, 4.0, 12.7 Hz); and 5.14ppm (1H, d, *J*=2.0Hz).

Mass Spectrum:
LREIMS: 214(M^+), 196, 183, 181, 170, 167, 157, 153, 139, and 127*m/e* (base peak).

TLC Data
R_f 0.35 on silica gel using benzene-ethyl acetate (1:1, v/v) as developing solvent.

References
G. A. Ellestad, W. J. McGahren, and M. P. Kunstmann; Structure of a New Fungal Lactone, LL-P880α, from an Unidentified *Penicillium* sp.; J. Org. Chem., Vol. 37, pp. 2045-2047(1972).

Y. Kimura, K. Katagiri, T. Inoue, and S. Tamura; Isolation and Biological Activity of Pestalotin, A Gibberellin Synergist from *Pestalotia cryptomeriaecola*; Agric. Biol. Chem., Vol. 35, pp. 1313-1314(1971).

Y. Kimura, K. Katagiri, and S. Tamura; Structure of Pestalotin, A New Metabolite from *Pestalotia cryptomeriaecola*; Tet. Lett., pp. 3137-3140(1971).

Y. Kimura and S. Tamura; Isolation and Structure of Pestalotin, A Gibberellin Synergist from *Pestalotia cryptomeriaecola*; Agric. Biol. Chem., vol. 36, pp. 1925-1930(1972).

Common/Systematic Name

LL-P880β

(6*S*,1'*S*,2'*R*)-4-Methoxy-5,6-dihydro-6-(1',2'-dihydroxypentyl)-2*H*-pyran-2-one

Molecular Formula/Molecular Weight

$C_{11}H_{18}O_5$; MW = 230.11542

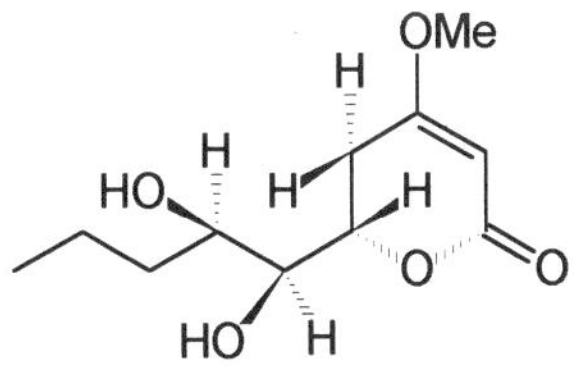

General Characteristics

White crystals from ethyl acetate-hexane; mp., 135.5-136°C; $[\alpha]_D^{25}$ -59.8° (*c*= 0.96, in MeOH).

Fungal Source

Penicillium sp.

Isolation/Purification

The fungal cultures were extracted with ethyl acetate. The extract was concentrated to a dark red solution, which, on being allowed to remain in the cold room for 2 days, yielded pure LL-P880β.

Spectral Data

UV:

λ_{max}^{MeOH} 234nm (ε=13,800)

IR:

(KBr) 1710 and 1624cm^{-1}.

CD:

$\Delta\epsilon_{246}$ -12.5, $\Delta\epsilon_{220}$ +6.6, shoulder at $\Delta\epsilon_{210}$ + 4.2(*c*=0.236, in MeOH).

^{1}H NMR:

($CDCl_3$) 0.93(-CH_2CH_3, t); 1.50((CH_2)$_2$$CH_3$, m); 2.36(Hc, dd, J_{cb}=17.5, J_{cd}=12.5Hz); 2.87(H_b, qq, J_{bd}=4Hz); 3.47(H_e, dd, J_{ef}=3.8, J_{ed}=4.2Hz); 3.78(H_d, m, hidden by OCH_3 signal); 3.80(OCH_3, s); 4.51(H_d, tt); and 5.16ppm (H_a, d, J_{ac} ~ 2Hz).

Mass Data:
Anal. calcd. for $C_{11}H_{18}O_5$: C, 57.38; H, 7.88; found: C, 57.02; H, 7.76.

Reference

W. G. McGahren, G. A. Ellestad, G. O. Morton, M. P. Kunstmann, and P. Mullen; A New Fungal Lactone. LL-P880β, and A New Pyrone, LL-P880γ, from *Penicillium* sp.; J. Org. Chem., Vol. 38, pp. 3542-3544(1973).

Common/Systematic Name

LL-P880γ

4-Methoxy-6-(1',2'-dihydroxypentenyl)-*2H*-pyran-2-one

Molecular Formula/Molecular Weight

$C_{11}H_{16}O_5$; MW = 228.09977

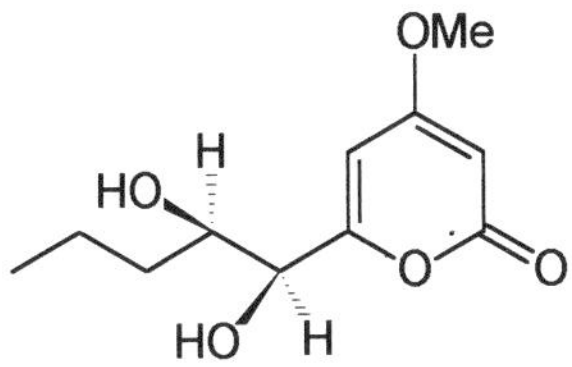

General Characteristics

Colorless needles from ethyl acetate; mp., 116-118°C; $[\alpha]_D^{20}$ -31° (*c*=1.0, in MeOH).

Fungal Source

Penicillium citreo-viride and *Penicillium* sp.

Isolation/Purification

The fungus was cultured statically in a malt extract medium containing 3% peptone at 24°C for 21 days. The culture broth was filtered and the filtrate was extracted with ethyl acetate at pH 2.0. The combined extracts were concentrated under reduced pressure below 40°C to give a dark brown residue which was dissolved in acetone and mixed with acid-washed Celite. After drying under reduced pressure, the impregnated Celite was charged onto a SilicAR CC-4 column and eluted with benzene-acetone mixtures, the content of acetone in the mixture being increased stepwise. The 5% and 10% acetone fractions were concentrated and the residual solid was recrystallized from ethyl acetate to afford colorless plates. The pale yellow solid obtained from the 15% and 20% acetone fractions was subjected to a silica gel column, which was eluted with benzene-ethyl acetate mixtures, the content of ethyl acetate being increased stepwise. The eluates with benzene containing more than 35% ethyl acetate fractions were evaporated to dryness to afford pale brown solids, which were recrystallized from ethyl acetate to give LL-P880γ as colorless needles.

Biological Activity

Acted as a gibberellin synergist using dwarf rice seeds (*Oryzae sativa*). LL-P880γ alone at dosages of 1.0 ~ 300mg/liter did not affect the growth of rice seedlings. When applied together with GA_3 at 1mg/liter, it enhanced the stimulative effect of GA_3 on the elongation of the second leaf sheath.

Spectral Data

UV:
λ_{max}^{MeOH} 225(sh, ε=3,500) and 280nm (8,400).

IR:
(KBr) 3300, 3100, 2690, 1735, 1660, 1580, and 1265cm^{-1}.

^{1}H NMR:
(pyridine-d_5) The signals at 5.50(1H, d, J=2.0Hz) and 6.58ppm (1H, d, J=2.0Hz) were due to olefinic protons and at 3.51ppm (3H, s) due to -OCH_3. The signals at 4.35(1H, m) and 4.62ppm (1H, d, J=3.0Hz) were assigned to C-2' and -1' protons, respectively. The presence of an *n*-propyl group was evident by the signals at 0.86(3H, s) and 1.75ppm (-CH_2-CH_2-, m).

Mass Spectrum:
LREIMS: 229(M^+ + 1), 211, 185, 155, 125, 110, 99, 83, 69*m/e* (base peak).

TLC Data
Silica gel thin-layer plates developed with benzene-ethyl acetate (3:7v/v), R_f=0.21.

References
Y. Kimura, T. Hamasaki, and H. Nakajima; Stereochemistry and Biological Activities of LL-P880γ, A Pestalotin Analogue, Produced by *Penicillium citreo-viride*; Agric. Biol. Chem., Vol. 50, pp. 1649-1650(1986).

Common/Systematic Name

3,5-Dihydroxydecanoic acid δ-lactone

Molecular Formula/Molecular Weight

$C_{10}H_{18}O_3$; MW = 186.12559

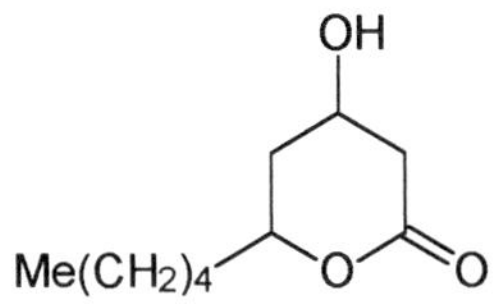

General Characteristics

Isolated as an oil; $[\alpha]_D^{25}$ + 24.7° (*c*=11.7, in $CHCl_3$).

Fungal Source

Cephalosporium recifei (NRRL 5161).

Isolation/Purification

Fungal cultures were extracted with hexane. The remaining hexane insoluble material was fractionated by silicic acid column chromatography (Bio Sil A), and eluted with ethyl ether-hexane (50:50, v/v) to yield 3,5-dihydroxydecanoic acid δ-lactone as an oil.

Spectral Data

IR:

(Neet) 3430 and 1720-1730cm^{-1}.

^{1}H NMR:

($CDCl_3$) 5.2-5.4(CH at junction of lactonic oxygen); 5.66(CH-OH); 6.94(OH); 7.36(doublet, CH_2 adjacent to carbonyl); and 9.12τ (C-CH_3).

Mass Data:

LREIMS: Highest detectable mass was at 168*m/e* (M^+ - H_2O), 97(-H_2O from 115*m/e*, 100%) with fragment ions at 115*m/e* (- hydrocarbon chain at C-5), and 97*m/e* (- H_2O from 115*m/e*; anal. calcd for $C_{10}H_{18}O_3$: C, 64.49; H, 9.74; found: C, 64.6; H, 9.54.

Reference

R. F. Vesonder, F. H. Stodola, and W. K. Rohwedder; Formation of the δ-Lactone of 3,5-Dihydroxydecanoic Acid by the Fungus *Cephalosporium recifei*; Can. J. Biochem., Vol. 50, pp. 363-365(1972).

Common/Systematic Name

Trichoacorenol

(1*S*,4*S*,5*S*,7*R*)-1-Isopropyl-4,8-dimethylspiro[4.5]dec-8-en-7-ol

Molecular Formula/Molecular Weight

$C_{15}H_{26}O$; MW = 222.19837

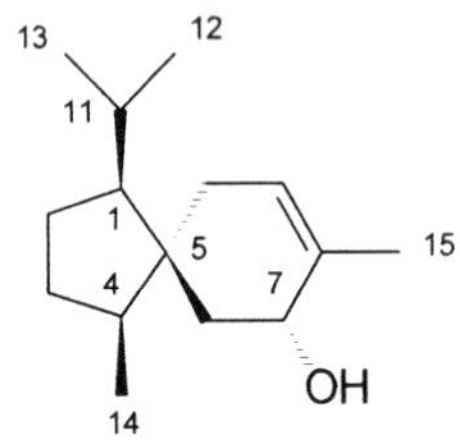

General Characteristics

Colorless amorphous solid; $[\alpha]_D$ 5.2° (*c*=0.12, in $CHCl_3$).

Fungal Source

Trichoderma koningii.

Isolation/Purification

The culture broth was extracted with butanol and the butanol extract subjected to silica gel column chromatography with a chloroform-methanol gradient (100:0 - 50:50) to give eight fractions. Fraction 2 was further purified by silica gel column chromatography with chloroform to give pure trichoacorenol.

Spectral Data

IR:

($CHCl_3$) 3600cm^{-1}

^{1}H NMR:

(acetone-d_6) 0.84(3H, d, *J*=7.0Hz, 14-H_3); 0.86; 0.93(each 3H, d, *J*=6.5Hz, 12-H_3, 13-H_3); 1.09(1H, m, 3-H); 1.27(1H, m, 1-H); 1.29(1H, dd, *J*=14.0, 10.0Hz, 6-H); 1.46(1H, tdd, *J*=11.5, 10.0, 4.0Hz, 2-H); 1.58(1H, ddq, *J*=10.0, 9.0, 6.5Hz, 4-H); 1.70(1H, m, 3-H); 1.70(1H, m, 11-H); 1.72(1H, m, 6-H); 1.73(3H, brs, 15-H_3); 1.74(1H, m, 2-H); 1.80(1H, br d, *J*=9.5Hz, 10-H); 2.10(1H, ddq, *J*=9.5, 5.5, 2.5Hz, 10-H); 3.51(1H, d, *J*=6.5Hz, 7-OH); 4.22(1H, br s, *J*=19.5Hz, 7-H); and 5.39ppm (1H, m, 9-H).

^{13}C NMR:

(acetone-d_6) 15.0(q, C-14); 90.0(q, C-15); 23.8, 24.3(each q, C-12, C-13); 27.0(t, C-2); 30.1(t, C-3); 31.5(d, C-11); 33.4(t, C-6); 36.7(t, C-10); 46.2(s, C-5); 48.0(d, C-4); 61.2(d, C-1); 68.6(d, C-7); 124.6(d, C-9); and 138.6ppm (s, C-8).

Mass Spectrum:

LREIMS: 222(M^+, 100), 207(M^+ - CH_3, 7), 204(M^+- H_2O, 4), 179(29), 166(20), 161(21), 151(70), and 138*m/e* (61). HREIMS: found 299.1970*m/e;* calcd for $C_{15}H_{26}O$, 222.1984.

Reference

Q. Huang, Y. Tezuka, Y. Hatanaka, T. Kikuchi, A. Nishi, and K. Tubaki; Studies on Metabolites of Mycoparasitic Fungi. III. New Sesquiterpene Alcohol from *Trichoderma koningii*; Chem. Pharm. Bull., Vol. 43, pp. 1035-1038(1995).

Common/Systematic Name
Aszonapyrone A

Molecular Formula/Molecular Weight
$C_{28}H_{40}O_5$; MW = 456.28757

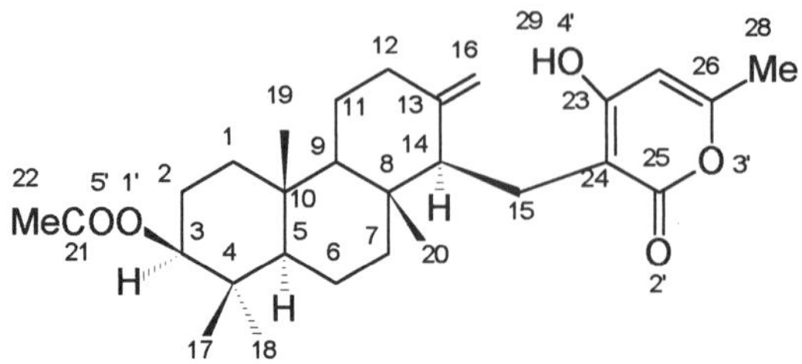

General Characteristics (Monomethyl ether)
Crystals from benzene-acetone; mp., 276-278°C.

Fungal Source
Mycelial mats of *Aspergillus zonatus* IFO 8817.

Biological Activity
Antibacterial activity.

Spectral Data (monomethyl ether of OH on carbon 23)

UV:
λ_{max}^{EtOH} 298nm (ϵ=8,600)

IR:
(KBr) 1740, 1700, 1690(sh), 1650, 1560, 1460, 1440, 1240, and 910cm^{-1}.

^{1}H NMR:
($CDCl_3$) 0.76(3H, s); 0.84(9H, s); 1.00~2.80(17H, m); 2.04(3H, s); 2.23(3H, s); 3.92(3H, s); 4.54(1H, dd, *J*=4.12Hz); 4.72(1H, s); 4.88(1H, s); and 6.06ppm (1H, s).

^{13}C NMR:
($CDCl_3$) 15.2(q); 16.4(q); 16.5(q); 19.0(t); 19.3(q); 20.1(q); 21.1(q); 23.7(t); 23.9(t); 28.1(q); 37.7(s); 38.0(s); 38.4(t); 38.5(t); 40.3(s); 40.3(t); 54.2(d); 55.6(d); 56.0(q); 60.1(d); 81.1(d); 95.1(d); 105.9(s); 106.2(t); 149.0(s); 160.7(s); 165.1(s); 165.5(s); and 170.6ppm (s).

Mass Spectrum:
470(M$^+$), 455, 411, 395, 154, and 153*m/e*.

References

Y. Katsube, Y.Kimura, T. Hamasaki, H. Nakajima, and A. Isogai; Structure of Aszonapyrone A Monomethyl Ether-1, a Derivative of Aszonapyrone A, Produced by *Aspergillus zonatus*; Agric. Biol. Chem.. Vol. 49, pp. 551-553(1985).

Y. Kimura, T. Hamasaki, A.Isogai, and H. Nakajima; Structure of Aszonapyrone A, a New Metabolite Produced by *Aspergillus zonatus*; Agric. Biol. Chem., Vol. 46, p. 1963(1982).

Common/Systematic Name
Oxysporone

Molecular Formula/Molecular Weight
$C_7H_8O_4$; MW = 156.04226

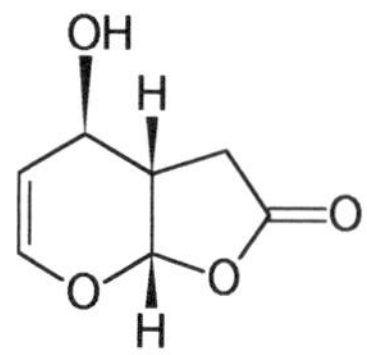

General Characteristics
$[\alpha]_D^{25}$ +39.5° (*c*= 0.9%, MeOH); gave a monoacetate derivative.

Fungal Source
Fusarium oxysporum IMI 211881 and *Pestalotia longiseta* PL8501 (IFO 32172) which causes tea gray blight, the most important disease of tea plantations in Japan.

Isolation/Purification
Separated on silica gel using ethyl acetate-petrol (2:3, v/v).

Biological Activity
Phytotoxin; caused leaf necrosis of cv. Yabukita at ca. 15ppm. Agent responsible for the antibiotic activity of earthworm casts used in some parts of Nigeria to treat chronic dysentery.

Spectral Data

UV:
λ_{max} 215(ϵ = 1277) and 284nm (417).

IR:
($CHCl_3$) 3600(OH), 1806(γ-alkoxy-γ-lactone), and 1654(C=C) cm^{-1}.

^1H NMR:
6.37(1H, *J*=6Hz, H-2); 5.06(1H, ddd, *J*=6, 5.5, 1.0Hz, H-3 coupled to H-2, H-4 and one of the high field protons possibly H-5); 5.82(1H, d, *J*=4.5Hz, H-6); 4.16(1H, dd, *J*=5.5 and 2.0Hz, H-4); and four high field protons between δ2 and 3, one of which was OH (exch. with D_2O).

^{13}C NMR:
175.3(C-8), 143.5(C-2), 100.1(C-3), 96.0(C-6), 60.0(C-4), 41.8(C-7), and 29.5ppm. (C-5); ($CDCl_3$) 174.5(C-8), 144.1(C-2), 99.8(C-3), 95.8(C-6), 60.3(C-4), 41.8(C-5), and 29.4ppm (C-7).

Mass Spectrum:
EIMS: 156(M^+, 11%), 128(12); 114(21); 110(19); 84(100), 73(73), and 55*m/e* (79).
HREIMS: 156.042*m/e*; calcd. For $C_7H_8O_4$, 156.042.

TLC Data
Silica gel; ethyl acetate-acetone (6:4, v/v); R_f=0.53.

References
E. K. Adesogan, and B. I. Alo; Oxysporone, A New Metabolite from *Fusarium oxysporum*; Phytochemistry, Vol. 18, pp. l886-1887(1979).

T. Nagata, and Y. Ando; Oxysporone, a Phytotoxin Isolated from the Tea Gray Bright Fungus *Pestalotia longiseta*; Agric. Biol. Chem., Vol. 53, p. 2811(1989).

Common/Systematic Name

Ascladiol

Molecular Formula/Molecular Weight

$C_7H_8O_4$; MW = 156.04226

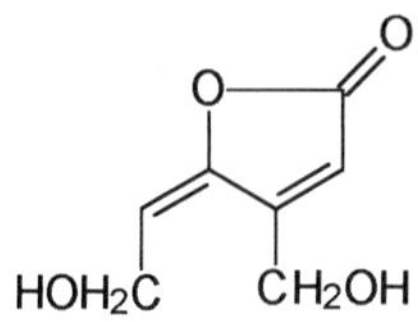

General Characteristics

Crystals from chloroform-acetone; mp., 65-66°C; very hygroscopic.

Fungal Source

Aspergillus clavatus (WF-38-11) isolated from wheat flour.

Isolation/Purification

Liquid culture was shaken with activated charcoal, which was centrifuged and extracted with acetone-water (4:1, v/v) followed by concentration and extraction with ether. The ether extract was chromatographed on silica gel using chloroform-acetone (50:2, v/v). The substance was crystallized from chloroform-acetone.

Biological Activity

Acute toxicity (I.P.) was about 1/4 times as strong as that of patulin.

Spectral Data

UV:

λ_{max}^{EtOH} 271nm

IR:

(KBr) 3300(associated OH), 1750(5-membered lactone-ring), and 1735cm^{-1} (5-membered lactone-ring).

^1H NMR:

(CD_3COCD_3) 6.29(1H, q); 5.87(1H, m); 4.74(4H, broad s); and 4.30ppm (2H).

Mass Data:

Found: C, 53.01, 53.40; H, 5.41, 5.29; calcd. for $C_7H_8O_4 \cdot H_2O$: 53.23; H, 5.23.

Reference

T. Suzuki, M. Takeda, and H. Tanabe; A New Mycotoxin Produced by *Aspergillus clavatus*; Chem. Pharm. Bull., Vol. 19, pp. 1786-1788(1971).

Common/Systematic Name

Anhydrosepedonin

Molecular Formula/Molecular Weight

$C_{11}H_{10}O_4$; MW= 206.05791

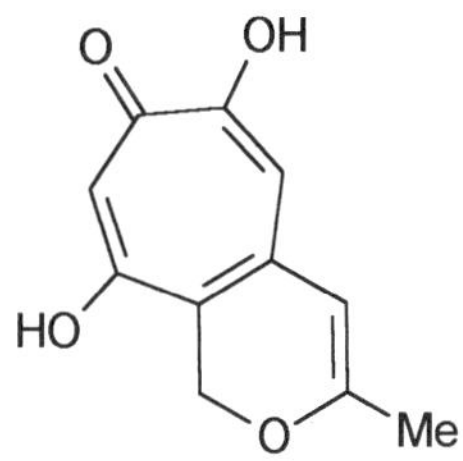

General Characteristics

Glistening, yellow needles; mp., 205 °C (decomp.); sublimed unchanged in high vacuum at 170 °C; Optically inactive. Insoluble in light petroleum (bp., 40-60 °C). Sparingly soluble in acetone, ethanol, methanol, ethyl acetate, dioxane, chloroform, and benzene. Dissolved slowly in aqueous $NaHCO_3$ but readily in aqueous NaOH, and also in conc. HCl from which it was precipitated unchanged on dilution with water. Gave a strong brown color with $FeCl_3$ solution. Anhydrosepedonin probably formed spontaneously in the culture filtrate, and during the isolation process, by chemical dehydration of sepedonin.

Fungal Source

Sepedonium chrysospermum isolated from *Boletus luridus.*

Isolation/Purification

Culture filtrate was acidified with HCl and extracted with ether. The extract was evaporated to dryness, and the resulting gum was triturated with acetone, giving an orange solid (mp., 190-197 °C) which was crystallized from absolute ethanol or glacial acetic acid.

Spectral Data

UV:

λ_{max}^{EtOH} 249(ϵ = 15750), 286(29400), 302(inflection, 18700), 354(inflection, 8270), 375(7250), and 386nm (8130).

IR:

(Nujol) 1660, 1607, 1465, 1430, 1370, 1340, 1310, 1230, 1170, 1140, 1105, 885, 835, and 725cm^{-1}.

^{1}H NMR:

(2N NaDH in D_2O) 6.58(1H of tropolone ring); 6.20(1H of tropolone ring at C-3);

^{1}H NMR:

(2N NaDH in D_2O) 6.58(1H of tropolone ring); 6.20(1H of tropolone ring at C-3); 5.49(1H, olefinic); 5.00(2H, methylene); and 1.87ppm (3H, methyl).

Mass Data:

Mass peaks higher than the expected molecular-ion peak at 206, probably due to the existence of oligomers in the solid state. After hydrogenation of the substance a mass spectrum showed expected molecular-ion peak at 206*m/e*; found: C, 64.0, 64.2; H, 5.52, 5.59; O, by direct estimation, 31.2, OCH_3, nil; CCH_3, 5.5; active H atoms in pyridine, 0.86%; titration equivalent to phenolphthalein, 204. $C_{11}H_{10}O_4$ required: C, 64.1; H, 4.89; O, 31.0; 1-CCH_3, 7.3; 2 active H atoms, 0.98%, equivalent titrating as a monobasic acid, 206.

Reference

P. V. Divekar, H. Raistrick, T. A. Dobson, and L. C. Vining; Studies in the Biochemistry of Microorganisms. Part 117. Sepedonin, a Tropolone Metabolite of *Sepedonium chrysospermum* Fries; Can. J. Chem., Vol. 43, pp. 1835-1848(1965).

Common/Systematic Name

Sepedonin

Molecular Formula/Molecular Weight

$C_{11}H_{12}O_5$; MW = 224.06847

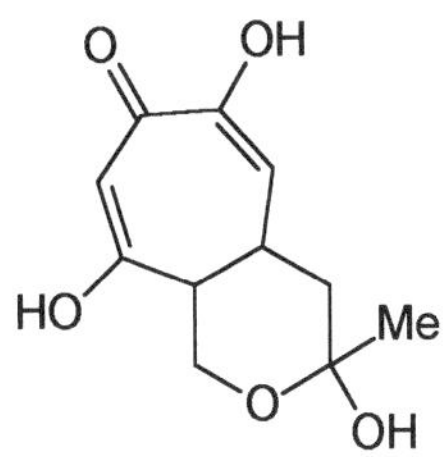

General Characteristics

Pale-yellow prisms from ethyl acetate; mp., 190-205°C (decomp.); $[\alpha]_D^{26}$ 0° (*c*=0.2, in EtOH). Drying at 80°C caused decomposition. Insoluble in petroleum ether, carbon tetrachloride, benzene, and dilute $NaHCO_3$. Poorly soluble in chloroform, ethyl acetate, acetone, and water. Fairly soluble in ethanol, pyridine, dioxane, dilute Na_2CO_3 or NaOH solution, also in NHCl giving a straw-colored solution which rapidly turned bright yellow. Gave a yellow solution in concentrated H_2SO_4 and deep-green color with $FeCl_3$ solution.

Fungal Source

Sepedonium chrysospermum isolated from *Boletus luridus*.

Isolation/Purification

The broth was acidified with HCl and extracted with ethyl acetate. The extract was crystallized after removal of the solvent and recrystallized several times from ethyl acetate to yield sepedonin.

Spectral Data

UV:

λ_{max}^{EtOH} 253(ϵ = 36800) and 327nm (6400).

IR:

(KBr) 3265, 2920, 2850, 1656(sh), 1600, 1573, 1525, 1462, 1378, 944, 870, 796, and 705cm^{-1}.

Mass Data:

Found in a sample dried in high vacuum at room temp.: C, 58.58; H, 5.63; OC_2H_5, 3.14%.

Reference

P. V. Divekar, H. Raistrick, T. A. Dobson, and L. C. Vining; Studies in the Biochemistry of Microorganisms. Part 117. Sepedonin, a Tropolone Metabolite of *Sepedonium chrysospermum* Fries; Can. J. Chem., Vol. 43, pp. 1835-1848(1965).

Common/Systematic Name

Alterporriol B

2,2'-Dimethoxy-4,6,4',5',6',7',8'-heptahydroxy-7,7'-dimethyl-(5'*S*,6'*R*,7'*S*,8'*R*)-5',6',7',8'-tetrahydro-1,1'-bianthraquinone

Molecular Formula/Molecular Weight

$C_{32}H_{26}O_{13}$; MW = 618.13734

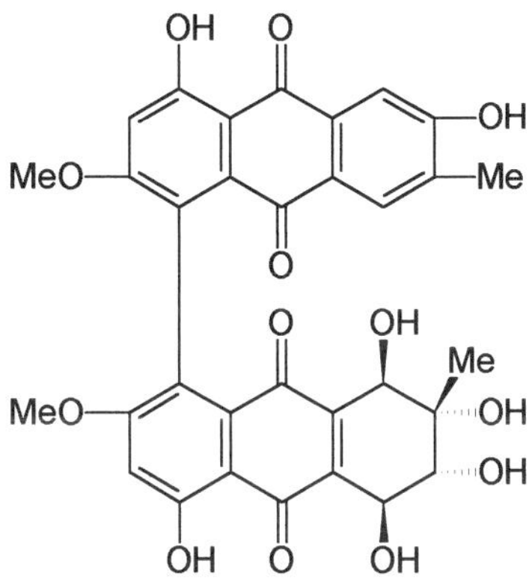

General Characteristics

Melting point >300°C; $[\alpha]_D^{25}$ -310° (*c*=0.05, in EtOH).

Fungal Source

Alternaria porri, a causal fungal agent of black spot disease of stone leek.

Isolation/Purification

The coloring matter obtained from an ethyl acetate soluble part of the culture liquid and mycelium was separated into 5% sodium bicarbonate soluble and insoluble parts and the latter was further separated into 5% sodium carbonate soluble and insoluble parts. Repeated silicic acid chromatography (Merck Kiesel Gel 60) of the 5% sodium carbonate soluble part with a solvent system of methanol-chloroform-formic acid (5:95:2, v/v/v) and preparative TLC (Merck Kiesel Gel GF_{254}) with a solvent system of chloroform-acetone-formic acid (200:100:1, v/v/v) gave a dark-red pigment named alterporriol B.

Spectral Data

IR:

(KBr) 3000-3700(broad OH), 1670, 1650(non-chelated C=O), 1640(chelated C=O), 1210(phenolic OH), 1160(tert-OH), and 1110ppm (sec-OH).

^{1}H NMR:

(d_4-acetic acid) 4.40(1H, s, H-8'); 4.01(1H, d, $J_{5',6'}$=7.8Hz, H-6); 4.91(1H, d, $J_{5',6'}$=7.8Hz, H-5'); 6.81(1H, s, H-4'); 6.81(1H, s, H-3'); 1.39(3H, s, C-7'-CH_3); 3.71(3H, s, C-2'-OCH_3); 7.55(1H, s, H-8); 7.60(1H, s, H-5); 6.82(1H, s, H-3); 6.82(1H, s, H-1); 2.26(3H, s, C-7-CH_3); and 3.72ppm (3H, s, C-2-OCH_3).

Mass Spectrum:

The field desorption mass spectrum of alterporriol B showed only a characteristic fragment ion *m/e* 582 ($C_{32}H_{22}O_{11}$). However, the molecular secondary ion mass spectrum showed fragment ions at 641(M^+ +Na) and 619(M^+ +H). The high resolution mass spectrum (In Beam EI) gave peaks at *m/e* 582.1161($C_{32}H_{22}O_{11}$, M^+ - $2H_2O$), 566.1204($C_{32}H_{22}O_{10}$, 582 +2H -H_2O), 551.1026($C_{31}H_{19}O_{10}$), 535.1000($C_{31}H_{29}O_9$, 566 -OCH_3), 284.0695($C_{16}H_{12}O_5$), 255.0646($C_{15}H_{11}O_4$, 284 -CHO), and 241.0941*m/e* ($C_{14}H_9O_4$).

Reference

R. Suemitsu, T. Sano, M. Yamamoto, Y. Anmoto, F. Morimatsu, and T. Nabeshima; Structural Elucidation of Alterporriol B, a Novel Metabolic Pigment Produced by *Alternaria porri* (Ellis) Ciferri; Agric. Biol. Chem., Vol. 48, pp. 2611-2613(1984).

Common/Systematic Name
LL-S491β

Molecular Formula/Molecular Weight
$C_{20}H_{26}O_5$; MW = 346.17802

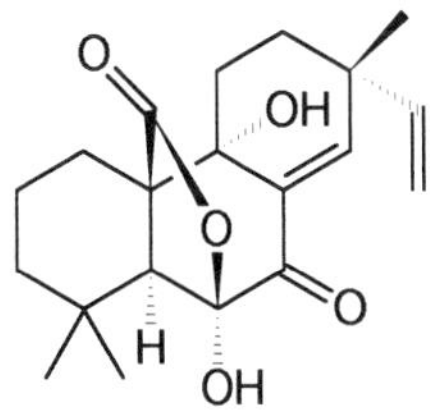

General Characteristics
Melting point, 180-185°C; $[\alpha]_D$ +112.4° (MeOH).

Fungal Source
Aspergillus chevalieri (Lederle culture S491).

Biological Activity
Antibacterial activity against certain gram-positive organisms. Strong antiprotozoal activity against *Tetrahymena pyriformis.*

Spectral Data

UV:
λ_{max}^{MeOH} 241nm (ε=5850); $\lambda_{max}^{MeOH+NaOH}$ 263nm (ε=4600) (reversible on addition of acid; α-diketone chromophore).

IR:
(KBr) 3500(-OH), 1755(γ-lactone), 1710(ketone), and 1620cm^{-1} (=bond).

CD:
Negative Cotton effect at 245nm with a Δε=3.7.

^{1}H NMR:
($CDCl_3$) 5.02-5.65(2H, 8-line pattern); 5.84(1H, 4-line system); 1.09(3H, s, 3C-CH_3); 1.20(6H, s, 3C-CH_3); and 1.90 and 5.25ppm (two 1H exchangeable, sharp s, two tertiary OH).

Reference
G. A. Ellestad, M. P. Kunstmann, P. Mirando, and G. O. Morton; Structures of Fungal Diterpene Antibiotics LL-S491β and γ; J. Amer. Chem. Soc., Vol. 94, pp. 6206 6208(1972).

Common/Systematic Name

LL-S491γ

Molecular Formula/Molecular Weight

$C_{20}H_{28}O_5$; MW= 348.19367

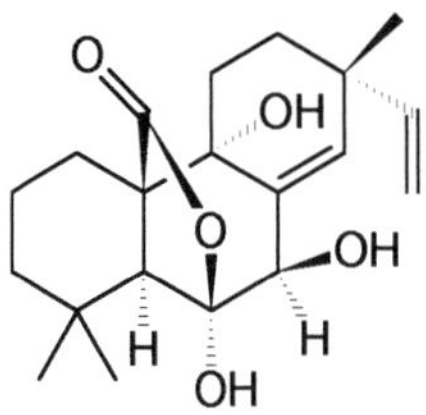

General Characteristics

Melting point, 190-195°C; $[\alpha]_D$ +63.3° (MeOH).

Fungal Source

Aspergillus chevalieri (Lederle culture S491).

Biological Activity

Antiviral activity against *Herpes simplex.* Strong antiprotozoal activity against *Tetrahymena pyriformis.*

Spectral Data

UV:

End absorption

IR:

(KBr) 1745cm-1 (γ-lactol)

CD:

Positive Cotton effect at 209nm with a Δϵ=8.3.

^{1}H NMR:

($CDCl_3$) 5.83(1, d, $J_{7,14}$=2.0Hz, H-14); and 4.30ppm (H-7).

Reference

G. A. Ellestad, M. P. Kunstmann, P. Mirando, and G. O. Morton; Structures of Fungal Diterpene Antibiotics LL-S491β and γ; J. Amer. Chem. Soc., Vol. 94, pp. 6206-6208 (1972).

Common/Systematic Name

cis-Sativenediol

Molecular Formula/Molecular Weight

$C_{15}H_{24}O_2$; MW = 238.19328

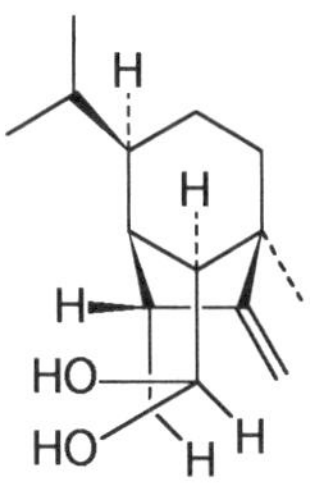

General Characteristics

Oil; $[\alpha]_D$ -119° (c=0.94, in $CHCl_3$).

Fungal Source

Cochliobolus setariae IFO 6635 and *Helminthosporium sativum*.

Isolation/Purification

Both species of fungi were cultured by shake culture on malt-dextrose medium. The culture broth was extracted with ethyl acetate, *cis*- and *trans*-sativenediol were isolated from the crude extract by silicic acid column chromatography and elution with 3% ethyl acetate in *n*-hexane. Final purification of these diols from each other was accomplished by treating the mixture with 2,2-dimethoxypropane and *p*-TsOH. Pure *trans*-sativenediol and an acetonide of the *cis*-diol were isolated from the reaction mixture, and the latter compound was subsequently hydrolyzed liberating *cis*-sativenediol in pure form.

Biological Activity

A plant growth promoter with gibberellin-like activity that promoted elongation of rice seedings; at concentrations of 3, 30, and 300μg/ml; *cis*-sativenediol promoted elongation at 77%, 176%, and 110% of control seedlings, respectively.

Spectral Data

IR:

(CCl_4) 3670 and 3360cm^{-1} (OH groups).

1H NMR:

($CDCl_3$) Two 3H doublets at 0.89, and 0.95ppm, J=7Hz, for an isopropyl group; a tertiary methyl (3H singlet at 1.08ppm); an exocyclic methylene (two 1H doublets at 4.66 and 4.98ppm); two vicinal hydroxyl groups (two 1H doublets at 3.68 and 4.08ppm, J=6Hz) which were decoupled to each other in double resonance experiments and shifted downfield (4.72 and 5.05ppm, respectively); acetylation to the diacetate resulted in two CH_3COO groups at 2.05ppm. These data were indicative of a dihydroxy derivative of sativene.

Mass Spectrum:
236*m/e* (M^+); diacetate derivative, 320*m/e* (M^+).

Reference

M. Nukina, H. Hattori, and S. Marumo; *cis*-Sativenediol, a Plant Growth Promotor, Produced by Fungi; J. Am. Chem. Soc., Vol. 97, pp. 2542-2543(1975).

Common/Systematic Name

trans-Sativenediol; Isosativenediol

Molecular Formula/Molecular Weight

$C_{15}H_{24}O_2$; MW = 236.17763

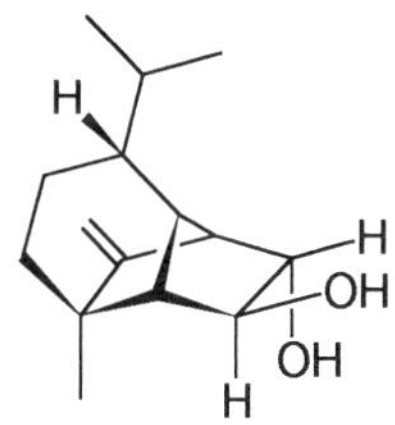

General Characteristics

Colorless needles; mp., 176°C; $[\alpha]_D^{27}$ -3.4° (*c*=1.4, in $CHCl_3$).

Fungal Source

Cochliobolus setariae IFO 6635, *Helminthosporium sativum* (reclassified as *Bipolaris sorokiniana,* which is known to cause seedling blight, foot and root rot, head and leaf spot of cereals and grasses), and *H. victoriae.*

Isolation/Purification

Fungi were cultured by shake culture on malt-dextrose medium. The culture broth was extracted with ethyl acetate, *cis-* and *trans*-sativenediol were isolated from the crude extract by silicic acid column chromatography and elution with 3% ethyl acetate in *n*-hexane. Final purification of these diols from each other was accomplished by treating the mixture with 2,2-dimethoxypropane and *p*-TsOH. Pure *trans*-sativenediol and an acetonide of the *cis*-diol were isolated from the reaction mixture, and the latter compound was subsequently hydrolyzed liberating *cis*-sativenediol in pure form.

Biological Activity

A plant growth inhibitor that inhibited elongation of rice seedlings only slightly; at concentrations of 3, 30, and 300μg/ml *trans*-sativenediol inhibited elongation 8%, 15%, and 16% of control seedlings, respectively.

Spectral Data

^{1}H NMR:

($CDCl_3$) Two 3H doublets at 0.95 and 1.05ppm, *J*=6.5Hz, for an isopropyl group; a tertiary methyl (3H singlet at 1.00ppm); an exocyclic methylene (two 1H doublets at 4.75 and 5.05ppm); two vicinal hydroxyl groups (two 1H of a slightly splitting broad singlet at 3.68 and 4.45ppm, *J*=<2Hz) which were decoupled from each other by double resonance experiments and shifted downfield to 4.70 and 4.92ppm, respectively); acetylation to the diacetate resulted in two CH_3COO groups at 1.99, and 2.07ppm.

Mass Spectrum:
218*m/e* (M^+ - H_2O); diacetate derivative, 320*m/e* (M^+).

References

F. Dorn and D. Arigoni; Zwei neue Sesquiterpene mit Isosativangerust aus *Helminthosporium sativum*: Derivatives of Isosativene from *Helminthosporium sativum*; Experientia, Vol. 31, pp. 753-754(1975).

M. Nukina, H. Hattori, and S. Marumo; *cis*-Sativenediol, a Plant Growth Promotor, Produced by Fungi; J. Am. Chem. Soc., Vol. 97, pp. 2542-2543(1975).

Common/Systematic Name
Isosativenetriol

Molecular Formula/Molecular Weight
$C_{15}H_{24}O_3$; MW = 252.17254

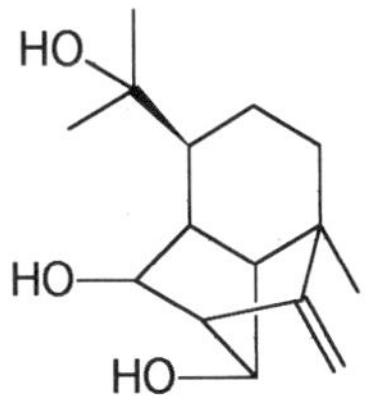

General Characteristics
Crystals; mp., 206°C; $[\alpha]_D^{25}$ +11° (EtOH).

Fungal Source
Helminthosporium sativum.

Biological Activity
Phytotoxin.

Reference
F. Dorn and D. Arigoni; Zwei neue Sesquiterpene mit Isosativangerust aus *Helminthosporium sativum*: Derivatives of Isosativene from *Helminthosporium sativum*; Experientia, Vol. 31, pp. 753-754(1975).

Common/Systematic Name

Patulin; Claviformin; Clavitin; Clavicin; Expansion; Penicidin; Mycoin; Leucopin; Tercinin
4-Hydroxy-4*H*-furo[3,2-c]pyran-2(6*H*)-one

Molecular Formula/Molecular Weight

$C_7H_6O_4$; MW = 154.02661

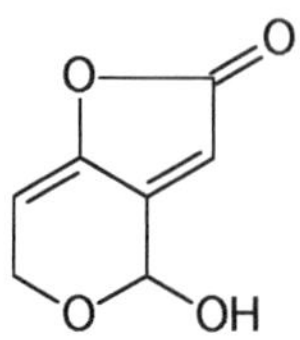

General Characteristics

Crystals from benzene; mp., 110-111°C; $[\alpha]_D$ 0°. Patulin is soluble in water, alcohols, acetone, ethyl acetate, chloroform; sparingly soluble in ethyl ether, benzene; insoluble in petroleum ether. It is somewhat unstable in polar solvents such as water and methanol and loses its biological potency in alkaline conditions. Acetyl derivative, prisms from methanol, mp., 118-120°C. Phenylhydrazone, orange plates, mp., 152-153°C. Oxime, crystals from water, mp., 152-153°C (dec.). Isopatulin, prisms from ethyl acetate, mp., 88-90°C.

Fungal Source

Penicillium expansum (P. leucopus), P. patulum Bainier (P. urticae, P. griseofulvum), P. claviforme, P. equinum (P. terrestre), P. novae-zeelandiae, P. lapidosum, P. granulatum (P. divergens), P. lanosum, P. melinfi, P. cyclopium, P. cyaneofulvum, P. roqueforh, Aspergillus clavatus, A. giganteus, A. terreus, and *Byssochlamys nivea (Gymnoascus* spp.*).*

Biological Activity

Patulin was originally isolated because of its antibiotic properties against gram-positive and gram-negative bacteria. However, it was soon found to be too toxic for use as an antimicrobial agent in humans. The LD_{50} of patulin in White Leghorn cockerels was 170mg/kg. Sublethal doses produced hemorrhaging in the digestive tract. LD_{50} (IP) in mice was 5mg/kg. Patulin has been reported to be carcinogenic and a potent mitotic toxin in addition to being toxic to many protozoa, fungi, mammals, plants, HeLa cells, and viruses. Patulin has been found contaminating apple juice and has been implicated in the death of cattle.

Spectral Data

UV:

λ_{max} 276nm (ϵ=16,600)

^1H NMR:
6.96(H-3); 5.90(H-4); 4.32(H-Sa); 4.66(H-Sb); 5.90(H-6); and 5.94ppm (4-OH).

^{13}C NMR:
168.5(C-2, s); 109.8(C-3, d); 88.4(C-4, d); 59.1(C-5, t); 107.9(C-6, d); 151.5(C-7, s); and 146.1ppm (C-8, s).

TLC Data
Adsorbent: silica gel G-HR; Solvent: toluene-ethyl acetate-formic acid, 5:4:1 v/v/v, R_f=0.58; Detection: brown spot after spraying with 50% ethanolic H_2SO_4 and heating at 100°C for 5 min.

GC Data
Solid support: Gas Chrom-Q, Liquid phase: 10% DC-200, Column temp: 175°C, Detector temp, 190°C, Injector temp, 190°C. Relative retention time (relative to 13-BHC and analyzed as TMS derivative): 0.56. Solid support: Gas Chrom-Q, 100-120 mesh; Liquid phases: 3% Dexsil 300, 3% OV-17 and 3% OV-25.

Reference
R. J. Cole, and R. H. Cox; Handbook of Toxic Fungal Metabolites; Academic Press, New York, pp. 937(1981).

Common/Systematic Name
Isopatulin

Molecular Formula/Molecular Weight
$C_7H_6O_4$; MW = 154.02661

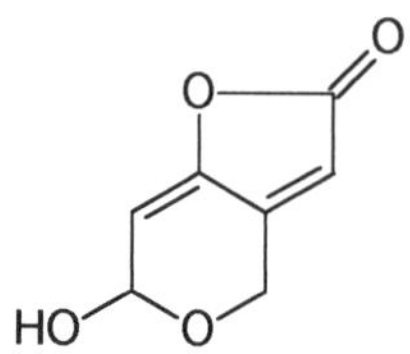

General Characteristics
Prisms from ethyl acetate; mp., 88-90°C.

Fungal Source
Penicillium urticae (2159k, cells).

Isolation/Purification
Acrylamide gel with fungi cells embedded in it was acidified and extracted with ethyl acetate. The extract was chromatographed on preparative TLC plates (chloroform-ethyl acetate-ethyl ether, 5:2:1, v/v/v). Isopatulin crystallized from ethyl acetate.

Spectral Data

UV:
λ_{max}^{MeOH} 269nm (ϵ=12800)

IR:
(Nujol) 3400, 1780, 1755, 1675, 1610, 1440, 1400, 1280, 1225, 1175, 1110, 1070, 1040, 1000, 980, 900, 845, 820, and 795cm^{-1}.

CD:
Optically inactive.

^{1}H NMR:
(CD_3COCD_3) 3.08(s, 5-OH); 4.74(dd, $J_{a\text{-}b}$=16.9, $J_{b\text{-}7}$=1.2Hz, 1-Hb); 5.04(dd, $J_{a\text{-}b}$=16.9, $J_{a\text{-}7}$=2.2Hz, 1-Ha; shifts of Ha and Hb may be interchanged); 5.72(m, 5-H); and 6.01ppm (m, 7-H).

^{13}C NMR:
(CD_3COCD_3) 58.2(t, 1-C); 90.4(d, 5-C); 109.0, 110.8(each d, for 4 and 7-C); 150.2(s, 2-C); 153.4(s, 3-C); and 169.6ppm (s, 8-C).

Mass Data:

154(M^+), 136, 126, 112, 97, 67, and 39*m/e*; found: C, 54.79; H, 4.15; calcd. for $C_7H_6O_4$: C, 54.55; H, 3.92.

TLC Data

Chloroform-ethyl acetate-ethyl ether (5:2:1, v/v/v); R_f=0.30. Detection: UV light.

Reference

J. Sekiguchi, G. M. Gaucher, and Y. Yamada; Biosynthesis of Patulin in *Penicillium urticae*: Identification of Isopatulin as a New Intermediate; Tetrahedron Lett., No. 1, pp. 41-42(1970).

Common/Systematic Name

Acrostalidic acid

Molecular Formula/Molecular Weight

$C_{16}H_{22}O_4$; MW = 278.15181

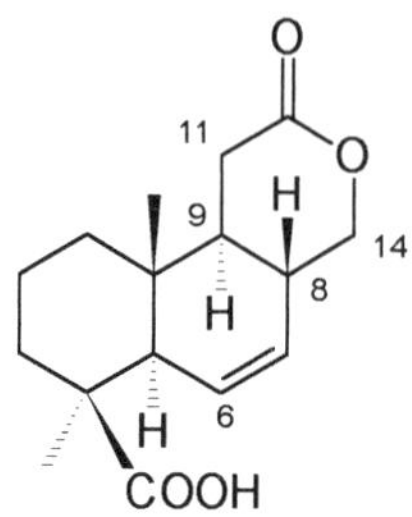

General Characteristics

Recrystallization from ethyl acetate gave crystals; mp., 210-211°C.

Fungal Source

Acrostalagmus sp. (NRRL 3481).

Isolation/Purification

The fermented mash was adjusted to pH 2 with 2N hydrochloric acid. The mash was filtered through Celite and the filter pad was washed with water. The filtrate was extracted with ethyl acetate, the organic layer was successively washed with water and saturated salt solution, dried (Na_2SO_4), and evaporated under reduced pressure to give an oily residue. This was washed with *n*-hexane, and the crude product was dissolved in ethyl acetate and precipitated with benzene. The supernatant was evaporated to give an oil. Sufficient acetone was added to dissolve the oil and silica gel was added to the solution. The solvent was removed under reduced pressure and the dry silica gel was placed on a silica gel column slurry packed in benzene. The column was eluted successively with benzene, benzene containing increasing amounts of chloroform, chloroform, and chloroform containing increasing amounts of ethyl acetate. From fractions eluted by chloroform-ethyl acetate, LL-Z1271α, LL-Z1271β, and LL-Z1271δ were isolated, respectively. Further column chromatography of fractions eluted by chloroform-ethyl acetate, (6:1 - 3:1) was carried out on silica gel by the usual method. The column was eluted successively with chloroform and chloroform containing an increasing proportion of ethyl acetate crude; acrostalidic acid was obtained from fractions eluted by chloroform-ethyl acetate (25:1). Recrystallization from ethyl acetate gave crystals of acrostalidic acid.

Biological Activity

Related to the antifungal metabolite LL-Z1271α. Therefore, may also be antifungal.

Spectral Data

IR:
(KBr) 3030cm^{-1} (olefinic); ($CHCl_3$) 1720 and 1690cm^{-1}.

^{1}H NMR:
($CDCl_3$); 2.03(H-5); 2.33(H-8); 1.57(H-9); 2.23(dd, *J*=18 and 12.5Hz, H=11β); 2.62(dd, *J*=18 and 5Hz, H-11α); 3.87(dd, *J*=11.5 and 10.5Hz, H-14α); 4.45(dd, *J*=10.5 and 5Hz, H-14β); and 10.1ppm (CO_2H).

^{13}C NMR:
($CDCl_3$) 54.7; 46.1; 43.1; 37.0; 35.5; 35.2(2 C); 30.6; 28.0; 18.9; and 10.7ppm.

Mass Spectrum:
278(M^+), 263, 260, 233(intense), 217, 173, 145, and 105*m/e* (base peak); high resolution 278.150(M^+; $C_{16}H_{22}O_4$ requires M, 278.152, 263.126($C_{15}H_{19}O_4$, M-CH_3), 233.153($C_{15}H_{21}O_2$, M-CO_2H), 217.116($C_{14}H_{17}O_2$), 173.132($C_{14}H_{17}$), 145.102($C_{11}H_{13}$); and 105.070*m/e* (C_8H_9).

Reference

M. Sato and H. Kakisawa; Structures of Three New C_{16} Terpenoids from an *Acrostalagmus* Fungus; J. Chem. Soc., Trans Perkin I, pp. 2407-2413(1976).

Common/Systematic Name

Acrostalic acid

Molecular Formula/Molecular Weight

$C_{16}H_{24}O_4$; MW = 280.16746

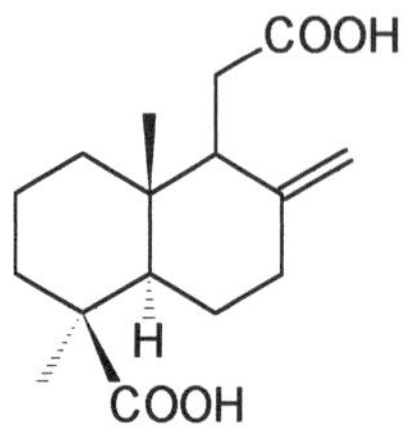

General Characteristics

Recrystallization from ethyl acetate gave crystals; mp., 219-220°C.

Fungal Source

Acrostalagmus sp. (NRRL – 3481).

Isolation/Purification

The fermented mash was adjusted to pH 2 with 2N hydrochloric acid. The mash was filtered through Celite and the filter pad was washed with water. The filtrate was extracted with ethyl acetate and the organic layer was successively washed with water and saturated salt solution, dried (Na_2SO_4), and evaporated under reduced pressure to give an oily residue. This was washed with *n*-hexane, and the crude product was dissolved in ethyl acetate and precipitated with benzene. The supernatant was evaporated to give an oil. Sufficient acetone was added to dissolve the oil and silica gel was added to the solution. The solvent was removed under reduced pressure and the dry silica gel was placed on a silica gel column slurry packed in benzene. The column was eluted successively with benzene, benzene containing increasing amounts of chloroform, chloroform, and chloroform containing increasing amounts of ethyl acetate. From fractions eluted by chloroform-ethyl acetate (12.5:1,v/v/v) crude acrostalic acid was obtained. It was recrystallized from ethyl acetate to give crystals.

Biological Activity

Related to the antifungal metabolite LL-Z1271α. Therefore it may also be antifungal.

Spectral Data

IR:

(KBr) 3080, 1645, and 890cm^{-1} (=CH_2).

^{1}H NMR:

(($(CD_3)_2CO$) 0.66 and 1.23ppm (s, 10- and 4-methyls).

Mass Spectrum:
280(M^+), 262, 234, and 109*m/e*; high resolution 280.169*m/e* (M^+); $C_{16}H_{24}O_4$ requires M, 280.167, 262.150($C_{16}H_{22}O_3$, M-H_2O), 234.159($C_{15}H_{22}O_2$, M-HCO_2H), and 109.107*m/e* (C_8H_{13}).

Reference

M. Sato and H. Kakisawa; Structures of Three New C_{16} Terpenoids from an *Acrostalagmus* Fungus; J. Chem. Soc., Trans Perkin I, pp. 2407-2413(1976).

Common/Systematic Name

Pimara-8(9),15-diene

Molecular Formula/Molecular Weight

$C_{20}H_{32}$; MW = 272.25040

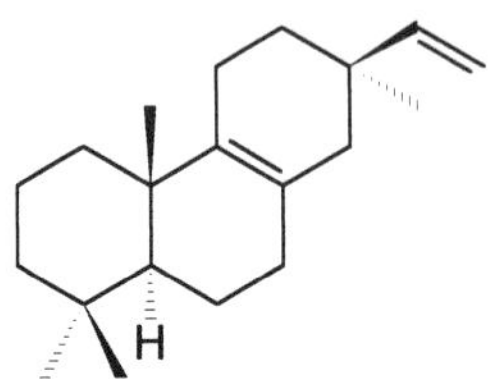

Fungal Source

Trichothecium roseum, Cephalosporium aphidicola, and *Nigrospora sphaerica.*

Isolation/Purification

Mycelium was extracted with acetone, acetone extract was concentrated *in vacuo* and extracted with ethyl acetate. The extract was dried, evaporated, and the residue adsorbed onto silica gel and chromatographed on alumina. Elution with light petroleum gave a mixture of hydrocarbons which was analyzed by GLC on carbowax.

Spectral Data

Mass Spectrum:

EIMS: 272(M^+), 257(base peak), 215, 201, 189, 187, 175, 161, 157, 147, 133, 121, 119, 109, 107, and 105*m/e*.

References

M. R. Adams and J. D. Bu'Lock; Biosynthesis of the Diterpene Antibiotic, Aphidicolin, by Radioisotope and ^{13}C-Nuclear Magnetic Resonance Methods; J. Chem. Soc. Chem. Commun., pp. 989-991(1975).

B. Dockerill and J. R. Hanson; Studies in Terpenoid Biosynthesis. Part 19. Formation of Pimara-8(9),15-diene by *Trichothecium roseum*; J. Chem. Soc., Perkin Trans I, pp. 324-327(1977).

Common/Systematic Name

Isoacrostalidic acid

Molecular Formula/Molecular Weight

$C_{15}H_{20}O_4$; MW = 264.13616

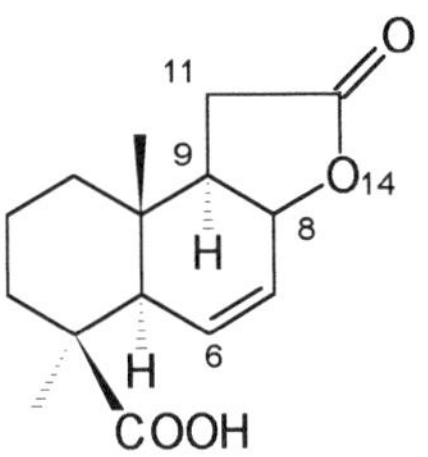

General Characteristics

Recrystallization from ethyl acetate gave crystals; mp., 204-206°C.

Fungal Source

Acrostalagmus sp. (NRRL – 3481).

Isolation/Purification

The fermented mash was adjusted to pH 2 with 2N hydrochloric acid. The mash was filtered through Celite and the filter pad was washed with water. The filtrate was extracted with ethyl acetate and the organic layer was successively washed with water and saturated salt solution, dried (Na_2SO_4), and evaporated under reduced pressure to give an oily residue. This was washed with *n*-hexane, and the crude product was dissolved in ethyl acetate and precipitated with benzene. The supernatant was evaporated to give an oil. Sufficient acetone was added to dissolve the oil and silica gel was added to the solution. The solvent was removed under reduced pressure and the dry silica gel was placed on a silica gel column slurry packed in benzene. The column was eluted successively with benzene, benzene containing increasing amounts of chloroform, chloroform, and chloroform containing increasing amounts of ethyl acetate. Fractions eluted by chloroform-ethyl acetate (50:1,v/v) followed by recrystallization from ethyl acetate giving needles of isoacrostalidic acid.

Biological Activity

Related to the antifungal metabolite LL-Z1271α. Therefore it may also be antifungal.

Spectral Data

IR:

(KBr) 3050cm^{-1} (olefinic); ($CHCl_3$) 1755cm^{-1}.

Mass Spectrum:

278*m/e* (M^+); high resolution 278.152*m*/e (M^+, $C_{16}H_{22}O_4$ requires M), 278.152, 263.132($C_{15}H_{19}O_4$, M-CH_3), 232.146($C_{15}H_{20}O_2$), 217.120($C_{14}H_{17}O_2$), 189.127, and 109.107*m/e* (C_8H_{13}).

Reference

M. Sato and H. Kakisawa; Structures of Three New C_{16} Terpenoids from an *Acrostalagmus* Fungus; J. Chem. Soc., Trans Perkin I, pp. 2407-2413(1976).

Common/Systematic Name
Deoxypatulinic acid
2,3-Dihydro-4-pyrone-5-acetic acid

Molecular Formula/Molecular Weight
$C_7H_8O_4$; MW = 156.04226

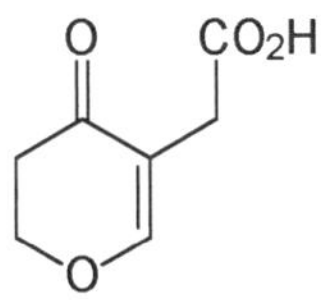

General Characteristics
Crystals; mp., 115-115.5°C; soluble in water (to give an acidic pH) and aqueous $NaHCO_3$ (with effervescence).

Fungal Source
Penicillium patulum (P. urticae), strain No. 562 from mixed feed.

Isolation/Purification
Extracted with ethyl acetate and chromatographed on silica gel. The substance was eluted with benzene-ethyl acetate (85:15 to 75:25, v/v); crystallized from isopropyl ether-*n*-hexane and toluene, followed by sublimation.

Biological Activity
Had no inhibitory effect (46mg/ml) on *Bacillus megaterium, B. subtilis, B. cereus, Staphylococcus aureus, S. epidermidis, Sarcina lutea, Micrococcus flavus,* and *Saccharomyces cerevisiae.*

Spectral Data

UV:
λ_{max}^{EtOH} 267nm (ϵ=7960)

IR:
($CHCl_3$) 3510(OH), 3000-2400(COOH), 1718(COOH dimer), 1672(conjugated C=O), and 1621cm^{-1} (C=C).

^{1}H NMR:
($CDCl_3$) 1.53(1H, s, exchanged with D_2O); 2.63(1H, s, COC=CHO); 5.48(2H, t, *J*=7Hz, CH_2O); 6.83(2H, s, $COCH_2C$=C); 7.34(2H, t, *J*=7Hz, COH_2); each triplet collapsed to a singlet on irradiation at the frequency of the other triplet.

Mass Data:

156(M^+), 112($M-CO_2$), 83, 60[$CH_2=C(OH)_2^+$], 55, 39, and 27*m/e*; found: C, 53.94; H, 5.16; calcd. for $C_7H_8O_4$: C, 53.85, H, 5.16%).

TLC Data

Silica gel F_{254}; toluene-ethyl acetate-formic acid (6:3:1, v/v/v); R_f= 0.22. Detection: dark spot under short-wavelength UV.

Reference

P. M. Scott, B. Kennedy, and W. van Walbeek; Desoxypatulinic Acid from a Patulin-Producing Strain of *Penicillium patulum*; Experientia, Vol. 28, p. 1252(1972).

Common/Systematic Name

Dihydropenicillic acid

Molecular Formula/Molecular Weight

$C_8H_{12}O_4$; MW = 172.07356

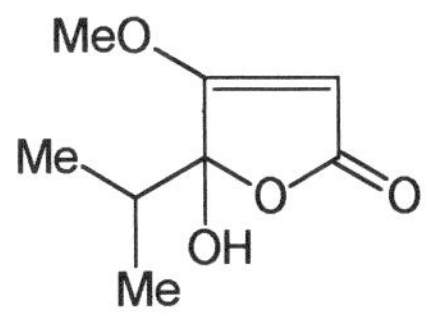

General Characteristics

Colorless needles from ethyl acetate-hexane; mp., 84°C. Hydrogenation with 5% palladium on charcoal gave penicillic acid.

Fungal Source

Unidentified fungus.

Isolation/Purification

Fungal culture acidified to pH 3.0 and extracted with ethyl acetate. The ethyl acetate extracts were concentrated and extracted several times with sodium bicarbonate solution. Concentration of the ethyl acetate solution produced crystals which were recrystallized from ethyl acetate-hexane to give colorless needles.

Spectral Data

UV:

Quite similar to that of penicillic acid.

IR:

Quite similar to that of penicillic acid.

^{1}H NMR:

0.88(3H, d, *J*=7.0Hz); 1.06(1H, d, *J*=7.0Hz); 2.18(1H, septet, *J*=7.0Hz); 3.90(3H, s, methoxy); and 5.02ppm (1H, vinyl proton).

Reference

T. Sassa, S. Hayakari, M. Ikeda, and Y. Miura; Plant Growth Inhibitors Produced by Fungi: Part I. Isolation and Identification of Penicillic Acid and Dihydropenicillic Acid; Agric Biol. Chem., Vol. 35, pp. 2130-2131(1971).

Common/Systematic Name

(-)-Longifolene

Molecular Formula/Molecular Weight

$C_{15}H_{24}$; MW = 204.18780

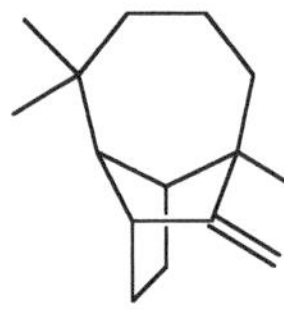

Fungal Source

Helminthosporium sativum, H. victoriae, and *H. setariae.*

References

F. Dorn and D. Argigoni; Ein Bicyclischer Abkommling von (-)-Longifolen aus *Helminthosporium sativum* und *H. victoriae*; Experientia, Vol. 30, pp. 851-852(1974).

F. Dorn and D. Argigoni; Sesquiterpenoide Metaboliten aus *Helminthosporium*-Arten: Struktur und Biosynthese; Chemica, Vol. 29, pp. 24-25(1975).

Common/Systematic Name

Victoxinine

Molecular Formula/Molecular Weight

$C_{17}H_{29}NO$; MW = 263.22491

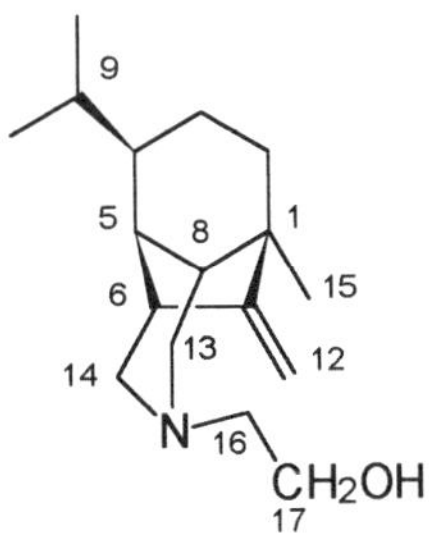

Fungal Source

Bipolaris sp. strain 36 (ATTC 64838), a fungal pathogen of Johnson Grass (*sorghum halepense* (L.) Pers.), *Helminthosporium victoriae*, the causal agent of Victoria blight of oats. Victoxinine has been reported from culture filtrates of *Helminthosporium sativum* (reclassified as *Bipolaris sorokiniana)*

Isolation/Purification

Purification of the crude fungal mixture was accomplished by flash column chromatography using Merck silica gel 60 and stepwise elution with *n*-hexane-Et_2O (90:10 to 30:70, v/v).

Biological Activity

Phytotoxic, inhibited root growth of *H. victoriae* susceptible and resistant oat plants at 75μg/ml. Victoxinine is a component of victorin, the host-specific toxin of *H. victoriae*.

Spectral Data

IR:

(KBr) 3464, 3063 and 1655cm^{-1}.

^1H NMR:

($CDCl_3$) 4.75(1H, s, H-12a); 4.59(1H, s, H-12b); 3.50(2H, m, H-17); 2.99(1H, dd, 3.0, 10.9Hz, H-13a); 2.81(1H, dd, 4.7, 9.3Hz, H-14a); 2.58(1H, br m, H-6); 2.52(1H, ddd, 4.7, 8 0, 12.6Hz, H-16a); 2.41(1H, dt, 4.4, 12.5Hz, H-16b); 2.34(1H, dd, 1.6, 10.9Hz, H-13b); 2.23(1H, d, 9.4Hz, H-14b); 1.82(1H, br s, H-5); 1.42(1H, m, H-8); 1.35(1H, m, H-9); 1.13(3H, s, H-15); 0.90(3H, d, 6.6Hz, H-10); 0.86(3H, d, 6.6, H-11). $J_{5,8}$, $J_{6,8}$, $J_{6,14}$ all = < 1Hz; $J_{6,14a}$=4.7Hz; $J_{8,13a}$=3.0Hz; $J_{8,13b}$=1.6Hz; $J_{13a,b}$=10.9Hz; $J_{14a,b}$=9.3Hz; $J_{16a,b}$=4.7Hz; $J_{16a,17b}$=8.0Hz; $J_{16b,17a}$=4.4Hz; and $J_{16b,17b}$=4.4Hz.

Mass Spectrum:
CIMS: (CH_4) 264(M^+, 100%), and 232*m/e* (39); HRMS: (parent ion not observed) $\{M - CH_2OH\}^+$; calcd for $C_{16}H_{22}N$ 232.2065, found 232.2066.

References

F. Dorn and D. Arigoni; Structure of Victoxinine; J. C. S. Chem. Commun., pp. 1342-1343(1972).

L. M. Peña-Rodriguez and W. S. Chilton; Victoxinine and Prehelminthosporolactone, Two Minor Phytoxic Metabolites Produced by *Bipolaris sp.*, A Pathogen of Johnson Grass; J. Natural Products, Vol. 52, pp. 899-901(1989).

R. B. Pringle; Compaative Biochemistry of the Phytopathogenic Fungus *Helminthosporium*. XVI. The Production of Victoxinine by *H. sativum* and *H. victoriae*; Can. J. Biochem., Vol. 54, pp. 783-787(1976).

R. B. Pringle and A. C. Braun; Isolation of Victoxinine from Cultures of *Helminthosporium victoriae*; Phytopathology, Vol. 50, pp. 324-325(1960).

Common/Systematic Name
Antibiotic A26771B

Molecular Formula/Molecular Weight
$C_{20}H_{30}O_7$; MW = 382.19915

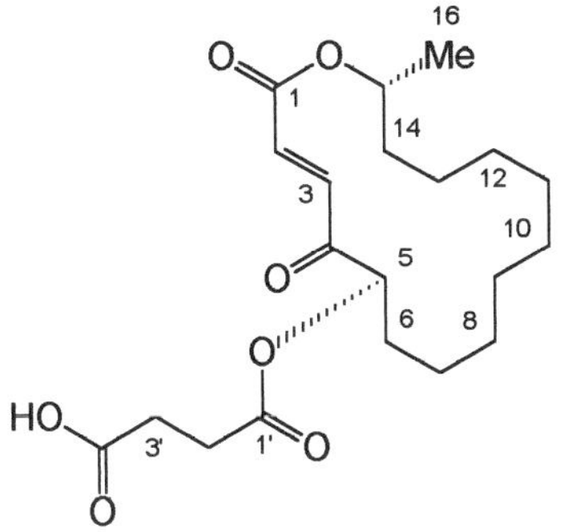

General Characteristics
Crystals from ethyl acetate-hexane; mp., 121-123°C; $[\alpha]_D$ -13.1° (*c*=0.65, in MeOH); $[\alpha]_D$ -14° (*c*=0.13, in MeOH).

Fungal Source
Penicillium turbatum (ATCC 28797).

Isolation/Purification
The fungal mycelium was filtered, dried, and extracted (3X) with ethyl acetate. The combined extracts were dried (Na_2SO_4), concentrated *in vacuo*, and chromatographed on preparative TLC plates with ethyl acetate-chloroform (2:1, v/v) to give a lipid fraction and crude antibiotic A26771B. The latter was recrystallized from ethyl acetate-hexane to give pure A26771B.

Biological Activity
Antibiotic

Spectral Data

IR:
(KBr) 3420, 1748, 1713, and 1701cm^{-1}.

^{1}H NMR:
($CDCl_3$) 1.13-1.60(15H, m); 1.29(3H, d, *J*=6.5Hz); 1.68(1H, m); 1.89(2H, m); 2.72(4H, m); 5.13(1H, m); 5.31(1H, br t, *J*=5.5Hz); 6.73(1H, d, *J*=15.5Hz); and 7.22ppm (1H, d, *J*=15.5Hz).

^{13}C NMR:

($CDCl_3$) C-16, 19.54; C-7, 22.08; C-13, 23.33; C-9, 26.75; C-10, 26.91; C-8, 27.05; C-12, 27.66; C-11, 27.77; C-2', 28.11; C-3', 28.41; C-6, 28.72; C-14, 34.34; C-15, 72.57; C-5, 77.32; C-2, 132.78; C-3, 135.59; C-1, 165.28; C-1', 171.85; C-4', 177.03; and C-4, 195.95ppm.

Mass Spectrum:

CIMS: (NH_3) 400($M^+ + NH_4^+$) and 381*m/e* ($M^+ - H^+$).

Reference

K. Arai, B. J. Rawlings, Y. Yoshizawa, and J. C. Vederas; Biosyntheses of Antibiotic A26771B by *Penicillium turbatum* and Dehydrocurvularin by *Alternaria cinerariae*: Comparison of Stereochemistry of Polyketide and Fatty Acid Enoyl Thiol Ester Reductases; J. Am. Chem. Soc., Vol. 111, pp. 3391-3399(1989).

Common/Systematic Name
IFO 6635 Aversion Factor

Molecular Formula/Molecular Weight
$C_{30}H_{46}O_4$; MW = 470.33961

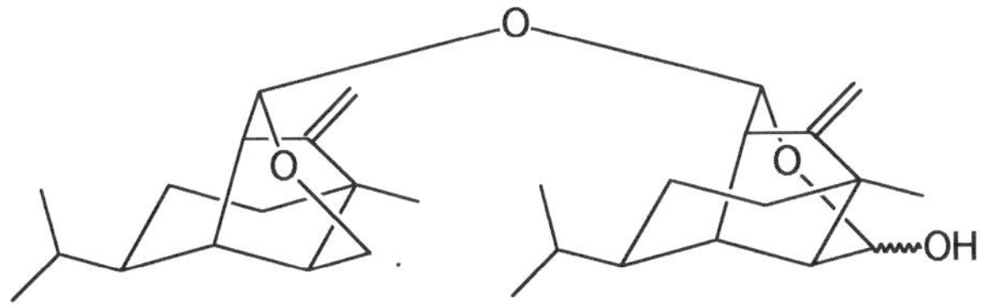

General Characteristics
A labilc oily substance that yields one molecule of prehelminthosporol and one molecule of prehelminthosporal on standing.

Fungal Source
Cochliobolus setariae strain IFO 6635.

Isolation/Purification
Isolated by silicic acid column chromatography eluted with ethyl acetate in *n*-hexane.

Biological Activity
Antibiotic that specifically causes inhibition of different strains of the same fungal species.

Spectral Data

IR:
(Acetate derivative) (CCl_4) 1760 (acetyl C=O)

^{1}H NMR:
($CDCl_3$)(acetate derivative) 5.81ppm (acetyl carbon as a broad singlet that sharpened upon irradiation at a frequency of 1.59 ppm, which was assigned to the vicinal methine proton); and 2.02ppm (acetyl methyl).

Mass Spectrum:
(Acetate derivative) 512(M^+), 235(45%), 219(100), and 277*m/e* (35).

Reference
M. Nukina and S. Marumo; Aversion Factors, Antibiotics among Different Strains of a Fungal Species: Aversion Factors of *Cochliobolus setariae;* Agric. Biol. Chem., Vol. 40, pp. 2121-2123(1976).

Common/Systematic Name

Penicillic acid
3-Methoxy-5-methyl-4-oxo-2,5-hexadienoic acid

Molecular Formula/Molecular Weight

$C_8H_{10}O_4$; MW = 170.05791

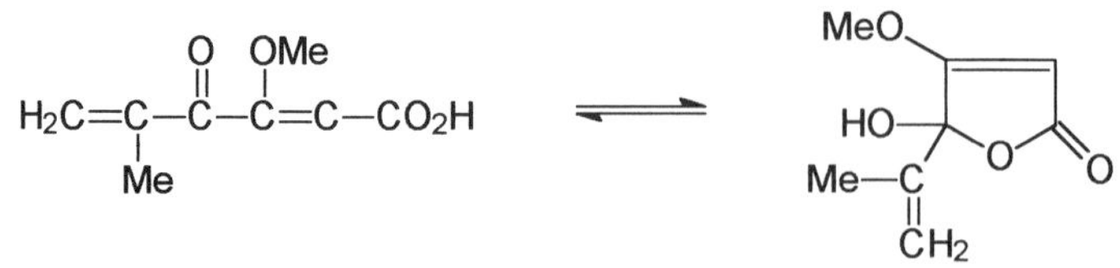

General Characteristics

Crystals from petroleum ether; mp., 86-87°C; from H_2O, 58-64°C; $[\alpha]_D = 0°$. Methyl ester, colorless prisms; mp., 35°C. Penicillic acid can occur as keto or lactone form.

Fungal Source

Penicillium lividum, P. puberulum, P. griseum, P. simplicissimum, P. cyclopium, P. thomii, P. roqueforti (P. suavolens), P. martensii, P. fenelliae, P. aurantio-virens, P. janthinellum, P. viridicatum, P. palitans, P. baarnense, P. madriti, P. lilacinum, P. canescens, P. chrysogenurn, P. olivino-viride, Aspergillus ochraceus, A. sulphureus, A. melleus (A. quercinus), A. sclerotiorum, A. alliaceus, A. ostianus and *Paecilomyces ehrlichii.*

Biological Activity

Subcutaneous injection of penicillic acid in mice at ca. 200-300mg/kg was lethal. The LD_{50} to mice dosed SC was 110mg/kg; IV was 250mg/kg. The LD_{50} for rabbits dosed SC was 100-200mg/kg. Penicillic acid has been shown to have antibacterial (mostly gram-negative), antiviral, antitumor, and cytotoxic effects.

Spectral Data

UV:

λ_{max}^{MeOH} 221nm (ϵ = 10,500)

^1H NMR:

5.09(H-2); 1.76(H-6); 5.18(H-7a); 5.44(H-7b); and 3.90ppm (H-8).

^{13}C NMR:

169.5(C-1, s); 89.6(C-2, d); 178.9(C-3, s); 102.1(C-4, s); 139.8(C-5, s); 16.9(C-6, q); 114.0(C-7, t); and 59.6ppm (C-8, q).

TLC Data

Adsorbent: Kieselgel G; Solvent: benzene-methanol-acetic acid, 18:1:1, v/v/v. R_f=0.25 (patulin has same R_f in this system). Detection: fluoresces under UV light after exposure to NH_3; yellow fluorescence with NH_3 and phenylhydrazine.

GLC Data

Solid support: Gas Chrom-Q liquid phase: 10% DC-200; Column temp:175°C; Detector temp: 190°C; Injector temp: 190°C; Relative retention time (relative to 13-BHC and were analyzed as TMS derivatives): 0.34; relative retention time for patulin in same system: 0.56. Solid support: Gas Chrom-Q, 100-102 mesh liquid phase: 3% Dexsil 300 3% OV-17 and 3% OV-25.

Reference

R. J. Cole, and R. H. Cox; Handbook of Toxic Fungal Metabolites; Academic Press, New York, pp. 520-524(1981).

Common/Systematic Name

5,5-Dimethyl-9-(2-hydroxyethyl)-8-hydroxymethyl-1-methylbicyclo[4.2.1]non-7-ene

Molecular Formula/Molecular Weight

$C_{15}H_{26}O_2$; MW = 238.19328

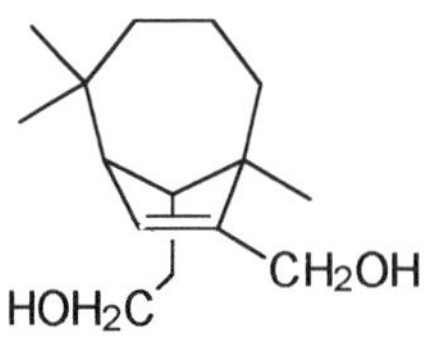

General Characteristics

Crystals; mp., 117°C; $[\alpha]_D^{25}$ = +3° (in $CHCl_3$); diacetate derivative, $[\alpha]_D^{25}$ = + 9°.

Fungal Source

Helminthosporium sativum and *H. victoriae.*

Isolation/Purification

Extracted from cultures with ethyl ether followed by silica gel column chromatography.

Spectral Data

^{1}H NMR:

($CDCl_3$) Signals for three methyl groups at 0.94(6H); 1.04(3H); 5.71(an olefinic proton); 3.67(m, 2H); 4.08(2H); and two D_2O exchangeable protons at 1.7ppm.

Mass Data:

Calcd. for $C_{15}H_{26}O_2$: C, 75.58%, H, 11.00%; found: C, 75.52%, H, 11.00%

Reference

F. Dorn and D. Arigoni; Ein Bicyclischer Abkommling von (-)-Longifolen aus *Helminthosporium sativum* und *H. victoriae*; Experientia, Vol. 30, pp. 851-852(1974).

Common/Systematic Name

Cynodontin

1,4,5,8-Tetrahydroxy-2-methylanthraquinone; 2,5,7-trihydroxyemodin

Molecular Formula/Molecular Weight

$C_{15}H_{10}O_6$; MW = 286.04774

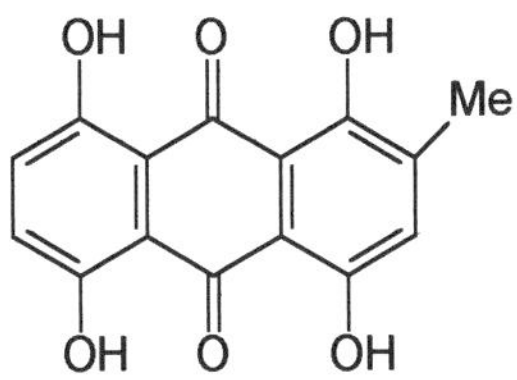

General Characteristics

Brown-bronze crystals (no melting point reported) from ethyl acetate.

Fungal Source

Curvularia lunata NRRL 2380, *C. geniculata, C. ramosa, C. trifolii, Drechslera halodes, D. spicifera, D. sorokiniana, D. dematioidea,* and *D. setariae.*

Isolation/Purification

Extracted and defatted with ethyl acetate; recrystallized from ethyl acetate.

Spectral Data

UV:

λ_{max}^{EtOH} 243, 297, 474sh, 484, 506sh, 517, 542, 554, and 603nm.

Mass Spectrum:

HREIMS: 286.04945*m/e* (M^+); calcd. for $C_{15}H_{10}O_6$, 286.04773.

Reference

G. W. van Eijk, and H. J. Roeymans; Cynodontin, The Tetrahydroxyanthraquinone of *Curvularia* and *Drechslera* spp.; Experientia,Vol. 10, pp. 1283-1284(1977).

Common/Systematic Name
Cyclonerodiol

Molecular Formula/Molecular Weight
$C_{15}H_{28}O_2$; MW = 240.20893

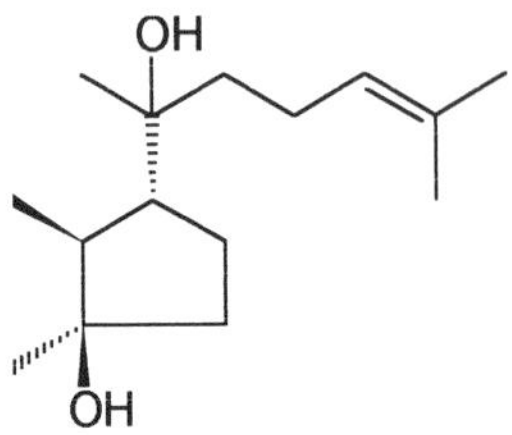

General Characteristics
Liquid.

Fungal Source
Gibberella fujikuroi, Trichoderma koningii isolated from *Diffenbachia* sp. (ATTC 46314), *Trichothecium* sp. *(T. roseum),* and *Fusarium culmorum.*

Isolation/Purification
Fungal cultures were extracted with acetone, the acetone reduced to an aqueous phase under vacuum, and partitioned twice with ethyl acetate. The ethyl acetate extract was chromatographed on a silica gel column eluted with benzene, ethyl ether, ethyl acetate, acetone and acetonitrile. The ethyl ether and ethyl acetate fractions were combined, reduced in volume, and further purified by silica gel column chromatography eluted with a linear gradient from benzene to acetone. The koninginin B fractions were reduced in volume and chromatographed on a column containing hydrated silica gel eluted with benzene-ethyl acetate (5:4, v/v). Biologically active fractions were pooled and chromatographed on C_{18} reverse phase silica gel with acetonitrile-water (1:1, v/v). Final purification was performed on a silica gel-silver nitrate column eluted with methylene chloride-acetone (9:1, v/v) to yield koninginin B and cyclonerodiol. Cyclonerodiol was separated from koninginin B on silica gel, eluting with ethyl acetate-benzene (9:1, v/v).

Biological Activity
The metabolite inhibited the growth of etiolated wheat coleoptiles 60% at 10^{-3}M but did not visibly affect them at 10^{-4} or -10^{-6}M.

Spectral Data

UV:
No UV absorption above 210nm.

IR:

(KBr) 3460(broad OH), 2965, 2025, 1446, 1375, 1285, 1200, 1143, 1110, 1025, 983, 916, 882, and $830cm^{-1}$.

^{1}H NMR:

(Acetone-d_6) 0.99(d, J=6.8Hz, H-1); 1.65-1.40(H-2); 1.79(m, H-4); 1.65-1.40(H-5); 1.65-1.40(H-6); 1.43(t, J=8.5Hz, H-8); 2.03(m, H-9); 5.10(br t, J=7.2Hz); 1.63*(s, H-12); 1.17*(s, H-13); 1.10*(s, H-14); 1.55*(s, H-15); and 2.88ppm (br s, OH).

* Assignments may be reversed.

^{13}C NMR:

(Acetone-d_6) 15.2(C-1); 44.8(C-2); 80.5(C-3); 41.6(C-4); 24.9(C-5); 55.1(C-6); 74.0(C-7); 41.4(H-8); 23.3(C-9); 125.9(C-10); 131.0(C-11); 25.7(C-12); 25.2(C-13); 26.4(C-14); and 17.5ppm (C-15).

TLC Data

Silica gel 60 thin-layer plates were developed with toluene-ethyl acetate-formic acid (5:4:1, v/v/v), R_f = 0.59-0.64; detected as a bright pink spot when sprayed with acidified anisaldehyde followed by heating.

References

H. G. Cutler, J. M. Jacyno, R. S. Phillips, R. L. VonTersch, P. D. Cole, and N. Montemurro; Cyclonerodiol from a Novel Source, *Trichoderma koningii*: Plant Growth Regulatory Activity; Agric. Biol. Chem., Vol. 55, pp. 243-244(1991).

S. Nozoe, M. Goi, and N. Morisaki; Structure of Cyclonerodiol; Tet. Lett., pp. 1293-1296(1970).

Common/Systematic Name

Cyperine

Molecular Formula/Molecular Weight

$C_{15}H_{16}O_4$; MW = 260.10486

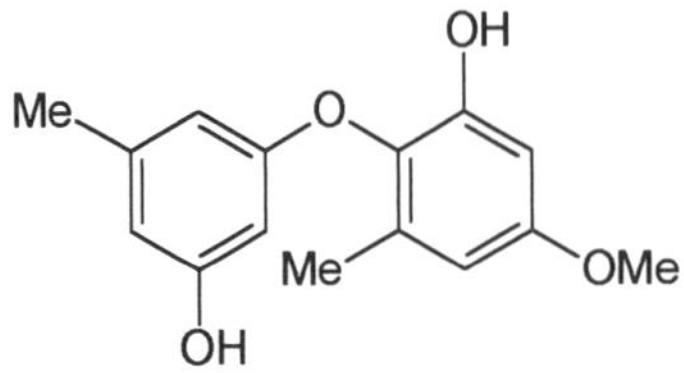

Fungal Source

Ascochyta cypericola, a new pathogen of purple nut sedge isolated in India near the center of the origin of the weed. Purple nut sedge (*Cyperus rotundus* L.) is one of the world's most notorious and ubiquitous weeds, reported throughout Africa, India, southern Asia, Australia, South and Central America, and southern United States.

Isolation/Purification

Three-week-old liquid cultures of *Ascochyta cypericola* were extracted with three equal volumes of ethyl acetate. This crude extract, which induced lesion formation on purple nut sedge, was purified by repeated bioassay-guided size exclusion chromatography (Sephadex LH-20). It yielded the extremely phytotoxic metabolite cyperine.

Biological Activity

Cyperine is an extremely active phytotoxin with modest host-selectivity within the genus *Cyperus*. *Cyperus rotundus*, the source of *A. cypericola*, exhibited the greatest sensitivity to cyperine among the species of *Cyperus* tested. The fungus was isolated from surface-sterilized diseased leaves of *C. rotundus*.

Spectral Data

IR:

A strong, broad IR absorption at 3400cm^{-1} suggested a phenolic moiety, and the absence of signals between 2000 and 1620cm^{-1} argued against carbonyl functionality.

^{1}H NMR:

The five ^{1}H NMR signals between 6.0 and 6.4ppm indicated the presence of five aromatic methines. The signals at 6.35 and 6.29ppm were *meta*-coupled to each other (J=2.7Hz), as were the three higher field signals(6.22, 6.12, and 6.02, J=2.1Hz). Methyl singlets at 2.03 and 2.18 indicated the presence of two aromatic methyl groups; a methyl singlet at 3.72ppm indicated a methoxyl substituent. There was benzylic coupling between the methyl singlet at 2.18 and the aromatic protons at 6.12 and

6.22ppm, and between the methyl singlet at 2.03 and the aromatic protons at 6.29 and 6.34ppm. Irradiation at 3.72ppm resulted in enhancement of the two aromatic protons at 6.34 and 6.29 (3.5 and 4.2%, respectively), which had been assigned to the tetra-substituted benzene ring. Irradiation of the methyl singlet at 2.10 resulted in enhancement (4.1%) of the aromatic proton at 6.34ppm.

Mass Spectrum:

HREIMS established the molecular formula of $C_{15}H_{16}O_4$, indicating a compound with eight sites of unsaturation.

Reference

A. Steerle, R. Upadhyay and G. Strobel; Cyperine, A Phytotoxin Produced by *Ascochyta cypericola*, A Fungal Pathogen of *Cyperus rotundus*; Phytochemistry, Vol. 30, pp. 2191-2192(1991).

Common/Systematic Name

Erinapyrone A

(2*S*)-2,3-Dihydro-6-hydroxymethyl-2-methyl-4*H*-pyran-4-one

Molecular Formula/Molecular Weight

$C_7H_{10}O_3$; MW = 142.06299

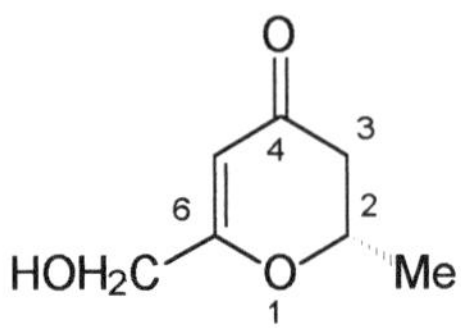

General Characteristics

A colorless oil.

Fungal Source

Hericium erinaceum, an edible mushroom.

Isolation/Purification

Culture filtrate from *H. erinaceum* was fractionated by solvent partition between water and ethyl acetate. Repeated silica gel column chromatography followed by HPLC with an ODS column gave purified erinapyrones A and B as colorless oils.

Biological Activity

Erinapyrone A exhibited cytotoxicity against HeLa cells; LD_{100} was 0.88mM.

Spectral Data

IR:

1608, 1655, and 3392cm^{-1}.

CD:

$\Delta\epsilon$ - 1.14 at 316nm (EtOH).

^{1}H NMR:

($CDCl_3$) 1.48(d, *J*=6.23Hz, H-methyl); 2.36(dd, *J*=16.85, 4.76, H-3eq); 2.40(dd, *J*=16.85, 12.07, H3ax); 4.18(d, *J*=15.93, H-6CH_2); 4.23(d, *J*=15.93Hz, H-6CH_2); 4.57(ddq, *J*=12.07, 4.76, 6.23, H-2); and 5.61ppm (s, H-5); Monoacetate: 1.40(d, *J*=6.59Hz, 2-methyl); 2.07(s, acetate); 2.37(dd, *J*=16.85, 4.13Hz, H-3eq); 2.42(dd, *J*=16.85, 13.92Hz, H-3ax); and 4.51ppm (ddq, *J*=13.92, 4.13, 6.59Hz, H-2);

^{13}C NMR:

($CDCl_3$) 20.4(C-2 methyl); 43.1 (C-3); 61.9 (C-2); 76.1 (C-6-CH_2); 102.4 (C-5); 174.9 (C-6);and 192.8ppm (C-4).

Mass Spectrum:
HREIMS: 142.0615*m/e* for $C_7H_{10}O_3$.

Reference

H. Kawagishi, M. R. Shirai, H. Sakamoto, S. Yoshida, F. Ojima, and Y. Ishiguro; Erinapyrones A and B from the Cultured Mycelia of *Hericium erinaceum*; Chemistry Letters, (Japan) pp. 2475-2476(1992).

Common/Systematic Name

Erinapyrone B

(2*R*)-2,3-Dihydro-2-hydroxymethyl-6-methyl-4*H*-pyran-4-one

Molecular Formula/Molecular Weight

$C_7H_{10}O_3$; MW = 142.06299

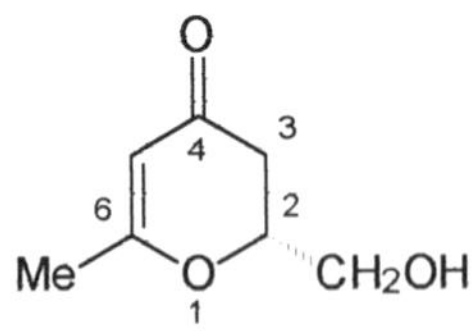

General Characteristics

A colorless oil.

Fungal Source

Hericium erinaceum, an edible mushroom.

Isolation/Purification

Culture filtrate from *H. erinaceum* was fractionated by solvent partition between water and ethyl acetate. Repeated silica gel column chromatography followed by HPLC with an ODS column gave purified erinapyrones A and B as colorless oils.

Biological Activity

Erinapyrone B exhibited cytotoxicity against HeLa cells; LD_{100} was 1.76mM.

Spectral Data

IR:

1606, 1655, and 3394cm^{-1}.

CD:

$\Delta\epsilon$ -0.56 at 315nm (EtOH).

^{1}H NMR:

($CDCl_3$) 2.01(s, H-6-methyl); 2.30(dd, *J*=16.85, 3.66, H-3eq); 2.61(dd, *J*=16.85, 14.29, H-3ax); 3.76(dd, *J*=12.46, 5.50, H-2-CH_2); 3.87(dd, *J*=12.46, 3.30Hz, H-2CH_2); 4.46(dddd, *J*=14.29, 3.66, 3.30, 5.50Hz, H-2); and 5.32ppm (s, H-5);

^{13}C NMR:

($CDCl_3$) 20.9(C-6 methyl); 36.7(C-3); 63.7(C-2-CH_2); 79.5(C-2); 104.9(C-5); 174.2(C-6); and 192.6ppm (C-4).

Mass Spectrum:
HREIMS: 142.0659*m/e* for $C_7H_{10}O_3$.

Reference

H. Kawagishi, M. R. Shirai, H. Sakamoto, S. Yoshida, F. Ojima, and Y. Ishiguro; Erinapyrones A and B from the Cultured Mycelia of *Hericium erinaceum*; Chemistry Letters, (Japan) pp. 2475-2476(1992).

Common/Systematic Name

Erinapyrone C

Molecular Formula/Molecular Weight

$C_8H_{10}O_5$; MW = 186.05282

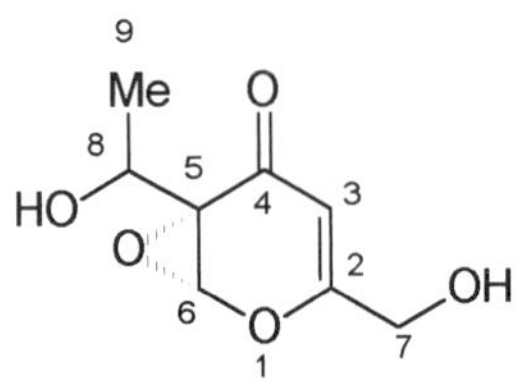

General Characteristics

Erinapyrone C was isolated as a solid; mp., 72°C; $[\alpha]_D^{25}$ + 141° (*c*=1.0, in MeOH).

Fungal Source

Hericium erinaceum (strain CBS 233.87).

Isolation/Purification

Fungal cultures were extracted with EtOAc containing 1% MeOH and the extracts were evaporated to give a mixture of crude metabolites. The crude mixture was chromatographed on a silica gel column using hexane-EtOAc (4:1, v/v) to yield hericenes A-C, fatty acids (palmitic, linoleic, and oleic acids); the residual material was then eluted with EtOAc containing MeOH (1%) to give erinapyrone C.

Spectral Data

UV:

λ_{max}^{EtOH} 280(ϵ = 9,700)

^{1}H NMR:

($CDCl_3$ - DMSO-d_6) H-3, 5.72(t, *J*=1.2 Hz); H-6, 5.74(s); H-7a, 4.16(ddd, *J*=16.7, 6.1, and 1.2Hz); H-7b, 4.07(ddd, *J*=16.7, 6.1 and 1.2Hz); H-8, 4.41(dq, *J*=6.1 and 6.4Hz); H-9, 1.28(q, *J*=6.4Hz); 7-OH, 5.40(t, *J*=6.1Hz); and 8-OH, 4.67ppm (d, *J*=6.1Hz).

^{13}C NMR:

($CDCl_3$ - DMSO-d_6) C-2, 170.61(s); C-3, 101.60(d); C-4, 189.47(s); C-5, 64.90(s); C-6, 80.98(d); C-7, 59.98(t); C-8, 60.68(t); and C-9, 18.17ppm (q).

Mass Data:
EIMS: 186(M^+), 168, and 155*m/e*; anal. found: C, 51.6; H, 5.4: calcd for $C_8H_{10}O_5$ C, 51.3; H, 5.7.

Reference
A. Arnone, R. Cardillo, G. Nasini, and O. Vajna De Pava; Secondary Mold Metabolites: Part 46. Hericenes A-C and Erinapyrone C, New Metabolites Produced by The Fungus *Hericium erinaceum*; J. Nat. Prod., Vol. 57, pp. 602-606(1994).

Common/Systematic Name

Asperflavin

(+)-3,4-Dihydro-3,6,9-trihydroxy-8-methoxy-3-methylanthracen-1(2*H*)-one

Molecular Formula/Molecular Weight

$C_{16}H_{16}O_5$; MW = 288.09977

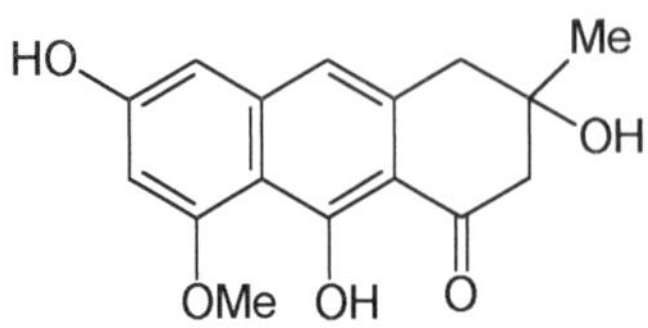

General Characteristics

Fractional crystallization from ethyl acetate gave citrine prisms; mp., 225-230°C, resetting to needles which decomposed at ca. 280°C without melting; $[\alpha]_D^{20}$ + 4° (*c*=0.3, in MeOH). It was insoluble in sodium hydrogen carbonate, but dilute solutions in sodium carbonate had an intense yellowish green fluorescence. In 2N sodium hydroxide it gave a yellow solution which rapidly became orange. Dilute solutions in organic solvents showed a green fluorescence. It gave an intense dark green color with iron chloride in ethanol. Sublimation at 160°C and 10^{-1} mmHg gave a solid which crystallized from ethyl acetate in needles (decomp. 290°C without melting). Asperflavin was unstable in light and air. Solutions in organic solvents slowly became dark green, while thin layers of silica impregnated with asperflavin rapidly changed color from citrine to green.

Fungal Source

An entomogenous strain of *Aspergillus flavus* (probably *A. nomius*), *A. repens,* and *Leucopaxillus tricolor.*

Isolation/Purification

Extracted with chloroform and purified by fractional crystallization from ethyl acetate solution after separation of 5'-hydroxyasperentin.

Spectral Data

UV:

λ_{max}^{MeOH} 230, 269, 317, 335, and 392nm (log ϵ = 4.13, 4.51, 3.64, 3.40, and 3.91).

IR:

(Nujol) 3490, 3350, 1632, and 1582cm^{-1}.

^{1}H NMR:

(CD_3OD) 3.25(10-H); 3.50, 3.57(5-, 7-H); 6.08(OMe); 7.00, 7.22(2-, 4-H_2); 8.64(3-Me); and 3.82τ (OH). [$(CD_3)_2CO$] 3.22(10-H); 3.38, 3.50(5-, 7-H); 6.10(OMe); 6.94, 7.18(2-, 4-H_2); 8.62(3-Me); -5.0, 3.72τ (OH).

Mass Data:

LREIMS: 288*m/e*; an abundant resonance-stabilized rearrangement ion at 230*m/e* (M^+ - C_3H_6O); anal. found C, 66.4; H, 5.75; $C_{16}H_{16}O_5$ requires C, 66.7; H, 5.6.

Reference

J. F. Grove; New Metabolic Products of *Aspergillus flavus*. Part II. Asperflavin, Anhydroasperflavin, and 5,7-Dihydroxy-4-methylphthalide; J. C. S. Perkin I, pp. 2406-2411(1972).

Common/Systematic Name

Diplodiol; Diplosporin

trans-6-Ethyl-5-hydroxy-3-hydroxymethyl-5,6,7,8-tetrahydrochromone; (5S,6X)-6-Ethyl-5-hydroxy-3-hydroxymethyl-5,6,7,8-tetrahydrobenzo[*b*]pyran-4-one

Molecular Formula/Molecular Weight

$C_{12}H_{16}O_4$; MW = 224.10486

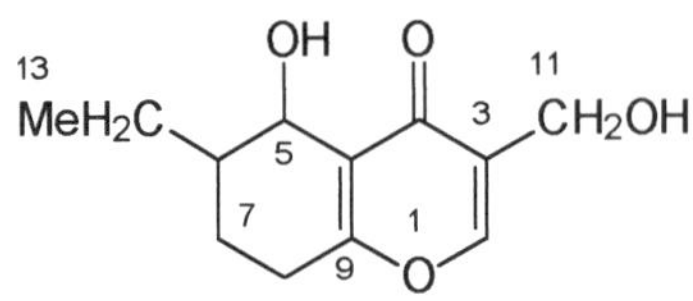

General Characteristics

Fine white crystals from benzene-acetone; mp., 47-48°C. Colorless crystals from benzene-hexane; mp., 83-84°C; $[\alpha]_D^{20}$ + 49.7° (*c*=1.01, in MeOH).

Fungal Source

Diplodia macrospora (ATTC 36896), a fungal pathogen of corn, causing ear- and stalk-rot in tropical and subtropical areas of the world and under humid conditions a severe leaf spot.

Isolation/Purification

Fungal cultures were extracted with acetone, reduced to an aqueous phase, and partitioned between ethyl acetate. The ethyl acetate extract was dried over anhydrous sodium sulfate, filtered, reduced in volume, and chromatographed on a silica gel column eluted with a series of solvents; benzene, ethyl ether, ethyl acetate, acetone, and methanol. The acetone fraction was further fractionated on a silica gel column eluted with a linear gradient from benzene to acetone. The diplodiol containing fractions were combined, reduced in volume, and diplodiol was crystallized as fine white needles.

Biological Activity

Toxic to day-old chicks dosed orally, LD_{50} 88mg/kg; not biologically active in the *Avena* coleoptile assay.

Spectral Data

UV:

λ_{max}^{EtOH} 217(log ϵ = 3.90) and 251nm (3.90); λ_{max} 214(log ϵ = 4.04), 252(4.08), and 256nm (4.07).

IR:

(KBr) 3400(broad, OH), 2960(methyl), 2930(methylene), 2870(methylene), 1653(strong, γ-pyrone), 1585(γ-pyrone), 1450, 1315, 1235, 1180 1125, and 1020cm^{-1}.

^{1}H NMR:
(CDCl$_3$) 7.79(H-2); 4.59(H-5); 1.40(H-6); 2.05, 1.75(H-7); 2.60(H-8); 4.46(H-11); 1.67(H-12); and 0.98ppm (H-13).

^{13}C NMR:
(CDCl$_3$) 152.05(C-2); 123.43(C-3); 180.29(C-4); 68.48(C-5); 41.22(C-6); 24.17(C-7); 22.96(C-8); 165.91(C-9); 126.80(C-10); 57.73(C-11); 26.73(C-12); and 11.28ppm (C-13) (Cutler *et al*); 152.0 (C-2); 127.0 (C-3); 179.9 (C-4);68.2 (C-5); 41.1 (C-6); 22.8 (C-7); 26.3 (C-8); 165.6 (C-9); 123.3 (C-10); 23.9 (C-11); 10.8 (C-12); and 57.2ppm (C-13) (Chalmers *et al*).

Mass Spectrum:
HREIMS: 224.1051(M^+, calculated mass for $C_{12}H_{16}O_4$, 224.1048), 208(M^+ - O), 206(M^+ - H_2O), 193(M^+ - CH_2OH), and 188*m/e* (M^+ - $2H_2O$).

TLC Data

R_f of diplodiol was 0.33-0.36 on silica gel-60 TLC plates using toluene-ethyl acetate-formic acid (5:4:1, v/v/v) as developing solvent.

References

A. A. Chalmers, C. P. Gorst-Allman, P. S. Steyn, R. Vleggaar, W. F. O. Marasas, and N. P. J. Kriek; The Structure of Diplosporin; S. Afr. J. Chem., Vol. 31, p. 111(1978).

H. G. Cutler, F. G. Crumley, R. H. Cox, R. J. Cole, J. W. Dorner, F. M. Latterell, and A. E. Rossi; Diplodiol: A New Toxin from *Diplodia macrospora*; Agric. Food Chem., Vol. 28, pp. 135-138(1980).

Common/Systematic Name

5-Deoxydiplosporin

Molecular Formula/Molecular Weight

$C_{12}H_{16}O_3$; MW = 208.10994

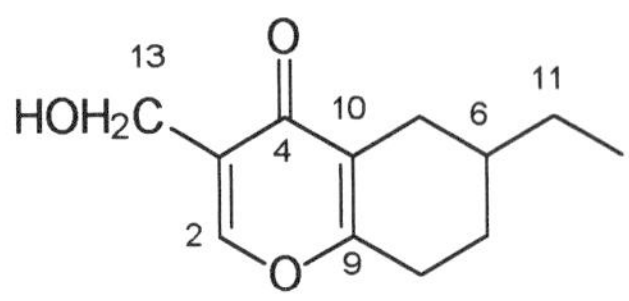

General Characteristics

Isolated as an oil; $[\alpha]_D^{20}$ + 55° (*c*=1.1, in MeOH).

Fungal Source

Diplodia macrospora (ATTC 36896), a fungal pathogen of corn, causing ear and stalk rot and under humid conditions a severe leaf spot.

Isolation/Purification

Fungal culture filtrates were extracted with chloroform, dried over anhydrous sodium sulfate, filtered, reduced in volume, and chromatographed on a silica gel column under pressure and eluted with chloroform-acetone-hexane (7:2:1, v/v/v). 5-Deoxydiplosporin was obtained as an oil from the first fractions to elute from the column.

Spectral Data

UV:

λ_{max}^{EtOH} 219(log ϵ = 3.72) and 258nm (3.73).

IR:

(KBr) 3400, 1660, 1610, 1590, and 1450cm^{-1}.

^{1}H NMR:

($CDCl_3$) 7.79(H-2); 4.59(H-5); 1.40H-6); 2.05, 1.75(H-7); 2.60(H-8); 4.46(H-11); 1.67(H-12); and 0.98ppm (H-13).

^{13}C NMR:

($CDCl_3$) 151.30(d, C-2); 125.8(s, C-3); 179.3(s, C-4); 27.4*(t, C-5); 34.6(d, C-6); 26.6*(t, C-7); 27.4(t, C-8); 164.4(s, C-9); 122.1(s, C-10); 28.4(t, C-11); 11.8(q, C-12); and 58.4ppm (t, C-13).

* Assignments may be reversed.

Mass Spectrum:
HREIMS: 208.1104*m/e* (M^+, calculated mass for $C_{12}H_{16}O_3$, 208.1099).

Reference

A. A. Chalmers, C. P. Gorst-Allman, P. S. Steyn, and R. Vleggaar; Biosynthesis of Diplosporin by *Diplodia macrospora*. Ring Formation Involving Methionine-derived Carbon Atoms; J. Chem. Soc. (Perkin Trans I), pp. 1481-1484(1979).

Common/Systematic Name

Betulin (Lup-20(29)-ene-3β,28-diol)

Molecular Formula/Molecular Weight

$C_{30}H_{50}O_2$; MW = 442.38108

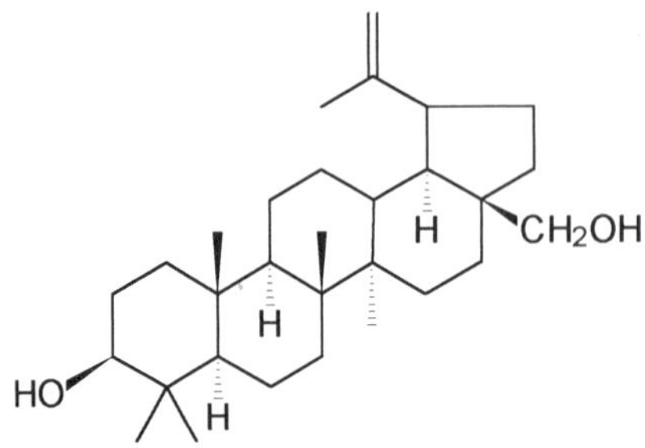

General Characteristics

Needles from methanol; mp., 250-251°C; $[\alpha]_D^{20}$ + 16° (*c*=0.3, in EtOH); diacetate derivative; mp., 216-217°C.

Fungal Source

The wood-rotting fungus, *Polyporus pinicola*. Betulin is an important constituent of the bark of birch trees and it has been suggested that the compound may be translocated from the bark directly into the fungus (in this case *Fomes fomentarius*). Although *Polyporus pinicola* is usually found on pines, betulin may arise in a similar manner or from small pieces of bark adhering to the fungus.

Spectral Data

IR:

(Nujol) 3279, 1658, 1265, and 915cm^{-1}.

^{1}H NMR:

($CDCl_3$) 0.78(3H, s, 24-CH_3); 0.85(3H, s, 25-CH_3); 0.99(6H, s, 23-, 27-CH_3); 1.04(3H, s, 26-CH_3); 1.40(2H, s, D_2O exchangeable); 1.70(3H, s, 29-CH_3); 3.25(1H, m, 3-H); 3.43; 3.75(2H, 28-CH_2); and 4.63, 4.73ppm (2H, 30-CH_3).

Mass Data:

LREIMS: 442(M^+, 91%), 411(82), 424(14), 234(44), 220(30), 203(62), 205(10), 207(85), and 189*m/e* (94.5). Diacetate, found: C, 81.0; H, 11.35%; calcd for $C_{30}H_{50}O_2$: C, 81.4; H, 11.4%; found: C, 77.4; H, 10.3%; calcd for $C_{34}H_{54}O_4$: C, 77.5; H, 10.3%.

References

R. C. Cambie; Chemistry of Fungi, 12. Betulin from *Polyporus pinicola* Fr.; New Zealand J. Science, Vol. 21, pp. 565-567(1978).

W. B. Turner and D. C. Aldridge; Fungal Metabolites II; Academic Press, New York, p. 330 (1983).

Secondary Metabolite Index

Molecular Formula Index

Molecular Weight Index

Fungal/Plant Source Index

R

S